Lecture Notes in Computer Science 7402

Commenced Publication in 1973
Founding and Former Series Editors
Gerhard Goos, Juris Hartmanis, and

Guohui Lin (Ed.)

Combinatorial Optimization and Applications

6th International Conference, COCOA 2012
Banff, AB, Canada, August 5-9, 2012
Proceedings

Volume Editor

Guohui Lin
University of Alberta
Department of Computing Science
Edmonton, Alberta T6G 2E8, Canada
E-mail: guohui@ualberta.ca

ISSN 0302-9743 e-ISSN 1611-3349
ISBN 978-3-642-31769-9 e-ISBN 978-3-642-31770-5
DOI 10.1007/978-3-642-31770-5
Springer Heidelberg Dordrecht London New York

Library of Congress Control Number: 2012941667

CR Subject Classification (1998): F.2, G.2.2, G.2, G.1.6, I.2.8, C.2

LNCS Sublibrary: SL 1 – Theoretical Computer Science and General Issues

Typesetting: Camera-ready by author, data conversion by Scientific Publishing Services, Chennai, India

Printed on acid-free paper

Springer is part of Springer Science+Business Media (www.springer.com)

Preface

This volume contains the papers presented at COCOA 2012: The 6th Annual International Conference on Combinatorial Optimization and Applications, held during August 5–9, 2012, in Banff, Alberta, Canada.

COCOA provides an annual forum for researchers working in the areas of combinatorial optimization and its applications. In addition to theoretical results, the conference is particularly focused on recent works on experimental and applied research of general algorithmic interest. Past COCOA conferences were held in Xi'an, China (2007), Newfoundland, Canada (2008), Huangshan, China (2009), Hawaii, USA (2010), and Zhangjiajie, China (2011).

There were 57 high-quality submissions. Using EasyChair, each submission was reviewed by at least 2, and on average 3.5, Program Committee members. The committee decided to accept 33 papers. The program also included one invited talk and one keynote talk. COCOA 2012 also organized a post-conference workshop on computational biology.

We wish to thank the authors for submitting their papers to the conference. We are grateful to the members of the Program Committee and the external referees for their work within demanding time constraints. We would like to thank the Alberta Innovates Technology Futures and the University of Alberta for their funding support. Members of the Omics Research Group in the Department of Computing Science at the University of Alberta, including Yi Shi, Ronghong Li, Yining Wang and Weitian Tong, were extremely helpful on the organizing side. Without their service, this conference would not have been successful.

August 2012 Guohui Lin

Organization

Program Committee

Zhipeng Cai	Georgia State University, USA
Zhi-Zhong Chen	Tokyo Denki University, Japan
Qi Cheng	University of Oklahoma, USA
Yongxi Cheng	Xi'an Jiaotong University, China
Annalisa De Bonis	Università degli Studi di Salerno, Italy
Donglei Du	University of New Brunswick, USA
Adrian Dumitrescu	University of Wisconsin-Milwaukee, USA
Stephane Durocher	University of Manitoba, Canada
Martin Fürer	Pennsylvania State University, USA
Carosten Gutwenger	Technische Universität Dortmund, Germany
Pinar Heggernes	University of Bergen, Norway
Iyad Kanj	DePaul University, USA
George Karakostas	McMaster University, Canada
Naoki Katoh	Kyoto University, Japan
Wonjun Lee	Korea University, Korea
Fei Li	George Mason University, USA
Guohui Lin	University of Alberta, Canada (Chair)
Bin Ma	University of Waterloo, Canada
Ian McQuillan	University of Saskatchewan, Canada
Peter Miltersen	Aarhus University, Denmark
Mitsunori Ogihara	University of Miami, USA
Hans Simon	Ruhr-Universität Bochum, Germany
Jack Snoeyink	University of North Carolina at Chapel Hill, USA
Daniel Stefankovic	University of Rochester, USA
Martin Strauss	University of Michigan, USA
Wing-Kin Sung	National University of Singapore
Zhiyi Tan	Zhejiang University, China
My Thai	University of Florida, USA
Weitian Tong	University of Alberta, Canada
Lusheng Wang	City University of Hong Kong, SAR China
Carola Wenk	University of Texas at San Antonio, USA
Boting Yang	University of Regina, Canada
Kaizhong Zhang	University of Western Ontario, Canada
Binhai Zhu	Montana State University, USA

Additional Reviewers

Ahmed, Mahmuda
Buchin, Kevin
Chen, Danny
Chen, Xujin
Cheng, Siyao
Dash, Sajal
Fink, Martin
Flatland, Robin
Gethner, Ellen
Han, Aram
Han, Qiaoming
Han, Xin
Li, Weiming
Liang, Zhewei
Liu, Yi
Mahajan, Meena
Mehrabi, Saeed
Sherette, Jessica
Shuai, Tianping
Skala, Matthew
Verma, Vishal
Yaroslavtsev, Grigory
Yu, Huiwen
Zey, Bernd
Zhang, Peng

Table of Contents

Load-Balanced Virtual Backbone Construction for Wireless Sensor Networks

Jing (Selena) He, Shouling Ji, Yi Pan, and Zhipeng Cai

Department of Computer Science,
Georgia State University, Atlanta, GA, USA
{jhe9,sji,pan,zcai}@cs.gsu.edu

Abstract. Virtual Backbones (VBs) are expected to bring substantial benefits to routing in Wireless Sensor Networks (WSNs). Connected Dominating Sets (CDSs) based VBs are competitive approaches among the existing methods used to establish VBs in WSNs. Most existing works focus on constructing Minimum-sized CDSs (MCDSs). However, few works consider the load-balance factor. In this work, the size and the load-balance factor are both taken into account when constructing VBs in WSNs. Specifically, three problems are investigated in the paper, namely, the MinMax Degree Maximal Independent Set (MDMIS) problem, the Load-Balanced Virtual Backbone (LBVB) problem, and the MinMax Valid-Degree non Backbone node Allocation (MVBA) problem. We claim that MDMIS and LBVB are NP-Complete and MVBA is NP-Hard. Moveover, approximation algorithms and comprehensive theoretical analysis of the approximation factors are presented in the paper.

1 Introduction

Wireless sensor networks (WSNs) are usually deployed for monitoring and controlling systems where human intervention is not desirable or feasible. Therefore, WSNs are widely used in many military and civilian applications such as battlefield surveillance, health care applications, environment and habitat monitoring, and traffic control [1]. Compared with traditional computer networks, WSNs have no fixed or pre-defined infrastructure as a hierarchical structure, resulting the difficulty to achieve routing scalability and efficiency [2]. To better improve the performance and increase the efficiency of routing protocols, a Connected Dominating Set (CDS) has become a well known approach to form a Virtual Backbone (VB) in WSNs. A Dominating Set (DS) is defined as a subset of nodes in a WSN such that each node in the network is either in the set or adjacent to some node in the set. If the induced graph by the nodes in a DS is connected, then this DS is called a CDS. The nodes in a CDS are called *backbone nodes*, otherwise, *non backbone nodes*. In a WSN with a CDS as its VB, non backbone nodes may forward their data only to their neighboring backbone nodes. With the help of a CDS, the average message burden of a WSN could be reduced so that routing becomes much easier and can adapt quickly to network topology changes [3]. In addition to routing protocols, a CDS-based VB has many other

G. Lin (Ed.): COCOA 2012, LNCS 7402, pp. 1–12, 2012.

applications in WSNs [4–7]. Clearly, the benefits of a CDS-based VB can be magnified by making its size smaller. In general, the smaller the CDS is, the less communication and storage overhead are incurred. Hence, it is desirable to build a Minimum-sized CDS (MCDS)-based VB. Since VBs are bring substantial benefits in WSNs, a huge amount of effort has been made to construct different CDS-based VBs for different applications, such as, a k-connect m-dominating CDS [8], a minimum routing cost CDS [9] or a bounded-diameter CDS [10].

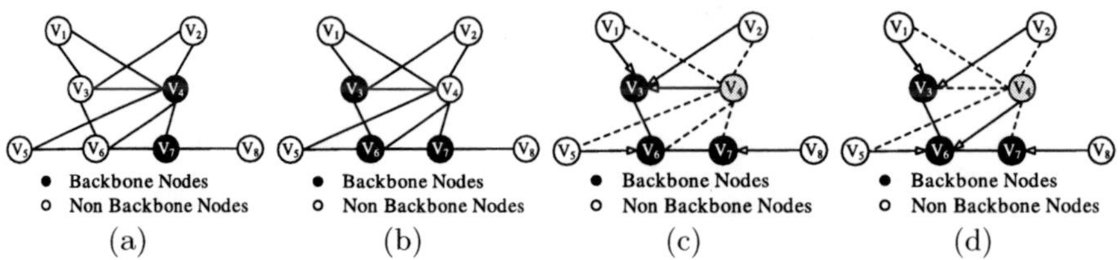

Fig. 1. Illustration of a regular VB vs a load balanced VB; and a regular Allocation vs a load-balanced allocation: (a) regular VB; (b) load-balanced VB; (c) regular allocation; (d) load-balanced allocation

Unfortunately, all the aforementioned works did not consider the *load-balance* factor when they construct a VB. For instance, when the MCDS-based VB is used in the network shown in Fig. 1(a), backbone node v_4 is adjacent to 5 different non backbone nodes, whereas, backbone node v_7 only connects to 2 non backbone nodes. If every non backbone node has the same amount of data to be transferred through the neighboring backbone node at a fixed data rate, then the number of neighboring non backbone nodes of each backbone node is a *potential* indicator of the traffic load on each backbone node. Hence, backbone nodes v_4 must deplete its energy much faster than backbone node v_7. A counter-example is shown in Fig. 1(b), the set $\{v_3, v_6, v_7\}$ is served as a VB. Compared with the VB constructed in Fig. 1(a), the numbers of neighboring non backbone nodes of all the backbone nodes in Fig. 1(b) are very similar. On the other hand, the criterion to allocate a non backbone node to a neighboring backbone node is also critical to balance traffic load on each backbone node. An illustration of the allocation schemes for non backbone nodes is depicted in Fig. 1(c) and (d), in which arrow lines represent that the non backbone nodes are allocated to the arrow pointed backbone nodes, while the dashed lines represent the communication links in the original network topological graph. Although the potential traffic load on each backbone node are evenly distributed in the VB constructed in Fig. 1(c) and (d), different allocation schemes for non backbone nodes might break the balance. In Fig. 1(c) and (d), only the gray non backbone node v_4 is adjacent to more than one backbone node. Allocating v_4 to different backbone nodes leads to distinct traffic load on the allocated backbone node. In Fig. 1(c), v_4 is allocated to backbone node v_3, while in Fig. 1(d), v_4 is allocated to backbone node v_6. Apparently, backbone node v_3 has more traffic load than backbone nodes v_6 and v_7 in Fig. 1(c). However, traffic loads are balanced among backbone nodes in

Fig. 1(d). Intuitively, compared with the WSN shown in Fig. 1(c), the VB and the allocation scheme for non backbone node v_4 shown in Fig. 1(d) can extend network lifetime notably.

To benefit from the CDS-based VB in WSNs and also take the load-balance factor into consideration, few attempts have been carried out to construct a VB in this manner [11,12]. In our previous work, we proposed a genetic-algorithm based method to build a load-balanced CDS (LBCDS) in WSNs. However, there is no performance ratio analysis in that paper. In this research, we first investigate how to construct an LBVB. It is well known that in graph theory, a Maximal Independent Set (MIS) is also a DS. MIS can be defined formally as follows: given a graph $G = (V, E)$, an Independent Set (IS) is a subset $I \subset V$ such that for any two vertex $v_1, v_2 \in I$, they are not adjacent, *i.e.*, $(v_1, v_2) \notin E$. An IS is called an MIS if we add one more arbitrary node to this subset, the new subset will not be an IS any more. Therefore, we construct an LBVB with two steps. The first step is to find a MinMax Degree MIS (MDMIS), and the second step is to make this MIS connected. Subsequently, we explore how to load-balancedly allocate non backbone nodes to backbone nodes, followed by comprehensive performance ratio analysis.

Particularly, our contributions mainly include three aspects as follows: 1) We first claim that the LBVB problem is NP-Complete. Hence, we solve LBVB with two steps. First, we propose an approximation algorithm to solve the MinMax Degree Maximal Independent Set (MDMIS) problem. It is shown that this algorithm yields a solution upper bounded by $O(\Delta \ln(n))OPT_{MDMIS}$, where OPT_{MDMIS} is the optimal result of MDMIS, Δ is the maximum node degree in the network, and n is number of sensors in a WSN. Subsequently, the minimum-sized set of nodes are found to make the MDMIS connected. The theoretical upper bound of the size of the constructed LBVB is analyzed in this paper as well. 2) We claim that the load-balancedly allocate non backbone nodes to backbone nodes problem is NP-Hard. Consequently, we present a randomized approximation algorithm, which produces a solution in which the traffic load on each backbone node is upper bounded by $O(\log^2(n))(OPT_{MVBA} + \frac{1}{\alpha^2})$ with probability $\frac{7}{8}$, where $\alpha = \log(n) + 3$, OPT_{MVBA} is the optimal result.

2 Problem Formulation

2.1 Network Model

We assume a static connected WSN and all the nodes in the WSN have the same transmission range. Hence, we model a WSN as an undirected graph $\mathbb{G} = (\mathbb{V}, \mathbb{E})$, where $\mathbb{V}$ is the set of n sensor nodes, denoted by v_i, where $1 \leq i \leq n$, i is called the node ID of v_i in the paper; $\mathbb{E}$ represents the link set $\forall\ u, v \in \mathbb{V}, u \neq v$, there exists a link (u, v) in $\mathbb{E}$ if and only if u and v are in each other's transmission range. In this paper, we assume links are undirected (bidirectional), which means two linked nodes are able to transmit and receive data from each other. Moreover, the degree of a node v_i is denoted by d_i, whereas Δ denotes the maximum degree in the network graph $\mathbb{G}$.

2.2 Problem Definition

As we mentioned in Section 1, we solve LBVB in two steps. The first step constructs a MinMax Degree Maximal Independent Set (MDMIS), and the second step selects additional nodes which together with the nodes in the MDMIS induce a connected topology *LBVB*. In this subsection, we first formally define the MDMIS problem, followed by the problem definition of LBVB.

Definition 21. *MinMax Degree Maximal Independent Set (MDMIS) Problem.* For a WSN represented by graph $\mathbb{G}(\mathbb{V}, \mathbb{E})$, the *MDMIS* problem is to find a node set $\mathbb{D} \subseteq \mathbb{V}$ such that:

1) $\forall u \in \mathbb{V}$ and $u \notin \mathbb{D}$, $\exists\ v \in \mathbb{D}$, such that $(u, v) \in \mathbb{E}$.
2) $\forall u \in \mathbb{D}$, $\forall v \in \mathbb{D}$, and $u \neq v$, such that $(u, v) \notin \mathbb{E}$.
3) There exists no proper subset or superset of $\mathbb{D}$ satisfying the above two conditions.
4) Minimize $\max\{d_i \mid \forall v_i \in \mathbb{D}\}$.

Taking the load-balance factor into consideration, we are seeking an MIS in which the maximum degree of the nodes in the constructed MIS is minimized. In other words, the potential traffic load on each node in the MIS is as balance as possible. Now, we are ready to define the LBVB problem.

Definition 22. *Load-Balanced Virtual Backbone (LBVB) Problem.* For a WSN represented by graph $\mathbb{G}(\mathbb{V}, \mathbb{E})$ and an *MDMIS* $\mathbb{D}$, the *LBVB* problem is to find a node set $\mathbb{C} \subseteq \mathbb{V} \backslash \mathbb{D}$ such that:

1) The induced graph $G[\mathbb{D} \bigcup \mathbb{C}]$ on $\mathbb{G}$ is connected.
2) Minimize $|\mathbb{C}|$, where $|\mathbb{C}|$ is the size of set $\mathbb{C}$.

For convenience, the nodes in the set $\mathbb{D}$ are called *independent nodes*, whereas, the nodes in the set $\mathbb{C}$ are called *MIS connectors*. Moreover, $\mathbb{B} = \mathbb{D} \bigcup \mathbb{C}$ is an *LBVB* of $\mathbb{G}$. Specifically speaking, $\forall v_i \in \mathbb{B}$, v_i is a *backbone node.*

Constructing an LBVB is a part of the work to balance traffic load on each backbone node. One more important task needs to be resolved is how to allocate non backbone nodes to its neighboring backbone nodes. The formal definition of the non backbone node allocation scheme are given as follows:

Definition 23. *Non Backbone Node Allocation Scheme ($\mathscr{A}$).* For a WSN represented by graph $\mathbb{G}(\mathbb{V}, \mathbb{E})$ and a VB $\mathbb{B} = \{v_1, v_2, \cdots, v_m\}$, we need to find m disjoint sets on $\mathbb{V}$, denoted by $\mathbb{A}(v_1), \mathbb{A}(v_2), \cdots, \mathbb{A}(v_m)$, such that:

1) Each set $\mathbb{A}(v_i)$ $(1 \leq i \leq m)$ contains exactly one backbone node v_i.
2) $\bigcup_{i=1}^{m} \mathbb{A}(v_i) = \mathbb{V}$, and $\mathbb{A}(v_i) \bigcap \mathbb{A}(v_j) = \emptyset$ $(1 \leq i \neq j \leq m)$.
3) $\forall v_u \in \mathbb{A}(v_i)$ $(1 \leq i \leq m)$ and $v_u \neq v_i$, such that $(v_u, v_i) \in \mathbb{E}$.

A Non Backbone Node Allocation Scheme *is:* $\mathscr{A} = \{\mathbb{A}(v_i) \mid \forall v_i \in \mathbb{B}, 1 \leq i \leq m\}$.

As we mentioned in Section 1, the potential traffic load indicator on each backbone node is the degree of the node, *i.e.*, d_i, for $\forall v_i \in \mathbb{B}$. However, d_i is not the actual traffic load. The actual traffic load only can be determined when a non backbone node allocation scheme $\mathscr{A}$ is decided. In other words, the number of allocated non backbone nodes is an indicator of the actual traffic load on each backbone node. According to this observation, we give the following definition:

Definition 24. *Valid Degree (d').* The *Valid Degree* of a backbone node v_i is the number of its allocated non backbone nodes, *i.e.*, $\forall v_i \in \mathbb{B}, d_i' = |\mathbb{A}(v_i)| - 1$, where $|\mathbb{A}(v_i)|$ represents the cardinality of the set $\mathbb{A}(v_i)$.

Finally, we are dedicated to find a load-balanced non backbone node allocation scheme $\mathscr{A}$, namely, the maximum *valid degree* of all the backbone nodes is minimized under $\mathscr{A}$.

Definition 25. *MinMax Valid-Degree non Backbone node Allocation (MVBA) Problem.* For a WSN represented by graph $\mathbb{G}(\mathbb{V}, \mathbb{E})$ and an *LBVB* $\mathbb{B} = \{v_1, v_2, \cdots, v_m\}$, the *MVBA* problem is to find a backbone allocation scheme $\mathscr{A}^*$, such that: the $\max\{d_i' \mid \forall v_i \in \mathbb{B})\}$ is minimized under $\mathscr{A}^*$.

3 Load Balanced Virtual Backbone Problem

In this section, we first introduce how to solve the MinMax Degree Maximal Independent Set (MDMIS) Problem. Since finding an MIS is a well-known NP-complete problem [13] in graph theory, LBVB is NP-complete as well. Next, we formulate the MDMIS problem as an Integer Nonlinear Programming (INP). Subsequently, we show how to obtain an $O(\Delta \ln(n))$ approximation solution by using Linear Programming (LP) relaxation techniques. Finally, we present how to find a minimum-sized set of MIS connectors to form an LBVB $\mathbb{B}$.

3.1 INP Formulation of MDMIS

Consider a WSN described by graph $\mathbb{G} = (\mathbb{V}, \mathbb{E})$. First we define the 1-Hop Neighborhood of a node v_i.

Definition 31. *1-Hop Neighborhood ($\mathbb{N}_1(v_i)$).* $\forall v_i \in \mathbb{V}$, the 1-Hop Neighborhood of node v_i is defined as: $\mathbb{N}_1(v_i) = \{v_j \mid v_j \in \mathbb{V}, e_{ij} = (v_i, v_j) \in \mathbb{E}\}$.

The physical meaning of 1-Hop Neighborhood is the set of the nodes that can be directly reached from node v_i.

Next we formally model the MDMIS problem as an Integer Nonlinear Program (INP).

DS Property Constraint. As we mentioned early, an MIS is also a DS. Hence, we should formulate the DS constraint for the MDMIS problem. For convenience, we assign a decision variable x_i for each sensor $v_i \in \mathbb{V}$, which is allowed to be 0/1 value. This variable sets to 1 *iff* the node is an independent node, *i.e.*, $\forall v_i \in \mathbb{D}, x_i = 1$. Otherwise, it sets to 0. The DS property states that each non independent node must reside within the 1-hop neighborhood of at least one independent node. We therefore have: $x_i + \sum_{v_j \in \mathbb{N}_1(v_i)} x_j \geq 1, \forall v_i \in \mathbb{V}$.

IS Property Constraint. Since the solution of the MDMIS problem is at least an IS, the IS property is also a constraint of MDMIS. The IS property indicates that no two independent nodes are adjacent, *i.e.*, $\forall v_i, v_j \in \mathbb{D}, (v_i, v_j) \notin \mathbb{E}$. In other words, we have: $\sum_{v_j \in \mathbb{N}_1(v_i)} x_i \cdot x_j = 0, \forall v_i \in \mathbb{V}$.

Consequently, the objective of the MDMIS problem is to minimize the maximum degree of all the independent nodes. We denote z as the objective of the MDMIS problem, *i.e.*, $z = \max_{v_i \in \mathbb{D}}(d_i)$. Mathematically, the MDMIS problem can be formulated as an integer nonlinear programming INP_{MDMIS} as follows:

$$\begin{aligned} \min \quad & z = \max\{d_i \mid \forall v_i \in \mathbb{D}\} \\ s.t. \quad & x_i + \sum_{v_j \in \mathbb{N}_1(v_i)} x_j \geq 1; \\ & \sum_{v_j \in \mathbb{N}_1(v_i)} x_i \cdot x_j = 0; \\ & x_i, x_j \in \{0,1\}, \ \forall v_i, v_j \in \mathbb{V}. \end{aligned} \qquad (INP_{MDMIS})$$

Since the *IS property constraint* is quadratic, the formulated integer programming INP_{MDMIS} is not linear. To linearize INP_{MDMIS}, the quadratic constraint is eliminated by applying the techniques proposed in [14]. More specifically, the product $x_i \cdot x_j$ is replaced by a new binary variable χ_{ij}, on which several additional constraints are imposed. As a consequence, we can reformulate INP_{MDMIS} exactly to an Integer Linear Programming ILP_{MDMIS} by introducing the following linear constraints:

$$\begin{aligned} & \sum_{v_j \in \mathbb{N}_1(v_i)} \chi_{ij} = 0; \\ & x_i \geq \chi_{ij}; \\ & x_j \geq \chi_{ij}; \\ & x_i + x_j - 1 \leq \chi_{ij}; \\ & \chi_{ij} \in \{0,1\}, \ \forall v_i, v_j \in \mathbb{V}. \end{aligned}$$

For convenience, we assign a random variable l_{ij} for each edge in the graph $\mathbb{G}$ modeled from a WSN, *i.e.*, $l_{ij} = \begin{cases} 1, & \textit{if } (v_i, v_j) \in \mathbb{E}. \\ 0, & \text{otherwise}. \end{cases}$ Thus, we obtain that $d_i = \sum_{v_j \in \mathbb{N}_1(v_i)} x_i l_{ij}, \forall v_i \in \mathbb{V}$. Moreover, by relaxing the conditions $x_j \in \{0,1\}$, and $\chi_{ij} \in \{0,1\}$ to $x_j \in [0,1]$, and $\chi_{ij} \in [0,1]$, correspondingly, we obtain the following relaxed linear programming LP^*_{MDMIS}:

$$\begin{aligned} \min \ & z = \max\{1, \max\{d_i = \sum_{v_j \in \mathbb{N}_1(v_i)} x_i l_{ij} \mid \forall v_i \in \mathbb{V}\}\} \\ s.t. \ & x_i + \sum_{v_j \in \mathbb{N}_1(v_i)} x_j \geq 1; \\ & \sum_{v_j \in \mathbb{N}_1(v_i)} \chi_{ij} = 0; \\ & x_i \geq \chi_{ij}; \\ & x_j \geq \chi_{ij}; \\ & x_i + x_j - 1 \leq \chi_{ij}; \\ & x_i, x_j, \chi_{ij} \in [0,1], \ \forall v_i, v_j \in \mathbb{V}. \end{aligned} \qquad (LP^*_{MDMIS})$$

3.2 Approximation Algorithm

Due to the relaxation enlarged the optimization space, the solution of LP^*_{MDMIS} corresponds to a lower bound to the objective of INP_{MDMIS}. Given an instance of MDMIS modeled by the integer nonlinear programming INP_{MDMIS}, the sketch of the proposed approximation algorithm (see Algorithm 1) is summarized as follows: first, solve the relaxed linear programming LP^*_{MDMIS} to get an optimal fractional solution, denoted by $(\mathbf{x}^*, z^*)$, where $\mathbf{x}^* =< x_1^*, x_2^*, \cdots, x_n^* >$, and then round x_i^* to integers $\widehat{x}_i$ according to five steps: 1) Sort sensor nodes by the x_i^* value (where $1 \leq i \leq n$) in the decreasing order (line 2). 2) Set all $\widehat{x}_i$ to be 0 (line 3-5). 3) Start from the first node in the sorted node array A (line 8). If there is no node been selected as an independent node in v_i's 1-hop neighborhood (line 11), then let $\widehat{x}_i = 1$ with probability $p_i = \max(x_i^*, \frac{1}{d_i})$ (line 12). 4) Repeat step 3) till reaching the end of array A (line 9 - 15). 5) Repeat step 3) and 4) for $3(\Delta + 1)\ln(n)$ times (line 7 - 17). Next the correctness of our proposed approximation algorithm (Algorithm 1) is proven, followed by the performance ratio analysis. Before showing the correctness of Algorithm 1, two important lemmas are derived as follows. The proofs of Lemma 1, Lemma 2, and Theorem 1 are omitted due to space limitation.

Algorithm 1. Approximation Algorithm for MDMIS

Require: A WSN represented by graph $\mathbb{G} = (\mathbb{V}, \mathbb{E})$; Node degree d_i.
1: Solve LP^*_{MDMIS}. Let $(\mathbf{x}^*, z^*)$ be the optimum solution, where $\mathbf{x}^* =< x_1^*, x_2^*, \cdots, x_n^* >$, $z^* = \max(1, \sum_{v_j \in \mathbb{N}_1(v_i)} x_i^* l_{ij})$.
2: Sort all the sensor nodes by the x_i^* value in the decreasing order. The sorted node ID i is stored in the array denoted by $A[n]$.
3: **for** $i = 1$ **to** n **do**
4:     $\widehat{x}_i = 0$.
5: **end for**
6: $counter = 0$.
7: **while** $counter \leq \beta$, where $\beta = 3(\Delta + 1)\ln(n)$ **do**
8:     $k = 0$.
9:     **while** $k < n$ **do**
10:         $i = A[k]$.
11:         **if** $\forall v_j \in \mathbb{N}_1(v_i)$, $\widehat{x}_j = 0$, **then**
12:             $\widehat{x}_i = 1$ with probability $p_i = \max(x_i^*, \frac{1}{d_i})$.
13:         **end if**
14:         $k = k + 1$.
15:     **end while**
16:     $counter = counter + 1$.
17: **end while**
18: **return** $(\widehat{\mathbf{x}}, \widehat{z} = \max(1, d_i = \sum_{v_j \in \mathbb{N}_1(v_i)} \widehat{x}_i l_{ij}))$.

Lemma 1. For a WSN represented by $\mathbb{G} = (\mathbb{V}, \mathbb{E})$, if a subset $\mathbb{S} \subseteq \mathbb{V}$ is a DS and meanwhile $\mathbb{S}$ is also an IS, then this subset $\mathbb{S}$ is an MIS of $\mathbb{G}$.

Lemma 2. The set $\mathbb{D} = \{v_i \mid \widehat{x_i} = 1, 1 \leq i \leq n\}$, where $\widehat{x_i}$ is derived from Algorithm 1, is a DS almost surely.

Theorem 1. The set $\mathbb{D} = \{v_i | \widehat{x_i} = 1, 1 \leq i \leq n\}$, where $\widehat{x_i}$ is derived from Algorithm 1, is an MIS.

From Theorem 1, the solution of our proposed approximation Algorithm 1 is an MIS. Subsequently, we analyze the approximation factor of Algorithm 1 in Theorem 2. We only provide proof sketch of Theorem 2 in the paper due to space limitation.

Theorem 2. Let OPT_{MDMIS} denote the optimal solution of the MDMIS problem. The proposed algorithm yields a solution of $O(\Delta \ln(n))OPT_{MDMIS}$.

Proof Sketch: The expected d_i of the independent node v_i found by Algorithm 1 is:

$$E[\sum_{v_j \in \mathbb{N}_1(v_i)} \widehat{x_i} l_{ij}] \leq \sum_{v_j \in \mathbb{N}_1(v_i)} (\beta x_i^*) E[l_{ij}] \leq \beta z^*. \tag{1}$$

The first inequality holds because the procedure, setting $\widehat{x_i} = 1$ with probability p_i, is repeated β times. By the union bound, we get $Pr[\widehat{x_i} = 1] = Pr[\bigcup_{t \leq \beta} \widehat{x_i} = 1 \textit{ at round } t] \leq \beta x_i^*$. This implies $E(\widehat{x_i}) \leq \beta x_i^*$. The last inequality follows from the fact that $\sum_{v_j \in \mathbb{N}_1(v_i)} x_i^* \cdot E[l_{ij}] \leq \max\{d_i \mid v_i \in \mathbb{D}\} = z^*$.

Applying the Chernoff bound, we obtain the bound:

$$Pr[\sum_{v_j \in \mathbb{N}_1(v_i)} \widehat{x_i} l_{ij} \geq (1+\mu)\beta z^*] \leq (\frac{e^{\mu}}{(1+\mu)^{1+\mu}})^{\beta z^*}. \tag{2}$$

for arbitrary $\mu > 0$. To simplify this bound, let $\mu = e - 1$, we get

$$Pr[\sum_{v_j \in \mathbb{N}_1(v_i)} \widehat{x_i} l_{ij} \geq (1+\mu)\beta z^*] \leq \frac{1}{n^3}. \tag{3}$$

The inequality holds since $z^* = \max\{1, \max\{d_i = \sum_{v_j \in \mathbb{N}_1(v_i)} x_i l_{ij} \mid \forall v_i \in \mathbb{V}\}\} \geq 1$. Applying the union bound, we get the probability that some independent node has a degree larger than $(1+\mu)\beta z^*$,

$$Pr[\widehat{z} \geq (1+\mu)\beta z^*] \leq n\frac{1}{n^3} = \frac{1}{n^2}. \tag{4}$$

$\sum_{n>0} \frac{1}{n^2}$ is a particular case of the Riemann Zeta function, then $\sum_{n>0} \frac{1}{n^2}$ is bound, *i.e.*, $\sum_{n>0} \frac{1}{n^2} < \infty$ by the result of the Basel problem. Thus, according to the Borel-Cantelli Lemma, $P[\widehat{z} \geq (1+\mu)\beta z^*] \sim 0$.

According to Lemma 2, and Inequality (4), we get

$$Pr[\text{some node is selected to be an independent node in 1-hop neighborhood} \\ \bigcap \widehat{z} \leq (1+\mu)\beta z^*] = 1 \cdot (1 - \tfrac{1}{n^2}) \sim 1, \text{ when } n \sim \infty, \text{and } \mu = e - 1.$$

□

3.3 Connected Virtual Backbone

To solve the LBVB problem, one more step is needed after constructing an MDMIS, which is to make the MDMIS connected. Next, we introduce how to find a minimum-sized set of MIS connectors to connect the MDMIS.

We first divide the MDMIS $\mathbb{D}$ into disjoint node sets according to the following criterion:

$$\mathbb{D}_0 = \{v_i \mid \forall v_i \in \mathbb{D} \text{ and } v_i \text{ has the minimized node ID among all the nodess in } \mathbb{D}\}$$

$$\mathbb{D}_\iota = \{v_i \mid v_i \in \mathbb{D}, \exists v_j \in \mathbb{D}_{\iota-1}, v_i \in \mathbb{N}_2(v_j), v_i \notin \bigcup_{k=0}^{\iota-1} \mathbb{D}_k\}$$

The independent node with smallest node ID is put into $\mathbb{D}_0$. Clearly, $|\mathbb{D}_0| = 1$. All the independent nodes in the 2-Hop Neighborhood of the nodes in $\mathbb{D}_{\iota-1}$ are put into $\mathbb{D}_\iota$. Hence, ι is called the *level* of an independent node. $\mathbb{D}_\iota$ represents the set of independent nodes of level ι in $\mathbb{G}$ with respect to the node in $\mathbb{D}_0$. Additionally, suppose the maximum level of an independent node is L. For each $0 \leq i \leq L-1$, let $\mathbb{S}_i$ be the set of the nodes adjacent to at least one node in $\mathbb{D}_i$ and at least one node in $\mathbb{D}_{i+1}$. Subsequently, compute a minimum-sized set of nodes $\mathbb{C}_i \subseteq \mathbb{S}_i$ cover node set $\mathbb{D}_{i+1}$. Let $\mathbb{C} = \bigcup_{i=0}^{L-1} \mathbb{C}_i$ and therefore $\mathbb{B} = \mathbb{D} \bigcup \mathbb{C}$ is a *Load Balanced Virtual Backbone* of the original graph $\mathbb{G}$.

We use the WSN shown in Fig. 2 (a) as an example to explain the construction process of an LBVB. In Fig. 2 (a), each circle represents a sensor node. As we mentioned early, the construction process consists of two steps. In the first step, it solves the MDMIS problem by Algorithm 1 to obtain $\mathbb{D}$ which is shown in Fig. 2 (b) by black circles. In $\mathbb{D}$, suppose v_i is the node with the smallest node ID. Then, the number besides each independent node is the level of that node with respect to v_i. In the second phase, we choose the appropriate MIS connectors ($\mathbb{C}$), shown by gray nodes in Fig. 2 (c), to connect all the nodes in $\mathbb{D}$ to form an LBVB ($\mathbb{B}$). Next, we analyze the number of backbone nodes $|\mathbb{B}|$ produced by our algorithm. The proof of Theorem 3 is omitted due to space limitation.

Theorem 3. The number of backbone nodes $|\mathbb{B}| \leq 2|\mathbb{D}|$.

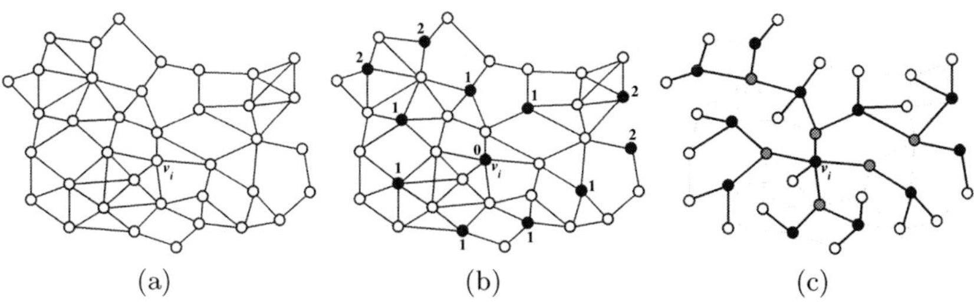

Fig. 2. Illustration of LBVB construction process

4 MinMax Valid-Degree Non Backbone Node Allocation

4.1 ILP Formulation of MVBA

According to Definition 25, the MVBA problem can be modeled by a binary problem with an linear objective functions, which is a known NP-Hard problem. In this subsection, we first model the MVBA problem as an ILP.

We define a binary variable b_i to indicate whether the sensor v_i is a backbone node or not. b_i sets to be 1 *iff* the sensor v_i is a backbone node. Otherwise, b_i sets to be 0 *iff* the sensor v_i is a non backbone node. Additionally, we assign a random variable a_{ij} for each edge connecting a backbone node v_i and a non backbone node v_j on the graph modeled from a WSN, *i.e.*,

$$a_{ij} = \begin{cases} 1, & \text{if non backbone node } v_j \text{ is allocated to backbone node } v_i. \\ 0, & \text{otherwise.} \end{cases}$$

Consequently, the MVBA problem can be formulated as an Integer Linear Programming ILP_{MVBA} as follows:

$$\begin{aligned} \min \quad & y = \max\{d'_i \mid \forall v_i \in \mathbb{B}\} \\ s.t. \quad & \sum_{v_i \in \mathbb{N}_1(v_j)} b_i a_{ij} = 1, \ \forall v_j \notin \mathbb{B}; \\ & a_{ij} \in \{0, 1\}. \end{aligned} \qquad (ILP_{MVBA})$$

The objective function y is the maximum valid degree (d') of all the backbone nodes. The first constraint states that each non backbone node can be allocated to only one backbone node, whereas the second constraint indicates that a_{ij} is a binary variable. By relaxing variable $a_{ij} \in \{0, 1\}$ to $a_{ij} \in [0, 1]$, we get the relaxed formulation which falls into a standard Linear Programming (LP) problem, denoted by LP^*_{MVBA} as follows:

$$\begin{aligned} \min \quad & y = \max\{1, \max\{ \sum_{v_j \in \mathbb{N}_1(v_i)} b_i a_{ij} \mid \forall v_i \in \mathbb{B}\}\} \\ s.t. \quad & \sum_{v_i \in \mathbb{N}_1(v_j)} b_i a_{ij} = 1, \ \forall v_j \notin \mathbb{B}; \\ & a_{ij} \in [0, 1]. \end{aligned} \qquad (LP^*_{MVBA})$$

Due to the relaxation enlarged the optimization space, the solution of LP^*_{MVBA} corresponds to a lower bound to the objective of ILP_{MVBA}.

4.2 Randomized Approximation Algorithm

Given an instance of MVBA modeled by the integer linear programming ILP_{MVBA}, the sketch of the randomized approximation algorithm (see Algorithm 2) is summarized as follows: first, solve the relaxed linear programming LP^*_{MVBA} to get an optimal fractional solution, denoted by $(\mathbf{a}^*, y^*)$, where $\mathbf{a}^* =< a^*_{11}, \cdots, a^*_{1n}, a^*_{21}, \cdots, a^*_{2n}, \cdots, a^*_{m1}, \cdots, a^*_{mn} >$, and then round a^*_{ij} to integers $\widehat{a_{ij}}$ by a random rounding procedure, which consists of four steps: 1) Set all $\widehat{a_{ij}}$ to be 0 (line 2). 2) Let $\widehat{a_{ij}} = 1$ with probability a^*_{ij} and execute this step

Algorithm 2. Approximation Algorithm for MVBA

Require: A WSN represented by graph $\mathbb{G} = (\mathbb{V}, \mathbb{E})$.
1: Solve LP^*_{MVBA}. Let $(\mathbf{a}^*, y^*)$ be the optimum solution.
2: $\widehat{a_{ij}} = 0$.
3: **while** $k \leq \alpha^2$, where $\alpha = \log(n) + 3$ **do**
4: $\quad \widehat{a_{ij}} = 1$ with probability a^*_{ij}
5: $\quad k = k + 1$
6: **end while**
7: **if** $((v_i, v_j) \in \mathbb{E})$ and $(v_i \in \mathbb{B}$ or $v_j \in \mathbb{B})$ **then**
8: $\quad \widehat{a_{ij}} = 1$ with probability $\frac{1}{\Delta}$.
9: **end if**
10: **repeat**
11: $\quad$ line 3 - 6
12: **until** $\sum_{v_i \in \mathbb{N}_1(v_j)} b_i \widehat{a_{ij}} = 1$
13: **return** $(\widehat{\mathbf{a}}, \widehat{y} = \max(1, \sum_{v_j \in \mathbb{N}_1(v_i)} b_i \widehat{a_{ij}}))$.

for α^2 times (line 3 - 6), where $\alpha = \log(n) + 3$. 3) Let $\widehat{a_{ij}} = 1$ with probability $\frac{1}{\Delta}$ (line 7). 4) To ensure $(\widehat{a_{ij}}, \widehat{y})$ is a feasible solution to ILP_{MVBA}, repeat steps 2) and 3) until every non backbone node is assigned a backbone node.

From the similar techniques we used in Theorem 2, we can obtain the approximation factor of Algorithm 2 is $O(\log^2(n))(OPT_{MVBA} + \frac{1}{\alpha^2})$ with probability $\frac{7}{8}$, when $\alpha = \log(n) + 3$. The proof is omitted due to space limitation.

5 Conclusion

In this paper, we address three fundamental problems of constructing a load-balanced VB in a WSN. More specifically, we solve the LBVB problem which is claimed to a NP-Complete problem with two steps. First, the MDMIS problem aims to find the optimal MIS such that the maximum degree of all the independent nodes is minimized. To solve this problem, a near optimal approximation algorithm is proposed, which yields an $O(\Delta \ln(n))$ approximation factor. Subsequently, the minimum-sized set of MIS connectors are found to make the MDMIS connected. The theoretical upper bound of the number of backbone nodes is analyzed in this paper as well. In the end, the MVBA problem is dedicated to allocate non backbone nodes to proper backbone nodes with an objective to minimize the maximum valid degree of all the backbone nodes, which is a NP-Hard problem. To solve this problem, we propose an approximation algorithm by using linear relaxing and random rounding techniques, which yields a solution of $O(\log^2(n))$ approximation factor of traffic load on each backbone node.

Acknowledgments. This research was supported in part by the National Science Foundation (NSF) under Grants CNS-1152001, CNS-0831634, and the 111 project of China under the grant No. 111-2-14.

References

1. Hadim, G., Mohamed, N.: Middleware Challenges and Approaches for Wireless Sensor Networks. IEEE Distributed Systems 7(3), 1–1 (2006)
2. Ni, S., Tseng, Y., Chen, Y., Sheu, J.: The Broadcast Storm Problem in a Mobile Ad Hoc Network. In: MOBICOM, pp. 152–162 (1999)
3. Das, B., Bharghavan, V.: Routing in Ad Hoc Networks Using Minimum Connected Dominating Sets. In: ICC (1997)
4. Ji, S., Li, Y., Jia, X.: Capacity of Dual-Radio Multi-Channel Wireless Sensor Networks for Continuous Data Collection. In: Infocom (2011)
5. Wan, P.J., Huang, S.C.-H., Wang, L., Wan, Z., Jia, X.: Minimumlatency aggregation scheduling in multihop wireless networks. In: MobiHoc (2009)
6. Yan, M., He, J., Ji, S., Li, Y.: Multi-Regional Query Scheduling in Wireless Sensor Networks with Minimum Latency. To appear in the Wireless Communications and Mobile Computing, WCMC (2012)
7. Cai, Z., Ji, S., He, J., Bourgeois, A.G.: Optimal Distributed Data Collection for Asynchronous Cognitive Radio Networks. In: ICDCS (2012)
8. Kim, D., Wang, W., Li, X., Zhang, Z., Wu, W.: A New Constant Factor Approximation for Computing 3-Connected m-Dominating Sets in Homogeneous Wireless Networks. In: INFOCOM (2010)
9. Ding, L., Gao, X., Wu, W., Lee, W., Zhu, X., Du, D.Z.: Distributed Construction of Connected Dominating Sets with Minimum Routing Cost in Wireless Networks. In: ICDCS (2010)
10. Kim, D., Wu, Y., Li, Y., Zou, F., Du, D.Z.: Constructing Minimum Connected Dominating Sets with Bounded Diameters in Wireless Networks. TPDS 20(2) (2009)
11. He, J., Ji, S., Pan, Y., Li, Y.: Greedy Construction of Load-Balanced Virtual Backbones in Wireless Sensor Networks. To appear in the Wireless Communications and Mobile Computing, WCMC (2012)
12. He, J., Ji, S., Yan, M., Pan, Y., Li, Y.: Load-Balanced CDS Construction in Wireless Sensor Networks Via Genetic Algorithm. International Journal of Sensor Networks (IJSNET) 11(3), 166–178 (2012)
13. Garey, M., Johnson, D.: Computers and Intractability: A Guide to the Theory of NP-Completeness. W. H. Freeman, New York (1983)
14. Gueyea, S., Michelonb, P.: A linearization framework for unconstrained quadratic (0-1) problems. Discrete Applied Mathematics 157, 1255–1266 (2009)

Maximum Matching in Multi-Interface Networks*

Adrian Kosowski[1], Alfredo Navarra[2], Dominik Pajak[1], and Cristina M. Pinotti[2]

[1] INRIA Bordeaux Sud-ouest, LaBRI, 33400 Talence, France
{adrian.kosowski,dominik.pajak}@labri.fr

[2] Dipartimento di Matematica e Informatica, Università degli Studi di Perugia, Italy
{alfredo.navarra,pinotti}@unipg.it

Abstract. In heterogeneous networks, devices can communicate by means of multiple wireless interfaces. By choosing which interfaces to switch on at each device, several connections might be established. That is, the devices at the endpoints of each connection share at least one active interface.

In this paper, we consider the standard matching problem in the context of multi-interface wireless networks. The aim is to maximize the number of parallel connections without incurring in interferences. Given a network $G = (V, E)$, nodes V represent the devices, edges E represent the connections that can be established. If node x participates in the communication with one of its neighbors by means of interface i, then another neighboring node of x can establish a connection (but not with x) only if it makes use of interface $j \neq i$. The size of a solution for an instance of the outcoming matching problem, that we call *Maximum Matching in Multi-Interface* networks (*3MI* for short), is always in between the sizes of the solutions for the same instance with respect to the standard matching and its induced version problems. However, we prove that *3MI* is *NP*-hard even for proper interval graphs and for bipartite graphs of maximum degree $\Delta \geq 3$. We also show polynomially solvable cases of *3MI* with respect to different assumptions.

1 Introduction

Wireless networks have been deeply considered as one of the most interesting topics from both practical and theoretical points of view. One of their more challenging characteristics is certainly related to the heterogeneity of the involved devices that might interact in order to exchange data. Wireless networks are, in fact, composed of devices with different capabilities like computational power, energy consumption, radio interfaces, supported communication protocols, and so forth. In this paper, we are interested in devices equipped with multiple interfaces (e.g., Bluetooth, WiFi, GPRS). Connections among devices might be accomplished by means of different communication networks according

* Research supported by the LaBRI under the "Project émergent" program.

G. Lin (Ed.): COCOA 2012, LNCS 7402, pp. 13–24, 2012.

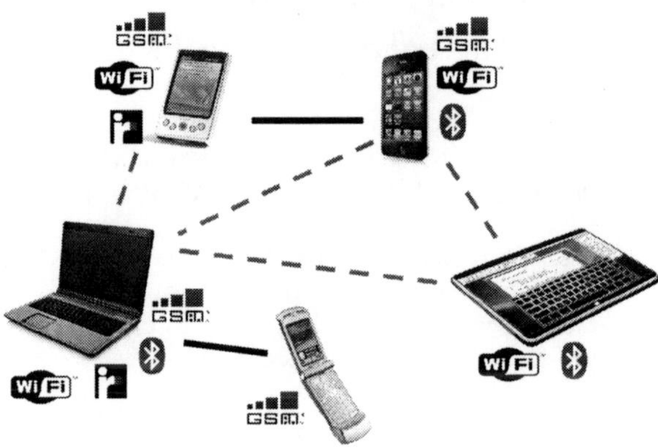

Fig. 1. An instance for *3MI* represented by a set of devices connected according to their proximities and their available interfaces. The black full edges represent a possible solution for *3MI* if they are correctly activated by means of suitable interfaces.

to connectivity and quality of service requirements. The selection of the most suitable interface for a specific connection might depend on various factors. Such factors include: its availability in specific devices, the required communication bandwidth, the cost (in terms of energy consumption) of maintaining an active interface, the available neighbors, and so forth.

We study communication problems in wireless networks supporting multiple interfaces. In the considered model, the input network is described by a graph $G = (V, E)$, where nodes V represent the set of wireless devices and E is the set of possible connections according to the devices' proximity and the available interfaces that they may share. Each $v \in V$ is associated with a set of available interfaces $W(v)$. The set of all the interfaces available in the network is then determined by $\bigcup_{v \in V} W(v)$; we denote the cardinality of this set by k. We say that a connection is satisfied (or covered) when the endpoints of the corresponding edge share at least one active interface. In this setting, we study the problem of establishing the maximum set of communication edges without incurring in interferences. We assume that two communications/edges do interfere if they share a node or if they are activated by means of the same interface and connected by one edge. Note that, in the latter case, i.e., when two edges at distance one are activated by means of the same interface, then also the edge in the middle is activated by means of the same interface, since its endpoints do share a common active interface. An example of such a behavior is shown in Figure 1. The two black full edges represent a possible solution to *3MI* if the black edge on the top is activated by means of the WiFi interface and the one on the bottom by means of the GSM interface. If both edges are activated by means of the GSM interface, then it will not represent a feasible solution since the connection between the laptop and the PDA on its top will be activated as well.

Related Work. Multi-interface wireless networks have been recently studied in a variety of contexts, usually focusing on the benefits of multiple radio devices of each node. Many basic problems of standard wireless network optimization can be reconsidered in such a setting [2], in particular, focusing on issues related to routing [10] and network connectivity [7,13]. The study of combinatorial problems on multi-interface wireless networks has originated from [6]. That paper, as well as [9,19], investigates the so called *Coverage* problem, where the goal is the activation of the minimum cost set of interfaces in such a way that all the edges of G are covered. *Connectivity* issues have been addressed in [1,9,21]. The goal becomes to activate the minimum cost set of interfaces in G in order to guarantee a path of communication between every pair of nodes. In [21], the attention has been devoted to the so called *Cheapest path* problem. This corresponds to the well-known shortest path problem but in the context of multi-interface networks. Bandwidth constraints have been addressed in [8], by studying flow problems on multi-interface networks.

A natural continuation on investigating such kind of networks is certainly to consider another basic problem like the matching. Given a graph $G = (V, E)$, a *maximum matching* $M \subseteq E$ in G is given by the largest set of pairwise disjoint edges (that is, no two edges in M share a common node). The maximum matching is well-known and it is polynomially solvable [12]. For a positive integer p, the *distance-p maximum matching* problem, asks for a maximum size set of edges whose pairwise distance[1] is at least p in G. Such a problem is in general *NP*-hard [4,5]. In particular, the special case of $p = 2$ is referred in the literature as the *maximum induced matching* problem, *MIM* for short, and it has been extensively studied. For instance, a solution for *MIM* on the underlying graph shown in Figure 1 would be given by only one edge. *MIM* has been shown to be *NP*-hard for various graph classes like bipartite graphs [5], line graphs (and hence including claw-free graphs) [20], and trapezoid graph [16]. Moreover, it is *APX*-complete in d-regular graphs, $d \geq 3$ [11], and a $d-1$-approximation algorithm has been provided for d-regular graphs, $d \geq 3$. The size of *MIM* on twinless graphs and planar graphs has been studied in [18], while for subcubic planar graphs in [17]. Polynomial time algorithms have been devised for chordal graphs [5], trees [22], and other special graphs. It is worth mentioning that, when $p = 3$, the problem is *NP*-hard [4] even for strongly chordal graphs.

Our Results. In this paper, we study the maximum matching problem in the context of multi-interface networks. The aim is to maximize the number of parallel connections without incurring in interferences. The main difference with the standard matching problem is that when an edge is selected, it is associated with an interface by which the connection is established. Hence, if an edge $e \in E$ is selected for the matching along with a communication interface i, then any other edge $e' \in E$ selected for the matching at distance 1 from e in G must be selected with interface $j \neq i$. We refer to this problem as the *Maximum Matching in Multi-Interface* networks (*3MI* for short). In general, we do also consider

[1] The distance between two edges is defined as the length (number of edges) of the shortest path connecting them.

Table 1. Complexity results achieved for *3MI*

Graph classes	k	Complexity
Complete	Unbounded	Optimally solvable in $O(n^{3/2}k^{3/2})$
Complete bipartite	Unbounded	Optimally solvable in $O(n^{3/2}k^{3/2})$
Bounded treewidth	Unbounded	Optimally solvable in polynomial time
Interval	Bounded	Optimally solvable in polynomial time
Unit interval	Unbounded	*NP*-hard
Bipartite, $\Delta \geq 3$	Bounded	*NP*-hard (from [5])
Claw-free	Bounded	*NP*-hard (from [20])

the total number k of available interfaces among the network as part of the input (*unbounded case*). However, sometimes k is differently specified as a fixed constant (*bounded case*). Clearly, the size of a solution to an instance G of *3MI* cannot be bigger than that of the solution provided by the standard maximum matching, and cannot be smaller than that of a solution provided by its induced version, i.e.:

$$\mathit{MIM}(G) \leq \mathit{3MI}(G) \leq \text{ Matching}(G).$$

In particular, when $k = 1$, *3MI* coincides with *MIM*. Hence, all the hardness results concerning *MIM* do also hold for *3MI*. However, it is worth mentioning that *MIM* has been proved to be polynomially solvable on chordal graphs while the distance-3 maximum matching, whose size is clearly smaller than *MIM*, is *NP*-hard for chordal graphs but polynomially solvable on strongly chordal graphs. In this paper, we prove that *3MI* is *NP*-hard on proper interval graphs (that are contained in the chordal graphs class), and we present some polynomial time algorithms for complete graphs, complete bipartite graphs, bounded treewidth graphs, and even for interval graphs but under the bounded case. Table 1 summarizes the obtained results.

2 Definitions and Notation

For a graph G, we denote by V its node set, by E its edge set, and by Δ its maximum node degree. Unless otherwise stated, the graph $G = (V, E)$ representing the network is assumed to be undirected, connected, and without multiple edges and loops. A global assignment of the interfaces to the nodes in V is given in terms of an appropriate interface assignment function W, as follows.

Definition 1. *A function* $W: V \to 2^{\{1,2,\ldots,k\}}$ *is said to* cover *graph* G *if for each* $\{u, v\} \in E$ *we have* $W(u) \cap W(v) \neq \emptyset$.

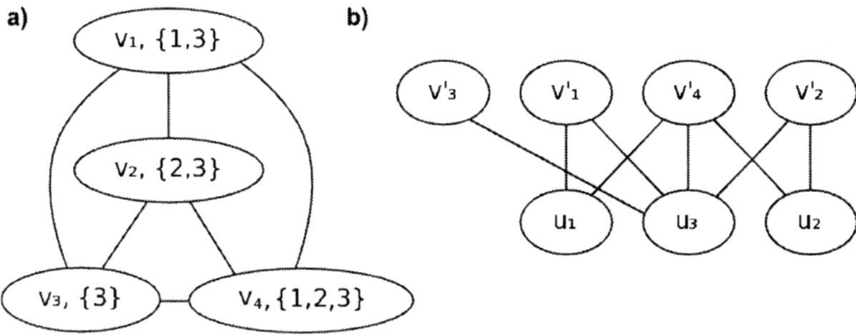

Fig. 2. a) Example of a clique G with allocation function $W(v_1) = \{1,3\}$, $W(v_2) = \{2,3\}$, $W(v_3) = \{3\}$, $W(v_4) = \{1,2,3\}$; b) The corresponding bipartite graph G'

The considered *3MI* optimization problem is formulated as follows.

3MI: Maximum Matching in Multi-Interface Networks

Input:	A graph $G = (V, E)$, a set of interfaces $I = \{1, 2, \ldots, k\}$, an allocation of available interfaces $W: V \to 2^I$ covering graph G.
Solution:	An allocation of active interfaces $W_A: V \to \{\emptyset, \{1\}, \{2\}, \ldots, \{k\}\}$, $W_A(v) \subseteq W(v)$ for all $v \in V$ such that for all $v \in V$ if $W_A(v) \neq \emptyset$, then there is exactly one neighbor of v, say w, with $W_A(w) = W_A(v)$.
Goal:	Maximize the number of edges $(v, w) \in E$ such that $W_A(v) = W_A(w) \neq \emptyset$.

Note that we do consider two variants of the above problem: the parameter k can be considered as part of the input (this is called the *unbounded case*), or k may be a fixed constant (the *bounded case*). When $k = 1$, i.e., a single interface is available, *3MI* coincides with the well-known induced matching. When each edge can be covered by means of an interface different from any other, *3MI* coincides with the standard maximum matching problem.

3 Polynomial Time Algorithms

In this section, we design polynomial time algorithms to solve *3MI* for various graph classes. Moreover, under the bounded case assumption, we polynomially solve *3MI* even for interval graphs.

Theorem 1. *3MI is solvable within $O(n^{3/2}k^{3/2})$ time for complete graphs.*

Proof. Consider the bipartite graph $G' = (V' \cup U, E')$, where one bipartite partition is a copy of the set of nodes of the complete graph, $V' = \{v' : v' \in V\}$, the other bipartite partition U corresponds to the set of available interfaces, $U = \{u_1, u_2, \ldots, u_k\}$, and the neighborhood of each node from V' is defined as the set of nodes representing interfaces available at the corresponding node of G, $E' = \{\{v', u_i\} : v' \in V, i \in W(v')\}$.

For any allocation of interfaces W on the complete graph $G = (V, E)$, an activation W_A is a solution to the *3MI* problem if and only if it is a function $W_A : V \to \{\emptyset, \{1\}, \{2\}, \ldots, \{k\}\}$ satisfying the following two properties:

(A) Each interface is either activated on exactly one pair of nodes or not at all, i.e., $|W_A^{-1}(\{i\})| \in \{0, 2\}$, for all $1 \leq i \leq k$;
(B) For all $v \in V$, only interfaces available at v may be used, i.e., $W_A(v) \subseteq W(v)$.

Consequently, there is a one-to-one correspondence between activations W_A which solve *3MI* in G and subsets of edges $F \subseteq E'$ in G' such that:

(1) $\deg_F(v') \leq 1$, for all $v' \in V'$,
(2) $\deg_F(u_i) \in \{0, 2\}$, for all $u_i \in U$.

For the sake of completeness, we note that the set of edges F corresponding to a given activation W_A is given by $F = \{\{v', u_i\} : v \in V, i \in W_A(v)\}$, and conversely, the activation W_A corresponding to such a set of edges F is given by the function $W_A(v) = \{i \in 1, \ldots, k : \{v', u_i\} \in F\}$, for all $v \in V$. Moreover, the number of vertices active in W_A is precisely equal to $2|F|$. Since the transformation between set F and function W_A can be performed in linear time, it now suffices to provide a polynomial time algorithm for finding a maximum-size set of edges $F \subseteq E'$ satisfying constraints (1) and (2). Next, form a graph G'^*, starting with graph G', and appending a triangle to each vertex $u_i \in U$, i.e., adding two new vertices, u_i^* and u_i^{**}, and three new edges, $\{u_i, u_i^*\}$, $\{u_i^*, u_i^{**}\}$, and $\{u_i^{**}, u_i\}$, to G'^*. Consider now the maximum subset of edges F^* of G'^* satisfying the following degree constraints in G'^*:

- $\deg_{F^*}(v') \leq 1$, for all $v' \in V'$,
- $\deg_{F^*}(u_i) = 2$ and $\deg_{F^*}(u_i^*) = \deg_{F^*}(u_i^{**}) = 1$, for all $u_i \in U$.

It follows by a local analysis in the triangles of G'^* that the size f of the maximum subset F in G', and the size f^* of the maximum subset F^* in G'^* satisfy the relation: $f^* = 2k + \frac{f}{2}$. Consequently, it suffices to find the maximum subset F^* in G'^*. Due to the nature of the imposed degree constraint for set F^*, this problem can be solved in $O(n^{3/2}k^{3/2})$ time by means of the algorithm for the maximum cardinality simple b-matching problem due to [14]. This determines the overall complexity of the algorithm. □

The next theorem characterizes *MIM* on complete bipartite graphs. Its proof proceeds by a slight modification of the argument used for complete graphs.

Theorem 2. *3MI is solvable in $O(n^{3/2}k^{3/2})$ time for complete bipartite graphs.*

Proof. Let $G = (V_1 \cup V_2, E)$ be the complete bipartite graph with bipartition $V = V_1 \cup V_2$. The proof proceeds by a slight modification of the argumentation for complete graphs.

We first consider the pair of bipartite graphs $G'^j = (V'^j \cup U^j, E'^j)$, $j \in \{1, 2\}$, where one bipartite partition is a copy of the chosen vertex partition of the original graph, $V'^j = \{v'^j : v \in V_j\}$, the other bipartite partition U^j corresponds to the set of available interfaces, $U^j = \{u^j_1, u^j_2, \ldots, u^j_k\}$, and the set of edges is given as $E'^j = \{\{v'^j, u^j_i\} : v \in V_j, i \in W(v)\}$.

Next, graph $G' = (V', E')$ is obtained by taking a copy of graph G'^1, a copy of graph G'^2, and connecting corresponding interfaces from U^1 and U^2 by edges; formally, $V' = V'^1 \cup U^1 \cup V'^2 \cup U^2$ and $E' = E'^1 \cup E'^2 \cup \{\{u^1_i, u^2_i\} : i \in \{1, \ldots, k\}\}$.

We now make the following characterization of functions $W_A : V \to \{\emptyset, \{1\}, \{2\}, \ldots, \{k\}\}$ which are valid activations for the instance W of *3MI* in a complete bipartite graph:

(A) Each interface is either activated on exactly one pair of nodes (one in V_1 and one in V_2) or not at all, i.e., $|W_A^{-1}(\{i\})| = 0$ or $|W_A^{-1}(\{i\}) \cap V_1| = |W_A^{-1}(\{i\}) \cap V_2| = 1$, for all $1 \le i \le k$;
(B) For all $v \in V$, only interfaces available at v may be used, i.e., $W_A(v) \subseteq W(v)$.

Let w denote the maximum possible number of nodes activated, taken over valid solutions to *3MI*. In view of the above characterization, consider subsets of edges $F \subseteq E'$ in graph G' which satisfy the following conditions:

(1) $\deg_F(v'^j) \le 1$, for all $v'^j \in V'^j$, $j \in \{1, 2\}$,
(2) $\deg_F(u) = 1$, for all $u \in U^1 \cup U^2$.

Let f denote the size of the maximum subset $F \subseteq E'$ satisfying the above constraints. By a local analysis of edge arrangements, we obtain that $f = k + \frac{w}{2}$, and knowing an appropriate maximum subset F, we can in linear time find the corresponding optimum activation W_A. As in the proof of Theorem 1, such a subset F can be found by directly applying the algorithm for the b-matching problem, leading to an optimal solution to *3MI* in $O(n^{3/2}k^{3/2})$ time. □

A wide class of problems, known as *ECC*, admit polynomial-time algorithms in graphs of bounded tree width using an approach due to Bodlaender (for a definition of the class *ECC* and details of the approach, the reader is referred to [3]). The next theorem shows that *3MI* $\in$ *ECC*.

Theorem 3. *3MI is solvable within polynomial time in the unbounded case when the input graph is a graph with bounded treewidth.*

Proof. Consider the underlying decision problem, denoted by $3MI_D$. We need to add one further bound $B \in \mathbb{Z}_0^+$ to *3MI* such that the problem will be to ask whether there exists a solution for *3MI* of size at least B. Note that, $3MI_D$ is equivalent to the following decision problem through a straightforward linear-time reduction of input data.

***Input*:** Graph $G = (V, E)$, function W, set $X = (I \cup \{\emptyset\}) \times V$, positive integer $B \le n$.
***Question*:** Do there exist function $f : V \to X$, where $f(\cdot) = (f_1(\cdot), f_2(\cdot))$, $f_1 \colon V \to I \cup \{\emptyset\}$, $f_2 \colon V \to V$, such that:
(1) $\forall_{v \in V} f_1(v) \subseteq W(v)$,
(2) $\forall_{(v,w) \in E}$ if $f_1(v) = f_1(w) \neq \emptyset$, then $f_2(v) = w$ and $f_2(w) = v$,
(3) the number of $v \in V$ such that $f_1(v) \neq \emptyset$ is at least B.

In the above, $f_1(v)$ describes the active interface at v (a value of $\emptyset$ means that the node does not turn on any interface), and $f_2(v)$ serves as a "pointer" to the neighbor of v which activates the same interface as v. If $f_1(v) \neq \emptyset$, the pair $\{v, f_2(v)\}$ belongs to the matching and the connection between this pair uses interface $f_1(v)$. Thus, any function f satisfying condition (2) yields some matching which is a solution to *3MI* in G. Thus, *3MI*$_D$ belongs to the class *ECC*. Since in [3] a polynomial time algorithm, based on dynamic programming, is given for problems of the class *ECC* for graphs with bounded treewidth, *3MI*$_D$ is polynomially solvable for graphs with bounded treewidth. A polynomial-time solution for *3MI* follows directly by considering different values of $B \in \{0, 1, \ldots, n\}$. □

The complexity of the above algorithm depends on the treewidth of the graph. We remark that when G is a tree, *3MI* can in fact be solved in $O(nk)$ time.

Finally, concerning the bounded case, the next theorem can be stated:

Theorem 4. *3MI is solvable within polynomial time for interval graphs in the bounded case.*

Proof. We recall that G is an interval graph if and only if the set of inclusion-wise maximal cliques of G can be ordered in a sequence $(M_1, M_2, \ldots, M_s)$ such that for any $v \in M_i \cap M_j$, where $i < j$, it is also the case that $v \in M_r$ for any r, $i \leq r \leq j$. Such an ordering of cliques can be performed in linear time (assuming a suitable data representation). For $i \leq r \leq s$, let $S_r = \bigcup_{i=1}^{r} M_r$, and let $M'_r = M_r \setminus S_{r-1}$ (where we assume $S_0 = \emptyset$).

We provide a polynomial-time algorithm for *3MI* using dynamic programming, and compute in the r-th step, $1 \leq r \leq s$, an array A_r indexed by all possible activations E_r of edges incident to at least one vertex of M_r, which may appear in some solution to *3MI* in graph $G[S_r]$. Then, $A_r[E_r]$ stores the size of the best solution to *3MI* in $G[S_r]$ having precisely activation E_r on the neighborhood of M_r. In any solution to *3MI* in $G[S_r]$, the number of vertices of M_r which have at least one interface activated cannot exceed $2k$, and each of these vertices can be used to activate at most one incident edge. Hence, the size of array A_r is bounded by $(|S_r| + 1)^{2k}(\Delta + 1)^{2k} = O(n^{4k})$.

In order to perform the $(r+1)$-th step, it suffices to consider all feasible sets E'_{r+1} of edges from $G[S_{r+1}]$, incident to at least one vertex from the set M'_{r+1}, which may be active in some solution to *3MI*. By a similar argument as before, there are $O(n^{4k})$ such sets to consider. We observe that any edge of $G[S_{r+1}]$, having one end-vertex in M'_{r+1} must have the other end-vertex either in M'_{r+1} or in M_r, by the properties of the clique ordering. Consequently, we build array A_{r+1} starting with an empty set, and for any pair of feasible activations E_r (indexing array A_r) and E'_{r+1}, modifying the array as follows:

- If E_r and E'_{r+1} contain a pair of edges which share an end-vertex, then no change is introduced to A_{r+1}.
- If E_r and E'_{r+1} contain a pair of edges which are activated using the same interface and which have at least one pair of end-vertices at distance one, then no change is introduced to A_{r+1}.

– Otherwise, let E_{r+1} be the union of activations E_r and E'_{r+1}, restricted to those edges which are incident to M_{r+1}. We then add a new element $A_{r+1}[E_{r+1}]{:=} A_r[E_r] + |E'_{r+1}|$ (or set $A_{r+1}[E_{r+1}]{:=} max\{A_{r+1}[E_{r+1}], A_r[E_r] + |E'_{r+1}|\}$, if an element already exists at this index).

Eventually, after step s, the maximum of values stored in A_s is the size of the optimal solution. A valid solution can be reconstructed using the set of data structures $\{A_1, \ldots, A_s\}$. Taking into account that $s \leq n$ and that each step of dynamic programming can be completed in $O(n^{8k+1})$ time, the overall runtime of the algorithm is bounded by $O(n^{8k+2})$. □

4 Hardness

In this section, we provide the main challenging but negative result for *3MI*. In fact, *3MI* turns out to be *NP*-hard even for proper interval graphs, while it is known that *MIM* is polynomially solvable for chordal graphs.

Theorem 5. *3MI is NP-hard even when for the class of proper interval graphs.*

Proof. We prove that the underlying decision problem $3MI_D$ obtained from *3MI* by adding the bound B (as described in the proof of Theorem 3), is in general *NP*-complete. The proof then proceeds by a reduction from the *3SAT* problem, which is known to be *NP*-complete [15].

	3SAT
Input:	A set U of variables, a collection C of clauses over U such that each $c \in C$ satisfies $\lvert c \rvert = 3$.
Question:	Is there a satisfying truth assignment for C?

Given an instance of *3SAT*, we construct an instance of $3MI_D$ provided by a proper interval graph G where we need to decide which interfaces to activate at each node, see Figure 3. For each variable $v \in U$ there is a chain of unit intervals (as shown in Figure 3 for the four variables x, y, z and w), each interval is associated with interfaces v, $-v$, and one further interface denoted as 1. Actually, all the nodes of G contain interface 1. Each clause $c \in C$ composed of three literals corresponds in G to 10 unit intervals vertically disposed as shown in the figure. There are two intervals associated per each literal l with interfaces $-l$, 1, and one further interface among the set $\{2, 3, 4\}$, a different one per pair. Plus, there are two intervals associated with interfaces 1, 2, 3, and 4, and other two intervals with only interface 1. The bound B is set to $(5 + 2|U|)|C|$.

($\Rightarrow$) Assume that *3SAT* has a positive answer, i.e., there exists a satisfying truth assignment for C. We show that also $3MI_D$ has a positive answer, i.e., there exists an activation of the available interfaces such that the number of edges in the obtained matching is at least $B = (5 + 2|U|)|C|$. For each variable v, if v is set to true in the solution of *3SAT*, then, along the chain of unit intervals corresponding to v we switch on interface v at the first two intervals, interface $-v$ at the second two intervals, again interface v at the third two intervals, and

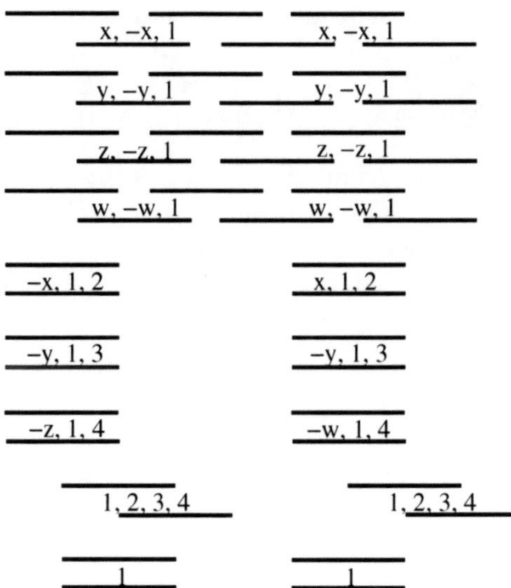

Fig. 3. The outcoming proper interval graph G from the transformation of clauses xyz and $-xyw$ to *3MI*$_D$. Interfaces hold by the intervals are specified between the lines.

so forth, always alternating between v and $-v$. If v is set to false in the solution of *3SAT*, then the activations of interfaces v and $-v$ in the chain corresponding to variable v would be still alternated but starting with $-v$. This provides a contribution of $2|U||C|$ activated edges to the final solution. Note that, in this way all the intervals corresponding to the clauses will be located below intervals that have activated interface v in the first case, and interface $-v$ in the second case. In the first case (in the second case, resp.), it follows that all the clauses containing the literal v ($-v$, resp.) have two intervals in their representation in G containing interface $-v$ (v, resp.). Hence, such intervals are allowed to switch on the interface corresponding to the negation of the appearing literal since this is not in conflict with the activation on the chain above. Since *3SAT* has been supposed to be solved, then there will always be an available interface for each clause that determines the connection of one pair of intervals among the 10 associated to the clause, as described above. This will contribute of $|C|$ edges to the final solution. Then, we activate interface 1 at all the bottom edges that hold only such interface, for a contribution of other $|C|$ activated edges. Finally, the remaining three pairs of intervals corresponding to each clause can be connected by means of interfaces 2, 3 and 4, hence contributing other $3|C|$ edges. In total, the obtained solution for *3MI*$_D$ is composed of $(5 + 2|U|)|C|$ edges.

($\Leftarrow$) Now, let us assume we have a positive answer to *3MI*$_D$. As graph G is composed of $2(5 + 2|U|)|C|$ nodes, the solution for *3MI*$_D$ must be composed of exactly $(5+2|U|)|C|$ edges, hence all the nodes participate to the matching. This means that nodes holding only interface 1 must be coupled among themselves. Any other node cannot use interface 1. For each generic clause $c = l_1l_2l_3$, the corresponding nodes holding only interfaces 1, 2, 3, and 4, must be connected among themselves by using one interface among the set $\{2, 3, 4\}$ since one of them do not share any available interface with other intervals. This means that among the other 6 intervals representing c, at least two of them must be connected among themselves by means of interface $-l_1$, or $-l_2$, or $-l_3$. Without loss of generality, let $-l_1$ be the interface used. It follows that on top of graph G, in the chain corresponding to variable $|l_1|$, the two intervals neighbors to the ones connected by means of interface $-l_1$, must be connected by means of interface l_1, as no other options are available. As "chain effect", the subsequent two intervals on the same chain (if any), and the previous two intervals on the same chain (if any) must make use of interface $-l_1$. The same arguments can be applied to each chain, and the corresponding truth assignment for the underlying *3SAT* problem is given by the interface used at the first two intervals of each chain. In fact, from the assumption that the solution for $3MI_D$ has size $(5 + 2|U|)|C|$, we have that all the intervals representing a clause have used at least one interface corresponding to the set of variables appearing in the instance of *3SAT*. That is, the assignment provided by the solution of $3MI_D$ is compatible with the satisfaction of all the clauses of *3SAT*. □

5 Conclusion

In this paper, we have considered the maximum matching problem in the context of multi-interface networks. As it has happened with other basic problems studied in the context of multi-interface networks, *3MI* turns out to be a very challenging problem. It is strictly related to the well-known maximum induced matching problem that has been extensively studied in the context of the standard graph theory. We have shown how the size of a solution for an instance of *3MI* is always in between the one for *MIM* and that for the maximum matching. Surprisingly, *3MI* has been shown to be harder than *MIM*. In fact, *MIM* is polynomially solvable on chordal graphs (and thus on interval graphs, which are a subfamily of the chordal graph), while *3MI* is *NP*-hard even for proper interval graphs. Moreover, polynomial time algorithms have been designed for some interesting graph classes, as reported in Table 1.

References

1. Athanassopoulos, S., Caragiannis, I., Kaklamanis, C., Papaioannou, E.: Energy-Efficient Communication in Multi-interface Wireless Networks. In: Královič, R., Niwiński, D. (eds.) MFCS 2009. LNCS, vol. 5734, pp. 102–111. Springer, Heidelberg (2009)
2. Bahl, P., Adya, A., Padhye, J., Walman, A.: Reconsidering wireless systems with multiple radios. SIGCOMM Comput. Commun. Rev. 34(5), 39–46 (2004)
3. Bodlaender, H.L.: Dynamic Programming on Graphs with Bounded Treewidth. In: Lepistö, T., Salomaa, A. (eds.) ICALP 1988. LNCS, vol. 317, pp. 105–118. Springer, Heidelberg (1988)
4. Brandstädt, A., Mosca, R.: On distance-3 matchings and induced matchings. Discrete Applied Mathematics 159, 509–520 (2011)

5. Cameron, K.: Induced matching. Discrete Applied Mathematics 24, 97–102 (1989)
6. Caporuscio, M., Charlet, D., Issarny, V., Navarra, A.: Energetic Performance of Service-oriented Multi-radio Networks: Issues and Perspectives. In: Proceedings of the 6th International Workshop on Software and Performance (WOSP), pp. 42–45. ACM Press (2007)
7. Cavalcanti, D., Gossain, H., Agrawal, D.: Connectivity in multi-radio, multi-channel heterogeneous ad hoc networks. In: Proceedings of the IEEE 16th International Symposium on Personal, Indoor and Mobile Radio Communications (PIMRC), pp. 1322–1326. IEEE (2005)
8. D'Angelo, G., Di Stefano, G., Navarra, A.: Bandwidth Constrained Multi-interface Networks. In: Černá, I., Gyimóthy, T., Hromkovič, J., Jefferey, K., Královič, R., Vukolić, M., Wolf, S. (eds.) SOFSEM 2011. LNCS, vol. 6543, pp. 202–213. Springer, Heidelberg (2011)
9. D'Angelo, G., Stefano, G.D., Navarra, A.: Minimize the maximum duty in multi-interface networks. Algorithmica 63(1–2), 274–295 (2012)
10. Draves, R., Padhye, J., Zill, B.: Routing in multi-radio, multi-hop wireless mesh networks. In: Proceedings of the 10th Annual International Conference on Mobile Computing and Networking (MobiCom), pp. 114–128. ACM (2004)
11. Duckworth, W., Manlove, D., Zito, M.: On the approximability of the maximum induced matching problem. Journal of Discrete Algorithms 3, 79–91 (2005)
12. Edmonds, J.: Paths, trees and flowers. Journal of Mathematics 17, 449–467 (1965)
13. Faragó, A., Basagni, S.: The effect of multi-radio nodes on network connectivity—a graph theoretic analysis. In: Proceedings of the IEEE International Workshop on Wireless Distributed Networks (WDM). IEEE (2008)
14. Gabow, H.N.: An efficient reduction technique for degree-constrained subgraph and bidirected network flow problems. In: Proceedings of the 15th Annual ACM Symposium on Theory of Computing (STOC), pp. 448–456. ACM (1983)
15. Garey, M.R., Johnson, D.S.: Computers and Intractability, A Guide to the Theory of NP-Completeness. W.H. Freeman and Company, New York (1979)
16. Golumbic, M.C., Lewenstein, M.: New results on induced matchings. Discrete Applied Mathematics 101, 157–165 (2000)
17. Kang, R.J., Mnich, M., Müller, T.: Induced Matchings in Subcubic Planar Graphs. In: de Berg, M., Meyer, U. (eds.) ESA 2010. LNCS, vol. 6347, pp. 112–122. Springer, Heidelberg (2010)
18. Kanj, I., Pelsmajer, M.J., Schaefer, M., Xia, G.: On the induced matching problem. Journal on Computer System Science 77, 1058–1070 (2011)
19. Klasing, R., Kosowski, A., Navarra, A.: Cost Minimization in Wireless Networks with a Bounded and Unbounded Number of Interfaces. Networks 53(3), 266–275 (2009)
20. Kobler, D., Rotics, U.: Finding maximum induced matchings in subclasses of claw-free and p5-free graphs, and in graphs with matching and induced matching of equal maximum size. Algorithmica 37, 327–346 (2003)
21. Kosowski, A., Navarra, A., Pinotti, M.: Exploiting Multi-Interface Networks: Connectivity and Cheapest Paths. Wireless Networks 16(4), 1063–1073 (2010)
22. Zito, M.: Induced Matchings in Regular Graphs and Trees. In: Widmayer, P., Neyer, G., Eidenbenz, S. (eds.) WG 1999. LNCS, vol. 1665, pp. 89–100. Springer, Heidelberg (1999)

Stretch Factor in Wireless Sensor Networks with Directional Antennae

Evangelos Kranakis*, Fraser MacQuarrie*, and Oscar Morales-Ponce**

School of Computer Science, Carleton University, Ottawa, Ontario, Canada

Abstract. Traditional study of wireless sensor networks has relied on the assumption that sensors transmit and receive using an omnidirectional antenna. There has been some recent study using a model where sensors transmit using a directional antenna. This study has focused on the problem of finding an optimal transmission range so that there exists an orientation of the antennae at each sensor which creates a strongly connected communication network. This is known as the Antenna Orientation Problem for Strong Connectivity. In this paper we examine a similar problem: we wish to optimize not only the transmission range, but also the hop-stretch factor of the communication network (in relation to the omnidirectional model). We refer to this as the Antenna Orientation Problem with Constant Stretch Factor. We present approximations to this problem for antennae with angles $\pi/2 \leq \phi \leq 2\pi$.

Keywords: Antenna Orientation Problem, Connectivity, Directional Antenna, Stretch Factor, Wireless Sensor Networks.

1 Introduction

A wireless sensor is a computational device which transmits and receives information using a radio antenna. The traditional study of wireless sensor networks (WSNs) assumes that the sensors employ omnidirectional antennae. In this model, a sensor (the sender) can transmit messages successfully to another sensor (the receiver) if and only if the Euclidean distance between the sender and receiver is less than or equal to the transmission range of the sender. Typically it is assumed that all sensors in a WSN have the same transmission range. This leads to an undirected communication graph which is a unit disk graph (UDG), where the unit is the transmission range of the sensors. This model has been studied extensively.

There is, however, no reason a sensor cannot use directional antennae to transmit and/or receive information. Recently, there has been some study in a directional WSN model where sensors receive information omnidirectionally, but transmit in a sector of angle (or beam width) ϕ with range r.

The use of directional rather than omnidirectional antennae has many possible advantages: longer ranges are achievable with the same amount of energy; different radiation patterns might lower interference in the network and lead to increased throughput;

* Supported in part by NSERC and MITACS grants.
** Supported by MITACS Postdoctoral Fellowship.

G. Lin (Ed.): COCOA 2012, LNCS 7402, pp. 25–36, 2012.

there may also be an increase in the security of a network by reducing the risk of eavesdropping. These possible advantages are of particular interest in connected networks. However not all connected networks are created equal. If replacing omnidirectional antenna with directed ones leads to a network with significantly longer paths between sensors, then any possible advantages may be offset by the increased path lengths.

Study of this directional antennae model has until now been focused on the Antenna Orientation Problem for Strong Connectivity. Solutions to this problem guarantee that a WSN with directional antennae can be strongly connected, but nothing more. The resulting communication graph could have shortest paths between sensors containing a very large number of hops. This may be unavoidable depending on the distribution of the sensors. It becomes a problem if these paths are very long compared to their counterparts in the omnidirectional model, i.e., if the sensors had omnidirectional antennae instead of directional antennae, would these paths remain fairly long? We examine a problem similar to the Antenna Orientation Problem, but which places a bound on the length of paths in the directional case compared to those in the omnidirectional case.

For P, a set of points, let $U(P)$ be the UDG on P where the unit is the longest edge of the MST of P. Let $G(P)$ be a (strongly) connected (di)graph on the vertices in P.

For any two vertices u, v in P, let $d_G(u,v)$ denote the minimum number of edges of any (directed) path from u to v in G. For the rest of the paper, we refer to a path with $d_G(u,v)$ edges as a *shortest path* from u to v. We will also refer to $d_G(u,v)$ as the *hop count* from u to v.

The hop-stretch factor of $G(P)$ on $U(P)$, denoted as $\tau_G(P)$, is defined as the maximum ratio $d_G(u,v)/d_U(u,v)$ among all pairs of vertices u, v in P.

$$\tau_G(P) = \max_{\forall u,v \in P} \frac{d_G(u,v)}{d_U(u,v)}.$$

The hop-stretch factor (hereafter referred to interchangeably as simply *stretch factor*) may depend on the size of the network. This raises an interesting problem when one attempts to construct a transmission network with minimum range for a given angle which guarantees a constant stretch factor.

> **Antenna Orientation Problem with Constant Stretch Factor.** Given a connected UDG $U(S)$ on a set of sensors S in the plane. Suppose the sensors have beam width $\phi \geq 0$. For a given hop stretch factor k, compute the minimum range, denoted by $r_k(U(S),\phi)$, so that an orientation of the antennae of the sensors of S creates a strongly connected communication digraph $G_\phi(S)$ such that $\tau_{G_\phi}(S) \leq k$.

1.1 Related Work

The Antennae Orientation Problem for Strong Connectivity was first proposed by Caragianis et al. [4]. They proved the Antenna Orientation Problem is NP-complete for angles less than $2\pi/3$. They also presented a polynomial algorithm for determining a solution for angles $\phi \geq \pi$. A similar problem was studied by Dobrev et al. [8], who studied the Antennae Orientation Problem when each sensor has more than one directional antenna. They proved the problem remains NP-Complete when each sensor has two directional antennae with sum of angles at most $9\pi/20$ even for a scaling range of 1.3. They also showed that when each sensor has $k \leq 5$ directional antennae, the upper bound on the range is bounded by $2\sin\left(\frac{\phi}{k+1}\right)$ times the optimal range, for every angle. A comprehensive survey of the antenna orientation problem is presented in [10]. In 3D, the problem was studied in [9] where the authors consider the case when each sensor has one directional antenna.

The Antenna Orientation Problem with Constant Stretch Factor was studied for the first time by Damian and Flatland [7]. They examined the particular cases when $\phi = \pi/2$ and $\phi = 2\pi/3$. They proved that $r_6(S,\pi/2) \leq 7$ and $r_5(S,2\pi/3) \leq 5$ respectively. Recently, Bose et al.[3] studied the case when $\phi \leq \pi/3$.

A distinct model is studied in [1,2] and [5] in which sensors both transmit and receive using the same directional antenna. In [1,5] it is proven that it is always possible to create a connected graph with angle $\phi \geq \pi/3$ and unbounded range. In [2] the authors also considered how to bound the range and stretch factor. They proved that a connected graph with stretch factor 8 can be constructed with angle $\pi/2$ and range $14\sqrt{2}$ times the range necessary to create a connected UDG on the set of sensors.

1.2 Outline and Results of the Paper

The strategies used for deriving our results rely on partitioning an arbitrary set of sensors into many small subsets. We then orient these subsets independently of each other and show that these orientations lead to our desired results.

In Section 2 we describe how we can orient and connect small groups of sensors. In Section 3 we address the Antenna Orientation Problem with Constant Stretch Factor. We provide a global algorithm for beam widths $\pi \leq \phi \leq 2\pi$, and a local algorithm for beam widths $\pi/2 \leq \phi < \pi$. All approximations are in relation to the longest edge of the MST of the sensors. This is a trivial lower bound on the optimal range, as it is a lower bound for strong connectivity. The summary of our results, along with existing results is shown in Table 1.

Table 1. Results for the Antenna Orientation Problem with Constant Stretch Factor

Beam Width	Approximation Ratio of r_s	Stretch Factor	Scope	Proof
$\frac{5\pi}{3} \leq \phi < 2\pi$ $\pi \leq \phi < \frac{5\pi}{3}$	1 $2\sin(\frac{\phi}{2})$	2	Global	Theorem 1
$\frac{\pi}{2} \leq \phi < \pi$	$4\cos(\frac{\phi}{2})+3$	3	Local	Theorem 2
$\phi = \frac{\pi}{3}$	$36\sqrt{2}$	10	Global	[3]
$\phi < \frac{\pi}{3}$	$4\sqrt{2}(\frac{7\pi}{\phi}-6)$	$\lceil 8\log(\frac{2\pi}{\phi})\rceil - 1$	Global	[3]

1.3 Preliminaries and Notation

A *sensor* is an object at a point in the plane. It is able to receive transmissions omnidirectionally. A sensor with an omnidirectional antenna is able to transmit in all directions up to a distance r called the *transmission range*. A sensor with a single directional antenna is able to transmit in a sector whose angle is referred to as the *beam width* of the antenna. This antenna may be initially facing any direction, but once *oriented* the antenna is fixed in this orientation. Any point that lies within the sector defined by the antenna, regardless of its distance from the sensor, is in the sensor's *line of sight*.

For the purposes of this paper, any sensors referred to are assumed to be sensors with a single directional antenna. Furthermore, the term *sensor* may be used interchangeably to mean the location of the sensor in the plane, the sensor object itself, or the vertex representing the sensor in a graph. We may also use the terminology "orienting a sensor" to mean orienting the antenna at a sensor.

We assume that any sensor, s, is able to determine its distance from other sensors, as well as the angle formed by any other two sensors with vertex s. While assuming that sensors have location awareness will satisfy these assumptions, it is not strictly necessary as sensors do not need access to a global co-ordinate system for our results. Furthermore, we assume that each sensor has the ability to communicate with all nearby sensors during the orientation process. This could be accomplished through the rotation of its directional antenna, or the use of its omnidirectional antenna to transmit as well as receive.

Let $D(a,r)$ denote the open disk centered at a with radius r and $D[a,r]$ denote the closed disk centered at a with radius r.

Definition 1 (Coverage). *Let a,b be sensors with range r. Sensor a* covers *sensor b if $b \in D[a,r]$ and b is within the line of sight of sensor a. This means that b will be a neighbour of a (although the reverse is not necessarily true). Sensor a* covers *area A if $\forall$ points $p \in A$, a sensor at p would be covered by a. A set of sensors S* covers *an area A if $\forall$ points $p \in A$, a sensor at p would be covered by at least one sensor $s \in S$*

Let a,b be two sensors. We say that sensor a can *reach* sensor b if a covers b, or a covers a sensor c which can reach b. In general terms, this means that a can reach b if there exists a directed path from a to b.

Definition 2 (k-orientation). *Let S be a set of sensors. An orientation of the antennae of S is a k-*orientation *if the directed communication graph $G(S)$ is strongly connected, and $\forall s \in S$:*

1. *$D[s,1]$ is covered by S, and*
2. *$\forall p \in D[s,1]$, the shortest path from s to a sensor covering p has length at most $k-1$ hops.*

2 Orientating Small Groups of Sensors

In this section we begin by showing how to merge k-orientations. We then relate k-orientations to stretch factor. We conclude the section by showing how we can orient various groups of small sensors to form k-orientations. These orientations will form the building blocks for our later results. Any omitted proofs can be found in the full paper.

2.1 Merging k-Orientations

Orienting small groups of sensors is the foundation for our results, however we first show how these small orientations can be put together to form an orientation for an entire graph.

Lemma 1. *Let S,T be two sets of sensors which have been oriented to form an i-orientation and a j-orientation, respectively. Suppose, without loss of generality, that $i \leq j$. If $\exists s_1 \in S, t_1 \in T$ such that s_1 covers t_1 and $\exists s_2 \in S, t_2 \in T$ such that t_2 covers s_2, then the orientations of S,T is a j-orientation of $S \cup T$.*

Proof. Let the orientations of S,T remain identical, so that the coverage and path length conditions hold for $S \cup T$. All that remains is to show that $S \cup T$ is strongly connected. This is trivial however, as both S and T are strongly connected and there is a path from S to T and vice versa.

2.2 Stretch Factor of k-Orientations

Lemma 2. *Let S be a set of sensors which have been oriented to form an i-orientation with the directed communication graph $G(S)$. The stretch factor of $G(S)$ on $UDG(S,1)$, $\tau_G(S) \leq i$.*

Proof. Assume there exists $G(S)$ such that $\tau_G(S) > i$. Therefore there must be two vertices $u,v \in S$ such that $d_G(u,v)/d_U(u,v) > i$. Since the ratio along the path in G from u to v is greater than i, this means that there must exist two vertices $a,b \in S$ such that $d_U(a,b) = 1$ and $d_G(a,b)/d_U(a,b) > i$. Therefore, $d_G(a,b) > i$. Note that since $d_U(a,b) = 1$, $b \in D[a,1]$ and vice versa. However, since $G(S)$ is the communication graph of an i-orientation of S, the shortest path between a and a sensor covering b cannot be more than $i-1$. Therefore, $d_G(a,b) \leq i$. This contradicts our assumption, and the lemma follows.

2.3 Forming k-Orientations from Small Groups of Sensors

Groups of 2 Sensors

Lemma 3. *Given two sensors u,v with beam width $\phi \geq \pi/2$. If the sensors are separated by Euclidean distance $\delta \geq 2\cos(\phi/2)$, then there exists a 2-orientation of u and v with transmission range $r = \delta + 1$.*

Proof. Suppose u,v are oriented as in Figure 1. Let c be the midpoint of the line segment uv. Each sensor beam pattern edge intersects with the intersection of the circles $D[u,1]$ and $D[v,1]$ at point i. This intersection forms a right angle triangle with c and u. Since the intersection lies on the boundary of $D[u,1]$, the hypotenuse of this triangle has length 1. Since uv bisects the sensor of beam width ϕ, the angle $\angle(ivc) = \phi/2$. Using simple trigonometry we calculate the length of the side $cv = \cos(\phi/2)$. Similarly, the side $uc = \cos(\phi/2)$, so the length of uv, $\delta = \cos(\phi/2)$.

Consider any point $p \in D[u,1]$. The farthest p can be from v is $\delta + 1$. Similarly, the farthest any point $q \in D[v,1]$ can be from u is $\delta + 1$. Therefore $r = \delta + 1$ is a sufficient transmission range to ensure that both $D[u,1]$ and $D[v,1]$ are covered.

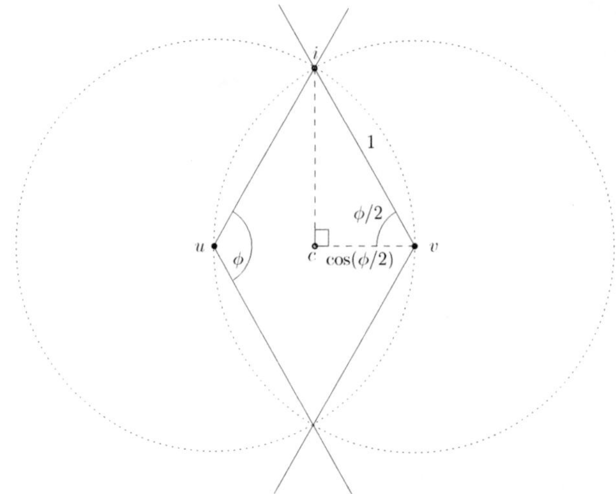

Fig. 1. Antenna orientation of two sensors with beam width $\frac{2\pi}{3} \leq \phi < \pi$

In the proposed antennae orientation, u covers v and vice versa, so each are connected by a path of length 1. Therefore this orientation of u and v is a 2-orientation.

A more specific version of the following result was first given in [6]. We derive a more general version and include the proof for completeness.

Lemma 4. *Given two sensors u, v with beam width $\phi \geq \pi$. Suppose the Euclidean distance between them is δ. There exists a 2-orientation of u and v with transmission range* $\max(1, \delta, \sqrt{1+\delta^2-2\delta\cos(\phi)})$.

Groups of 3 Sensors

Lemma 5. *Given three sensors u, v and w with beam width $\phi \geq \pi/2$. If two of the sensors are separated by Euclidean distance $\delta \geq 2\cos(\phi/2)$, there is a 3-orientation of $\{u, v, w\}$ with transmission range $r = \delta + 1$.*

Lemma 6. *Given a set S of $n \geq 3$ sensors with beam width $\phi \geq \pi$. Suppose $\exists c \in S$ such that the maximum distance between c, and any other sensor $s \in S - \{c\}$ is δ. If all the sensors $s \in S - \{c\}$ are contained within a sector centered at c with angle ϕ, there is a 2-orientation of S with transmission range $r =$* $\max(1, \delta, \sqrt{1+\delta^2-2\delta\cos(\phi)})$.

Groups of 4 Sensors

Lemma 7. *Given a set S of four sensors with beam width $\phi \geq \pi/2$. Suppose the maximum Euclidean distance between any two sensors in S is δ. There is a 3-orientation of S with transmission range $r = \delta + 1$.*

Proof. Let us first consider the case where the sensors have infinite transmission range. Label the sensors $t, u, v, w \in S$ such that the greatest pairwise Euclidean distance between sensors occurs between t and w. We then obtain one of two cases (Figure 2).

Case (a): Both sensors u and v are on the same side of the line tw. We orient the antennae of t and w to cover the half plane above (and including) the line tw. We then orient the antennae of u and v to cover the half plane below tw. Since t and w cover the half plane above tw, and u and v cover the half plane below tw, the entire plane is covered. Since t and w are the two sensors farthest apart, both u and v must lie between t and w. Therefore, t covers u, v, and w. Similarly w covers t, u and v. The sensors u and v each cover one of t and w. Therefore t and w cover each other, and u and v each cover one of t or w and are covered by both. This means the induced graph is strongly connected.

Case (b): u and v are on opposing sides of the line tw. We label u as the point whose projection on the line tw is closest to t. We can then orient the antennae of t and w so that they cover the entire plane except the shaded areas in Fig 2(b). The antenna of u (similarly v) can then be oriented to cover the shaded area adjacent to w (t) as well as sensors v, w (t, u). Collectively, the sensors t, u, v and w cover the entire plane. In this orientation, t covers u, u covers w, w covers v, and v covers t. Therefore the sensors are strongly connected.

In both cases, the antennae are oriented so that they cover the plane, and so that the sensors are strongly connected. Now consider any point $p \in D[t,1] \cup D[u,1] \cup D[v,1] \cup D[w,1]$. The farthest p can be from any of t, u, v, or w is $\delta + 1$. Therefore a transmission range of $r = \delta + 1$ is sufficient to obtain a k-orientation in both cases.

Furthermore, an examination of the orientations reveals that the shortest path between any two of the sensors is at most 2. Therefore these are both 3-orientations. Since we always obtain one of the two cases, the proof follows.

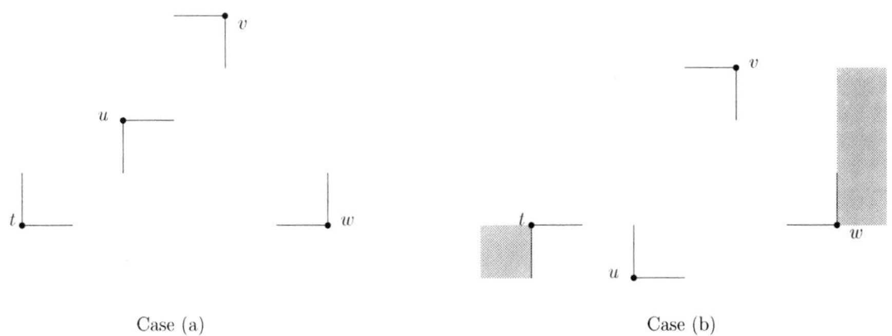

Fig. 2. Antenna orientation of four sensors with beam width $\phi \geq \frac{\pi}{2}$

Corollary 1. *Given a set S of $n \geq 4$ sensors with beam width $\phi \geq \pi/2$. Suppose the maximum Euclidean distance between any two sensors in S is δ. There is a 3-orientation of S with transmission range $r = \delta + 1$.*

3 Orienting Antennae with Constant Stretch Factor

In this section we will present some approximations for the Antenna Orientation Problem with Constant Stretch Factor. These approximations will have transmission ranges dependant on the beam width of the sensors, the desired stretch factor, as well as whether the orientation is to be found with local or global information.

3.1 Orienting Antennae of Beam Width $\phi \geq \pi$ with Constant Stretch Factor

Theorem 1. *Given a connected UDG $U(S)$ on a set of sensors S each with one directional antenna of beam width $\phi \geq \pi$. There exists an antennae orientation of S with range* $\max(1, 2\sin(\phi/2))$ *which creates a connected transmission network $G_\phi(S)$ such that* $\tau_{G_\phi}(S) \leq 2$.

Proof. For any graph H, denote $C(H)$ as the vertices of the convex hull of each of the connected components of H. We then define the following hierarchical structure, Q: $Q_0(V_0, E_0) = U(S)$, $Q_{k+1}(V_{k+1}, E_{k+1}) = Q_k[V_k - C(Q_k)]$ (the subset of Q_k induced by $V_n - C(Q_n)$). Intuitively, every iteration of the structure is the previous graph with the convex hulls of its connected components peeled away. We note that every iteration is a proper subset of the previous: $Q_{k+1} \subset Q_k$, unless $Q_{k+1} = Q_k = \emptyset$. The construction of such a hierarchical structure is illustrated in Figure 3.

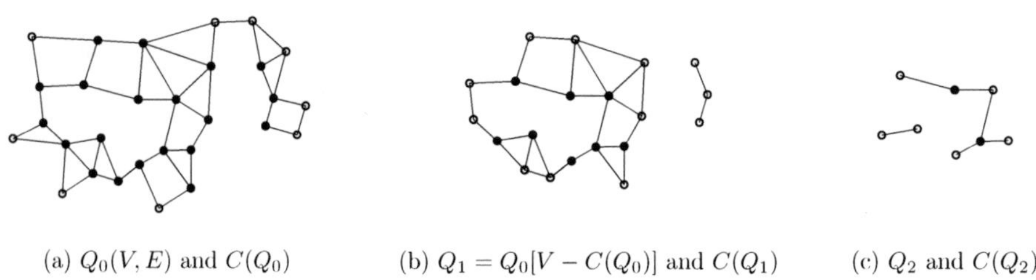

(a) $Q_0(V, E)$ and $C(Q_0)$ (b) $Q_1 = Q_0[V - C(Q_0)]$ and $C(Q_1)$ (c) Q_2 and $C(Q_2)$

Fig. 3. Construction of Q for a given graph $G(V, E)$. $C(Q_i)$ denoted by hollow points

We want to use induction to show that we can find a suitable orientation for some Q_i and that given an orientation for any Q_k we can find an orientation for Q_{k-1}. We do this by ensuring that each sensor is either: *oriented* so that it part of a k-orientation, or the sensor is *convex*. We say that a sensor s is convex in a graph H if s is on the convex hull of $N_H(s) \cup \{s\}$, where $N_H(s)$ is the set of neighbours of s in H. Intuitively, s is convex in H if a line can be drawn through s such that all neighbours of s in H are on one side of the line. Since the sensors have beam width $\phi \geq \pi$, the antennae of convex sensors can be positioned so that they cover all their neighbours. Therefore if all sensors in S are either oriented or convex, there exists an orientation of S which is strongly connected.

Let i be the smallest value such that $Q_i = Q_{i+1} = \emptyset$. Consider Q_{i-1}. It must contain at least one sensor, and all sensors are on the convex hull of their connected components, so they are all convex. Therefore we have a valid orientation for Q_{i-1}.

Consider now Q_{i-2}. We know that all the sensors in $C(Q_{i-2})$ are convex. What about the remaining sensors? None have yet been oriented, some may remain convex, but some may not. It is the sensors that are no longer convex which prevent an orientation from being achieved. For each sensor s which is no longer convex, there must exist at least one sensor in $C(Q_{i-1})$ which is a neighbour of s in $U(S)$ (otherwise s would still be convex). Suppose each such unoriented, non-convex sensor selects one of its neighbours in $C(Q_{i-1})$ and requests to orient with it. Let us now consider each sensor in $C(Q_{i-1})$. A sensor may receive no requests to orient, in which case it remains convex. It may receive a single request, in which case the two sensors orient themselves according to Lemma 4 to form a 2-orientation. It may also receive multiple requests, in which case the sensors can orient themselves according to Lemma 6 to form a 2-orientation. After each sensor of $C(Q_{i-1})$ has taken the appropriate action, we now have the case where all sensors are either oriented or convex, so we have a valid orientation for Q_{i-2}.

Consider now Q_{k-1}. Assume that we have a valid orientation for Q_k. We know that all the sensors in $C(Q_{k-1})$ are convex. What about the remaining sensors? Some sensors may have already been oriented, some may remain convex, but some may not. Using the same method as above, we can orient all the unoriented, non-convex sensors with sensors in $C(Q_{k-1})$ to achieve a valid orientation of Q_{k-1}.

Since we have proven we can find an orientation for Q_{i-2}, and since we can find a valid orientation for Q_{k-1} given a valid orientation for Q_k, we know that we can find a valid orientation $G_\phi(S)$ for $Q_0 = U(S)$.

All sensors in the orientation $G_\phi(S)$ are either part of a 2-orientation or convex. Sensors in a 2-orientation can reach all their neighbours in $U(S)$ is at most two hops. Convex sensors can directly reach all their neighbours in $U(S)$. This means that $\tau_{G_\phi}(S) \leq 2$.

3.2 Orienting Antennae of Beam Width $\pi/2 \leq \phi < \pi$ with Constant Stretch Factor

Definition 3. *We define the annulus graph $A(P,r,R)$ on a set of points P as the straight line graph where two points a,b at distance d are connected if and only if $r \leq d \leq R$.*

Theorem 2. *Given a connected UDG, $U(S,1)$, on a set of sensors S each with beam width $\pi/2 \leq \phi < \pi$. There exists an antenna orientation of S with range at most $4\cos(\phi/2) + 3$ which creates a strongly connected communication graph $G_\phi(S)$ such that $\tau_{G_\phi}(S) \leq 3$. This communication graph can be constructed in constant time.*

Proof. Let $\mathcal{A} = A(S, 2\cos(\phi/2), 2\cos(\phi/2)+1)$ be an annulus graph on S.

Claim. Let u be any point in S and v the farthest point in S from u. If $d(u,v) \geq 2\cos(\phi/2)$, then the degree of u in A, $d_{\mathcal{A}}(u)$ is at least 1.

Proof. Assume there exist two points u and v such that $d(u,v) \geq 2\cos(\phi/2)$ and $d_{\mathcal{A}}(u) = 0$. Since U is connected we can always find a path $P = u = u_0, u_1, \ldots u_k = v$. Observe that $d(u_i, u_{i+1}) \leq 1$ for all i. Therefore, at least one vertex in the path $u_1, \ldots u_k = v$ is at distance between $2\cos(\phi/2)$ and $2\cos(\phi/2)+1$ since P crosses the annulus of u. This contradicts the assumption, therefore $d_{\mathcal{A}}(u) > 0$.

Given two sensors u,v in S, we say that they form a *2-group* if $d(u,v) \geq 2\cos(\phi/2)$. By Lemma 3 there exists a 2-orientation of the antennae at u and v with angle ϕ and range $d(u,v)+1$.

Given three sensors u,v,w in S where $\delta = \max(d(u,v),d(u,w),d(w,v))$. We say that u,v,w form a *3-group* if $\delta \geq 2\cos(\phi/2)$. By Lemma 5 there exists a 3-orientation of u,v,w with range $\delta+1$.

We say that two points u,v are *close* if $d(u,v) \leq 4\cos(\phi/2)+2$. Given $S' \subseteq S$, we say that S' is a *4-strong subset* if there exists four sensors $u,v,w,x \in S'$ such that $\forall a,b \in S'$, a is close to b. By Corollary 1 we can find a 3-orientation of u,v,w and x with range $4\cos(\phi/2)+3$.

If no two vertices are distance at least $2\cos(\phi/2)$ apart, either U is a 4-strong set, or there are three or fewer sensors in U. If U is a 4-strong set, we can orient it according to Corollary 1. If U consists of only three sensors u,v,w, then they can be oriented so that u covers v, v covers w and w covers u. If there are two or fewer sensors, the orientation is trivial. In the rest of the proof we assume that there exist two vertices separated by distance at least $2\cos(\phi/2)$.

Let M be a maximal matching of $\mathcal{A}$. Consider the following geometric graph $G=(S,E)$ where $\{a,b\} \in E$ if and only if a is an unmatched sensor and b is the nearest matched sensor to a.

Claim. For each edge $\{a,b\}$ in G, $d(a,b) \leq 2\cos(\phi/2)+1$

Proof. From the first claim we know that each point has degree at least one in $\mathcal{A}$. Therefore, a point is only unmatched if all its neighbours in $\mathcal{A}$ are matched. Thus, $d(a,b) \leq 2\cos(\phi/2)+1$.

Let $\{u,v\}$ be any edge in M and $N_G(u)$ denote the neighbours of u in G. Since M is a matching, clearly u,v are not incident to any other edge in M. From our definition of G, $v \notin N_G(u)$. Furthermore, $\forall a,b \in M, N_G(a) \cap N_G(b) = \emptyset$. Since every sensor in U is incident to at least one edge in either M or G, the previous conditions mean that we can partition the graph based on the edges in M. For each edge $\{u,v\}$ in M, we define a subset $S'_{\{u,v\}} = \{u,v\} \cup N_G(u) \cup N_G(v)$. Each subset is non-empty since it must contain u,v. As mentioned previously, every sensor will be part of one and only one subset. This is therefore a valid partition. We will show that each subset can be oriented to form a 3-orientation.

Without loss of generality assume that $|N_G(u)| \leq |N_G(v)|$. There are now multiple cases we may encounter.

Case 1: $|N_G(u)| = |N_G(v)| = 0$.
In this case, $S'_{\{u,v\}}$ forms a 2-group.

Case 2: $|N_G(u)| = 0$ **and** $|N_G(v)| \geq 1$.
If $|N_G(v)| = 1$, $S'_{\{u,v\}}$ forms a 3-group. Otherwise, it is a 4-strong subset.

Case 3: $|N_G(u)| = 1$ **and** $|N_G(v)| \geq 1$.
Let $x \in N_G(u)$. If $d(x,v) \leq 2\cos(\phi/2)+1$, $S'_{\{u,v\}}$ is a 4-strong subset. If not, we consider three possible cases:

- $|N_G(v)| = 1$. Let $y \in N_G(v)$. If $d(y,u) \leq \cos(\phi/2) + 1$, $S'_{\{u,v\}}$ is a 4-strong subset. Otherwise, $\{x,v\}$ forms a 2-group and $\{y,u\}$ forms a 2-group. If two groups are formed, we remove $S'_{\{u,v\}}$ from the partition and add $\{x,v\}$ and $\{y,u\}$.
- $|N_G(v)| = 2$. Let y and z be the elements in $N_G(v)$. If $\max(d(y,u), d(z,v)) \leq 2\cos(\phi/2) + 1$, $S'_{\{u,v\}}$ is a 4-strong subset. Otherwise, $\{x,v\}$ forms a 2-group and $\{u,y,z\}$ forms a 3-group. If two groups are formed, we remove $S'_{\{u,v\}}$ from the partition and add $\{x,v\}$ and $\{u,y,z\}$.
- $|N_G(v)| \geq 3$. Let z,w be the nearest sensor of v in $N_G(v) \setminus \{y\}$. $N_G(v) \cup \{u\}$ is a 4-strong subset and $\{x,v\}$ forms a 2-group. We remove $S'_{\{u,v\}}$ from the partition and add $N_G(v) \cup \{u\}$ and $\{x,v\}$.

Case 4: $|N_G(u)| = 2$ and $|N_G(v)| \geq 2$.
If there are two sensors x,y in $N_G(u) \cup N_G(v)$ that are at distance at most $2\cos(\phi/2) + 1$ of u and v, $S'_{\{u,v\}}$ is a 4-strong subset. Otherwise, $N_G(u) \cup \{v\}$ forms a 3-group. If $d_G(v) = 2$, $N_G(v) \cup \{u\}$ forms a 3-group, otherwise it is a 4-strong subset. If two groups are formed remove $S'_{\{u,v\}}$ from the partition and add $N_G(v) \cup \{v\}$ and $N_G(v) \cup \{u\}$.

Case 5: $|N_G(u)| \geq 3$ and $|N_G(v)| \geq 3$.
$N_G(u) \cup \{u\}$ is a 4-strong subset and $N_G(v) \cup \{v\}$ is a 4-strong subset. We remove $S'_{\{u,v\}}$ from the partition and add $N_G(v) \cup \{u\}$ and $N_G(v) \cup \{v\}$.

Once we have oriented every subset in the partition, we observe that they now all consist of 2-groups, 3-groups and 4-strong subsets. As mentioned previously, we know that we can form 3-orientations for each of these. Therefore we now have a partition of the sensors of U such that each sensor is a part of a 3-orientation. Note that the transmission range required to orient any of the sensor groups was always less than or equal to $2\cos(\phi/2) + 3$. What is left to show is that U is a 3-orientation. Consider any edge $\{a,b\} \in U$. Suppose a,b are not in the same 3-orientation. Since $b \in D[a,1]$, there is a sensor in a's 3-orientation which covers b. Similarly, there is a sensor in b's 3-orientation which covers a. Therefore, by Lemma 1 these 3-orientations can be combined to form a larger 3-orientation of which both a and b are members. Since U is connected, this process can be repeated until all sensors are a part of the same 3-orientation. Therefore there must exist some orientation $G_\phi(S)$ of the sensors of S which is a 3-orientation. Therefore by Lemma 2, there must exist some orientation $G_\phi(S)$ of the sensors of S such that $\tau_{G_\phi}(S) \leq 3$.

Regarding the complexity, a maximal matching can be constructed in constant time [11] and each other step is local.

One may ask whether we can improve our result by considering the annulus graph $G = A(P, 2\cos(\phi/2), 2\cos(\phi/2) + 1 - \varepsilon)$. However, we cannot guarantee minimum degree greater than zero on G and consequently the properties of the unmatched vertices do not hold.

4 Conclusion

In this paper we have examined issues relating to connectivity in the directional antenna model. There remain unanswered questions relating to this problem. Can approximations can be found for angles between $\pi/2$ and $\pi/3$? Can tighter bounds on range and/or stretch factor be found? How does the Antenna Orientation Problem with Constant Stretch Factor relate to the Antenna Orientation Problem for Strong Connectivity? How would multiple antennae per sensor affect connectivity? How does the problem change if Euclidean stretch factor is considered instead of hop-stretch factor?

The properties of this model may be of particular interest for questions such as: How would routing work? What level of sender and receiver interference would be expected? These are interesting questions and are worthy of study.

References

1. Ackerman, E., Gelander, T., Pinchasi, R.: Ice-creams and wedge graphs. Arxiv preprint arXiv:1106.0855 (2011)
2. Aschner, R., Katz, M.J., Morgenstern, G.: Symmetric connectivity with directional antennas. Arxiv preprint arXiv:1108.0492 (2011)
3. Bose, P., Carmi, P., Damian, M., Flatland, R., Katz, M., Maheshwari, A.: Switching to directional antennas with constant increase in radius and hop distance. In: Proceedings of Workshop on Algorithms and Data Structures, pp. 134–146 (2011)
4. Caragiannis, I., Kaklamanis, C., Kranakis, E., Krizanc, D., Wiese, A.: Communication in Wireless Networks with Directional Antennae. In: Proceedings of 20th ACM Symposium on Parallelism in Algorithms and Architectures (SPAA 2008), Munich, Germany, June 14-16, pp. 344–351 (2008)
5. Carmi, P., Katz, M.J., Lotker, Z., Rosén, A.: Connectivity guarantees for wireless networks with directional antennas. Comput. Geom. Theory Appl. 44(9), 477–485 (2011)
6. Damian, M., Flatland, R.: Connectivity of graphs induced by directional antennas. Arxiv preprint arXiv:1008.3889 (2010)
7. Damian, M., Flatland, R.: Spanning properties of graphs induced by directional antennas. In: Electronic Proc. 20th Fall Workshop on Computational Geometry. Stony Brook University, Stony Brook (2010)
8. Dobrev, S., Kranakis, E., Krizanc, D., Opatrny, J., Ponce, O.M., Stacho, L.: Strong Connectivity in Sensor Networks with Given Number of Directional Antennae of Bounded Angle. In: Wu, W., Daescu, O. (eds.) COCOA 2010, Part II. LNCS, vol. 6509, pp. 72–86. Springer, Heidelberg (2010)
9. Kranakis, E., Krizanc, D., Modi, A., Morales Ponce, O.: Maintaining connectivity in 3D wireless sensor networks using directional antennae. In: 25th IEEE International Parallel & Distributed Processing Symposium (IPDPS 2011), May 16–20. IEEE Press, Anchorage (2011)
10. Kranakis, E., Krizanc, D., Morales, O.: Maintaining connectivity in sensor networks using directional antennae. In: Nikoletseas, S., Rolim, J. (eds.) Theoretical Aspects of Distributed Computing in Sensor Networks. Springer (2010)
11. Wiese, A., Kranakis, E.: Local Maximal Matching and Local 2-Approximation for Vertex Cover in UDGs (Extended Abstract). In: Coudert, D., Simplot-Ryl, D., Stojmenovic, I. (eds.) ADHOC-NOW 2008. LNCS, vol. 5198, pp. 1–14. Springer, Heidelberg (2008)

On the Minimum Diameter Cost-Constrained Steiner Tree Problem

Wei Ding[1,⋆] and Guoliang Xue[2]

[1] Zhejiang Water Conservancy and Hydropower College
Hangzhou, Zhejiang 310000, China
dingweicumt@163.com

[2] Department of Computer Science and Engineering
Arizona State University, Tempe, AZ 85287-8809, USA
xue@asu.edu

Abstract. Given an edge-weighted undirected graph $G = (V, E, c, w)$ where each edge $e \in E$ has a cost $c(e) \geq 0$ and another weight $w(e) \geq 0$, a set $S \subseteq V$ of terminals and a given constant $\mathrm{C}_0 \geq 0$, the aim is to find a minimum diameter Steiner tree whose all terminals appear as leaves and the cost of tree is bounded by C_0. The diameter of tree refers to the maximum weight of the paths connecting two different leaves in the tree. This problem is called the minimum diameter cost-constrained Steiner tree problem, which is NP-hard even when the topology of the Steiner tree is fixed. In this paper, we deal with the fixed-topology restricted version. We prove the restricted version to be polynomially solvable when the topology is not part of the input and propose a weakly fully polynomial time approximation scheme (weakly FPTAS) when the topology is part of the input, which can find a $(1+\epsilon)$–approximation of the restricted version problem for any $\epsilon > 0$ with specific characteristic.

Keywords: Minimum diameter, cost-constrained Steiner tree, weakly fully polynomial time approximation scheme, fixed topology.

1 Introduction

The *Steiner minimum tree* (SMT) problem in graphs asks for a minimum length connected subgraph of the given graph spanning a set of given *terminals*, which has many applications in a variety of fields [9,14], such as communication networks, computational biology. The problem has been proved to be NP-hard in the strong sense [11] and admitted a number of approximation algorithms with a constant performance ratio [9,19,28]. In the paper, we are concerned with a variant of SMT, called *terminal Steiner tree* (TeST), which requires all terminals as its leaves. Since Lin and Xue proposed the definition of TeST and proved the problem of finding a minimum length TeST to be NP-complete and MAX SNP-hard [16], many researchers have put interests onto this problem in past a decade and devised some constant performance ratio approximation algorithms [3,8,10,16,17].

⋆ Correspondence author.

G. Lin (Ed.): COCOA 2012, LNCS 7402, pp. 37–48, 2012.

1.1 Related Works

A computer or communication network is frequently modeled as an undirected edge-weighted graph $G = (V, E, c, w)$ [2], where $c(e) \geq 0$ and $w(e) \geq 0$ represent the cost and another weight on e respectively for each edge $e \in E$. Given a subset $s \subset V$ of terminals, let $T = (U, F, c, w)$ be a TeST in G spanning S. Clearly $U \subseteq V$ and $F \subseteq E$. The *cost* of T is the sum of its all costs of edge, denoted as $c(T) = \sum_{e \in F} c(e)$. For any two different nodes $u, v \in U, u \neq v$, let $p_T(u, v)$ denote the unique u-to-v path of T and $w(p_T(u, v))$ denote the *weight* of $p_T(u, v)$, similarly $w(p_T(u, v)) = \sum_{e \in p_T(u,v)} w(e)$. Specially we call $p_T(u, v)$ a *leaf-to-leaf path* when $u, v \in S, u \neq v$. The maximum weight of leaf-to-leaf path of T is called the *diameter* of T, denoted as $w(T) = \max_{u,v \in S, u \neq v} \sum_{e \in p_T(u,v)} w(e)$. A leaf-to-leaf path of T with its weight equal to the diameter of T is called a *diameter leaf-to-leaf path* of T, denoted by $p(T)$. The number of edges on a path p of T is called *hop count* of p, denoted by $h(p)$. It is noted that the diameter of tree involved in this paper is the maximum weight of path connecting two different leaves instead of the maximum hop count between leaves [4,12,20].

The *diameter-constrained Steiner tree* (DCST) is a kind of extended Steiner tree with its diameter no more than a given constant D_0. The objective of the *minimum cost diameter-constrained Steiner Tree problem* (MCDCSTP) is to find a DCST with its cost minimized from the given graph. Marathe *et al.* [18] studied the generalized MCDCSTP and proved it NP-hard to approximate the generalization within a logarithmic factor, presented a $(1+\gamma)\lceil \log |S| \rceil$–approximation algorithm for a fixed $\gamma > 0$ and proved the weight of any leaf-to-leaf path in the approximation is at most $2\lceil \log |S| \rceil \mathrm{D}_0$. When the solution tree is restricted as a TeST, Ding *et al.* [6] studied the fixed topology version of MCDCSTP, namely *minimum cost diameter-constrained realization of objective tree problem* (MCDCRP), and gave a fully polynomial time approximation scheme (FPTAS). The *cost-constrained Steiner tree* (CCST) is another kind of extension of Steiner tree with its cost bounded by a given constant C_0. The *minimum diameter cost-constrained Steiner Tree problem* (MDCCSTP) aims to find a CCST with a minimum diameter from the given graph. In [18], Marathe *et al.* studied the generalization of MDCCSTP and proved its inapproximability, designed an approximation algorithm with a performance ratio of $2\lceil \log |S| \rceil$ and proved the cost of the approximation is no more than $(1+\gamma)\lceil \log |S| \rceil \mathrm{C}_0$ for a fixed $\gamma > 0$.

1.2 Our Contribution

In this paper, we are the first one to propose the *weakly fully polynomial time approximation scheme* (weakly FPTAS). For ease of comparison, we present the definitions of the classical FPTAS and weakly FPTAS as follows.

Definition 1. *Given an input $\mathcal{I}$, the algorithm $\mathcal{A}_\epsilon$ is a fully polynomial time approximation scheme (FPTAS) provided that it can find an approximation $\mathcal{A}_\epsilon(\mathcal{I})$ satisfying $\mathcal{A}_\epsilon(\mathcal{I}) \leq (1+\epsilon)OPT(\mathcal{I})$ for any $\epsilon > 0$ within a polynomial time in $\frac{1}{\epsilon}$ and $|\mathcal{I}|$. In addition, $\mathcal{A}_\epsilon(\mathcal{I}) \to OPT(\mathcal{I})$ when $\epsilon \to 0$.*

Definition 2. *Given an input $\mathcal{I}$, the algorithm $\mathcal{A}_\Sigma$ is a weakly fully polynomial time approximation scheme (weakly FPTAS or WFPTAS) provided that it can find an approximation $\mathcal{A}_\Sigma(\mathcal{I})$ satisfying $\mathcal{A}_\Sigma(\mathcal{I}) \leq (1+\epsilon)OPT(\mathcal{I})$ for any $\epsilon > 0$ with specific characteristic Π, i.e., $\epsilon \in \Sigma(\Pi, \mathcal{I}) = \{\epsilon > 0 : \epsilon$ satisfies Π for $\mathcal{I}\}$, within a polynomial time in $\frac{1}{\epsilon}$ and $|\mathcal{I}|$. Also, $\mathcal{A}_\Sigma(\mathcal{I}) \to OPT(\mathcal{I})$ when $\epsilon \to 0$.*

According to Definition 1, a classical FPTAS can find a $(1+\epsilon)$–approximation of the problem definitely in a fully polynomial time for any real number $\epsilon > 0$, see [6,13,15,21,23,26,27]. However, Definition 2 implies that a weakly FPTAS can find a $(1+\epsilon)$–approximation only for some available values of ϵ with specific characteristic. Simply speaking, whether a weakly FPTAS can find a $(1+\epsilon)$–approximation or not depends on the value of ϵ.

In the paper, we focus on the fixed topology restricted version of MDCCSTP, called the *minimum diameter cost-constrained realization of objective tree problem* (MDCCRP) (formally defined in Sect. 2.2). Here, we give its rough definition as follows: given an undirected edge-weighted complete graph $G = (V, E, c, w)$, a subset $S \subset V$ of terminals, a given constant $C_0 \geq 0$, a sample TeST $\mathcal{T} = (\mathcal{V}, \mathcal{E})$, we let $\mathcal{T}$–ROT denote a realization of $\mathcal{T}$ in G and aim to find a minimum diameter $\mathcal{T}$–ROT as $T = (U, F, c, w)$ subject to the constraint of $c(T) \leq C_0$. The major contribution of this paper is to design a weakly FPTAS for MDCCRP.

The rest of this paper is organized as follows. In Sect. 2, we define MDCCRP formally as well as its decision version, prove its NP-hardness, and make some fundamental preliminaries. In Sect. 3, we present a pseudo-polynomial-time algorithm for MCDCRP. In Sect. 4, we design a weakly FPTAS for MDCCRP. In Sect. 5, we present some concluding remarks on this paper.

2 Preliminaries

2.1 Illustrate the Realization of Objective Tree

In [6], Ding *et al.* formally proposed the definition of realization of objective tree in the given graph although the rudiment of such realization has been introduced by Wang *et al.* in [24] and Xue *et al.* in [26] independently. Now let us recall the realization of objective tree.

Let $\mathcal{J}$ represent an input as follows: an undirected edge-weighted complete graph $G = (V, E, c, w)$ in which every edge $e \in E$ has a real-number cost $c(e) \geq 0$ and an integer weight $w(e) \geq 0$, a subset $S \subset V$ of terminals and a sample TeST $\mathcal{T} = (\mathcal{V}, \mathcal{E})$.

Given a $\mathcal{J}$, we always take $\mathcal{T}$ as a rooted tree since any unrooted tree can be transformed into a rooted tree by designating its any nonleaf vertex as the root. Let $\mathcal{T}$–ROT denote a realization of objective tree $\mathcal{T}$ and $\mathcal{T}(\alpha)$–ROT denote a realization of $\mathcal{T}(\alpha)$ where $\mathcal{T}(\alpha)$ is the *subtree* of $\mathcal{T}$ rooted at α for any vertex $\alpha \in \mathcal{V}$. The $\mathcal{T}$–ROT consists of all realizations of edge of $\mathcal{T}$. In essence, for any edge $\{\alpha, \beta\} \in \mathcal{E}$, the realization of $\{\alpha, \beta\}$ in G is determined by the mappings of two endpoints α and β to two vertices of G. That is, the realization of $\mathcal{T}$ is determined by all mappings of vertex of $\mathcal{T}$. It is noted that each *leaf* (i.e., a

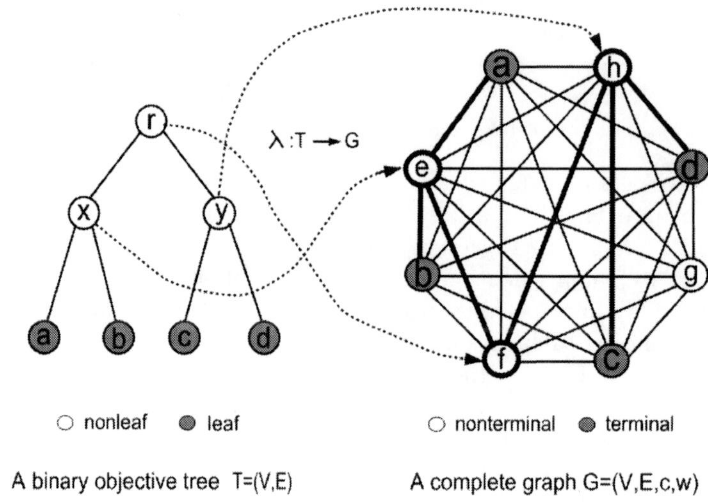

Fig. 1. Illustrating a realization of a binary objective tree

terminal) $s \in S \subset \mathcal{V}$ is required to be mapped to a fixed vertex $s \in S \subset V$ and each *nonleaf* $\alpha \in \mathcal{V} \setminus S$ to a *nonterminal* $v \in V \setminus S$. Consequently, the key task is to map all nonleaves of $\mathcal{T}$ to some nonterminals of G, which forms an *mapping function*, denoted as $\lambda : \mathcal{V} \setminus S \to V \setminus S$ where $\lambda(\alpha) = v$ means that $\alpha \in \mathcal{V} \setminus S$ is mapped to $v \in V \setminus S$. As an example in Fig. 1, the left-hand graph is a given binary objective tree $\mathcal{T} = (\mathcal{V}, \mathcal{E})$ and the right-hand graph is a given graph $G = (V, E, c, w)$ where all weights on edges are not marked for ease of view. Three nonleaves x, y, r of $\mathcal{T}$ are mapped to three nonterminals e, h, f of G respectively, and then $\{x, a\}, \{x, b\}, \{y, c\}, \{y, d\}, \{r, x\}, \{r, y\}$ of $\mathcal{T}$ are mapped to $\{e, a\}, \{e, b\}, \{h, c\}, \{h, d\}, \{f, e\}, \{f, h\}$ of G respectively. These form a realization of $\mathcal{T}$ in G distinguished by bold edges. Evidently, $\mathcal{T}$–ROT has a same tree topology as $\mathcal{T}$.

An objective tree $\mathcal{T}$ always can be taken as a rooted tree by designating its any nonleaf as its root, as well transformed into a binary tree using the method in [6,22]. All vertices of $\mathcal{V}$ can be labeled by numbers $1, \ldots, |\mathcal{V}|$ in the order of from bottom to root and from left to right on a level amongst $\mathcal{T}$. Clearly $|\mathcal{V}|$ represents the root ν of $\mathcal{T}$. Let $|S| = k, |V| = n$, and then $|\mathcal{V} \setminus S| = k - 1, |V \setminus S| = n - k$. Let the maximum hop count between two different leaves in $\mathcal{T}$ be m.

According to above analysis, there exist three kinds of $\mathcal{T}$–ROT in all. We call $\mathcal{T}$–ROT *degenerate* if different nonleaves of $\mathcal{T}$ are mapped to a same nonterminal of G and *undegenerate* otherwise. Furthermore, we call a degenerate $\mathcal{T}$–ROT *strongly degenerate* if two adjacent vertices of $\mathcal{T}$ are mapped to a same nonterminal of G and *weakly degenerate* otherwise. In practice, the strongly degenerate $\mathcal{T}$–ROT is invalid and easy to be avoided by letting $w(v, v) = \infty$ additionally for each vertex v of G.

2.2 The Problems and NP-Hardness

Problem 1. Given an input $\mathcal{J}$ and a real number $C_0 \geq 0$, the *minimum diameter cost-constrained realization of objective tree problem* (MDCCRP) aims to find a minimum diamter $\mathcal{T}$–ROT as $T = (U, F, c, w)$ in G subject to the constraint that $c(T) \leq C_0$.

The decision version of MDCCRP is formally defined as follows.

Problem 2. Given an input $\mathcal{J}$, a real number $C_0 \geq 0$, and a constant $\mathcal{B} \geq 0$, is there a $\mathcal{T}$–ROT as $T = (U, F, c, w)$ in G subject to the constraint that $c(T) \leq C_0$ such that $w(T) \leq \mathcal{B}$?

Theorem 1. *Problem 2 is NP-hard.*

Theorem 1 implies that the undegenerate version of MDCCRP is NP-hard (its hardness is no less than that of Hamilton Path Problem) when $\mathcal{T}$ is part of the input. Given any $\mathcal{T}$, the number of all available undegenerate $\mathcal{T}$–ROT's is $\binom{|V\setminus S|}{|\mathcal{V}\setminus S|} \cdot (|\mathcal{V} \setminus S|)! = \binom{n-k}{k-1} \cdot (k-1)! = \frac{(n-k)!}{(n-2k+1)!} = \Omega(n^{k-1})$. Since we can compute its diameter and cost by using DF (depth-first search) procedure to travel it for a given TeST spanning S, the time complexity of computing the undegenerate version of MDCCRP is $O(kn^{k-1})$. Thus the undegenerate version of MDCCRP is polynomially solvable when $\mathcal{T}$ is not part of the input but fixed. Fortunately, the weakly degenerate $\mathcal{T}$–ROT arises frequently in practice and more representative than the undegenerate one. Therefore, we focus on the weakly degenerate version of MDCCRP in the rest of this paper where the objective tree is a rooted binary tree.

Problem 3. Given an input $\mathcal{J}$ and a real number $D_0 \geq 0$, the *minimum cost diameter-constrained realization of objective tree problem* (MCDCRP) aims to find a minimum cost $\mathcal{T}$–ROT as $T = (U, F, c, w)$ in G subject to the constraint that $w(T) \leq D_0$.

3 A Pseudo-polynomial-Time Algorithm for MCDCRP

In this section, we present a pseudo-polynomial-time algorithm for MCDCRP, which can compute the minimum cost of $\mathcal{T}$–ROT with a diameter of no more than η, for given $\eta \geq 0$. Let $w(\mathcal{T})$ denote the minimum diameter of $\mathcal{T}$–ROT with a cost of no more than C_0. On basis of this algorithm, we devise another algorithm described as Algorithm 1, which can decide whether $w(\mathcal{T}) > \eta$ or $w(\mathcal{T}) \leq \eta$ for MDCCRP in a pseudo polynomial time.

Given an objective tree $\mathcal{T}$, we use the bottom-up dynamic programming to realize it. For any $\mathcal{T}$–ROT as $T = (U, F, c, w)$, we refer to the *radius* of T as the maximum weight of $p_T(s, r), s \in S$ in T provided that $\lambda(\nu) = r$. For any $\alpha \in \mathcal{V}$, when $\lambda(\alpha) = v$, we use $C[\alpha, v, \mathcal{D}, \mathcal{R}]$ to denote the minimum cost of all $\mathcal{T}(\alpha)$–ROT's with a diameter bounded by $\mathcal{D}$ and a radius bounded by $\mathcal{R}$.

For all $\mathcal{D} = 0, 1, \ldots, \eta$ and $\mathcal{R} = 0, 1, \ldots, \mathcal{D}$, when $\alpha \in \mathcal{V}$ is a leaf of $\mathcal{T}$, considering that $\mathcal{T}(\alpha)$ has a single vertex, we initialize $C[s, s, \mathcal{D}, \mathcal{R}] = 0$ if $\alpha = s \in S$. When $\alpha \in \mathcal{V}$ is a nonleaf of $\mathcal{T}$, it is required that $\lambda(\alpha) = v \in V \setminus S$. We show a recurrence equation for computing $C[\alpha, v, \mathcal{D}, \mathcal{R}]$ in Theorem 2. Moreover, we set $C[\alpha, v, \mathcal{D}, \mathcal{R}] = \infty$ when either $\mathcal{D} < 0$ or $\mathcal{R} < 0$. It is noted that $\lambda(\beta) = \lambda(\alpha)$ if α is a dummy vertex for β in the resulting binary tree $\mathcal{T}^{\mathrm{B}}$, as well the cost and weight on $\{x, y\}$ are both set to zero if $\{x, y\}$ is an artificial edge in $\mathcal{T}^{\mathrm{B}}$. Here, for any edge $\{u, v\} \in E$, we let $c(u, v)$ and $w(u, v)$ denote the cost and weight on $\{u, v\}$ respectively.

Theorem 2. *For all $\mathcal{D} = 0, 1, \ldots, \eta$ and $\mathcal{R} = 0, 1, \ldots, \mathcal{D}$, when $\lambda(\alpha) = v$, we can compute $C[\alpha, v, \mathcal{D}, \mathcal{R}]$ by*

$$C[\alpha, v, \mathcal{D}, \mathcal{R}] = \min_{\substack{\mathcal{R}_l + \mathcal{R}_r \leq \mathcal{D}, \\ 0 \leq \mathcal{R}_l, \mathcal{R}_r \leq \mathcal{R}}} \left\{ \min_{v_l \in V \setminus S} \{C[\alpha_l, v_l, \mathcal{D}, \mathcal{R}_l - w(v_l, v)] + c(v_l, v)\} \right. \\ \left. + \min_{v_r \in V \setminus S} \{C[\alpha_r, v_r, \mathcal{D}, \mathcal{R}_r - w(v_r, v)] + c(v_r, v)\} \right\}. \tag{1}$$

We can accelerate the computation of $C[\alpha, v, \mathcal{D}, \mathcal{R}]$ using the following way. Let $C^{\mathbb{D}}[\alpha, v, \mathcal{D}, \mathcal{R}]$ denote the minimum cost of all $\mathcal{T}(\alpha)$–ROT's with diameter $\mathcal{D}$ and a radius of no more than $\mathcal{R}$, and $C^{\mathbb{D}}_{\mathbb{R}}[\alpha, v, \mathcal{D}, \mathcal{R}]$ denote the minimum cost of all $\mathcal{T}(\alpha)$–ROT's with diameter $\mathcal{D}$ and radius $\mathcal{R}$. We have

- $C[\alpha, v, \mathcal{D}, \mathcal{R}] = \min\{C[\alpha, v, \mathcal{D} - 1, \mathcal{R}], C^{\mathbb{D}}[\alpha, v, \mathcal{D}, \mathcal{R}]\}$;
- $C^{\mathbb{D}}[\alpha, v, \mathcal{D}, \mathcal{R}] = \min\{C^{\mathbb{D}}[\alpha, v, \mathcal{D}, \mathcal{R} - 1], C^{\mathbb{D}}_{\mathbb{R}}[\alpha, v, \mathcal{D}, \mathcal{R}]\}$.

When α is the root ν, we compute $\min_{\mathcal{R}=0,1,\ldots,\mathcal{D}} \min_{v \in V \setminus S} C[\nu, v, \mathcal{D}, \mathcal{R}]$ for all $\mathcal{D} = 0, 1, \ldots, \eta$, the value of which is just the minimum cost of $\mathcal{T}$–ROT with diameter bounded by η. Above discussions form a pseudo-polynomial-time algorithm for finding a minimum cost $\mathcal{T}$–ROT with diameter bounded by η for MCDCRP, see the analysis of Step_2 in the proof of Theorem 3. By investigation, we discover that either none or some of them are no more than C_0 in all values of $\min_{\mathcal{R}=0,1,\ldots,\mathcal{D}} \min_{v \in V \setminus S} C[\nu, v, \mathcal{D}, \mathcal{R}], \mathcal{D} = 0, 1, \ldots, \eta$. Let

$$w(\mathcal{T}, \eta) = \min \left\{ \mathcal{D} \in \{0, 1, \ldots, \eta\} : \min_{\mathcal{R}=0,1,\ldots,\mathcal{D}} \min_{v \in V \setminus S} C[\nu, v, \mathcal{D}, \mathcal{R}] \leq \mathrm{C}_0 \right\}. \tag{2}$$

If the former occurs, we set $w(\mathcal{T}, \eta) = \infty$ and output NO. Else if the latter occurs, we record $w(\mathcal{T}, \eta)$ and output YES. This leads us to Algorithm 1, the step_2 of which uses the dynamic programming approach [6,7,26] for deciding whether $w(\mathcal{T}) > \eta$ or $w(\mathcal{T}) \leq \eta$ for MDCCRP. Theorem 3 shows that the time complexity of Algorithm 1 is also pseudopolynomial.

Theorem 3. *Given an MDCCRP where G has n vertices and k terminals, the time complexity of Algorithm 1 is $O(kn^2\eta^3)$. Furthermore, we infer $w(\mathcal{T}) \leq \eta$ if the output is* YES *and $w(\mathcal{T}) > \eta$ if the output is* NO.

Algorithm 1. Pseudo-polynomial-time algorithm for MDCCRP.

Input: An input $\mathcal{J}$, two positive integers C_0 and η.
Output: Either YES together with $w(\mathcal{T}, \eta)$ or NO.

```
Step_1 for {∀α ∈ 𝒱; ∀v ∈ V; 𝒟 = 0 to η; ℛ = 0 to η} do
            Initialize C[α, v, 𝒟, ℛ] := 0;
       endfor
Step_2 for α = 1 to |𝒱| do
            if α ∈ S then break;
            else for {∀v ∈ V \ S; 𝒟 = 0 to η; ℛ = 0 to 𝒟} do
                    Compute C[α, v, 𝒟, ℛ] by Eq. (1);
                 endfor
Step_3 w(𝒯, η) := min{ 𝒟 ∈ {0, 1, ..., η} : min_{ℛ=0,1,...,𝒟} min_{v∈V\S} C[ν, v, 𝒟, ℛ] ≤ C0 };
       if w(𝒯, η) = ∞ then output NO; else output YES;
       When the answer is YES, a minimum diameter 𝒯-ROT with cost
       bounded by C0 can be traced out top-down from ν of 𝒯.
```

4 A Weakly FPTAS for MDCCRP

In this section, we apply standard technique of scaling and rounding to design a weakly FPTAS for MDCCRP based on Algorithm 1, see [6,13,15,21,26,27]. To prepare for the weakly FPTAS, we need several auxiliary algorithms which are used as subroutines in the weakly FPTAS. In the following subsections, we will present these in detail.

4.1 Auxiliary Graphs

Let w_θ be the *scaled edge weight function* such that $w_\theta(e) = w(e) \times \theta$ for each $e \in E$. Then we construct an auxiliary graph $G_\theta = (V, E, c, w_\theta)$ which is the same as $G = (V, E, c, w)$ except that the weight $w(e)$ is changed to $w_\theta(e)$ for each $e \in E$. Correspondingly, we have $w_\theta(\mathcal{T}, \eta)$ and $w_\theta(\mathcal{T})$. Let $w_{\lfloor\theta\rfloor}(e)$ be the *scaled-rounding edge weight function* such that $w_{\lfloor\theta\rfloor}(e) = \lfloor w(e) \times \theta \rfloor$ for each $e \in E$. Similarly, we construct $G_{\lfloor\theta\rfloor} = (V, E, c, w_{\lfloor\theta\rfloor})$. Let T, T_θ and $T_{\lfloor\theta\rfloor}$ be a minimum diameter $\mathcal{T}$–ROT with cost bounded by C_0 in G, G_θ and $G_{\lfloor\theta\rfloor}$ respectively. Lemma 1 follows immediately.

Lemma 1. $w_\theta(T) = w_\theta(T_\theta) \leq w_\theta(T_{\lfloor\theta\rfloor})$, $w_{\lfloor\theta\rfloor}(T_{\lfloor\theta\rfloor}) \leq w_{\lfloor\theta\rfloor}(T)$.

For ease of presentation, we also use $\lfloor\theta\rfloor$ to represent the edge scaled-rounding operation. Each $\lfloor\theta\rfloor$ results in a $G_{\lfloor\theta\rfloor}$. Let $p(T)$ be a diameter leaf-to-leaf path of T in G and $p(T_{\lfloor\theta\rfloor})$ be a diameter leaf-to-leaf path of $T_{\lfloor\theta\rfloor}$ in $G_{\lfloor\theta\rfloor}$. Clearly, $w(T) = w(p(T))$ and $w_\theta(T) = w_\theta(p(T))$. All θ such that $p(T)$ is also a diameter leaf-to-leaf path of T in $G_{\lfloor\theta\rfloor}$ form a set, formulated as Eq. (3), and all θ such that $p(T_{\lfloor\theta\rfloor})$ in $G_{\lfloor\theta\rfloor}$ is also a diameter leaf-to-leaf path of $T_{\lfloor\theta\rfloor}$ in G, formulated as Eq. (4), and otherwise. An input $\mathcal{J}$ together with $\mathbf{C}_0$ are represented as I.

$$\Theta_1(I) = \{\theta : \ w(T) = w(p(T)), w_{\lfloor\theta\rfloor}(T) = w_{\lfloor\theta\rfloor}(p(T))\}\,, \tag{3}$$

$$\Theta_2(I) = \{\theta : w(T_{\lfloor\theta\rfloor}) = w(p(T_{\lfloor\theta\rfloor})), w_{\lfloor\theta\rfloor}(T_{\lfloor\theta\rfloor}) = w_{\lfloor\theta\rfloor}(p(T_{\lfloor\theta\rfloor}))\} . \quad (4)$$

Lemma 2. *For any $\theta > 0$, we infer that $\theta \in \Theta_1(I)$ if θ satisfies that $w(p(T)) - w(p_1) \geq \frac{h(p(T))}{\theta}$ for any leaf-to-leaf path p_1 on T, and $\theta \in \Theta_2(I)$ if θ satisfies that $w_{\lfloor\theta\rfloor}(p(T_{\lfloor\theta\rfloor})) - w_{\lfloor\theta\rfloor}(p_2) \geq h(p_2)$ for any leaf-to-leaf path p_2 on $T_{\lfloor\theta\rfloor}$.*

As a counterexample, given a graph G and a T, T has two different leaf-to-leaf paths $p' = e_1e_2e_3e_4e_5e_6$ with $w(e_1) = \ldots = w(e_6) = 3$ and $p'' = e_7e_8$ with $w(e_7) = 8, w(e_8) = 9$. Let $p(T) = p'$ in G. We have $w(p') = w(e_1)+\ldots+w(e_6) = 18$ and $w(p'') = w(e_7)+w(e_8) = 17$. Set $\theta_1 = 3.3$, then $\lfloor\theta_1\rfloor$ to G generates $G_{\lfloor\theta_1\rfloor}$ with $w_{\lfloor\theta_1\rfloor}(e_1) = \ldots = w_{\lfloor\theta_1\rfloor}(e_6) = \lfloor 3 \times 3.3\rfloor = 9$ and $w_{\lfloor\theta_1\rfloor}(e_7) = \lfloor 8 \times 3.3\rfloor = 26, w_{\lfloor\theta_1\rfloor}(e_8) = \lfloor 9 \times 3.3\rfloor = 29$. We have $w_{\lfloor\theta_1\rfloor}(p') = w_{\lfloor\theta_1\rfloor}(e_1)+\ldots+w_{\lfloor\theta_1\rfloor}(e_6) = 54$ and $w_{\lfloor\theta_1\rfloor}(p'') = w_{\lfloor\theta_1\rfloor}(e_7) + w_{\lfloor\theta_1\rfloor}(e_8) = 55$. Therefore, p' is not a diameter leaf-to-leaf path of T in $G_{\lfloor\theta_1\rfloor}$. As a result, $\theta_1 = 3.3 \notin \Theta_1(I)$. Likewise, we can get a counterexample of $\theta_2 \notin \Theta_2(I)$.

Theorem 4. *Given any I of MDCCRP, it is certain that every $\theta \geq 2m$ satisfies that $\theta \in \Theta_1(I) \cap \Theta_2(I)$.*

4.2 Polynomial Time Approximate Testing

Given a real number $\mathrm{W} > 0$, deciding whether $w(\mathcal{T}) > \mathrm{W}$ or $w(\mathcal{T}) < \mathrm{W}$ for MDCCRP is NP-hard. However, for any given constant $\epsilon > 0$, we can decide whether $w(\mathcal{T}) > \mathrm{W}$ or $w(\mathcal{T}) < (1+\epsilon) \times \mathrm{W}$ in a fully polynomial time using the standard technique of scaling and rounding [6,13,15,21,26,27]. The technique plays an important role in our weakly FPTAS. This approximate testing is described as TEST, see Algorithm 2.

Algorithm 2. TEST(W, ϵ).
Input: An input $\mathcal{J}$, two positive constants C_0 and W, and a positive real number $\epsilon \in (0, \frac{1}{\mathrm{W}}]$. **Output:** Either YES or NO.
Step_1 Set $\theta := \frac{2m}{\mathrm{W}\times\epsilon}$; $w_{\lfloor\theta\rfloor}(e) := \lfloor w(e) \times \theta\rfloor$ for each $e \in E$; Set $\eta := \lfloor \mathrm{W} \times \theta\rfloor$; Step_2 Replace G by $G_{\lfloor\theta\rfloor}$ into $\mathcal{J}$; Apply Algorithm 1; if $w_{\lfloor\theta\rfloor}(\mathcal{T}, \eta) \leq \eta$ then output YES; else output NO;

Theorem 5. *Given a $\mathcal{J}$, three constants $\mathrm{W} > 0, \mathrm{C}_0 \geq 0$ and $0 < \epsilon \leq \frac{1}{\mathrm{W}}$, we infer $w(\mathcal{T}) > \mathrm{W}$ if* TEST$(\mathrm{W}, \epsilon) =$ NO *and $w(\mathcal{T}) < (1+\epsilon) \times \mathrm{W}$ if* TEST$(\mathrm{W}, \epsilon) =$ YES. *In addition, the worst-case time complexity of* TEST(W, ϵ) *is $O(\frac{kn^2m^3}{\epsilon^3})$.*

4.3 Weakly Fully Polynomial Time Approximation Schemes

In this subsection, we will present a weakly fully polynomial time approximation scheme (abbreviated to weakly FPTAS or WFPTAS), described as Algorithm 3, and an improved weakly FPTAS based on Algorithm 3, described as Algorithm 4. There exists some constant Ω such that both of Algorithm 3 and 4 can find a $(1+\epsilon)$-approximation of MDCCRP for a subset of available values of $\epsilon > \Omega$ with specific characteristic of $\theta = \theta(\epsilon) \in \Theta_1(I) \cap \Theta_2(I)$, and find one definitely for all $\epsilon \leq \Omega$.

Algorithm 3. $\mathsf{WFPTAS}_1(\epsilon)$.
Input: An input $\mathcal{J}$, three positive constants LB, UB and C_0, and a positive real number $\epsilon \in (0, \frac{1}{2\times\mathsf{LB}}]$.
Output: A cost-constrained $\mathcal{T}$–ROT as T^{A} such that $w(T^{\mathrm{A}}) < (1+\epsilon)\times w(\mathcal{T})$.
Step_1 Set $\theta := \frac{m}{\mathsf{LB}\times\epsilon}$; $w_{\lfloor\theta\rfloor}(e) := \lfloor w(e)\times\theta\rfloor$ for each $e \in E$; Set $\eta := \lfloor \mathsf{UB}\times\theta\rfloor$;
Step_2 Replace G by $G_{\lfloor\theta\rfloor}$ into $\mathcal{J}$; Apply Algorithm 1;

Theorem 6. *Given an input $\mathcal{J}$, a known lower bound* LB *and an upper bound* UB *on $w(\mathcal{T})$, and two constants $\mathrm{C}_0 \geq 0$ and $0 < \epsilon \leq \frac{1}{2\times\mathsf{LB}}$,* $\mathsf{WFPTAS}_1(\epsilon)$ *can compute a $\mathcal{T}$–ROT as T^{A} such that $w(T^{\mathrm{A}}) < (1+\epsilon)\times w(\mathcal{T})$ in $O(\frac{kn^2m^3\times\mathsf{UB}^3}{\epsilon^3\times\mathsf{LB}^3})$ time, provided that we get a scaled-rounding weight function $w_{\lfloor\theta\rfloor}(e) = \lfloor w(e)\times\theta\rfloor$ and set $\eta = \lfloor\mathsf{UB}\times\theta\rfloor$ using $\theta = \frac{m}{\mathsf{LB}\times\epsilon}$.*

Theorem 6 shows that the time complexity of WFPTAS_1 is related to the ratio $\frac{\mathsf{UB}}{\mathsf{LB}}$. As in [6,13,26], we can reduce the time complexity by initializing LB and UB as easily computable values and then using the bisection method to reduce the ratio. An initial value of LB can be computed as follows. Use the method in [5] to compute the minimum diameter of $\mathcal{T}$–ROT ignoring the bound on cost, which occupies $O(n^3 + k(n-k)^3)$ time. We can take this minimum as the initial value of LB. If $\mathrm{W} < \mathsf{LB}$ (sufficient but unnecessary condition), there is no $\mathcal{T}$–ROT with cost bounded by C_0. An initial value of UB can be computed as follows. Find the maximum edge weight in G, which occupies $O(\log n)$ time. We can take $m \times \max_{e\in E} w(e)$ as the initial value of UB.

Let $\zeta = \frac{1}{\mathsf{UB}}$, we apply the bisection method to drive $\frac{\mathsf{UB}}{\mathsf{LB}}$ down to some number below $2\times(1+\zeta)$. Suppose that our lower bound LB and upper bound UB satisfy that $\frac{\mathsf{UB}}{\mathsf{LB}} > 2\times(1+\zeta)$. Let $\mathrm{W} = (\frac{\mathsf{LB}\times\mathsf{UB}}{1+\zeta})^{\frac{1}{2}}$. If $\mathsf{TEST}(\mathrm{W},\zeta) = \mathsf{NO}$ then W is a new lower bound and UB is also an upper bound on $w(\mathcal{T})$. If $\mathsf{TEST}(\mathrm{W},\zeta) = \mathsf{YES}$ then $(1+\zeta)\times\mathrm{W}$ is a new upper bound and LB is also a lower bound on $w(\mathcal{T})$. Therefore, the ratio of the new upper bound over the new lower bound will be always no more than $(\frac{\mathsf{UB}}{\mathsf{LB}}\times(1+\zeta))^{\frac{1}{2}}$ since $\theta \in \Theta_1(I)\cap\Theta_2(I)$ always follows from $\theta = \frac{2m}{\mathrm{W}\times\zeta} \geq 2m$ by Theorem 4. Above process is called an *iteration*. Such an iteration can be accomplished in a fully polynomial time (according to Theorem 5). Moreover, $\frac{\mathsf{UB}}{\mathsf{LB}}$ will be reduced to a number below 4 in $\log\mathcal{S}$ iterations ($\mathcal{S}$ is the input size of the given instance), see [6,26]. Above analysis leads to our WFPTAS_2, described as Algorithm 4. The time complexity of Algorithm 4 is shown in Theorem 7.

Algorithm 4. $\mathsf{WFPTAS}_2(\epsilon)$.

Input: An input $\mathcal{J}$, two positive constants C_0 and ϵ.
Output: A cost-constrained $\mathcal{T}$–ROT as T^A such that $w(T^A) < (1+\epsilon) \times w(\mathcal{T})$.

Step_1 Set both LB and UB to their initial values as mentioned above;
  if $0 < \epsilon \leq \frac{1}{2\times \mathsf{UB}}$ then
    Let $\zeta := \frac{1}{\mathsf{UB}}$, goto Step_2;
  else Return;
  endif
Step_2 if $\mathsf{UB} \leq 2(1+\zeta) \times \mathsf{LB}$ then
    goto Step_3;
  else
    $\mathrm{W} := (\frac{\mathsf{LB}\times \mathsf{UB}}{1+\zeta})^{\frac{1}{2}}$;
    if $\mathsf{TEST}(\mathrm{W}, \zeta) = \mathsf{NO}$, then set $\mathsf{LB} := \mathrm{W}$;
    if $\mathsf{TEST}(\mathrm{W}, \zeta) = \mathsf{YES}$, then set $\mathsf{UB} := (1+\zeta) \times \mathrm{W}$;
    goto Step_2;
  endif
Step_3 Set $\theta := \frac{m}{\mathsf{LB}\times\epsilon}$; $w_{\lfloor\theta\rfloor}(e) := \lfloor w(e) \times \theta \rfloor$ for each $e \in E$; Set $\eta := \lfloor \mathsf{UB} \times \theta \rfloor$;
  Replace G by $G_{\lfloor\theta\rfloor}$ into $\mathcal{J}$; Apply Algorithm 1;

Theorem 7. *Given an input $\mathcal{J}$, two constants $C_0 \geq 0$ and $\epsilon > 0$, if there is a cost-constrained $\mathcal{T}$–ROT,* $\mathsf{WFPTAS}_2(\epsilon)$ *will find a cost-constrained $\mathcal{T}$–ROT as T^A such that $w(T^A) < (1+\epsilon) \times w(\mathcal{T})$. Furthermore, the time complexity of* $\mathsf{WFPTAS}_2(\epsilon)$ *is $O(\frac{kn^2m^3}{\epsilon^3} \times \log \mathcal{S})$, where $\mathcal{S}$ is the input size of the given instance.*

5 Concluding Remarks

In this paper, we are concerned with MDCCRP in a complete graph and have presented two weakly FPTAS's. The major difference between a classical FPTAS and a weakly FPTAS is that, given a real number $\epsilon > 0$, the former can find a $(1+\epsilon)$–approximation of problem definitely regardless of the value of ϵ while the latter indefinitely. Theorems 5, 6 and 7 reflect a common fact that whether their corresponding results hold or not relies on $\theta \in \Theta_1(I) \cap \Theta_2(I)$ or not, that is, essentially on the value of ϵ. The characteristic of $\theta = \theta(\epsilon) \in \Theta_1(I) \cap \Theta_2(I)$ is represented as Π. As a whole, for a subset of available values of ϵ with specific characteristic of admitting Π, WFPTAS can find a $(1+\epsilon)$–approximation of problem. The subset can be formulated as

$$\Sigma(\Pi, I) = \{\epsilon > 0 : \theta = \theta(\epsilon) \in \Theta_1(I) \cap \Theta_2(I)\} \ . \tag{5}$$

Theorem 4 shows that every $\theta \geq 2m$ satisfies $\theta \in \Theta_1(I) \cap \Theta_2(I)$, i.e., there exists a constant Ω such that every $0 < \epsilon \leq \Omega$ satisfies $\theta = \theta(\epsilon) \geq 2m$, which ensures that $w(T^A) \to w(\mathcal{T})$ when $\epsilon \to 0$. However, only a subset of available values of ϵ satisfies $\theta = \theta(\epsilon) \in \Theta_1(I) \cap \Theta_2(I)$ when $\epsilon > \Omega$. This paper gives $\Omega = \theta^{-1}(2m)$. It is also of interests to improve the value of Ω.

Given a real number $\epsilon > 0$, $\mathsf{LB}_{[0]}$ and $\mathsf{UB}_{[0]}$ as discussed in Sect. 4.3, when $0 < \epsilon \leq \frac{1}{2\times\mathsf{UB}_{[0]}}$, WFPTAS_1 can find a $(1+\epsilon)$–approximation of MDCCRP in $O(\frac{kn^2m^3}{\epsilon^3} \times \Delta_1)$ time where $\Delta_1 = (\mathsf{UB}_{[0]}/\mathsf{LB}_{[0]})^3$ and WFPTAS_2 can find a $(1+\epsilon)$–approximation of MDCCRP in $O(\frac{kn^2m^3}{\epsilon^3} \times \Delta_2)$ time where $\Delta_2 = \lceil \log(\log \mathsf{UB}_{[0]} - \log \mathsf{LB}_{[0]}) \rceil$. From the theoretical point of view, since $\Delta_2 \ll \Delta_1$, the time complexity of WFPTAS_2 is quite lower than that of WFPTAS_1.

References

1. Apostolopoulos, G., Guerin, R., Kamat, S., Tripathi, S.: Quality of service based routing: a performance perspective. In: Proc. ACM SigComm 1998, pp. 17–28 (1998)
2. Bondy, J.A., Murty, U.S.R.: Graph Theory with Application. Macmillan, London (1976)
3. Chen, Y.H.: An Improved Approximation Algorithm for the Terminal Steiner Tree Problem. In: Murgante, B., Gervasi, O., Iglesias, A., Taniar, D., Apduhan, B.O. (eds.) ICCSA 2011, Part III. LNCS, vol. 6784, pp. 141–151. Springer, Heidelberg (2011)
4. Deo, N., Abdalla, A.: Computing a Diameter-Constrained Minimum Spanning Tree in Parallel. In: Bongiovanni, G., Petreschi, R., Gambosi, G. (eds.) CIAC 2000. LNCS, vol. 1767, pp. 17–31. Springer, Heidelberg (2000)
5. Ding, W.: Many-to-Many Multicast Routing Under a Fixed Topology: Basic Architecture, Problems and Algorithms. In: First International Conference on Networking and Distributed Computing (ICNDC 2010), pp. 128–132 (2010)
6. Ding, W., Lin, G., Xue, G.: Diameter-Constrained Steiner Trees. Discrete Mathematics, Algorithms and Applications 3(4), 491–502 (2011)
7. Ding, W., Xue, G.: A Linear Time Algorithm for Computing a Most Reliable Source on a Tree Network with Faulty Nodes. Theor. Comput. Sci. 412, 225–232 (2011)
8. Drake, D.E., Hougrady, S.: On Approximation Algorithms for the Terminal Steiner Tree Problem. Info. Proc. Lett. 89, 15–18 (2004)
9. Du, D., Hu, X.: Steiner Tree Problems in Computer Communication Networks. World Scientific Publishing Co. Pte. Ltd., Singapore (2008)
10. Fuchs, B.: A Note on the Terminal Steiner tree Problem. Info. Proc. Lett. 87, 219–220 (2003)
11. Garey, M.R., Johnson, D.S.: Computers and Intractability: A Guide to the Theory of NP-Completeness. Freeman, San Francisco (1979)
12. Gouveia, L., Magnanti, T.L.: Network Flow Models for Designing Diameter-Constrained Minimum Spanning and Steiner Trees. Networks 41, 159–173 (2003)
13. Hassin, R.: Approximation Schemes for the Restricted Shortest Path Problem. Math. of Oper. Res. 17, 36–42 (1992)
14. Hwang, F.K., Richards, D.S., Winter, P.: The Steiner Tree Problem. Annals of Disc. Math., vol. 53. North-Holland, Amsterdam (1992)
15. Ibarra, O., Kim, C.: Fast Approximation Algorithms for the Knapsack and Sum of Subset Problems. J. ACM 22(4), 463–468 (1975)
16. Lin, G., Xue, G.: On the Terminal Steiner Problem. Info. Proc. Lett. 84, 103–107 (2002)

17. Martineza, F.V., Pinab, J.C.D., Soares, J.: Algorithm for Terminal Steiner Trees. Theor. Comput. Sci. 389, 133–142 (2007)
18. Marathe, M.V., Ravi, R., Sundaram, R., Ravi, S.S., Rosenkrantz, D.J., Hunt III, H.B.: Bicriteria Network Design Problems. J. Algorithms 28(1), 142–171 (1998)
19. Robins, G., Zelikovsky, A.: Improved Steiner Tree Approximation in Graphs. In: Proc. of the 11th Annual ACM-SIAM Symposium on Discrete Algorithm (SODA 2000), pp. 770–779 (2000)
20. dos Santos, A.C., Lucena, A., Ribeiro, C.C.: Solving Diameter Constrained Minimum Spanning Tree Problems in Dense Graphs. In: Ribeiro, C.C., Martins, S.L. (eds.) WEA 2004. LNCS, vol. 3059, pp. 458–467. Springer, Heidelberg (2004)
21. Sahni, S.: General Techniques for Combinatorial Approximations. Oper. Res. 35, 70–79 (1977)
22. Tamir, A.: An $O(pn^2)$ Algorithm for the p-Median and Related Problems on Tree Graphs. Oper. Res. Lett. 19, 59–64 (1996)
23. Vazirani, V.V.: Approximation Algorithms. Springer, Berlin (2001)
24. Wang, L., Jia, X.: Note Fixed Topology Steiner Trees and Spanning Forests. Theor. Comput. Sci. 215(1-2), 359–370 (1999)
25. Wang, Z., Crowcroft, J.: Quality-of-service routing for supporting multimedia applications. IEEE Journal on Selected Areas in Communications 14, 1228–1234 (1996)
26. Xue, G., Xiao, W.: A Polynomial Time Approximation Scheme for Minimum Cost Delay-Constrained Multicast Tree under a Steiner Topology. Algorithmica 41(1), 53–72 (2004)
27. Xue, G., Zhang, W., Tang, J., Thulasiraman, K.: Polynomial Time Approximation Algorithms for Multi-Constrained QoS Routing. IEEE/ACM Trans. Netw. 16, 656–669 (2008)
28. Zelikovsky, A.: An $\frac{11}{6}$-Approximation Algorithm for the Network Steiner Problem. Algorithmica 9(5), 463–470 (1993)

The Edge-Centered Surface Area of the Arrangement Graph

Eddie Cheng[1], Ke Qiu[2], and Zhizhang Shen[3]

[1] Dept. of Mathematics and Statistics, Oakland University, USA
echeng@oakland.edu

[2] Dept. of Computer Science, Brock University, Canada
kqiu@brocku.ca

[3] Dept. of Computer Science and Technology, Plymouth State University, USA
zshen@plymouth.edu

Abstract. We suggest the notion of the surface area centered at an edge for an interconnection network, which generalizes the usual notion of surface area of a network centered at a vertex. Following an elementary approach, we derive an explicit expression of the edge-centered surface area of the arrangement graph.

Keywords: Edge-centered surface area, arrangement graph, combinatorial studies, interconnection networks.

1 Introduction

Given a vertex v in a graph G, a question one may ask is *how many vertices are at distance i from v, $i \in [0, D(G)]$*, where $D(G)$ stands for the diameter of G. This quantity is referred to in the literature, among others, as the *surface area with radius i centered at v* [6].

The surface area of a (di)graph can find several applications in network performance evaluation, e.g., in computing various bounds for the problem of k-neighborhood broadcasting[5] and in identifying spanning trees [9]. As a result, this surface area problem has been studied for a variety of graphs, including the rotator graph, the star graph, the k-ary n-cube, the (n, k)-star graph, and the arrangement graph (For the solution to this problem for the aforementioned and other graphs, readers are referred to[6,10] and the references cited within.)

In this paper, we study a related question: given a reference edge (v, w) in a graph G, *how many vertices are at distance i from (v, w)*, $i \in [0, D(G)]$. We refer to this quantity as the *surface area with radius i centered at (v, w)*, denoted as $B_G(v, w, i)$ in this paper. We will drop G from this notation, and the others, when the context is clear. We also use $(B_G(v, w, 0), B_G(v, w, 1), \ldots, B_G(v, w, D(G)))$ to refer to the *(v, w)-centered surface area sequence of G*.

This notion of edge-centered surface area of a graph is clearly an immediate generalization of the above vertex-centered surface area, thus interesting in its own right as a combinatorial problem. Moreover, it is recently suggested that

G. Lin (Ed.): COCOA 2012, LNCS 7402, pp. 49–60, 2012.

the surface area centered at a path of length 2 is directly related to the conditional diagnosability of interconnection networks [11]. Thus, the study of the edge-centered surface area provides a starting point to this unexplored territory.

Let G be a simple and connected graph, and let (v, w) be an edge in G. It is clear that $B_G(v, w, 0) = 2$, since the only such vertices of distance 0 from (v, w) are v and w themselves. In general, let $i \in [1, D(G)]$, for a vertex u, if $d(u, v) = i$, but $d(u, w) < i$, then the distance from u to (v, w) would be strictly less than i. On the other hand, the very existence of (v, w) implies $d(u, w) \leq i + 1$. These facts and a symmetric consideration lead to the following definition.

Definition 1. *Let G be a simple and connected graph, and let (v, w) be an edge of G. For $i \in [1, D(G)]$,*

$$B_G(u, v, i) = E_G(v, w, i) + S_G(v, w, i) + L_G(v, w, i), \tag{1}$$

where

$$E_G(v, w, i) = |\{u|(d(u, v) = i) \wedge (d(u, w) = d(u, v))\}|, \tag{2}$$
$$S_G(v, w, i) = |\{u|(d(u, v) = i) \wedge (d(u, w) = d(u, v) + 1)\}|, \text{ and} \tag{3}$$
$$L_G(v, w, i) = |\{u|(d(u, v) = i + 1) \wedge (d(u, w) = d(u, v) - 1)\}|. \tag{4}$$

We note that, for $u, v, w \in V(G)$, if (v, w) is an edge in G and $d(u, v) = d(u, w) \geq 1$, then any pair of shortest paths from u to v and w, respectively, together with (v, w), induces an odd cycle in G. Since no bipartite graph contains an odd cycle, the following result is immediate.

Proposition 1. *Let G be a bipartite graph. For all $(v, w) \in G$, and $i \in [1, D(G)]$, $E_G(v, w, i) = 0$.*

The rest of the paper proceeds as follows: in the next section, after a brief discussion of the arrangement graph, we explore the relevant structures of its vertices. We then derive the edge-centered surface area of the arrangement graph in Section 3, and those of some of the other structures in Section 4, which also concludes this paper.

2 Arrangement Graph and Its Vertex Structures

The arrangement graph [4] preserves many of the nice properties of the well-known star graph [1]: both vertex and edge symmetric, a small diameter, a hierarchical structure, simple minimum routing, and various fault tolerance features; while bringing a solution to the scalability issue associated with the star graph by providing flexibility, with an additional parameter, in choosing an appropriate graph for a network with a given size.

Let $n \geq 2$, $\langle n \rangle = \{1, 2, \ldots, n\}$, and $k \in [1, n)$, an *arrangement graph, $A_{n,k}(V, E)$*, $A_{n,k}$ for short, is defined as follows: V is the collection of all the k-permutations, $p_1 p_2 \cdots p_k$, over $\langle n \rangle$, and for all $u, v \in V$, $(u, v) \in E$ if and only if u and v differ in exactly one position. We call $e_k = 12 \ldots k$ the *identity vertex* of $A_{n,k}$, all the symbols in $[1, k]$ *internal symbols,* and those in $(k, n]$ *external.* We also call all positions in $[1, k]$ *internal positions*, and those in $(k, n]$ *external.* Fig. 1 shows $A_{4,2}$, where 12 is adjacent to 13 and the external symbol 3 occurs in the internal position 2 of node 13.

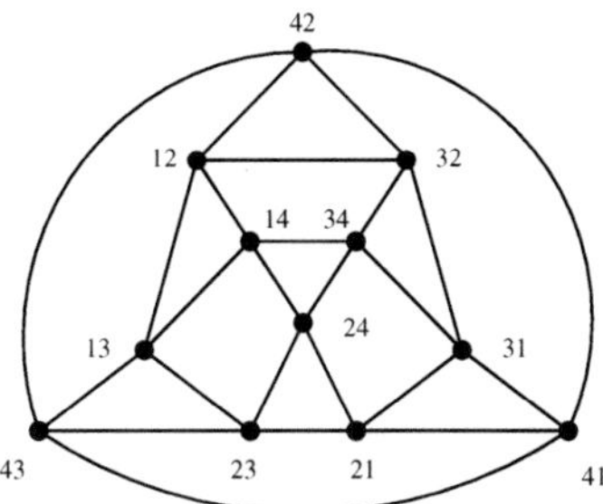

Fig. 1. $A_{4,2}$

It is pointed out in [4,10,2] that a vertex $u(\neq e_k)$ in $A_{n,k}$, as a partial permutation, can be converted to an *extended permutation* on $\langle n \rangle$ and then be factored as a collection of cycles, referred to as the *cycle structure* of u, denoted by $\mathcal{C}(u)$ in this paper. Such a cycle structure is unique except for the order of these cycles. Furthermore, for $u \in A_{n,k}$, $\mathcal{C}(u)$ with $b(u)$ symbols consists of $g_I(u)$ non-trivial internal cycles[1] of $b_I(u)$ symbols and $g_E(u)$ non-trivial external cycles of $b_E(u)$ symbols. Each *non-trivial internal cycle* contains at least two internal symbols and no external symbol and each *non-trivial external cycle* contains exactly one external symbol and at least one internal symbol.

For example, given $z = 6351792 \in A_{9,7}$, we first convert z to $z' = 6351792\underline{84}$ on $\langle 9 \rangle$ as follows: Since 8 does not occur in z, we let $z'_8 = 8$. On the other hand, since $z_6 = 9, z_1 = 6$, and $z_4 = 1$, but 4 does not occur in z, we define $z'_9 = 4$. Then, it is easy to see that $\mathcal{C}(z')$, always denoted by $\mathcal{C}(z)$, is $(9; 4, 1, 6)(2, 3, 5, 7)$, where $(9; 4, 1, 6)$ is a non-trivial external cycle and 9 its external symbol, as indicated with ';', $(2, 3, 5, 7)$ is a non-trivial internal cycle, and, incidentally, 8 is a fixed point of $\mathcal{C}(z)$. Thus, $g_I(z) = g_E(z) = 1$, $b_I(z) = b_E(z) = 4$, and $b(z) = 8$.

Let $u \in A_{n,k}$, $\mathcal{C}(u)$ be its cycle structure, its distance to e_k can be expressed as follows [4,2]:

$$d(u, e_k) = b(u) + g_I(u) - g_E(u). \tag{5}$$

For example, $d(z, e_7) = 8$, and a shortest path from z to e_7 is given as follows:

$$6351792\underline{84} \overset{(4,9)}{\to} 6354792\underline{81} \overset{(1,9)}{\to} 1354792\underline{86} \overset{(6,9)}{\to} 1354762\underline{89} \overset{(7,8)}{\to} 1354768\underline{29} \overset{(2,8)}{\to} 1254768\underline{39} \overset{(3,8)}{\to} 1234768\underline{59} \overset{(5,8)}{\to} 1234568\underline{79} \overset{(7,8)}{\to} 1234567\underline{89},$$

[1] When a cycle contains exactly one symbol, we call it a *trivial cycle,* and the symbol a *fixed point* of $\mathcal{C}(u)$.

where, with "$u \overset{(p_1,p_2)}{\rightarrow} v$", we obtain an extended permutation v by applying a transposition (p_1, p_2) to another extended permutation u.

The vertex-centered surface area problem of $A_{n,k}$ has been studied and solved in [2]. When investigating the edge-centered surface area of $A_{n,k}$, since this graph is edge symmetric, we can choose any edge as the reference edge of the sought surface area. Let $u \in A_{n,k}$, we define $\varphi(u)$ by replacing 1 with $k+1$, if 1 occurs in u, and/or replacing $k+1$ with 1, if $k+1$ occurs in u. For example, in $A_{4,2}$, $\varphi(13) = 31$, $\varphi(32) = 12$ since only 2 occurs in (32), and $\varphi(24) = 24$, since neither 1 nor 3 occurs in 24. Clearly, $\varphi(e_k) = (k+1)2\cdots k$, and $\varphi(\varphi(e_k)) = e_k$. Moreover, since φ is an automorphism of $A_{n,k}$, for $u \in A_{n,k}$, $d(u, \varphi(e_k)) = d(\varphi(u), e_k)$. We henceforth adopt $(e_k, \varphi(e_k))$ as the reference edge.

We are now ready to characterize, for any vertex $u \in A_{n,k}$, the relationship between $d(u, e_k)$ and $d(u, \varphi(e_k))$, in terms of its cycle structure.

Theorem 1. *Let u be a vertex of $A_{n,k}$, then*

1. *if $k+1$ occurs in a trivial external cycle in $\mathcal{C}(u)$, and 1 is a fixed point of $\mathcal{C}(u)$, then $d(u, \varphi(e_k)) = d(u, e_k) + 1$;*
2. *if $k+1$ occurs in a trivial external cycle and 1 occurs in a non-trivial internal cycle in $\mathcal{C}(u)$, then $d(u, \varphi(e_k)) = d(u, e_k) - 1$;*
3. *if $k+1$ occurs in a trivial external cycle and 1 occurs in a non-trivial external cycle in $\mathcal{C}(u)$, then $d(u, \varphi(e_k)) = d(u, e_k)$;*
4. *if $k+1$ occurs in a non-trivial external cycle and 1 is a fixed point of $\mathcal{C}(u)$, then $d(u, \varphi(e_k)) = d(u, e_k) + 1$;*
5. *if $k+1$ occurs in a non-trivial external cycle and 1 occurs in a non-trivial internal cycle in $\mathcal{C}(u)$, then $d(u, \varphi(e_k)) = d(u, e_k) - 1$;*
6. *if $k+1$ and 1 occur in the same non-trivial external cycle, $(k+1, \ldots, 1)$, in $\mathcal{C}(u)$, then $d(u, \varphi(e_k)) = d(u, e_k) - 1$;*
7. *if $k+1$ and 1 occur in the same non-trivial external cycle, $(k+1, \ldots, 1, A)$, $A \neq \epsilon$, in $\mathcal{C}(u)$, then $d(u, \varphi(e_k)) = d(u, e_k) + 1$;*
8. *if $k+1$ and 1 occur in two different non-trivial external cycles in $\mathcal{C}(u)$, then $d(u, \varphi(e_k)) = d(u, e_k)$.*

Proof: We prove the first case when $k+1$ occurs in a trivial external cycle, and 1 occurs in a trivial internal cycle. Since $u_1 = 1$, by definition, $\varphi(u)_1 = k+1$. On the other hand, since $k+1$ does not occur in u, 1 does not occur in $\varphi(u)_1 \cdots \varphi(u)_k$. By definition, $\varphi(u)'_{k+1} = 1$. This leads to an extra external cycle $(k+1; 1)$, containing two extra symbols, in $\mathcal{C}(\varphi(u))$. Since none of the other cycles is involved, by Eq. 5,

$$\begin{aligned} d(\varphi(u), e) &= b(\varphi(u)) + g_I(\varphi(u)) - g_E(\varphi(u)) = (b(u) + 2) + g_I(u) - (g_E(u) + 1) \\ &= d(u, e) + 1. \end{aligned}$$

The other cases can be similarly proved. □

We use Table 1 to demonstrate Theorem 1 in terms of $A_{4,2}$, where the Case column refers to the case index as given in the Theorem 1. We note one case is missing, i.e., when $k+1$ occurs in an external cycle and 1 occurs in an internal cycle in $\mathcal{C}(u)$, then $d(u, \varphi(e_k)) = d(u, e_k) - 1$. Indeed, in $A_{4,2}$, if the symbol 3 occurs in a non-trivial external cycle, containing at least one internal symbol, which cannot be 1, thus, 2 has to be associated with 3. Then, as a non-trivial internal cycle has to contain at least two internal symbols but only the symbol 1 is remaining, no vertex in $A_{4,2}$ could fall into Case 5.

Table 1. $(e_k, \varphi(e_k))$-centered surface area for $A_{4,2}$

u	u'	$\mathcal{C}(u)$	$\varphi(u)$	$d(u, e_k)$	$d(u, \varphi(e_k))$	$i = d(u, (e_k, \varphi(e_k)))$	Case
12	$12\underline{34}$	$\emptyset$	$32\underline{14}$	1	0	0	1
32	$32\underline{14}$	$(3;1)$	$12\underline{34}$	0	1	0	6
14	$14\underline{32}$	$(4;2)$	$34\underline{12}$	2	1	1	1
42	$42\underline{31}$	$(4;1)$	$42\underline{31}$	1	1	1	3
13	$13\underline{24}$	$(3;2)$	$31\underline{24}$	2	1	1	4
34	$34\underline{12}$	$(3;1)(4;2)$	$14\underline{32}$	1	2	1	6
31	$31\underline{24}$	$(3;2,1)$	$13\underline{24}$	1	2	1	6
21	$21\underline{34}$	$(1,2)$	$23\underline{14}$	2	3	2	2
41	$41\underline{32}$	$(4;2,1)$	$43\underline{21}$	2	2	2	3
24	$24\underline{31}$	$(4;1,2)$	$24\underline{31}$	2	2	2	3
23	$23\underline{14}$	$(3;1,2)$	$21\underline{34}$	3	2	2	7
43	$43\underline{21}$	$(3;2)(4;1)$	$41\underline{32}$	2	2	2	8

3 Derivation of the Surface Area Expression

We now proceed to derive the surface area of the arrangement graph, $A_{n,k}$, centered at the edge $(e_k, \varphi(e_k))$. We start with the following result, which is immediate by Theorem 1, Eqs. 2, 3 and 4.

Corollary 1. *Let* $u \in A_{n,k}$, *for* $i \in [1, D(A_{n,k})]$,

$$E_{A_{n,k}}(e_k, \varphi(e_k), i) = |\{u|d(u, e_k) = i, \text{ and } \mathcal{C}(u) \text{ falls into either Case 3 or 8 }\}|.$$

$$S_{A_{n,k}}(e_k, \varphi(e_k), i) = |\{u|d(u, e_k) = i, \text{ and } \mathcal{C}(u) \text{ falls into either Case 2, 5, or 6 }\}|.$$

$$L_{A_{n,k}}(e_k, \varphi(e_k), i) = |\{u|d(u, e_k) = i+1, \text{ and } \mathcal{C}(u) \text{ falls into either Case 1, 4, or 7}\}|.$$

We notice that the distance formula, i.e., Eq. 5, holds even if we allow trivial external cycles in the cycle structure, since any such a cycle contains exactly one external symbol, thus is counted in the $b(u)$ term exactly once, which is canceled out in the $g_E(u)$ term. We thus slightly overload the quantity $g_E(u), u \in A_{n,k}$, by using it to refer to the number of all the non-trivial external cycles in u, plus the *one containing* $k+1$, referred to as *the* $\langle k+1 \rangle$*-cycle* henceforth, which *may be trivial.* We can then combine, e.g., Cases 3 and 8 to one case where $k+1$ and 1 occur in two separate external cycles, and the $\langle k+1 \rangle$-cycle may be trivial. Similarly, Cases 1 and 4, and Cases 2 and 5, can be so combined, as well. As a result, we only need to discuss five, instead of eight, cases in our analysis: the combined cases of 1 and 4 (Case 1+4), that of 2 and 5 (Case 2+5), that of 3 and 8 (Case 3+8), Case 6, and Case 7.

We are ready to enumerate cycle structures by constructing them in the following steps: select the needed external symbols (ES); select internal symbols for these external cycles (IE); construct these external cycles (EC); and select internal symbols and construct the non-trivial internal cycles (IC). We discuss these steps in terms of the above five cases:

ES: We need to select g_E external symbols for the g_E external cycles. For all the cases, the external symbol $k+1$ occurs in a potentially trivial external cycle, thus one of the chosen. Besides $k+1$, we need to choose $g_E - 1$ external symbols from the range of $(k, n]$ in $\binom{n-k-1}{g_E-1}$ ways.

For all the cases, $1 \leq g_E \leq n-k$, except Case 3+8, where the two symbols 1 and $k+1$ have to occur in two different external cycles, thus, $2 \leq g_E \leq n-k$.

IE: We need to select internal symbols for these external cycles. For Cases 1+4, and 2+5, the symbol 1 does not occur in any external cycle, thus we need to choose $b_E - g_E$ internal symbols out of the range $(1, k]$, excluding 1, in $\binom{k-1}{b_E-g_E}$ ways. On the other hand, for Cases 3+8, 6, and 7, the symbol 1 does occur in an external cycle, thus, we choose $b_E - g_E - 1$ such symbols out of $(1, k]$, again excluding 1, in $\binom{k-1}{b_E-g_E-1}$ ways, and then add 1 into such a chosen set to have a total of $b_E - g_E$ symbols.

For Cases 1+4, 2+5 and 3+8, the symbol $k+1$ may occur in a trivial external cycle. Since any non-trivial external cycle contains at least two symbols, and a trivial cycle contains at least one symbol, for these cases, $b_E \geq 2(g_E - 1) + 1 = 2g_E - 1$; while for Cases 6, and 7, no external cycle can be trivial, thus, $b_E \geq 2g_E$.

Similarly, for Cases 1+4 and 2+5, 1 does not occur in any external cycle, thus, $b_E - g_E \leq k-1$, i.e., $b_E \leq g_E + k - 1$; while for Cases 3+8, 6 and 7, $b_E \leq g_E + k$.

EC: When constructing these external cycles, we first select l_E internal symbols out of the $b_E - g_E$ internal symbols as chosen in Step IE to construct the potentially trivial $\langle k+1\rangle$-cycle together with the external symbol $k+1$.

For Cases 1+4, and 2+5, the symbol 1 does not occur in an external cycle, thus not one of the chosen in Step IE. As a result, we simply select l_E internal symbols out of $b_E - g_E$ chosen internal symbols in $\binom{b_E-g_E}{l_E}$ ways. Since for these two cases, the $\langle k+1\rangle$-cycle can be trivial, $l_E \geq 0$. Moreover, since the $\langle k+1\rangle$-cycle contains exactly $l_E + 1$ symbols, and each of the remaining $g_E - 1$ external cycles contains at least two symbols, $b_E \geq l_E + 1 + 2(g_E - 1) = l_E + 2g_E - 1$, i.e., $l_E \leq b_E - 2g_E + 1$.

For Case 3+8, the symbol 1 has to stay in an external cycle different from the $\langle k+1\rangle$-cycle, thus, we have to chose l_E external symbols intended for the $\langle k+1\rangle$-cycle out of $b_E - g_E - 1$ internal symbols, excluding 1, in $\binom{b_E-g_E-1}{l_E}$ ways. The bounds of l_E for this case are the same as those for the above two cases. We note there are $b_E - g_E - l_E$ internal symbols left for the other $g_E - 1$ non-trivial external cycles, after putting back the symbol 1.

It is clear that for these three combined cases, assuming $k+1$ sits at the very first position in its cycle, any permutation of these l_E internal symbols leads to a unique $\langle k+1\rangle$-cycle.

For Cases 6, and 7, the symbol 1 has to be in the same external cycle as $k+1$ does, so, we choose l_E-1 internal symbols out of b_E-g_E-1, in $\binom{b_E-g_E-1}{l_E-1}$ ways, then add back the symbol 1, l_E symbols in total. For Case 6, $l_E-1\geq 0$, i.e., $l_E\geq 1$; while, for Case 7, we have to insert at least one number in between $k+1$ and 1, hence, $l_E-1\geq 1$, i.e., $l_E\geq 2$. The upper bound of l_E is the same for both cases, i.e., b_E-2g_E+1.

We note that, for Case 6, where 1 is required to occur in the last position, while $k+1$ occurs in the first place, any permutation of those chosen l_E-1 internal symbols will lead to a unique $\langle k+1\rangle$-cycle. On the other hand, for Case 7, the symbol 1 can't sit in the last position. We thus have to choose such a symbol out of these chosen l_E-1 symbols in l_E-1 ways, then any permutation of the remaining l_E-2 symbols, together with 1, a total of l_E-1 symbols, lead to a unique $\langle k+1\rangle$-cycle.

We then have to distribute the remaining $b_E-g_E-l_E$ internal symbols to the other g_E-1 external cycles, in which the associated external symbol sits there as the very first symbol. Although the order of these external cycles does not matter, that of those internal symbols in each and every of those cycles does. In general, to insert $q(\geq 1)$ symbols into $r\in[1,k)$ blocks, each containing at least one symbol, and, while the order of these blocks is not important, that of those symbols within these blocks is, we first order those q symbols in $q!$ ways, then, for each such a permutation, insert $r-1$ "slashes" to separate them while making sure that no two slashes are adjacent to each other. It is clear that we can select $r-1$ positions for these slashes, out of a total of $q-1$ possible positions. Thus, for all the above five cases, there are $\mathbf{p}(b_E-g_E-l_E,g_E-1)$ ways to distribute the remaining internal symbols to the remaining non-trivial external cycles, where

$$\mathbf{p}(1,1)=1,$$
$$\forall q\geq 1, r\in[1,q), \mathbf{p}(q,r)=q!\binom{q-1}{r-1}. \tag{6}$$

IC: We will now wrap up the construction by selecting $b_I=b-b_E$ internal symbols, out of the remaining internal symbols, to construct the g_I internal cycles, each of which contains at least two internal symbols.

For Case 1+4, symbol 1 is a fixed point in $\mathcal{C}(u)$. It was thus not chosen for any of the external cycles, and it should not be chosen for any of the internal cycles, either. Thus, we have to choose those b_I symbols out of $k-1-(b_E-g_E)$, excluding 1, in $\binom{k-1-b_E+g_E}{b_I}$ ways.

For Case 2+5, symbol 1 has to appear in a non-trivial internal cycle, thus, not chosen in Step ES. What we will do now is to choose b_I-1 symbols out of $k-1-(b_E-g_E)$ symbols, excluding 1, then throw 1 into this chosen set of b_I symbols.

Finally, for Cases 3+8, 6 and 7, symbol 1 has already been reserved for a non-trivial external cycle, we simply choose these b_I internal symbols out of the remaining $k - (b_E - g_E)$ symbols in $\binom{k-b_E+g_E}{b_I}$ ways.

For any such a chosen set of b_I symbols, we are going to construct g_I non-trivial internal cycles, each of which contains at least 2 symbols. The general quantity of $\mathbf{d}(q,r)$, *the number of ways of factoring n distinct symbols into r non-trivial cycles,* is discussed in [8, §4.4]. Based on Eqs. 4.9 and 4.18 [8]: for $q \geq 2r \geq 1$,

$$\mathbf{d}(q,r) = \sum_{j=0}^{q} (-1)^{q+r-j} \binom{q}{j} s(q-j, r-j). \tag{7}$$

In the above, $s(\_,\_)$ stands for the signless Stirling numbers of the first kind, which can be represented as an explicit formula itself [6, Eqs. 5 and 6].

Since each internal cycle contains at least two symbols, by Eq. 5, $2g_I \leq b_I = b - b_E = d - g_I + g_E - b_E$, i.e., $3g_I \leq d + g_E - b_E$. We note that, for cycle structures $\mathcal{C}(u)$ falling into $E_{A_{n,k}}(e_k, \varphi(e_k), i)$ and $S_{A_{n,k}}(e_k, \varphi(e_k), i)$, by Corollary 1, $d(u, e_k) = i$; and for those falling into $L_{A_{n,k}}(e_k, \varphi(e_k), i)$, $d(u, e_k) = i + 1$.

Furthermore, for cycle structures falling into Cases 1+4, 3+8, 6, and 7, where the symbol 1 either occurs in a trivial internal cycle, or an external cycle, we just use the chosen b_I symbols to construct g_I internal cycles. Based on the above discussion, for those falling into Cases 1+4, and 3+8, there are $\mathbf{d}(i - g_I + g_E - b_E, g_I)$ ways to construct the internal cycles, where $3g_I \leq i + g_E - b_E$; and for those falling into Case 6, there are $\mathbf{d}(i + 1 - g_I + g_E - b_E, g_I)$ ways to construct such cycles, where $3g_I \leq i + 1 + g_E - b_E$.

For Case 2+5 where the symbol 1 is intended for a non-trivial internal cycle, we have to select l_I symbols out of $b_I - 1$ symbols, excluding 1, to form this internal cycle with 1. For each permutation of such a chosen set of l_I symbols, there is a unique internal cycle for 1 and these l_I symbols. We then use the remaining $b_I - l_I - 1$ symbols to form the other $g_I - 1$ internal cycles in $\mathbf{d}(b_I - l_I - 1, g_I - 1)$ ways. Clearly, $1 \leq l_I \leq b_I - 1$, and for this case, $b_I = b - b_E = i + 1 - g_I + g_E - b_E$.

Finally, for Cases 1+4, 3+8, 6 and 7, $g_I \geq 0$, and for Case 2+5, $g_I \geq 1$. We also notice that, for Cases 1+4, 6, and 7, $g_E + g_I \geq 1$, and for the other cases, $g_E + g_I \geq 2$. Since $\binom{n}{k} = 0$, whenever $k < n$, for Cases 1+4, 3+8, and 7, $k - b_E + g_E \geq i - g_I - b_E + g_E$, thus, $g_I \geq i + 1 - k$, and for the other two cases, $g_I \geq i - k$.

We use Tables 2, 3, and 4 to summarize our findings, where the IC term for Case 2+5 is given as follows:

$$IC(2+5) = \sum_{l_I=1}^{i-g_I+g_E-b_E} \binom{k-1-b_E+g_E}{i-g_I+g_E-b_E} \binom{i-g_I+g_E-b_E}{l_I} (l_I)!\mathbf{d}(i-g_I+g_E-b_E-l_I, g_I-1). \tag{8}$$

Table 2. Results by Cases (I)

Case	ES	Range of g_E	IE	Range of b_E
1+4	$\binom{n-k-1}{g_E-1}$	$[1, n-k]$	$\binom{k-1}{b_E-g_E}$	$[2g_E-1, g_E+k-1]$
2+5	$\binom{n-k-1}{g_E-1}$	$[1, n-k]$	$\binom{k-1}{b_E-g_E}$	$[2g_E-1, g_E+k-1]$
3+8	$\binom{n-k-1}{g_E-1}$	$[2, n-k]$	$\binom{k-1}{b_E-g_E-1}$	$[2g_E-1, g_E+k]$
6	$\binom{n-k-1}{g_E-1}$	$[1, n-k]$	$\binom{k-1}{b_E-g_E-1}$	$[2g_E, g_E+k]$
7	$\binom{n-k-1}{g_E-1}$	$[1, n-k]$	$\binom{k-1}{b_E-g_E-1}$	$[2g_E, g_E+k]$

Table 3. Results by Cases (II)

Case	EC	Range of l_E
1+4	$l_E!\binom{b_E-g_E}{l_E}\mathbf{p}(b_E-g_E-l_E, g_E-1)$	$[0, b_E-2g_E+1]$
2+5	$l_E!\binom{b_E-g_E}{l_E}\mathbf{p}(b_E-g_E-l_E, g_E-1)$	$[0, b_E-2g_E+1]$
3+8	$l_E!\binom{b_E-g_E-1}{l_E}\mathbf{p}(b_E-g_E-l_E, g_E-1)$	$[0, b_E-2g_E+1]$
6	$(l_E-1)!\binom{b_E-g_E-1}{l_E-1}\mathbf{p}(b_E-g_E-l_E, g_E-1)$	$[1, b_E-2g_E+1]$
7	$(l_E-1)(l_E-1)!\binom{b_E-g_E-1}{l_E-1}\mathbf{p}(b_E-g_E-l_E, g_E-1)$	$[2, b_E-2g_E+1]$

To summarize, let $R(Q(C))$ stand for the range of $Q \in \{g_E, b_E, l_E, g_I\}$ and $C \in \{1+4, 2+5, 3+8, 6, 7\}$, for $i \in [1, D(A_{n,k})-1]$,

$$
\begin{aligned}
&E_{A_{n,k}}(e_k, \varphi(e_k), i) \\
&= \sum_{R(g_E(3+8))} \sum_{R(b_E(3+8))} \sum_{R(l_E(3+8))} \sum_{R(g_I(3+8))} ES(3+8)IE(3+8)EC(3+8)IC(3+8) \\
&= \sum_{g_E=2}^{n-k} \sum_{b_E=2g_E-1}^{g_E+k} \sum_{l_E=0}^{b_E-2g_E+1} \sum_{\max\{0, i-k, 2-g_E\}}^{\left\lfloor \frac{i-b_E+g_E}{3} \right\rfloor} \binom{n-k-1}{g_E-1}\binom{k-1}{b_E-g_E-1}\binom{b_E-g_E-1}{l_E} \\
&\quad l_E!\mathbf{p}(b_E-g_E-l_E, g_E-1)\binom{k-b_E+g_E}{i-g_I+g_E-b_E}\mathbf{d}(i-g_I+g_E-b_E, g_I).
\end{aligned} \tag{9}
$$

For $i \in [0, D(A_{n,k})]$,

$$
\begin{aligned}
&S_{A_{n,k}}(e_k, \varphi(e_k), i) \\
&= \sum_{C\in\{2+5,6\}} \sum_{R(g_E(C))} \sum_{R(b_E(C))} \sum_{R(l_E(C))} \sum_{R(g_I(C))} ES(C)IE(C)EC(c)IC(C),
\end{aligned} \tag{10}
$$

and, for $i \in [0, D(A_{n,k})]$,

$$
\begin{aligned}
&L_{A_{n,k}}(e_k, \varphi(e_k), i) \\
&= \sum_{C\in\{1+4,7\}} \sum_{R(g_E(C))} \sum_{R(b_E(C))} \sum_{R(l_E(C))} \sum_{R(g_I(C))} ES(C)IE(C)EC(C)IC(C).
\end{aligned} \tag{11}
$$

Table 4. Results by Cases (III)

Case	IC	Range of g_I
1+4	$\binom{k-1-b_E+g_E}{i-g_I+g_E-b_E}\mathbf{d}(i-g_I+g_E-b_E, g_I)$	$\max\{0, , i-k, 1-g_E\}, \frac{i-b_E+g_E}{3}$
2+5	Eq. 8	$\max\{1, i+1-k, 2-g_E\}, \frac{i+1-b_E+g_E}{3}$
3+8	$\binom{k-b_E+g_E}{i-g_I+g_E-b_E}\mathbf{d}(i-g_I+g_E-b_E, g_I)$	$\max\{0, i-k, 2-g_E\}, \frac{i-b_E+g_E}{3}$
6	$\binom{k-b_E+g_E}{i+1-g_I+g_E-b_E}\mathbf{d}(i+1-g_I+g_E-b_E, g_I)$	$\max\{0, i+1-k, 1-g_E\}, \frac{i+1-b_E+g_E}{3}$
7	$\binom{k-b_E+g_E}{i-g_I+g_E-b_E}\mathbf{d}(i-g_I+g_E-b_E, g_I)$	$\max\{0, i-k, 1-g_E\}, \frac{i-b_E+g_E}{3}$

Readers might notice that $E_{A_{n,k}}(e_k, \varphi(e_k), i)$ is not given in the above for $i = 0$, or $i = D(A_{n,k})$. Indeed, we may conclude that $E_{A_{n,k}}(e_k, \varphi(e_k), i) = 0$, for these two values of i, for the following reasons: Let $u \in A_{n,k}$, by Eq. 5, $d(u, e_k) = b(u) + g_I(u) - g_E(u)$. Assume that $d(u, e_k) = 0$, then[2] $\mathcal{C}(u) = \emptyset$, in particular, it contains no external cycle. On the other hand, when $d(u, e_k) = D(A_{n,k})$, since $d(u, e_k)$ is maximized, $g_E(u) = 0$, i.e., $\mathcal{C}(u)$ necessarily contains no non-trivial external cycle. In either case, by Theorem 1, $d(\varphi(u), e_k) \neq d(u, e_k)$.

By Eq. 1, we have the following central result of this paper.

Theorem 2. *The edge-centered surface area of $A_{n,k}$ with radius $i \in [0, D(A_{n,k})]$, centered at $(e_k, \varphi(e_k))$, is given as follows:*

$$B_{A_{n,k}}(e_k, \varphi(e_k), 0) = 2,$$

and, for $i \in [1, D(A_{n,k})]$,

$$B_{A_{n,k}}(e_k, \varphi(e_k), i) = E_{A_{n,k}}(e_k, \varphi(e_k), i) + S_{A_{n,k}}(e_k, \varphi(e_k), i) + L_{A_{n,k}}(e_k, \varphi(e_k), i),$$

where $E_{A_{n,k}}(e_k, \varphi(e_k), i)$, $S_{A_{n,k}}(e_k, \varphi(e_k), i)$, and $L_{A_{n,k}}(e_k, \varphi(e_k), i)$, are given in Eqs. 9, 10, and 11, respectively.

One can then easily find out that the $(12, 32)$-*centered surface area sequence of* $A_{4,2}$ is $(2, 5, 5, 0)$, consistent with Table 1.

An upper bound of the total number of terms as contained in Eq. 10, an explicit-form expression of $S_{A_{n,k}}(e_k, \phi(e_k), i)$, can be estimated as follows, when factorial is considered as a basic operation:

$$TT_S(n, k) = \sum_{g_E=1}^{n-k} \sum_{b_E=2g_E-1}^{g_E+k} \sum_{l_E=0}^{b_E-2g_E+1} \sum_{l_I=1}^{b_I} \sum_{g_I=1}^{\frac{b-b_E}{2}} 1.$$

[2] Assume for some u, $b(u) + g_I(u) = g_E(u)$. Since every non-trivial internal (external) cycle contains at least two symbols, by definition, we have $g_E(u) - g_I(u) = b(u) = b_E(u) + b_I(u) \geq 2g_E(u) + 2g_I(u)$, i.e., $g_E(u) + 3g_I(u) \leq 0$. Thus, we have to conclude $g_E(u) = g_I(u) = 0$, as neither can be negative.

Since $g_E \geq 1$, $b_I = b - b_E$, $b \leq n$, $g_E \leq n - k$, we have $1 - 2g_E \leq 0$, $b_I \leq n - b_E$, and $g_E + k \leq n$, which leads to the following estimation:

$$TT_S(n,k) = O\left(n^4(n-k)\right).$$

Similarly, for Eqs. 9 and 11, we have the following results:

$$TT_E(n,k) = TT_L(n,k) = O\left(n^3(n-k)\right).$$

Therefore, these explicit form expressions all contain a polynomial number of terms, thus computationally feasible. It is certainly straightforward to come up with a computer program to calculate $B_{A_{n,k}}(e_k, \varphi(e_k), i)$, $i \in [0, D(A_{n,k})]$, based on the result as given in Theorem 2. For example, Table 5 shows $B_{A_{8,k}}(e_k, \varphi(e_k), i)$, $k \in [1,7]$, $i \in [0, D(A_{8,k})]$. These sample results agree with what have been directly derived via a BFS search.

Table 5. Sample data for $A_{8,k}(i)$

	i										
k	0	1	2	3	4	5	6	7	8	9	10
1	2	6	0	0	0	0	0	0	0	0	0
2	2	17	37	0	0	0	0	0	0	0	0
3	2	24	112	194	4	0	0	0	0	0	0
4	2	27	171	576	847	57	0	0	0	0	0
5	2	26	184	828	2,260	2,980	434	6	0	0	0
6	2	21	145	740	2,690	6,390	7,988	2,089	95	0	0
7	2	12	72	390	1,640	5,220	11,538	14,628	6,188	630	0

Since $A_{n,k}$ is edge symmetric, this edge-centered surface area result as given in Theorem 2 holds for any reference edge.

4 Concluding Remarks

In this paper, we proposed the notion of the edge-centered surface area of a structure, and derived an explicit expression of the edge-centered surface area of the general arrangement graph by following an elementary approach. Readers are referred to [3] for a solution of the same problem following a generating function approach.

The class of arrangement graphs are relatively general in the sense that several interesting interconnection structures are isomorphic to certain arrangement graphs. Thus, various topological properties of the arrangement graph, including the just discovered one related to the edge-centered surface area, also hold in such graphs.

As is pointed out in [4, §1] that S_n, the well-studied star graph of n dimensions [1], is isomorphic to $A_{n,n-1}$. Theorem 2 thus provides the edge-centered surface area result for S_n, when setting $k = n - 1$. For example, the last row in Table 5 gives the edge-centered surface area for S_8.

Incidentally, $E_{A_{n,n-1}}(e_k, \varphi(e_k), i) = 0$, as the range of g_E for Case 3+8 becomes $[2, 1]$. Indeed, since S_n is a bipartite graph, by Proposition 1, $E_{S_n}(v, w, i) = 0, i \in [1, D(S_n)]$.

The class of the alternating group graphs is proposed in [7]. When compared with the star graph of the same dimension, an alternating group graph has half the vertices, but nearly twice the degree. It is both vertex and edge symmetric, and performs better than the star graph and close to the hypercube, as far as the contention problem is concerned. It is also known that, for all $n \geq 3$, AG_n is isomorphic to $A_{n,n-2}$, where, we take $(v, \varphi(v)) = ((n-1)n12 \cdot (n-2), n1(n-1)2 \cdots (n-2))$ as the reference edge for AG_n. Hence, Theorem 2 also provides the edge-centered surface area result for $AG_n, n \geq 3$, when setting $k = n - 2$. As an example, the second last row, i.e., the one for $k = 6$, as shown in Table 5 provides the edge-centered surface area for AG_8.

References

1. Akers, S.B., Krishnamurthy, B.: A group theoretic model for symmetric interconnection networks. IEEE Trans. on Computers 38, 555–566 (1989)
2. Cheng, E., Qiu, K., Shen, Z.: On deriving explicit formulas of the surface areas for the arrangement graphs and some of the related graphs. International Journal of Computer Mathematics 87, 2903–2914 (2010)
3. Cheng, E., Qiu, K., Shen, Z.: A generating function approach to the edge surface area of the arrangement graphs. To appear in the Computer Journal
4. Day, K., Tripathi, A.: Arrangement graphs: a class of generalized star graphs. Information Processing Letters 42, 235–241 (1992)
5. Fertin, G., Raspaud, A.: k-Neighbourhood broadcasting. In: 8th International Colloquium on Structural Information and Communication Complexity, pp. 133–146. Carleton Scientific, Ontario (2001)
6. Imani, N., Sarbazi-Azad, H., Akl, S.G.: On some combinatorial properties of the star graph. In: 2005 International Symposium on Parallel Architecture, Algorithms and Networks, pp. 58–65. IEEE Computer Society, California (2005)
7. Jwo, J.S., Lakshmivarahan, S., Dhall, S.K.: A new class of interconnection networks based on the alternating graph. Networks 23, 315–326 (1993)
8. Riordan, J.: An Introduction to Combinatorial Analysis. Wiley, New York (1980)
9. Sarbazi-Azad, H., Ould-Khaoua, M., Mackenzie, L.M., Akl, S.G.: On some properties of k-ary n-cubes. In: Eighth International Conference on Parallel and Distributed Systems, pp. 517–524. IEEE Computer Society, California (2001)
10. Shen, Z., Qiu, K., Cheng, E.: On the surface area of the (n, k)-star graph. Theoretical Computer Science 410, 5481–5490 (2009)
11. Stewart, I.: A general technique to establish the asymptotic conditional diagnosability of interconnection networks (2011) (manuscript)

On Zero Forcing Number of Permutation Graphs

Eunjeong Yi

Texas A&M University at Galveston, Galveston, TX 77553, USA
yie@tamug.edu

Abstract. *Zero forcing number*, $Z(G)$, of a graph G is the minimum cardinality of a set S of black vertices (whereas vertices in $V(G)\backslash S$ are colored white) such that $V(G)$ is turned black after finitely many applications of "the color-change rule": a white vertex is converted black if it is the only white neighbor of a black vertex. Zero forcing number was introduced and used to bound the minimum rank of graphs by the "AIM Minimum Rank – Special Graphs Work Group". Let G_1 and G_2 be disjoint copies of a graph G and let $\sigma : V(G_1) \to V(G_2)$ be a permutation. Then a *permutation graph* $G_\sigma = (V, E)$ has the vertex set $V = V(G_1) \cup V(G_2)$ and the edge set $E = E(G_1) \cup E(G_2) \cup \{uv \mid v = \sigma(u)\}$. It is readily seen that $1+\delta(G) \leq Z(G_\sigma) \leq n$, if G is a graph of order $n \geq 2$; here $\delta(G)$ is the minimum degree of G. We give examples showing that $|Z(G) - Z(G_\sigma)|$ can be arbitrarily large. Further, we characterize permutation graphs G_σ satisfying $Z(G_\sigma) = n$ for a graph G that is a nearly complete graph, a complete k-partite graph, a cycle, and a path, respectively, on n vertices.

Keywords: zero forcing set, zero forcing number, permutation graph, generalized prism, nearly complete graph, complete k-partite graph, cycle, path.

1 Introduction

Let $G = (V(G), E(G))$ be a finite, simple, and undirected graph of order $|V(G)| = n \geq 2$. For a given graph G and $S \subseteq V(G)$, we denote by $\langle S \rangle$ the subgraph induced by S. For a vertex $v \in V(G)$, the *open neighborhood of* v is the set $N_G(v) = \{u \mid uv \in E(G)\}$, and the *closed neighborhood of* v is the set $N_G[v] = N_G(v) \cup \{v\}$. The *degree* $\deg_G(v)$ of a vertex $v \in V(G)$ is the number of edges incident with the vertex v in G. We denote by $\delta(G)$ the *minimum degree* of a graph G. We denote by K_n, C_n, and P_n the complete graph, the cycle, and the path on n vertices, respectively. For other terminologies in graph theory, refer to [7].

The notion of a zero forcing set, as well as the associated zero forcing number, of a simple graph was introduced by the "AIM Minimum Rank – Special Graphs Work Group" in [1] to bound the minimum rank of associated matrices for numerous families of graphs. Let each vertex of a graph G be given one of two colors, "black" and "white" by convention. Let S denote the (initial) set of black vertices of G. The *color-change rule* converts the color of a vertex from white to black if the white vertex u_2 is the only white neighbor of a black vertex

G. Lin (Ed.): COCOA 2012, LNCS 7402, pp. 61–72, 2012.

u_1; we say that u_1 forces u_2, which we denote by $u_1 \to u_2$. And a sequence, $u_1 \to u_2 \to \cdots \to u_i \to u_{i+1} \to \cdots \to u_t$, obtained through iterative applications of the color-change rule is called a *forcing chain.* Note that, at each step of the color change, there may be two or more vertices capable of forcing the same vertex. The set S is said to be *a zero forcing set* of G if all vertices of G will be turned black after finitely many applications of the color-change rule. The *zero forcing number* of G, denoted by $Z(G)$, is the minimum of $|S|$, as S varies over all zero forcing sets $S \subseteq V(G)$.

Since its introduction by the aforementioned "AIM group", zero forcing number has become a graph parameter studied for its own sake, as an interesting invariant of a graph. In [8], the authors studied the number of steps it takes for a zero forcing set to turn the entire graph black; they named this new graph parameter the *iteration index* of a graph: from the "real world" modeling (or discrete dynamical system) perspective, if the initial black set is capable of passing a certain condition or trait to the entire population (i.e. "zero forcing"), then the iteration index of a graph may represent the number of units of time (anything from days to millennia) necessary for the entire population to acquire the condition or trait. Independently, Hogben et al. studied the same parameter (iteration index) in [15], which they called *propagation time.* It's also noteworthy that physicists have independently studied the zero forcing parameter, referring to it as the *graph infection number*, in conjunction with the control of quantum systems (see [4], [5], and [19]). More recently in [10] and [11], the authors initiated a comparative study between metric dimension and zero forcing number for graphs. For more articles and surveys pertaining to the zero forcing parameter, see [2], [3], [8], [9], [12], [13], [14], [17], [18].

Chartrand and Harary [6] introduced a "permutation graph", which is also called a "generalized prism".

Definition 1. *Let G_1 and G_2 be disjoint copies of a graph G, and let $\sigma : V(G_1) \to V(G_2)$ be a permutation. A* permutation graph *$G_\sigma = (V, E)$ consists of the vertex set $V = V(G_1) \cup V(G_2)$ and the edge set $E(G) = E(G_1) \cup E(G_2) \cup \{uv \mid v = \sigma(u)\}$.*

It is readily seen that $1 + \delta(G) \leq Z(G_\sigma) \leq n$, if G is a graph of order $n \geq 2$. In this paper, we investigate the zero forcing number of permutation graphs. First, we give examples showing that $|Z(G) - Z(G_\sigma)|$ can be arbitrarily large. Second, we characterize permutation graphs G_σ satisfying $Z(G_\sigma) = n$ for a graph G that is a nearly complete graph, a complete k-partite graph, a cycle, and a path, respectively, on n vertices. Further, we prove that the zero forcing number of any complete k-partite graph of order $n \geq 3$, which is not the complete graph, is $n - 2$.

2 $Z(G)$ versus $Z(G_\sigma)$

The *path cover number* $P(G)$ of G is the minimum number of vertex disjoint paths, occurring as induced subgraphs of G, that cover all the vertices of G.

Next, we recall the definition that is stated in [16]. A graph G is a graph of *two parallel paths* if there exist two independent induced paths of G that cover all the vertices of G and such that G can be drawn in the plane in such a way that the two paths are parallel and the edges (drawn as segments, not curves) between the two paths do not cross. A simple path is not considered to be such a graph. A graph that consists of two connected components, each of which is a path, is considered to be such a graph.

Theorem 1. *[1], [2], [18]*

(a) [2] For any graph G, $P(G) \leq Z(G)$.
(b) [1] For any tree T, $P(T) = Z(T)$.
(c) [18] For any unicyclic graph G, $P(G) = Z(G)$.

Theorem 2. *[3] For any graph G, $Z(G) \geq \delta(G)$.*

Theorem 3. *[1] For any graphs G and H, $Z(G\Box H) \leq \min\{Z(G)|V(H)|, Z(H)|V(G)|\}$, where $G\Box H$ denotes the Cartesian product of G and H.*

Theorem 4. *Let G be a connected graph of order $n \geq 2$. Then*

(a) [10], [18] $Z(G) = 1$ if and only if $G = P_n$,
(b) [18] $Z(G) = 2$ if and only if G is a graph of two parallel paths,
(c) [10], [18] $Z(G) = n - 1$ if and only if $G = K_n$.

Theorem 5. *[9] Let G be any graph. Then*

(a) For $v \in V(G)$, $Z(G) - 1 \leq Z(G - \{v\}) \leq Z(G) + 1$.
(b) For $e \in E(G)$, $Z(G) - 1 \leq Z(G - e) \leq Z(G) + 1$.

Theorem 6. *[18] Let G be a graph with cut-vertex $v \in V(G)$. Let $W_1, W_2, \ldots, W_k$ be the vertex sets for the connected components of $\langle V(G)\backslash\{v\}\rangle$, and for $1 \leq i \leq k$, let $G_i = \langle W_i \cup \{v\}\rangle$. Then $Z(G) \geq [\sum_{i=1}^{k} Z(G_i)] - k + 1$.*

If G is a graph of order 2, then $Z(G_\sigma) = 2$ for any permutation σ. So, we only consider a graph G of order $n \geq 3$ for the rest of the paper. Noting that $Z(G_1)$ forms a zero forcing set for G_σ, together with Theorem 2, we have the following

Corollary 1. *Let G be a graph of order $n \geq 3$, and let $\sigma : V(G_1) \to V(G_2)$ be a permutation. Then $1 + \delta(G) \leq Z(G_\sigma) \leq n$.*

A graph G is *strongly regular* with parameters (n, k, α, β) if $|V(G)| = n$, G is k-regular (i.e., the degree of each vertex in G is k), every pair of adjacent vertices has α common neighbors, and every pair of non-adjacent vertices has β common neighbors.

Proposition 1. *[1] If G is a strongly regular graph, then $Z(G) \geq \left\lfloor \frac{|V(G)|}{2} \right\rfloor$.*

Remark 1. The Petersen graph $\mathcal{P}$ (see Fig. 1) is strongly regular; thus, $Z(\mathcal{P}) = 5$ by Corollary 1 and Proposition 1.

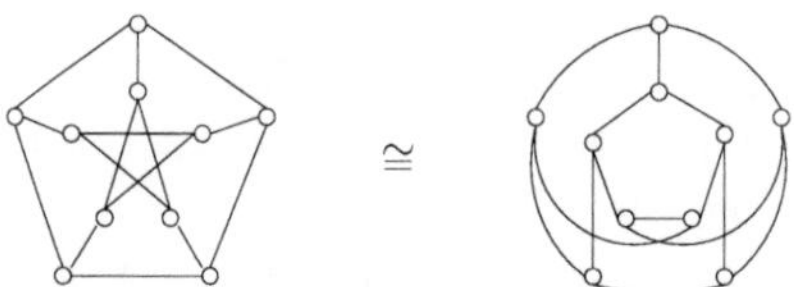

Fig. 1. The Petersen graph $\mathcal{P}$ with $Z(\mathcal{P}) = 5$

Remark 2. There exists a permutation graph G_σ such that $Z(G_\sigma) - Z(G)$ can be arbitrarily large; take $G = mP_2$ (m copies of P_2) and $\sigma = id$, the identity, for $m \geq 1$ (see (A) of Fig. 2), and notice that $Z(G) = m$ and $Z(G_\sigma) = 2m$. For another permutation graph, with a connected graph G, satisfying that $Z(G_\sigma) - Z(G)$ can be arbitrarily large, see (B) of Fig. 2; notice that $Z(G) = m + 1$ by (b) of Theorem 1, and $Z(G_\sigma) \geq 2m$ since at least a vertex in each $B_i \setminus \{u_i, v_i\}$ must belong to any zero forcing set of G_σ (otherwise, each of u_i and v_i has two white neighbors in B_i).

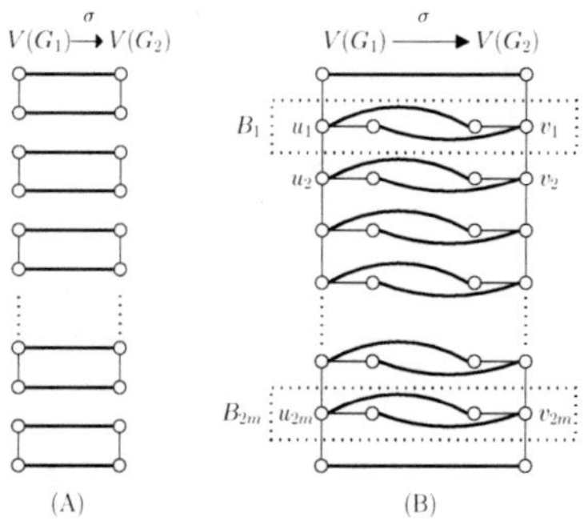

Fig. 2. Examples showing that $Z(G_\sigma) - Z(G)$ can be arbitrarily large

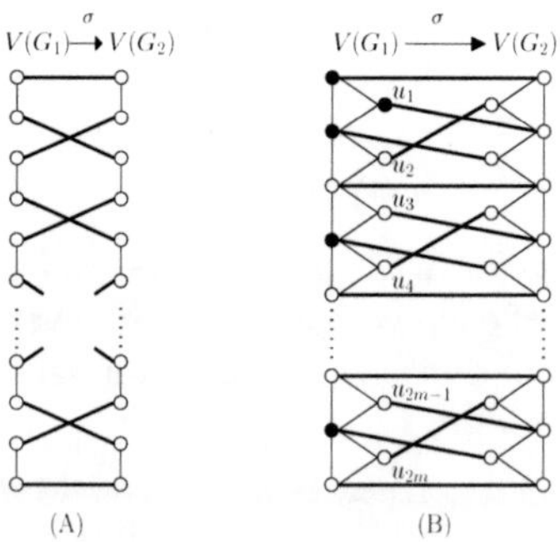

Fig. 3. Examples showing that $Z(G) - Z(G_\sigma)$ can be arbitrarily large

Remark 3. There exists a permutation graph G_σ such that $Z(G) - Z(G_\sigma)$ can be arbitrarily large; take $G = mP_2$ and $G_\sigma \cong C_{2m}$ for $m > 2$ (see (A) of Fig. 3), and notice that $Z(G) = m$ and $Z(G_\sigma) = 2$. For another permutation graph, with a connected graph G, satisfying that $Z(G) - Z(G_\sigma)$ can be arbitrarily large, see (B) of Fig. 3; notice that $Z(G) \geq 2m + 1$ by successively applying Theorem 6, and $Z(G_\sigma) \leq m + 2$ since the solid vertices form a zero forcing set for G_σ.

3 Zero Forcing Number of Permutation Graphs on Nearly Complete Graphs and on Complete *k*-Partite Graphs

We define a "nearly complete graph" to be a graph G of order $n \geq 3$ with $\delta(G) = n - 2$.

Theorem 7. *[17] Let $G = K_n$ be the complete graph of order $n \geq 3$, and let $\sigma : V(G_1) \to V(G_2)$ be a permutation. Then $Z(G_\sigma) = n$.*

Theorem 8. *Let G be a nearly complete graph of order $n \geq 4$, and let $\sigma : V(G_1) \to V(G_2)$ be a permutation. Then $Z(G_\sigma) = n$.*

Proof. Let S be a zero forcing set for G_σ; notice that $|S| \geq n-1$ by Corollary 1. Without loss of generality, assume that $u_1 \in S \cap V(G_1)$ forces first. If $\deg_{G_1}(u_1) = n-1$, then $|S| \geq n$. Next, we consider the case $\deg_{G_1}(u_1) = n-2$. Suppose that $u_1u_2 \notin E(G_1)$ for some $u_2 \in V(G_1)$; notice that $\deg_{G_1}(u_1) = \deg_{G_1}(u_2) = n-2$. If $S = N_{G_1}[u_1]$ with $|S| = n-1$, then $u_1 \to \sigma(u_1)$. Since each vertex $u_i \in S \setminus \{u_1\}$ $(3 \leq i \leq n)$ has two white neighbors u_2 and $\sigma(u_i)$, no vertex in $S \setminus \{u_1\}$ can force. And, noting that $n \geq 4$, $\sigma(u_1)$ has at least two white neighbors in G_2, and thus $\sigma(u_1)$ cannot force any vertex in G_σ. So, $|S| \geq n$. Therefore, in each case, $Z(G_\sigma) = n$ by Corollary 1. □

Next, we consider the zero forcing number of a complete k-partite graph $G = K_{a_1,a_2,\ldots,a_k}$ and its permutation graph G_σ, where $k \geq 2$.

Proposition 2. *For $k \geq 2$, let $G = K_{a_1,a_2,\ldots,a_k}$ be a complete k-partite of order $\sum_{i=1}^k a_i = n \geq 3$. If $G \ncong K_n$, then $Z(G) = n - 2$.*

Proof. For $k \geq 2$, let $G = K_{a_1,a_2,\ldots,a_k}$ be a complete k-partite of order $\sum_{i=1}^k a_i = n \geq 3$, and let $G \ncong K_n$. Let $V(G)$ be partitioned into k-partite sets $V_1, V_2, \ldots, V_k$, where $|V_i| = a_i$ $(1 \leq i \leq k)$. Since $G \ncong K_n$, $Z(G) \leq n - 2$ by (c) of Theorem 4. Next, we show that $Z(G) \geq n-2$. Let S be a zero forcing set for G. Suppose that, for some $a_i \geq 2$, $v_i \in S \cap V_i$ forces first (If $a_i = 1$, then $\deg_G(v_i) = n-1$, and thus $|S| \geq n-1$). First, notice that $|S \cap (V(G) \setminus V_i)| \geq n - a_i - 1$; otherwise, v_i has at least two white neighbors in $V(G) \setminus V_i$. Second, notice that $|S \cap V_i| \geq a_i - 1$; otherwise, each vertex in $V(G) \setminus V_i$ has two white neighbors in V_i, and no vertex in $V_i \setminus \{v_i\}$ can be forced by any vertex in V_i. So, $Z(G) \geq n-2$. Thus, $Z(G) = n-2$. □

Theorem 9. *For $k \geq 2$, let $G = K_{a_1,a_2,...,a_k}$ be a complete k-partite graph of order $n = \sum_{i=1}^{k} a_i \geq 3$, where $a_k \geq a_j$ $(1 \leq j \leq k-1)$; let $V(G_1)$ be partitioned into k-partite sets $V_1, V_2, ..., V_k$, and let $V(G_2)$ be partitioned into k-partite sets $V'_1, V'_2, ..., V'_k$, where $a_i = |V_i| = |V'_i|$ $(1 \leq i \leq k)$. Let $\sigma : V(G_1) \to V(G_2)$ be a permutation. Then $Z(G_\sigma) < n$ if and only if either V_k satisfies $|\sigma(V_k) \cap V'_k| \geq 2$ and $|\sigma(V_k) \cap (\cup_{i=1}^{k-1} V'_i)| = (\sum_{i=1}^{k-1} a_i) - 1$, or V'_k satisfies $|\sigma^{-1}(V'_k) \cap V_k| \geq 2$ and $|\sigma^{-1}(V'_k) \cap (\cup_{i=1}^{k-1} V_i)| = (\sum_{i=1}^{k-1} a_i) - 1$.*

Proof. For $k \geq 2$, let $G = K_{a_1,a_2,...,a_k}$ be a complete k-partite graph of order $n = \sum_{i=1}^{k} a_i \geq 3$, where $a_k \geq a_{k-1} \geq \ldots \geq a_1 \geq 1$. Let $V(G_1)$ be partitioned into k-partite sets $V_1, V_2, ..., V_k$, and let $V(G_2)$ be partitioned into k-partite sets $V'_1, V'_2, ..., V'_k$, where $a_i = |V_i| = |V'_i|$ $(1 \leq i \leq k)$. For each i $(1 \leq i \leq k)$, let $V_i = \{u_{i,1}, u_{i,2}, \ldots, u_{i,a_i}\}$, and let $V'_i = \{u'_{i,1}, u'_{i,2}, \ldots, u'_{i,a_i}\}$. Let S be a zero forcing set for G_σ. We consider two cases.

Case 1: Either V_k satisfies $|\sigma(V_k) \cap V'_k| \geq 2$ and $|\sigma(V_k) \cap (\cup_{i=1}^{k-1} V'_i)| = (\sum_{i=1}^{k-1} a_i) - 1$, or V'_k satisfies $|\sigma^{-1}(V'_k) \cap V_k| \geq 2$ and $|\sigma^{-1}(V'_k) \cap (\cup_{i=1}^{k-1} V_i)| = (\sum_{i=1}^{k-1} a_i) - 1$. Notice that $a_k \geq (\sum_{i=1}^{k-1} a_i) + 1$ in this case. Without loss of generality, we may assume that $|\sigma(V_k) \cap V'_k| \geq 2$ and $|\sigma(V_k) \cap (\cup_{i=1}^{k-1} V'_i)| = (\sum_{i=1}^{k-1} a_i) - 1$. Further, we may assume that $(\cup_{i=1}^{k-1} V'_i) \setminus \sigma(V_k) = \{u'_{1,1}\}$ and $\sigma(V_k) \cap V'_k = \{u'_{k,1}, u'_{k,2}\}$, by relabeling if necessary. Then $S = V(G_1) \setminus \{\sigma^{-1}(u'_{k,2})\}$ forms a zero forcing set (It suffices to show that all vertices of G_1 are turned black after finitely many applications of the color-change rule on S.): (i) after one global application of the color change-rule, vertices in $[(\cup_{i=1}^{k-1} V'_i) \setminus \{u'_{1,1}\}] \cup \{u'_{k,1}\} \subseteq V(G_2)$ are turned black; (ii) $u'_{k,1} \to u'_{1,1}$; (iii) $\sigma^{-1}(u'_{1,1}) \to \sigma^{-1}(u'_{k,2})$. Thus, $Z(G_\sigma) < n$. (See Fig. 4, where the solid vertices form a zero forcing set for G_σ.)

Case 2: V_k satisfies $|\sigma(V_k) \cap V'_k| \leq 1$ or $|\sigma(V_k) \cap (\cup_{i=1}^{k-1} V'_i)| \neq (\sum_{i=1}^{k-1} a_i) - 1$, and V'_k satisfies $|\sigma^{-1}(V'_k) \cap V_k| \leq 1$ or $|\sigma^{-1}(V'_k) \cap (\cup_{i=1}^{k-1} V_i)| \neq (\sum_{i=1}^{k-1} a_i) - 1$.

Subcase 2.1: $S \cap V(G_1) = \emptyset$ or $S \cap V(G_2) = \emptyset$. Without loss of generality, assume that $S \cap V(G_2) = \emptyset$. Suppose $S = V(G_1) \setminus \{u_{i,x}\}$ for some i $(1 \leq i \leq k)$; then $u_{i,s} \to \sigma(u_{i,s})$, where $1 \leq s \leq a_i$ and $s \neq x$. In order for a vertex in G_2 to force the vertex $\sigma(u_{i,x})$, there must exist a vertex, say $\sigma(u_{i,s^*}) \in V'_t$ for $s^* \neq x$ and for some t $(1 \leq t \leq k)$, in G_2 such that all vertices in $(V(G_2) \setminus V'_t) \setminus \{\sigma(u_{i,x})\}$ are turned black after one global application of the color-change rule, where $\sigma(u_{i,x}) \notin V'_t$; otherwise, each vertex $u_{\alpha,j} \in V(G_1) \setminus V_i$ has two white neighbors, $u_{i,x}$ and $\sigma(u_{\alpha,j})$, and each vertex $\sigma(u_{i,s})$ has at least two white neighbors in G_2. That is, $\sigma(V_i \setminus \{u_{i,x}\}) \supseteq [(V(G_2) \setminus V'_t) \setminus \{\sigma(u_{i,x})\}] \cup \sigma(u_{i,s^*})$, and thus $a_i - 1 \geq n - a_t \iff a_i + a_t \geq n + 1$, which is impossible. Thus, S of cardinality $n-1$ fails to be a zero forcing set for G_σ, and hence $Z(G_\sigma) \geq n$. By Corollary 1, $Z(G_\sigma) = n$.

Subcase 2.2: $S \cap V(G_1) \neq \emptyset$ and $S \cap V(G_2) \neq \emptyset$. Suppose that a vertex in $V_i \subseteq V(G_1)$ forces first. Then $|S \cap V(G_1)| \geq n - a_i + \ell$, where $1 \leq \ell = |S \cap V_i| < a_i$.

In order for a vertex in G_2 to be able to force, $|S \cap V(G_2)| \geq n - a_j - \ell + 1$. So, $|S| \geq n + 1 + (n - a_i - a_j)$. If $i \neq j$, then $|S| \geq n + 1$. If $i = j$ and $i \neq k$, then $|S| \geq n+1$. So, we only need to consider when $i = j = k$; then $|S| \geq 2(n-a_k)+1$. If $a_k \leq \frac{1}{2}(n + 1)$, then $|S| \geq n$; thus, $Z(G_\sigma) = n$ by Corollary 1. So, suppose that $a_k > \frac{1}{2}(n+1)$; then $a_k \geq (\sum_{i=1}^{k-1} a_i) + 2$. First, notice that $|\sigma(V_k) \cap V_k'| \leq 1$ is impossible, since $a_k \geq (\sum_{i=1}^{k-1} a_k) + 2$. Next, we consider $|\sigma(V_k) \cap (\cup_{i=1}^{k-1} V_i')| \neq (\sum_{i=1}^{k-1} a_i)-1$. Let $|S \cap V_k| = \ell = \ell_1 + \ell_2$ such that $|\sigma(S \cap V_k) \cap (\cup_{i=1}^{k-1} V_i')| = \ell_1$ and $|\sigma(S \cap V_k) \cap V_k'| = \ell_2$. Assume that $|S| < n$. In order for vertices in V_k' to be able to force vertices in G_1 so that $V(G_1)$ turns black after finitely many applications of the color-change rule, $|S \cap (\cup_{i=1}^{k-1} V_i')| = (\sum_{i=1}^{k-1} a_i) - \ell_1$ and $|S \cap V_k'| \geq a_k - \ell - 1$. So, $|S| \geq (\sum_{i=1}^{k-1} a_i) + \ell + (\sum_{i=1}^{k-1} a_i) - \ell_1 + a_k - \ell - 1 = n + (\sum_{i=1}^{k-1} a_i) - \ell_1 - 1$. If $\sum_{i=1}^{k-1} a_i \geq \ell_1 + 1$, then $|S| \geq n$, contradicting the assumption. If $\sum_{i=1}^{k-1} a_i = \ell_1$, then $|S \cap V_k'| \geq a_k - \ell$; otherwise, after applying the color-change rule on S as long as possible, either (i) two or more vertices in V_k are still white or (ii) one vertex in V_k is still white and each vertex $u_{i,r} \in \cup_{i=1}^{k-1} V_i$ has two white neighbors, $\sigma(u_{i,r})$ and the one white vertex in V_k. Thus, $|S| \geq (\sum_{i=1}^{k-1} a_i) + \ell + a_k - \ell = n$. So, $Z(G_\sigma) \geq n$ in each case, and thus $Z(G_\sigma) = n$ by Corollary 1. □

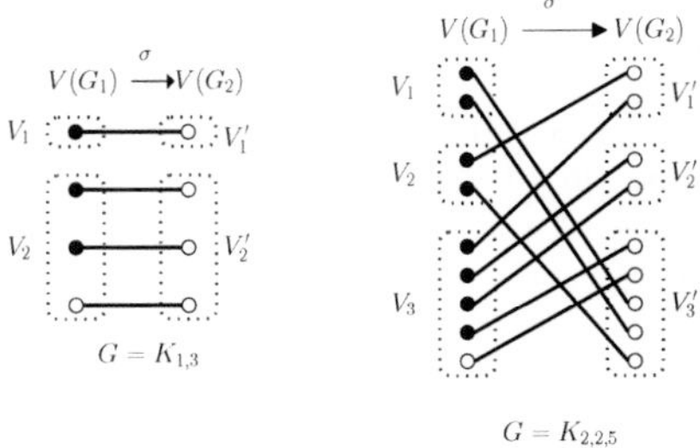

Fig. 4. Permutation graphs on complete k-partite graphs with $Z(G_\sigma) < |V(G)|$

4 Zero Forcing Number of Permutation Graphs on Cycles

In this section, we characterize permutation graphs achieving the lower or upper bounds of Corollary 1 when G is a cycle C_n on $n \geq 3$ vertices. Let $V(G_1) = \{u_i \mid 1 \leq i \leq n\}$ and let $E(G_1) = \{u_i u_{i+1} \mid 1 \leq i \leq n-1\} \cup \{u_1 u_n\}$; similarly, let $V(G_2) = \{v_i \mid 1 \leq i \leq n\}$ and let $E(G_2) = \{v_i v_{i+1} \mid 1 \leq i \leq n-1\} \cup \{v_1 v_n\}$. By $v_i < v_j$, we mean that $i < j$.

Proposition 3. *[1] For $s \geq 3$ and $t \geq 2$, $Z(C_s \square P_t) = \min\{s, 2t\}$.*

Theorem 10. *Let $G = C_n$ be the cycle of order $n \geq 3$, and let $\sigma : V(G_1) \rightarrow V(G_2)$ be a permutation. Then*

(a) $Z(G_\sigma) = 3$ if and only if $n = 3$ (for any σ);

(b) $Z(G_\sigma) = n$ *if and only if* $n = 3$ *or* $n = 4$ *(for any* σ*) or* $G_\sigma \cong \mathcal{P}$*, the Petersen graph.*

Proof. Let $G = C_n$ be the cycle of order $n \geq 3$; then $Z(G_\sigma) \geq 3$ by Corollary 1. If $n = 3$, then $Z(G_\sigma) \leq 3$ by Corollary 1; thus $Z(G_\sigma) = 3$ for any permutation σ. If $n = 4$, then $G = C_4$ is a nearly complete graph since $\delta(C_4) = 2$; thus, by Theorem 8, $Z(G_\sigma) = 4$ for any permutation σ.

Next, we consider $n \geq 5$. One can easily verify that $Z(G_\sigma) > 3$ for any permutation σ. Let $\ell(G_\sigma)$ denote the maximum over all a such that there exists an i with $\langle\{\sigma(u_i), \sigma(u_{i+1}), \ldots \sigma(u_{i+a})\}\rangle \cong P_{a+1}$, where $\sigma(u_i), \sigma(u_{i+1}), \ldots, \sigma(u_{i+a})$ form a monotone sequence. And, let $H = \langle V(G_2) \setminus \{\sigma(u_1), \sigma(u_2), \sigma(u_3)\}\rangle$. We consider three cases.

Case 1: $\ell(G_\sigma) \geq 2$. Without loss of generality, we may assume that $\sigma(u_1) = v_1$, $\sigma(u_2) = v_2$, and $\sigma(u_3) = v_3$. Then $S = V(G_1) \setminus \{u_2\}$ forms a zero forcing set for G_σ: (i) after one global application of the color-change rule, each vertex in $V(G_\sigma) \setminus \{u_2, v_1, v_2, v_3\}$ are turned black; (ii) $v_4 \to v_3$ and $v_n \to v_1$; (iii) $u_1 \to u_2$ and $v_1 \to v_2$. So, $Z(G_\sigma) < n$.

Case 2: $\ell(G_\sigma) = 1$. Without loss of generality, we may assume that $\sigma(u_1) = v_1$ and $\sigma(u_2) = v_2$. We will show that $S = V(G_1) \setminus \{u_2\}$ forms a zero forcing set for G_σ. We consider two subcases.

Subcase 2.1: H contains P_2. Suppose that $\{v_3, v_4, \ldots, v_t\} \subseteq V(H)$. Notice (i) after one global application of the color-change rule, each vertex in $V(G_\sigma) \setminus \{u_2, v_1, v_2, \sigma(u_3)\}$ are turned black; (ii) $v_t \to \sigma(u_3)$ and $v_n \to v_1 \to v_2 \to u_2$. A similar argument works for other cases. So, S is a zero forcing set for G_σ, and thus $Z(G_\sigma) < n$.

Subcase 2.2: H consists of isolated vertices. This implies that $n = 5$ and that $\sigma(u_3) = v_4$. If $\sigma(u_5) = v_5$, then G_σ satisfy the condition of Case 1. So, $\sigma(u_5) = v_3$ and $\sigma(u_4) = v_5$ (see (A) of Fig. 5); one can easily check that $V(G_1) \setminus \{u_1\}$ forms a zero forcing set for G_σ, and thus $Z(G_\sigma) < 5$.

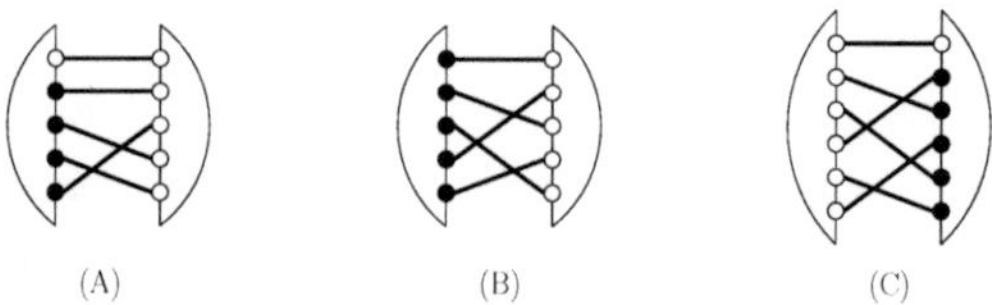

Fig. 5. G_σ for $G = C_n$ such that $\ell(G_\sigma) \leq 1$ and H consists of isolated vertices

Case 3: $\ell(G_\sigma) = 0$. Without loss of generality, let $\sigma(u_1) = v_1$, and we may assume that $v_1 < \sigma(u_2) < \sigma(u_3)$ by relabeling if necessary. We consider two subcases.

Subcase 3.1: H contains P_2. Suppose that $\{v_2, v_3, \ldots, v_t\} \subseteq V(H)$. Then $S = V(G_1) \setminus \{u_2\}$ forms a zero forcing set for G_σ: (i) after one global application of the color-change rule, each vertex in $V(G_\sigma) \setminus \{u_2, v_1, \sigma(u_2), \sigma(u_3)\}$ are turned black; (ii) $v_2 \to v_1$ and $v_t \to \sigma(u_2)$; (iii) $u_1 \to u_2$; (iv) $u_3 \to \sigma(u_3)$. A similar argument works for other cases. So, S is a zero forcing set for G_σ, and thus $Z(G_\sigma) < n$.

Subcase 3.2: H consists of isolated vertices. This implies that $n \leq 6$, and that $\sigma(u_2) = v_3$ and $\sigma(u_3) = v_5$. First, suppose $n = 5$. If $\sigma(u_4) = v_4$, then $\sigma(u_3)\sigma(u_4) \in E(G_2)$, contradicting the hypothesis that $\ell(G_\sigma) = 0$. So, $\sigma(u_4) = v_2$ and $\sigma(u_5) = v_4$ (see (B) of Fig. 5); since $G_\sigma \cong \mathcal{P}$, the Petersen graph, $Z(G_\sigma) = 5$ by Remark 1. Second, suppose $n = 6$. Then $\sigma(u_4) = v_2$, $\sigma(u_5) = v_6$, and $\sigma(u_6) = v_4$ (see (C) of Fig. 5); since $S = V(G_2) \setminus \{v_1\}$ forms a zero forcing set for G_σ, $Z(G_\sigma) < 6$. □

5 Zero Forcing Number of Permutation Graphs on Paths

In this section, we characterize permutations graphs achieving the lower or upper bounds of Corollary 1 when G is a path P_n on $n \geq 3$ vertices. Let $V(G_1) = \{u_i \mid 1 \leq i \leq n\}$ and let $E(G_1) = \{u_i u_{i+1} \mid 1 \leq i \leq n-1\}$; similarly, let $V(G_2) = \{v_i \mid 1 \leq i \leq n\}$ and let $E(G_2) = \{v_i v_{i+1} \mid 1 \leq i \leq n-1\}$. By $v_i < v_j$, we mean that $i < j$.

Proposition 4. *[1] For* $s, t \geq 2$, $Z(P_s \Box P_t) = \min\{s, t\}$.

Theorem 11. *Let* $G = P_n$ *be the path of order* $n \geq 3$, *and let* $\sigma : V(G_1) \to V(G_2)$ *be a permutation. Then*

(a) $Z(G_\sigma) = 2$ *if and only if* $G_\sigma \cong P_n \Box P_2$;
(b) $Z(G_\sigma) = n$ *if and only if (i)* $n = 3$ *and* $G_\sigma \ncong P_3 \Box P_2$, *or (ii)* $n = 4$ *and* G_σ *is isomorphic to (B) of Fig. 7.*

Proof. Let $G = P_n$ be the path of order $n \geq 3$. Part (a) of the present theorem follows by (b) of Theorem 4. So, it remains to prove part (b). Notice that $Z(G_\sigma) \geq 2$ by Corollary 1. First, we consider $n = 3$; notice that $Z(G_\sigma) \leq 3$ by Corollary 1. There are two non-isomorphic permutation graphs (see Fig. 6). If G_σ is isomorphic to (A) of Fig. 6, then $Z(G_\sigma) = 2$ by Proposition 4. If G_σ is isomorphic to (B) of Fig. 6, then $Z(G_\sigma) = 3$ by (b) of Theorem 4.

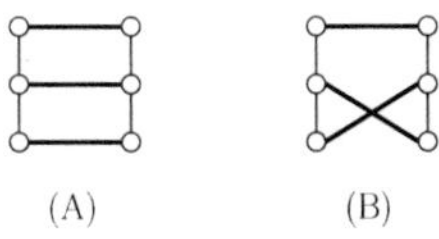

Fig. 6. Two non-isomorphic G_σ for $G = P_3$

Next, we consider $n \geq 4$. We consider two cases.

Case 1: $\{\sigma(u_1), \sigma(u_n)\} \cap \{v_1, v_n\} \neq \emptyset$. We may assume that $\sigma(u_1) = v_1$, by relabeling if necessary. We will show that $S = V(G_1) \setminus \{u_1\}$ forms a zero forcing set for G_σ, and thus $Z(G_\sigma) < n$. If $d_{G_2}(v_1, \sigma(u_2)) \geq 3$, then (i) after one global application of the color-change rule on S, $V(G_\sigma) \setminus \{u_1, v_1, \sigma(u_2)\}$ are turned black; (ii) $v_2 \to v_1 \to u_1$; (iii) $u_2 \to \sigma(u_2)$. If $d_{G_2}(v_1, \sigma(u_2)) = 2$ (notice that $\sigma(u_2) = v_3$), then (i) after one global application of the color-change rule on S, $V(G_\sigma) \setminus \{u_1, v_1, v_3\}$ are turned black; (ii) $v_4 \to v_3$; (iii) $u_2 \to u_1 \to v_1$. If $d_{G_2}(v_1, \sigma(u_2)) = 1$ (notice that $\sigma(u_2) = v_2$), then (i) after one global application of the color-change rule on S, $V(G_\sigma) \setminus \{u_1, v_1, v_2\}$ are turned black; (ii) $v_3 \to v_2 \to v_1 \to u_1$. In each case, S is a zero forcing set for G_σ with $Z(G_\sigma) < n$.

Case 2: $\{\sigma(u_1), \sigma(u_n)\} \cap \{v_1, v_n\} = \emptyset$. We may assume that $\sigma(u_1) < \sigma(u_2)$, by relabeling if necessary. If $d_{G_2}(\sigma(u_1), \sigma(u_2)) \geq 3$, then (i) after one global application of the color-change rule on $S = V(G_1) \setminus \{u_1\}$, $V(G_\sigma) \setminus \{u_1, \sigma(u_1), \sigma(u_2)\}$ are turned black; (ii) if $\langle\{\sigma(u_1), v_s, v_{s+1}, \ldots, v_t, \sigma(u_2)\}\rangle$ is an induced path in G_2, then $v_s \to \sigma(u_1) \to u_1$ and $v_t \to \sigma(u_2)$. If $d_{G_2}(\sigma(u_1), \sigma(u_2)) = 2$, then (i) after one global application of the color-change rule on $S = V(G_1) \setminus \{u_1\}$, $V(G_\sigma) \setminus \{u_1, \sigma(u_1), \sigma(u_2)\}$ are turned black; (ii) if $\langle\{v_s, \sigma(u_1), v_t, \sigma(u_2)\}\rangle \cong P_4$ in G_2, then $v_s \to \sigma(u_1) \to u_1$; (iii) $v_t \to \sigma(u_2)$. So, suppose $d_{G_2}(\sigma(u_1), \sigma(u_2)) = 1$, and we consider two subcases.

Subcase 2.1: $d_{G_2}(\sigma(u_1), \sigma(u_2)) = 1$ and $\sigma(u_2) \neq v_n$. Then (i) after one global application of the color-change rule on $S = V(G_1) \setminus \{u_1\}$, $V(G_\sigma) \setminus \{u_1, \sigma(u_1), \sigma(u_2)\}$ are turned black; (ii) if $\langle\{\sigma(u_1), \sigma(u_2), v_t\}\rangle \cong P_3$ in G_2 with $\sigma(u_1) < \sigma(u_2) < v_t$, then $v_t \to \sigma(u_2) \to \sigma(u_1) \to u_1$.

Subcase 2.2: $d_{G_2}(\sigma(u_1), \sigma(u_2)) = 1$ and $\sigma(u_2) = v_n$. This implies that $\sigma(u_1) = v_{n-1}$. First, we consider $n \geq 5$. We will show that $S = V(G_1) \setminus \{u_2\}$ forms a zero forcing set for G_σ, and thus $Z(G_\sigma) < n$. If $d_{G_2}(v_{n-1}, \sigma(u_3)) \geq 3$, then (i) after one global application of the color-change rule on S, $V(G_\sigma) \setminus \{u_2, \sigma(u_3), v_{n-1}, v_n\}$ are turned black; (ii) if $\langle\{\sigma(u_3), v_s, v_{s+1}, \ldots, v_{n-1}\}\rangle$ is an induced path in G_2, then $v_s \to \sigma(u_3)$ and $v_{n-2} \to v_{n-1} \to v_n \to u_2$. If $d_{G_2}(v_{n-1}, \sigma(u_3)) = 2$ (notice that $\sigma(u_3) = v_{n-3}$), then (i) after one global application of the color-change rule on S, $V(G_\sigma) \setminus \{u_2, v_{n-3}, v_{n-1}, v_n\}$ are turned black; (ii) $v_{n-4} \to v_{n-3}$; (iii) $v_{n-2} \to v_{n-1} \to v_n \to u_2$. If $d_{G_2}(v_{n-1}, \sigma(u_3)) = 1$ (notice that $\sigma(u_3) = v_{n-2}$), then (i) after one global application of the color-change rule on S, $V(G_\sigma) \setminus \{u_2, v_{n-2}, v_{n-1}, v_n\}$ are turned black; (ii) $v_{n-3} \to v_{n-2} \to v_{n-1} \to v_n \to u_2$. So, for each case, S is a zero forcing set for G_σ with $Z(G_\sigma) < n$ for $n \geq 5$. Second, we consider $n = 4$. In this case, G_σ is isomorphic to (A) or (B) of Fig. 7. If G_σ is isomorphic to (A) in Fig. 7, one can easily check that $\{u_1, u_2, u_3\}$ forms a zero forcing set for G_σ, and thus $Z(G_\sigma) < 4$. If G_σ is isomorphic to (B) of Fig. 7, no three vertices form a zero forcing set for G_σ, and thus $Z(G_\sigma) = 4$. □

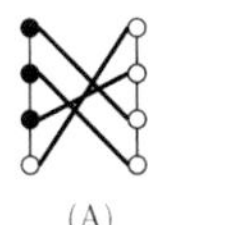

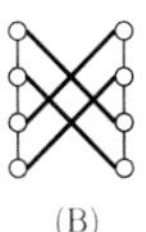

Fig. 7. Two non-isomorphic G_σ for $G = P_4$, where $\sigma(u_1) = u_3$ and $\sigma(u_2) = v_4$

6 Open Problems

We conclude this paper with some open problems. Let G be a graph, and let $\sigma : V(G_1) \to V(G_2)$ be a permutation.

(1) Some permutation graphs achieving the upper bound of Corollary 1 are shown in sections 3, 4, and 5. Can we characterize permutation graphs G_σ satisfying $Z(G_\sigma) = |V(G)|$?
(2) A characterization of graphs G with $Z(G) = 2$ is given in [18]. Can we characterize graphs G for which $Z(G) = |V(G)| - 2$?

Acknowledgement. The author wishes to thank anonymous referees for valuable comments and suggestions, which improved the paper.

References

1. (AIM Minimum Rank - Special Graphs Work Group) Barioli, F., Barrett, W., Butler, S., Cioabă, S.M., Cvetković, D., Fallat, S.M., Godsil, C., Haemers, W., Hogben, L., Mikkelson, R., Narayan, S., Pryporova, O., Sciriha, I., So, W., Stevanović, D., van der Holst, H., Vander Meulen, K., Wehe, A.W.: Zero forcing sets and the minimum rank of graphs. Linear Algebra Appl. 428/7, 1628–1648 (2008)
2. Barioli, F., Barrett, W., Fallat, S.M., Hall, H.T., Hogben, L., Shader, B., van den Driessche, P., van der Holst, H.: Zero forcing parameters and minimum rank problems. Linear Algebra Appl. 433, 401–411 (2010)
3. Berman, A., Friedland, S., Hogben, L., Rothblum, U.G., Shader, B.: An upper bound for the minimum rank of a graph. Linear Algebra Appl. 429, 1629–1638 (2008)
4. Burgarth, D., Giovannetti, V.: Full Control by Locally Induced Relaxation. Phys. Rev. Lett. 99, 100501 (2007)
5. Burgarth, D., Maruyama, K.: Indirect Hamiltonian identification through a small gateway. New J. Phys. 11, 103019 (2009)
6. Chartrand, G., Harary, F.: Planar permutation graphs. Ann. Inst. H. Poincare (Sect. B) 3, 433–438 (1967)
7. Chartrand, G., Zhang, P.: Introduction to Graph Theory. McGraw-Hill, Kalamazoo (2004)
8. Chilakamarri, K., Dean, N., Kang, C.X., Yi, E.: Iteration Index of a Zero Forcing Set in a Graph. Bull. Inst. Combin. Appl. 64, 57–72 (2012)
9. Edholm, C.J., Hogben, L., Hyunh, M., LaGrange, J., Row, D.D.: Vertex and edge spread of zero forcing number, maximum nullity, and minimum rank of a graph. Linear Algebra Appl. 436, 4352–4372 (2012)

10. Eroh, L., Kang, C.X., Yi, E.: A Comparison between the Metric Dimension and Zero Forcing Number of Trees and Unicyclic Graphs (submitted)
11. Eroh, L., Kang, C.X., Yi, E.: A Comparison between the Metric Dimension and Zero Forcing Number of Line Graphs (submitted)
12. Eroh, L., Kang, C.X., Yi, E.: On Zero Forcing Number of Graphs and Their Complements (submitted)
13. Fallat, S.M., Hogben, L.: The minimum rank of symmetric matrices described by a graph: A survey. Linear Algebra Appl. 426, 558–582 (2007)
14. Fallet, S.M., Hogben, L.: Variants on the minimum rank problem: A survey II. arXiv:1102.5142v1
15. Hogben, L., Huynh, M., Kingsley, N., Meyer, S., Walker, S., Young, M.: Propagation time for zero forcing on a graph (preprint)
16. Johnson, C.R., Loewy, R., Smith, P.A.: The graphs for which the maximum multiplicity of an eigenvalue is two. Linear Multilinear Algebra 57, 713–736 (2009)
17. Kang, C.X., Yi, E.: On Zero Forcing Number of Functigraphs (submitted)
18. Row, D.D.: A technique for computing the zero forcing number of a graph with a cut-vertex. Linear Algebra Appl., doi:10.1016/j.laa.2011.05.012
19. Severini, S.: Nondiscriminatory propagation on trees. J. Phys. A: Math. Theor. 41, 482002 (2008)

Complexity Results for the Empire Problem in Collection of Stars

Basile Couetoux[1], Jérome Monnot[3,2], and Sonia Toubaline[4]

[1] Laboratoire d'Informatique Fondamentale, Faculté des Sciences de Luminy, 163, av. de Luminy F-13288 Marseille cedex 9, France
Basile.Couetoux@lif.univ-mrs.fr
[2] PSL, Université Paris-Dauphine, LAMSADE, Place du Maréchal de Lattre de Tassigny, 75775 Paris Cedex 16, France
[3] CNRS, UMR 7243
monnot@lamsade.dauphine.fr
[4] Department of Security and Crime Science, UCL Jill Dando Institute, University College London, 35 Tavistock Square, London, WC1H 9EZ, UK
s.toubaline@ucl.ac.uk

Abstract. In this paper, we study the Empire Problem, a generalization of the coloring problem to maps on two-dimensional compact surface whose genus is positive. Given a planar graph with a certain partition of the vertices into blocks of size r, for a given integer r, the problem consists of deciding if s colors are sufficient to color the vertices of the graph such that vertices of the same block have the same color and vertices of two adjacent blocks have different colors. In this paper, we prove that given a 5-regular graph, deciding if there exists a 4-coloration is **NP**-complete. Also, we propose conditional **NP**-completeness results for the Empire Problem when the graph is a collection of stars. A star is a graph isomorphic to $K_{1,q}$ for some $q \geq 1$. More exactly, we prove that for $r \geq 2$, if the $(2r-1)$-coloring problem in $2r$-regular connected graphs is **NP**-complete, then the Empire Problem for blocks of size $r+1$ and $s = 2r-1$ is **NP**-complete for forests of $K_{1,r}$. Moreover, we prove that this result holds for $r = 2$. Also for $r \geq 3$, if the r-coloring problem in $(r+1)$-regular graphs is **NP**-complete, then the Empire Problem for blocks of size $r+1$ and $s = r$ is **NP**-complete for forests of $K_{1,1} = K_2$, i.e., forest of edges. Additionally, we prove that this result is valid for $r = 2$ and $r = 3$. Finally, we prove that these results are the best possible, that is for smallest value of s or r, the Empire Problem in these classes of graphs becomes polynomial.

Keywords: Empire Problem, Coloring in regular graphs, **NP**-completeness, Forests of stars.

1 Introduction

Graph coloring problem is an important optimization problem because scheduling problems appearing in real-life situations may often be modeled as graph coloring problems (see [1,7,13]). For instance, scheduling problems involving only incompatibility constraints correspond to the classical vertex coloring problem. A k-coloration of a graph

G. Lin (Ed.): COCOA 2012, LNCS 7402, pp. 73–82, 2012.

$G = (V, E)$ is a mapping $c : V \to \{1, \dots, k\}$ such that $c(u) \neq c(v)$ for all $[u, v] \in E$. It is well known that, given an integer k, deciding if a graph admits a k-coloration is a **NP**-complete problem if $k \geq 3$ and polynomial otherwise [6]. The coloring problem consists in finding the minimum k such that G is k-colorable, this number is called the chromatic number of G denoted $\chi(G)$.

Using the Brooks' theorem, it is well known that a connected graph with maximum degree Δ, for any $\Delta \geq 3$, is Δ-colorable, except for $K_{\Delta+1}$, and such a coloration can be found within polynomial time. So, an open question concerning the coloration is to know the complexity of the k-coloring problem in $(k + 1)$-regular graphs and such a result will help narrowing down the gap between **P** and **NP**-complete classes for the coloration with respect to the maximum degree of the graph. Dailey [3] proved that the 3-coloring problem in 4-regular graph is **NP**-complete, even if the graph is planar. To our best knowledge, no result exist for $k > 3$. We prove in this paper that the 4-coloring problem in 5-regular graph is **NP**-complete.

The Empire Problem is a generalization of the coloring problem to maps on two-dimensional compact surfaces whose genus is positive. In graph theory terminology, the Empire Problem can be described as follows: given an integer r and a planar graph $G = (V, E)$, what is the minimum number s of colors needed to color a map in which each country has r colonies? A country and its colonies should be colored the same, and no pair of distinct colonies (belonging to two different countries) are adjacent. Each country is called block (sometimes a block is also called empire). We denote the decision version of the Empire Problem by s-COL_r.

In 1890, Heawood [8] conjectured that at most $6r$ colors are necessary to color every instance of the Empire Problem in the plane, where blocks are of size at most r. He could only proved the conjecture for $r = 2$. In 1981, Taylor [5] proved Heawood's conjecture for $r = 3$ and $r = 4$. Later, in 1984 this conjecture has completely been proved by Jackson and Ringel [9] for every $r \geq 2$. For random graphs, better bounds exist. For instance, McGrae and Zito [10] proved that for every fixed integer $r > 1$ there exists a positive integer $s_r = O(\frac{r}{\log r})$ such that $s_r < s \leq 2r$ asymptotically almost surely for a random n-vertex tree. Later, Coper, McGrae and Zito [2] improved this result and showed that for every fixed integer $r > 1$, $s_r \leq s \leq s_r + 6$. Furthermore, if $r > \frac{2(s_r-1)^2}{2s_r-3} \log s_r$ then $s_r + 1 \leq s \leq s_r + 6$.

From a deterministic point of view, the complexity of the Empire Problem in several classes of planar graphs, has been mainly studied by McGrae and Zito in two recent papers [12,11]. The case $r = 1$ corresponds to the coloring problem in planar graphs. Conversely, for any s, r, the complexity of s-COL_r in G is equivalent to the complexity of the s-coloring problem in the reduced graph $R_r(G)$, where $R_r(G)$ is obtained from G by contracting each country into a distinct node. In particular, s-COL_r is polynomial as soon as $s \leq 2$. Hence, the complexity of the Empire Problem seems interesting only when we deal with very particular classes of planar graphs. For instance in [12], the authors give a full dichotomy theorem for trees. More exactly, McGrae and Zito proved that for fixed positive integers r and s, with $r \geq 2$, s-COL_r in trees is **NP**-complete, if $3 \leq s \leq 2r - 1$ and polynomial otherwise. For general planar graphs, McGrae and Zito [12] showed that for fixed positive integers r and s, with $r \geq 2$, s-COL_r is **NP**-complete if $3 \leq s \leq 6(r - 1)$. They proved also in [11] that s-COL_r is **NP**-complete if $s < 7$ for

$r = 2$ and $s < 6r - 3$ for $r \geq 3$. Finally, for linear forests, i.e., collection of induced paths, McGrae and Zito [12] proved on the one hand that $(2r-1)$-COL_r is polynomial for paths of length at most $2r - 1$, for fixed positive integer $r \geq 2$, and on the other hand, they showed that s-COL_r is **NP**-complete for any fixed positive integers r and s, with $3 \leq s < r$, if the paths have an arbitrary length. In [11], the values of r, s have been improved to any $r \geq 2$ and $3 \leq s \leq 2r - \sqrt{2r + \frac{1}{4}} + \frac{3}{2}$, but the length of the paths remain arbitrary large. Here, as corollary of Theorem 2, we strengthen this result by proving that 3-COL_3 is **NP**-complete, even for linear forests of length exactly 2, i.e., collection of disjoint $K_{1,2}$.

A related problem is the selective graph coloring problem. It consists of selecting one vertex per empire (not all the empire) in such a way that the chromatic number of the resulting induced selection is the smallest possible. In [4], some complexity results for various classes of graphs are given. Hence, the decision version of the selective graph coloring problem can be viewed as the restriction of the Empire Problem to select one vertex per empire.

This paper is organized as follows. In Section 2 we introduce some definition and notation. We show in Section 3 that the 4-coloring problem in 5-regular graph is **NP**-complete. We present in Section 4 some polynomial results and conditional **NP**-completeness results for s-COL_r for special classes of graphs and for given values of s and r. Indeed, we prove in this section that for $r \geq 2$, if the $(2r-1)$-coloring problem in $2r$-regular connected graphs is **NP**-complete, then $(2r-1)$-COL_{r+1} is **NP**-complete for forests of $K_{1,r}$. For $r = 2$, the result is valid without conditions. We prove that for $r \geq 2$ and $s \geq 2r$, s-COL_{r+1} is polynomial for forests of $K_{1,r}$, showing that the previous conditional **NP**-complete result will be the best possible for forests of $K_{1,r}$. Also, for any $r \geq 3$, if the r-coloring problem in $(r+1)$-regular graphs is **NP**-complete, then r-COL_{r+1} is **NP**-complete for graphs of disjoint edges. For $r = 3, 4$, the result is valid without conditions. Moreover, we have that r-COL_r is polynomial in forest of edges. Finally, we conclude by some discussions and perspectives in Section 5.

2 Definitions

All graphs in this paper are finite, simple and loopless. Let $G = (V, E)$ be a graph. An edge between u and v will be denoted $[u, v]$. For a vertex $v \in V$, let $N_G(v)$ denote the set of vertices in G that are adjacent to v, i.e., the neighbors of v and the degree of v is $d_G(v) = |N_G(v)|$. A graph $G = (V, E)$ is r-regular if $\forall v \in V$, $d_G(v) = r$. For any $V' \subseteq V$, $G[V']$ is the graph induced by the set of vertices V', i.e., the graph obtained from G by deleting the vertices of $V - V'$ and all edges incident to at least one vertex of $V - V'$. If the graph is directed we denote the set of oriented edges $\overrightarrow{E}$ and (u, v) is the arc from u to v. For a directed graph, let $N^+_{\overrightarrow{G}}(v) = \{u \in V : (v, u) \in \overrightarrow{E}\}$ be the outgoing neighbors of v and $N^-_{\overrightarrow{G}}(v) = \{u \in V : (u, v) \in \overrightarrow{E}\}$ be the incoming neighbors of v. Finally, $d^+_{\overrightarrow{G}}(v) = |N^+_{\overrightarrow{G}}(v)|$ and $d^-_{\overrightarrow{G}}(v) = |N^-_{\overrightarrow{G}}(v)|$.

A simple graph $G = (V, E)$ is called Eulerian if it has a cycle (called Eulerian cycle) which visits each edge exactly once. The famous Euler's theorem asserts that a graph $G = (V, E)$ is Eulerian iff it is connected and the degree of all its vertices is even. An independent set in a graph $G = (V, E)$ is a set $S \subseteq V$ of pairwise nonadjacent vertices. Alternatively, a k-coloration of $G = (V, E)$ can be viewed as a partition of V into k independent sets.

We denote by nG the disjoint union of n copies of a graph G. As usual P_n denotes the path induced by n vertices. The length of a path is the number of its edges. The complete bipartite graph $K_{1,q}$ is also called a q-star, ie., a root adjacent to q leafs. For more graph definitions and notations see [14].

In this paper we will be interested in the following two problems. Let $r \geq 3$ be a fixed integer.

r-COLORING in $(r+1)$-regular graph
Input: A $(r+1)$-regular connected graph $G = (V, E)$
Output: Deciding if there is an assignment of vertices to r colors such that no edge is monochromatic (no two adjacent vertices share the same color).

THE EMPIRE PROBLEM (denoted s-COL$_r$)
Input: A planar graph (not necessary connected) $G = (V, E)$ on pr vertices; a partition $\mathcal{V} = (V_1, \ldots, V_p)$ of V where $\forall i = 1, \ldots, p$, $|V_i| = r$.
Output: Deciding if there is an assignment of vertices to r colors such that no edge between V_i and V_j with $1 \leq i < j \leq p$ is monochromatic and all vertices in each V_i have the same colors.

Without loss of generality, we can assume that the partition $\mathcal{V}$ of the Empire Problem is a coloring. For an instance of s-COL$_r$ formed by a graph G, we denote $R_r(G)$ its reduced graph. $R_r(G)$ is obtained from G by contracting each empire to a distinct pseudo-vertex and by adding an edge between a pair of pseudo-vertices if there exists in G at least one edge between one vertex of the first empire of this pair and one vertex of the second empire of this pair.

3 The 4-Coloring Problem in 5-Regular Graphs

We study in this section the complexity of the k-coloring problem in $(k+1)$-regular graphs. We prove in the following theorem that the problem is **NP**-complete for $k = 4$.

Theorem 1. 4-COLORING *in 5-regular connected graphs is* ***NP****-complete.*

Proof: We propose a polynomial reduction from 3-COLORING in 4-regular connected graphs, proved to be **NP**-complete in [3]. Let $G = (V, E)$ be a 4-regular connected graph on n vertices instance of the 3-coloring problem. We construct a 5-regular connected graph H in the following way: it contains two copies of G where copies of the $i-th$ vertex are v_i, v_i' and a gadget F depicted in Figure 1. This gadget contains $n-3$ copies $H_1, \ldots, H_{n-3}$ of a same graph, n special vertices $u_1, \ldots, u_n$ where for $i = 1, \ldots, n-4$, u_i is in copy H_i while the last four vertices $u_{n-3}, \ldots, u_n$ are in

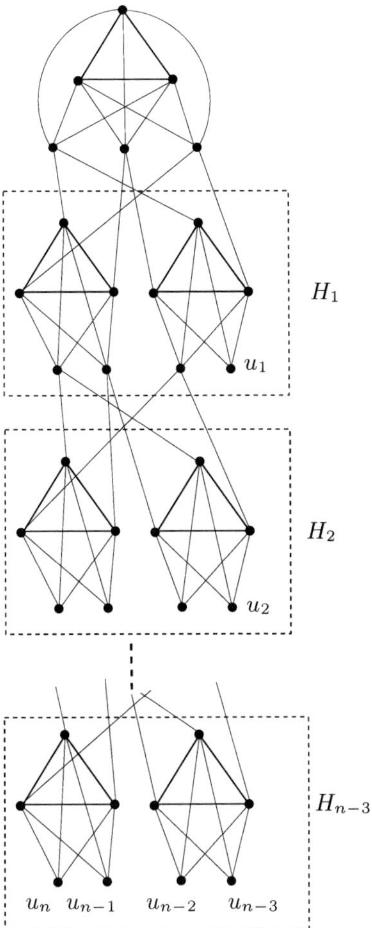

Fig. 1. The gadget F

H_{n-3}. In F, all vertices have a degree 5 except special vertices $u_1, \ldots, u_n$ of degree 3. Finally, each vertex u_i is linked to v_i and v_i'. Hence, H is a 5-regular connected graph and this construction can be done within polynomial time.

We claim that G is 3-colorable iff H is 4-colorable.

Actually, by pointing out that F is 4-colorable and the special vertices $u_1, \ldots, u_n$ must have the same color in any 4-coloration, the result follows. □

4 Complexity Results for the Empire Problem

We propose in this section some conditional **NP**-completeness results as well as polynomial results for the Empire Problem when the graph is a collection of stars.

Theorem 2. *Let $r \geq 2$ be an integer. If $(2r-1)$-COLORING in $2r$-regular connected graphs is **NP**-complete, then $(2r-1)$-COL$_{r+1}$ is **NP**-complete in $nK_{1,r}$.*

Proof: We propose a polynomial reduction from $(2r-1)$-COLORING in $2r$-regular connected graphs.

Let $r \geq 2$. Consider an instance I of $(2r-1)$-COLORING formed by a $2r$-regular connected graph $G = (V, E)$ with $V = \{1, \ldots, n\}$. We define an orientation on the edges of E according to an Eulerian cycle in G. This cycle exists since G is connected and all vertices of V have an even degree (see Figure 2 for an illustration when $r = 2$).

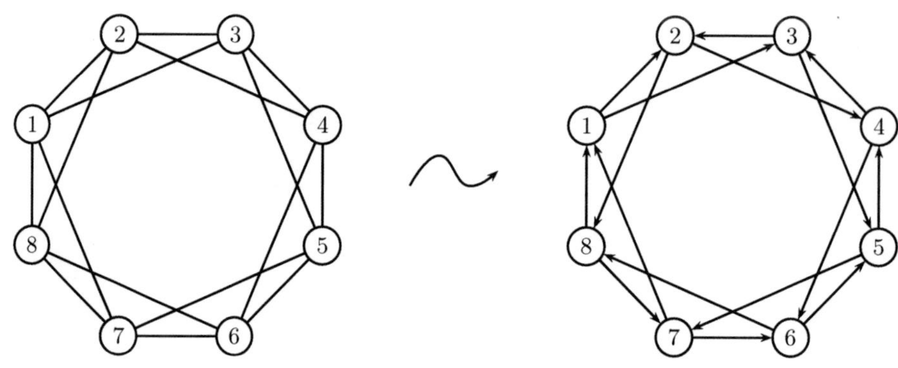

Fig. 2. Case $r = 2$. Orienting the edges of a 3-regular connected graph G according to an Eulerian cycle.

Denote by $\overrightarrow{G} = (V, \overrightarrow{E})$ the corresponding digraph. For $i \leq n$, let $e_\ell(i)$, for $\ell = 1, \ldots, r$, be the arcs incoming in vertex i in $\overrightarrow{E}$. We have $\overrightarrow{E} = \{e_\ell(i), \ell = 1, \ldots, r : i \leq n\}$ because G is $2r$-regular. Let $\overrightarrow{G}_i = (V, \overrightarrow{E_i})$, where $\overrightarrow{E_i} = \{e_\ell(j), \ell = 1, \ldots, r : j \leq i\}$. By construction $\overrightarrow{E_0} = \emptyset$.

We construct an instance I' of $(2r-1)$-COL$_{r+1}$ problem formed by a graph $G' = (V', E')$ as follows (see Figure 3 for the case $r = 2$ and the 3-regular graph G given in Figure 2). We associate to each vertex $i \in V$, $r+1$ copies in V', for $p = 0, \ldots, r$, denoted i^p, which form an empire of size $r+1$ in G'. Let $N^-_{\overrightarrow{G}}(i) = \{h_1, \ldots, h_r\}$ be the incoming neighbors vertices of i. For each vertex i, for $i = 1, \ldots, n$, noting $e_\ell(i) = (h_\ell, i)$, for $\ell = 1, \ldots, r$, we add to G' the edges $[h_\ell^{d^+_{\overrightarrow{G}_{i-1}}(h_\ell)+1}, i^0]$, for $\ell = 1, \ldots, r$. Since $d^+_{\overrightarrow{G}}(i) = r, \forall i \in V$, we have $d_{G'}(i^p) \leq r, \forall i \leq n$ and $\forall p = 0, \ldots, r$. Indeed, by construction :

- $d_{G'}(i^0) = r$, because $d_{G'}(i^0)$ corresponds to the r arcs $e_\ell(i)$, for $\ell = 1, \ldots, r$, the only r arcs incoming in vertex i in $\overrightarrow{G}$.
- $d_{G'}(i^p) = 1$, for $p = 1, \ldots, r$, because by construction i^p has an edge incident if and only if $d^+_{\overrightarrow{G}_{j_p-1}}(i) = p-1$ at the iteration j_p. Since $d^+_{\overrightarrow{G}}(i) = r$, the iteration j_p exists.

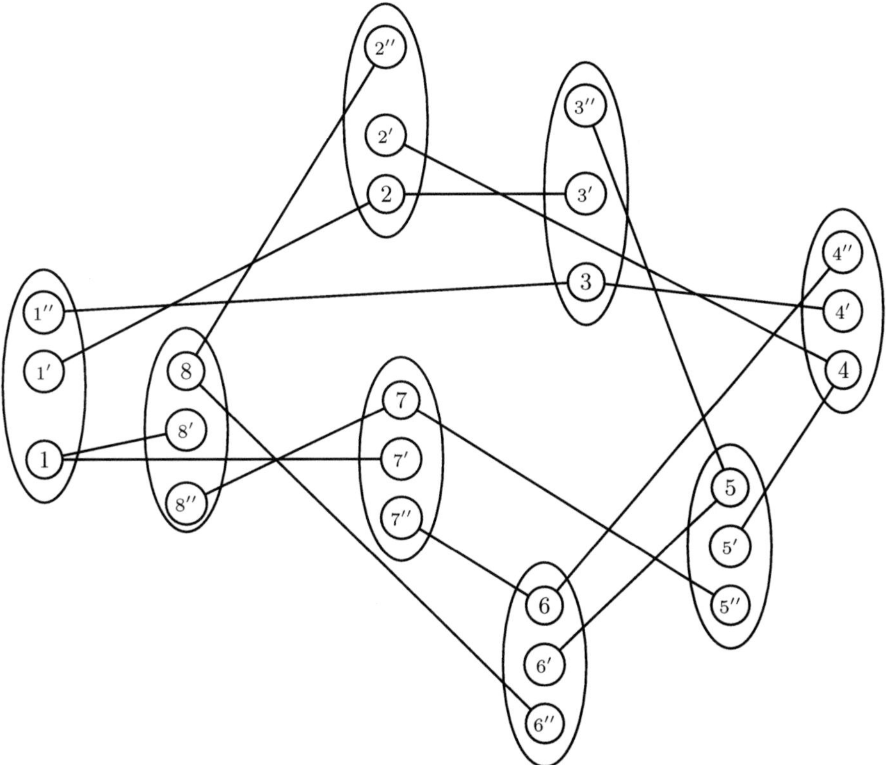

Fig. 3. Construction of G' from G when $r = 2$

Thus, since by construction $d_{G'}(i^0) = r$ and $d_{G'}(i^p) = 1$, for $p = 1, \ldots, r$ and $i = 1, \ldots, n$, G' is a $nK_{1,r}$ and this construction can be done within polynomial time. We remark that $R_{r+1}(G') = G$.

We show that G is $(2r-1)$-colorable if and only if G' is $(2r-1)$-colorable.

Suppose first that G is $(2r-1)$-colorable. We assign to vertices i^p in V', for $p = 0, \ldots, r$, the same color as i in V, for $i = 1, \ldots, n$. Hence, the coloring defines a truth assignment of vertices of G' to $2r-1$ colors. Therefore G' is $(2r-1)$-colorable.

Suppose now that G' is $(2r-1)$-colorable. Since $R_{r+1}(G') = G$, we have that G is $(2r-1)$-colorable by assigning to every vertex $i \in V$ the same color as its copies in V'. □

In [12,11], McGrae and Zito have obtained, on the one hand that for any fixed integer $k \geq 1$, $\lceil \frac{2kr}{k+1} \rceil$-Col$_r$ in collection of paths of length at most k is polynomial (p. 182 in [12]. Indeed, it is a corollary of a more general result in sparse graphs). For instance, $\lceil \frac{4r}{3} \rceil$-Col$_r$ in collection of paths of length at most 2 are polynomial. On the other hand, they proved that s-Col$_r$ is **NP**-complete for any $r > s \geq 3$ in linear forests, i.e., collection of paths of arbitrary length (Theorem 3 in [12]) and (in [11],

Theorem 6), they have strengthen this result to any $r \geq 2$ and $3 \leq s \leq 2r - \sqrt{2r + 1/4} + 3/2$. Here, we deduce from Theorem 2 that 3-COL$_3$ is **NP**-complete in collection of paths of length exactly 2.

Corollary 1. *3-COL$_3$ is **NP**-complete in $nK_{1,2} = nP_3$.*

Proof: Since 3-COLORING in 4-regular connected graphs has been proved **NP**-complete in [3], the result follows from Theorem 2 with $r = 2$. □

Now, we prove that the result given in Theorem 2 is the best possible according the parameter s.

Proposition 1. *Let r and s be two integers such that $r \geq 2$ and $s \geq 2r$. s-COL$_{r+1}$ is polynomial in $nK_{1,r}$.*

Proof: Let G be a $nK_{1,r}$. G is a planar graph containing no induced subgraph of average degree larger than $\frac{2r}{r+1}$. According to Theorem 1 of [11], s-COL$_r$ is polynomial for planar graphs containing no induced subgraph of average degree larger than $\frac{s}{r}$. Since, $s \geq 2r$ we have that $\frac{s}{r} \geq \frac{2r}{r+1}$. Therefore, s-COL$_{r+1}$ is polynomial in $nK_{1,r}$ for $r \geq 2$ and $s \geq 2r$. □

Theorem 3. *If r-COLORING in $(r+1)$-regular graphs is **NP**-complete, then r-COL$_{r+1}$ is **NP**-complete in $nK_{1,1} = nK_2 = nP_2$.*

Proof: We propose a polynomial reduction from r-COLORING defined on $(r + 1)$-regular graph.

Consider an instance I of r-COLORING formed by a $(r + 1)$-regular graph $G = (V, E)$ with $V = \{1, \ldots, n\}$ and $E = \{e_1, \ldots, e_m\}$. We construct an instance I' of r-COL$_{r+1}$ problem formed by a graph $G' = (V', E')$ as follows. We associate to each vertex $i \in V$, $r + 1$ copies i^{e_h} in V' with $e_h \in E_G(i)$ where $E_G(i)$ are the subset of edges incident to i in G. These $r + 1$ copies form an empire of size $r + 1$ in G'. For each edge $e_\ell = [i, j] \in E$, we add the edge $[i^{e_\ell}, j^{e_\ell}]$ in G'.

Since $d_G(i) = r + 1$ and each edge is exactly incident to two vertices, we have that $d_{G'}(i^{e_\ell}) = 1, \forall i^{e_\ell} \in V'$. Thus G' is isomorphic to nK_2 and this construction can be done within polynomial time. We remark that $R_{r+1}(G') = G$.

We show that G is r-colorable if and only if G' is r-colorable.

Suppose first that G is r-colorable. We assign to vertices i^{e_h} in V', for $e_h \in E_G(i)$, the same color as i in V, for $i = 1, \ldots, n$. This coloring constitutes a truth assignment of vertices of G' to r colors. Therefore G' is r-colorable.

Suppose now that G' is r-colorable. Since $R_{r+1}(G') = G$, we have that G is r-colorable by assigning to every vertex $i \in V$ the same color as its copies in V'. □

From [12], we know that s-COL$_r$ in collection of disjoint edges and isolated vertices is polynomial for $s \geq r$. Here, we deduce that this result is the best possible for $r = 3, 4$.

Corollary 2. *r-COL$_{r+1}$ is **NP**-complete in nK_2 for $r = 3, 4$.*

Proof: Since 3-COLORING in 4-regular graphs and 4-COLORING in 5-regular graphs have been proved **NP**-complete in [3] and in Theorem 1 respectively, the result follows from Theorem 3 with $r = 3$ and $r = 4$. □

Proposition 2. *Let $r \geq 1$ be integer. r-COL$_r$ is polynomial in forest of edges.*

Proof: In [12,11], McGrae and Zito showed that $\left\lceil \frac{2kr}{k+1} \right\rceil$-COL$_r$ can be decided in polynomial time for forests of paths of length at most k. Thus, for $k = 1$ we have that r-COL$_r$ is polynomial in forest of edges. □

5 Conclusion

We are interested in this paper to some coloration problems. First, we gave a partial answer to the open question about the complexity of the k-coloring problem in $(k+1)$-regular graphs, or more generally in graphs of maximum degree $k+1$. To the best of our knowledge, this question has never been raised in the literature, although the case $k = 3$ was solved in 1980 by Dailey [3]. Here, we have continued the investigation of this problem for the particular case of $k = 4$. Indeed, we showed that 4-COLORING in 5-regular connected graphs is **NP**-complete. We are not able to solve the cases $k \geq 5$, but we think that this question is important to better understand the complexity of the coloring problem. Based on this open problem, we have proposed some complexity results on the Empire Problem in sparse planar graphs. More exactly, we studied the Empire Problem a generalization of the k-coloring problem for which we proposed some conditional **NP**-completeness results for collection of stars. Hence, we proved that if $(2r-1)$-COLORING in $2r$-regular connected graphs is **NP**-complete, then $(2r-1)$-COL$_{r+1}$ is **NP**-complete for forests of $K_{1,r}$. This result holds for $r = 2$ and strengthens the results obtained by McGrae and Zito in [12,11] for s-COL$_r$ in forests of paths of arbitrary length. Furthermore, we proved that if r-COLORING in $(r+1)$-regular connected graphs is **NP**-complete, then r-COL$_{r+1}$ is **NP**-complete for forests of edges. This results is valid for $r = 2$ and $r = 3$. Also, we showed that these results are best possible, that is for smallest values of s and r, the Empire Problem in these classes of graphs becomes polynomial.

Looking for the approximation, we show that the optimization version of the Empire Problem with blocks of size r, noted MIN COL$_r$, is $\frac{4}{3}$-approximable for forests of paths of length at most 2. Indeed, if $R(G)$ is bipartite then G is 2-empire colorable. Otherwise, in [12,11], the authors showed that $\left\lceil \frac{2kr}{k+1} \right\rceil$-COL$_r$ can be decided in polynomial time for forests of paths of length at most k. Since G is not bipartite, G is at least 3-colorable which means that the value of an optimal coloration of G is larger than 3. Therefore, MIN COL$_r$ is $\frac{4}{3}$-approximable for forests of paths of length at most 2. It is then interesting to study the approximation of MIN COL$_r$ more generally for forests of stars. Using Proposition 1, we trivially get that MIN COL$_r$ is $\frac{2r}{3}$-approximable in $nK_{1,r}$. Thus, an interesting perspective is to try to improve this approximation ratio.

Another perspective is to study the complexity and approximation of the Empire Problem for other classes of sparse planar graphs with small average degree like nK_3 and nK_4.

References

1. Al-Mouhamed, M., Dandashi, A.: Graph coloring for class scheduling. In: IEEE/ACS Internation Conference on Computer Systems and Applications (AICCSA), pp. 1–4 (2010)
2. Cooper, C., McGrae, A.R.A., Zito, M.: Martingales on Trees and the Empire Chromatic Number of Random Trees. In: Kutyłowski, M., Charatonik, W., Gębala, M. (eds.) FCT 2009. LNCS, vol. 5699, pp. 74–83. Springer, Heidelberg (2009)
3. Dailey, D.P.: Uniqueness of colorability and colorability of planar 4-regular graphes are NP-complete. Discrete Mathematics 30(3), 289–293 (1980)
4. Demange, M., Monnot, J., Pop, P., Ries, B.: Selective graph coloring in some special classes of graphs. In: Proceedings of 2nd International Symposium on Combinatorial Optimization, ISCO 2012 (to appear in LNCS, 2012)
5. Gardner, M.: M-Pire Maps. In: The last recreations. Hydras, Eggs and Other Mathematical Mystifications, pp. 85–100. Spring-Verlag New York, Inc. (1997)
6. Garey, M.R., Johnson, D.S.: Computers and Intractability: A Guide to the Theory of NP-Completeness. W. H. Freeman (1979)
7. Giaro, K., Kubale, M., Obszarski, P.: A graph coloring approach to scheduling of multiprocessor tasks on dedicated machines with availability constraints. Discrete Applied Mathematics 157(17), 3625–3630 (2009)
8. Heawood, P.J.: Map colour theorem. Quarterly Journal of Pure and Applied Mathematics 24, 332–338 (1890)
9. Jackson, B., Ringel, G.: Solution of heawood's empire problem in the plane. Journal für die Reine und Angewandte Mathematik 347, 146–153 (1984)
10. McGrae, A.R.A., Zito, M.: Colouring Random Empire Trees. In: Ochmański, E., Tyszkiewicz, J. (eds.) MFCS 2008. LNCS, vol. 5162, pp. 515–526. Springer, Heidelberg (2008)
11. McGrae, A.R.A., Zito, M.: The complexity of the empire colouring problem. CoRR, abs/1109.2162 (2011)
12. McGrae, A.R.A., Zito, M.: Empires Make Cartography Hard: The Complexity of the Empire Colouring Problem. In: Kolman, P., Kratochvíl, J. (eds.) WG 2011. LNCS, vol. 6986, pp. 179–190. Springer, Heidelberg (2011)
13. Ries, B.: Complexity of two coloring problems in cubic planar bipartite mixed graphs. Discrete Applied Mathematics 158(5), 592–596 (2010)
14. West, D.B.: Introduction to Graph Theory, 2nd edn. Prentice Hall (2001)

Hamiltonian Paths and Cycles in Planar Graphs

Sudip Biswas[1], Stephane Durocher[2,⋆],
Debajyoti Mondal[2], and Rahnuma Islam Nishat[3]

[1] Department of Computer Science, Louisiana State University
[2] Department of Computer Science, University of Manitoba
[3] Department of Computer Science, University of Victoria
sudipid@gmail.com, {durocher,jyoti}@cs.umanitoba.ca,
rnishat@cs.uvic.ca

Abstract. We examine the problem of counting the number of Hamiltonian paths and Hamiltonian cycles in outerplanar graphs and planar graphs, respectively. We give an $O(n\alpha^n)$ upper bound and an $\Omega(\alpha^n)$ lower bound on the maximum number of Hamiltonian paths in an outerplanar graph with n vertices, where $\alpha \approx 1.46557$ is the unique real root of $\alpha^3 = \alpha^2 + 1$. For any positive integer $n \geq 6$, we define an outerplanar graph G, called a ZigZag outerplanar graph, such that the number of Hamiltonian paths starting at a single vertex in G is the maximum over all possible outerplanar graphs with n vertices. Finally, we prove a 2.2134^n upper bound on the number of Hamiltonian cycles in planar graphs, which improves the previously best known upper bound 2.3404^n.

1 Introduction

Counting of combinatorial objects is a fundamental problem in combinatorics. Given a graph G with n vertices, a straightforward approach to count the number of Hamiltonian paths in G is to use a naive backtracking algorithm that enumerates all possible paths in G. Since the problem of determining whether any Hamiltonian path exists in a given graph is NP-hard [7], determining their exact number is also NP-hard.

Much research effort has been devoted to counting as well as bounding the number of Hamiltonian paths and Hamiltonian cycles in graphs [1,3,4] and various classes of graphs, such as cubic graphs [8,6], grid graphs [3] and planar graphs [2]. The currently best known upper and lower bounds on the number of Hamiltonian cycles in planar graphs are established by Buchin et al. [2], which are 2.3404^n and 2.0845^n, respectively. They also gave a 2.8927^n upper bound and a 2.4262^n lower bound on the number of simple cycles in planar graphs. Recently, de Mier and Noy [5] proved that the number of simple cycles in an outerplanar graph is $\Theta(1.502837^n)$.

Although there exists a polynomial-time algorithm to determine the number of Hamiltonian paths in the graphs with bounded treewidth [9], finding a tight

⋆ Work of the author is supported in part by the Natural Sciences and Engineering Research Council of Canada (NSERC).

G. Lin (Ed.): COCOA 2012, LNCS 7402, pp. 83–94, 2012.

upper bound on that number is a non-trivial task (e.g., counting Hamiltonian paths in a rectangular grid of small width [3]). On the other hand, we can find a fairly tight upper bound on the number of Hamiltonian paths in an outerplanar graph G with n vertices by simply solving a recurrence formula as follows. Let a be a vertex of G (without loss of generality assume that G is maximal). We can define the number of Hamiltonian paths of G starting at a recursively, as shown in Figures 1(a)–(c). The numbers k_1 and k_2 represent the number of vertices (that are not shown explicitly) in the corresponding shaded regions. A partial Hamiltonian path starting at a can be extended along the path shown in bold, where the vertices already visited are shown in gray, the current vertex is shown in white. The vertices still to be visited either lie in the dark gray region or are shown in black. Consequently, the number of Hamiltonian paths starting at a is

$$\begin{aligned} T(n) = \quad & \max\{T(n-k_2-2)+T(n-k_2-3)+T(n-k_1-3)+T(n-k_1-2), \\ & T(n-1)+T(n-3), T(n-k_1-2)+T(n-k_2-2)\}, \end{aligned}$$

which is dominated by $T(n-1)+T(n-3)$ and hence bounded by $O(1.46557^n)$. This also suggests that the number of Hamiltonian paths of an outerplanar graph is maximized when the graph has low maximum degree.

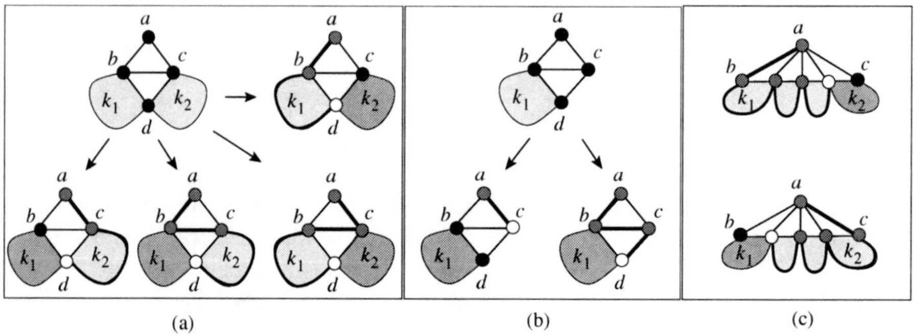

Fig. 1. Counting Hamiltonian Paths. Illustration for the cases when (a) degree(a)=2 and $k_1, k_2 > 0$, (b) degree(a)=2 and $k_2 = 0$, and (c) degree(a)> 2 and $k_1, k_2 \geq 0$.

We give a combinatorial proof for an $O(n\alpha^n)$ upper bound and an $\Omega(\alpha^n)$ lower bound on the maximum number of Hamiltonian paths in an outerplanar graph with n vertices, where $\alpha \approx 1.46557$ is the unique real root of $\alpha^3 = \alpha^2 + 1$. Our proof relies on graph transformation. We show that given a maximal outerplanar graph G with n vertices and a vertex x in G, one can insert/delete constant number of vertices and edges to obtain another combinatorially different maximal outerplanar graph G' with n vertices such that the maximum number of Hamiltonian paths starting at some vertex y in G' is at least as large as the maximum number of Hamiltonian paths in G that start at x. If we apply such a transformation repeatedly, then within $3n/2$ steps we can find an outerplanar graph G'' such that the number of Hamiltonian paths starting at a vertex in G'' is maximum over all the outerplanar graphs with n vertices. Contrary to proofs using recurrence relations, this proof helps characterize some of the structural properties of outerplanar graphs. Furthermore, we prove a 2.2134^n upper bound on the number of Hamiltonian cycles in planar graphs, which improves the previously best known upper bound 2.3404^n and reduces the previous gap between the upper and lower bound for the exponential growth from 0.46 to 0.13.

2 Preliminaries

Let G be a graph with n vertices. By $V(G)$ and $E(G)$ we denote the set of vertices and the set of edges in G, respectively. By $|V(G)|$ we denote the number of vertices of G, i.e., $|V(G)| = n$. By (u, v) we denote an edge between the vertices u and v. Let G be a graph and let G' be a subgraph of G. By $G - G'$ we denote the graph obtained by deleting all the vertices of G' from G. A *separating pair* of G is a pair of vertices $\{x, y\}$ whose deletion disconnects G. If x and y are neighbors, then the pair is called a *separating edge*.

A graph is *outerplanar* if it has a planar embedding with all its vertices on the outer face. An outerplanar graph is *maximal* if the addition of any edge violates outerplanarity. Let G be a maximal outerplanar graph with $n > 3$ vertices and let $\{x, y\}$ be a separating edge of G. Then deletion of the vertices x and y from G will give two connected components G' and G''. We call the subgraphs $G - G'$ and $G - G''$ the *split graphs* with respect to $\{x, y\}$. By $\langle u_1, u_2, \ldots, u_k\rangle$ we denote a simple path of k vertices. We now have the following fact.

Fact 1. *Let G be a maximal outerplanar graph with n vertices. For any Hamiltonian path $\langle v_1, v_2, \ldots, v_n\rangle$ in G, the edge (v_1, v_2) must be an outer edge of G. Let (u, v) be an outer edge of G. Then the Hamiltonian path that starts at u and ends at v is unique and lies along the outer face of G.*

It is straightforward to design a backtracking algorithm based on Fact 1 that takes a maximal outerplanar graph G (a fixed combinatorial plane embedding of G) and a vertex x of G as input and then enumerates all the Hamiltonian paths of G that start at vertex x. Starting at x such an algorithm constructs a Hamiltonian path incrementally by visiting the unvisited vertices one after another. At each vertex the algorithm can have at most two choices to move forward to the next vertex and at each forward phase the algorithm is guaranteed to produce a new Hamiltonian path. Once a Hamiltonian path is produced, the algorithm backtracks to find a vertex that can initiate a forward move that has not been taken yet. If there is no such vertex, then the algorithm terminates. We will use this idea of enumerating Hamiltonian paths in our counting technique.

3 Hamiltonian Paths in Outerplanar Graphs

In this section we give an $O(n1.47^n)$ upper bound on the number of Hamiltonian paths in an outerplanar graph with n vertices. Since the addition of an edge in a graph does not decrease the number of Hamiltonian paths, it suffices to consider only maximal outerplanar graphs.

Let G be a maximal outerplanar graph and let v be a vertex of G. By $h(G)_v$ we denote the number of Hamiltonian paths in G starting at vertex v. If the number of Hamiltonian paths starting from v in G is the maximum over all vertices of G, then we say v is an *ace vertex* of G. By $\mathcal{N}(G)$ we denote the number of Hamiltonian paths in G starting at an ace vertex of G. In the following we give an outline of our proof technique.

Step 1: Let $\mathcal{S}_n$ be the set of all maximal outerplanar graphs of n vertices, where $n \geq 6$, whose weak dual is a path. We prove that there exists a graph $G \in \mathcal{S}_n$, such that $\mathcal{N}(G)$ is the maximum over all possible maximal outerplanar graphs of n vertices. See Theorem 1.
Step 2: We then identify such a graph G and refer to that graph as a ZigZag graph. See Theorem 2.
Step 3: Finally, we identify an ace vertex v in G. We give an $O(1.47^n)$ upper bound on $h(G)_v$. Consequently, we obtain an $O(n1.47^n)$ upper bound on the number of Hamiltonian paths in any outerplanar graph of n vertices. See Theorem 3.

Let G be a maximal outerplanar graph with n vertices. Then the *weak dual T of G* is a binary tree which has a vertex for each bounded face of G, and two vertices in T are adjacent if the corresponding faces in G share an edge. Let f be a face in G. Then the node in T that corresponds to the face f is the *dual node* of f. We can define the weak dual T as a rooted ordered binary tree as follows. If $n = 3$, then T contains a single node which is the root r. Otherwise, $n > 3$ and we take any vertex of degree one as the root r of T. Observe that r has only one child v. By convention, we set v to be the left child of r. For any node $u \neq r$ in T, let the parent of u be w and let (a, b) be the common edge of the two faces of G that correspond to the vertices u and w. Let the vertices on the triangular face corresponding to u be a, b, c in clockwise order. Let f and f' be the triangular faces (if any) other than abc that contain the edges (a, c) and (b, c), respectively. Then the dual nodes of f and f' are the left and right children of u, respectively.

We now prove the correctness of Step 1. For any maximal outerplanar graph G with $n \geq 6$ vertices we construct another maximal outerplanar graph G' with n vertices such that $\mathcal{N}(G') \geq \mathcal{N}(G)$ and the number of vertices of degree three in the weak dual of G' is less than the number of vertices of degree three in the weak dual of G.

We first examine the properties of Hamiltonian paths in G. Let the number of vertices of degree three in T be x, where $x \geq 1$. Let abc be a face of G such that no edge of abc is an outer edge. Then the dual node v of abc must be a vertex of degree three in T. See Figure 2. Let G_1 and G_2 be the two split graphs of G with respect to the separating edge $\{a, b\}$, where G_1 contains the vertex c. Let p be any vertex in G_2 other than a, b. Since $\{a, b\}$ is a separating edge in G, any Hamiltonian path starting at p must contain a subpath, which is a Hamiltonian path of $G_1, G_1 - \{a\}$ or $G_1 - \{b\}$.

For convenience, we redefine T as an ordered rooted tree, where the root corresponds to some face in G_2. We now compute $h(G_1)_a, h(G_1)_b$, and then $h(G_1 - \{b\})_a$, $h(G_1 - \{a\})_b$. We need the following lemma, whose proof is omitted due to space constraints.

Lemma 1. *Let G be a maximal outerplanar graph and let (a, b) be an outer edge of G. Then there exists a Hamiltonian path in $G - \{a\}$ starting from b that ends at a vertex of degree two in G.*

Let G_L (resp., G_R) be the subgraph of G that contains the left (resp., right) subtree of v as its weak dual. We now compute $h(G_1)_a$ and $h(G_1)_b$ considering the following cases.

(a) The Hamiltonian paths that start at a, visit the vertices in G_L along the outer face ending at c, and then visit G_R starting at c.
(b) The Hamiltonian paths that start at a, visit the vertices in G_R along the outer face starting at b and ending at c, and then visit the vertices in $G_L - \{a, c\}$.
(c) The Hamiltonian paths that start at b, visit the vertices in G_R along the outer face ending at c, and then visit the vertices in G_L starting at c.
(d) The Hamiltonian paths that start at b, visit the vertices in G_L along the outer face starting at a and ending at c, and then visit the vertices in $G_R - \{b, c\}$.

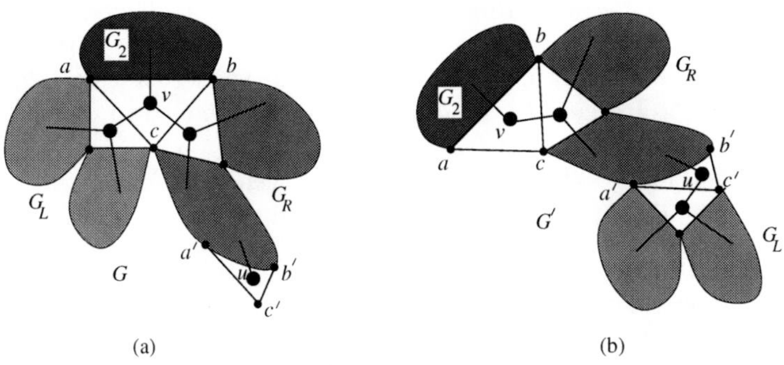

Fig. 2. Illustration for (a) G, and (b) G'

Therefore, $h(G_1)_a = h(G_R)_c + h(G_L - \{a\})_c$ and $h(G_1)_b = h(G_L)_c + h(G_R - \{b\})_c$.

In the following we construct the graph G'. By Lemma 1, at least one Hamiltonian path in $G_R - \{b\}$ starts from c and ends at a vertex of degree two in $G_R - \{b\}$. Let that vertex be c' and let the vertex just before c' on that path be a'. Let u be the dual node of the face $a'b'c'$ of G. See Figure 2(a). Take a copy of G, remove all the vertices of G_L other than a and c from that copy. Let the resulting graph be X. Now take a copy of G_L and merge the vertices a and c of G_L with the vertices a' and c' of X, respectively, and then remove any resulting multi-edges. We denote the resulting graph by G'. See Figure 2(b). Let the weak dual of G' be T'. Observe that the construction of G' can be described by an operation on T as follows: remove the left subtree of v and add that subtree as a subtree of u. We call this operation *child swap*. Since u and v are vertices of degree two in T', the number of vertices of degree three in T' is $x-1$.

Let G'_1 be the subgraph of G', where the weak dual of G'_1 is the subtree of T' rooted at v. Observe that the two split graphs of $G'_1 - \{a\}$ with respect to $\{a', c'\}$ consist of a copy of G_L and a copy of G_R. For simplicity, we use the same notation, i.e., G_L and G_R, to denote those split graphs. We now compute $h(G'_1)_a$ and $h(G'_1)_b$ considering the following cases.

(a) The paths that start at a and then visit the vertices in G_R starting at c. For every such path that does not end at c', we replace the subpath $\langle a', c' \rangle$ (resp., $\langle c', a' \rangle$) with the outer face of G_L starting at a' and ending at c' (resp., starting at c' and ending at a'). For every path that ends at c', we extend that path along the outer face of $G_L - \{a'\}$. Therefore, the number of such Hamiltonian paths is at least $h(G_R)_c$.

(b) The paths that start at a and then visit the vertices in G_R starting at b. For every such path we replace the subpath $\langle a', c' \rangle$ (resp., $\langle c', a' \rangle$) with the outer face of G_L starting at a' and ending at c' (resp., starting at c' and ending at a'). If we visit c after b, then by construction of G', at least one Hamiltonian path in $G_R - \{b\}$ that starts from c must end at c'. Recall that the path visits a' just before c'. Therefore, we can take the sequence a to a' of that path and extend it in $h(G_L)_{a'}$ ways. Otherwise, we start from b and visit the outer face of $G'_1 - \{a\}$ ending at c. Therefore, the number of Hamiltonian paths is at least $h(G_R - \{b\})_c - 1 + h(G_L)_{a'} + 1 = h(G_R - \{b\})_c + h(G_L)_{a'}$.

(c) The paths that start at b, visit a, then visit the vertices in $G_R - \{b\}$ starting at c. As in Case 2, i.e., (b), these paths can be extended to at least $h(G_R - \{b\})_c - 1 + h(G_L)_{a'}$ Hamiltonian paths in G'_1.

(d) The Hamiltonian path that starts at b and then visits the outer face of G'_1 ending at a. This Hamiltonian path is unique by Fact 1.

Before the child swap operation we relabel the vertices a, c of G in the following way so that after child swap $h(G_L)_{a'} \geq h(G_L)_{c'}$ holds. If $h(G_L)_a < h(G_L)_c$, we swap the labels of the vertices a and c. In the case when we do not change labels, $h(G_L)_{a'} = h(G_L)_a > h(G_L - \{a\})_c$. Otherwise, $h(G_L)_{a'} = h(G_L)_c > h(G_L)_a > h(G_L - \{a\})_c$. Therefore, $h(G'_1)_a \geq h(G_R)_c + h(G_R - \{b\})_c + h(G_L)_{a'} \geq h(G_1)_a$, and $h(G'_1)_b \geq h(G_R - \{b\})_c + h(G_L)_{a'} \geq h(G_1)_b$.

Similarly, we can compute that $h(G_1 - \{a\})_b = h(G_L - \{a\})_c$, $h(G_1 - \{b\})_a = h(G_R - \{b\})_c$, $h(G'_1 - \{a\})_b \geq h(G_R - \{b\})_c + h(G_L)_{a'}$, and $h(G'_1 - \{b\})_a \geq h(G_R - \{b\})_c$. Therefore, $h(G'_1 - \{b\})_a \geq h(G_1 - \{b\})_a$ and $h(G'_1 - \{a\})_b \geq h(G_1 - \{a\})_b$.

Recall that for any vertex $p \in V(G_2 - \{a, b\})$, any Hamiltonian path starting at p must contain a subpath, which is a Hamiltonian path of $G_1, G_1 - \{a\}$ or $G_1 - \{b\}$. We have proved that in each of these cases, the number of such subpaths in G' is greater than or equal to the number of such subpaths in G. Therefore, for any vertex $p \in V(G_2 - \{a, b\})$, $h(G')_p \geq h(G)_p$. We use the above technique to prove the following theorem.

Theorem 1. *For any positive integer $n \geq 6$, there exists an outerplanar graph G such that the weak dual of G is a path and $\mathcal{N}(G)$ is the maximum over all possible outerplanar graphs of n vertices.*

Proof (Outline). Let Y be a graph whose weak dual is not a path and $\mathcal{N}(Y)$ is the maximum over all possible outerplanar graphs of n vertices. Suppose for a contradiction that there is no graph G whose weak dual is a path and $h(G)_x \geq \mathcal{N}(Y)$, for some vertex x in G.

Let T be the weak dual of Y. Since T is not a path, there is at least one node v of degree three in T. Let abc be the face of Y that corresponds to v, where the vertices a, b, c appear on the face abc in clockwise order. Let S_{ab} be the set of outer vertices between a and b in clockwise order on the outer face of Y. Define sets S_{bc} and S_{ca} in a similar way. Let w be an ace vertex of Y.

If w belongs to S_{ab}, S_{bc} or S_{ca}, then by the child swap operation we can construct a graph Y' from Y such that $h(Y')_w \geq \mathcal{N}(Y)$ and the number of vertices of degree three in the weak dual of Y' is one less than that of T. Otherwise, $w \in \{a, b, c\}$. Also in this case, we can prove that $h(Y')_w \geq \mathcal{N}(Y)$; a detailed proof is omitted due to space constraints. We apply the process repeatedly on the resulting graph to construct a graph G whose weak dual is a path and $h(G)_x \geq \mathcal{N}(Y)$, for some vertex x in G, a contradiction. □

We now prove the correctness of Step 2. Let G be a maximal outerplanar graph with $n \geq 3$ vertices and let T be its weak dual. Let T be a path $\langle r, u_1, u_2, \ldots, u_{n-3} \rangle$ rooted at r. By definition, u_1 is the left child of r. For each i, $1 \leq i \leq n-4$, assume that u_{i+1} is the left child of u_i if i is even, and right child of u_i otherwise. We then call G a *ZigZag outerplanar graph.* See Figure 3(d). Let G' be another outerplanar graph of n vertices such that the weak dual T' of G' is a path $v_1, v_2, \ldots, v_{n-2}$ rooted at v_1. If G' is not a ZigZag outerplanar graph, then there is a subpath $\langle v_{i-1}, v_i, v_{i+1} \rangle$, $1 < i < n-2$, such that either each of v_i and v_{i+1} is the left child of their parents, or both of them are the respective right children of their parents. We call such a subpath a *repeated ancestry.* Without loss of generality, suppose that both of v_i and v_{i+1} are the respective left children of their parents and the child of v_{i+1} is a right child, if any. We construct a graph G'' from G', by applying one of the following *flip* operations such that the number of repeated ancestries in the weak dual T'' of G'' is at least one less than the number of repeated ancestries in T'. See Figures 3(a)–(c).

Child Flip: Let Y be the subgraph of G' that contains the subtree rooted at v_i as its weak dual. This operation takes a copy of G' and replaces the subgraph Y with a mirror copy of Y. This construction can be described by T' as follows: for each node y in the subtree rooted at v_i, this operation flips the left-right order of the child of y. See Figure 3(b), where $v = v_i$ and $w = v_{i+1}$.

Parent Flip: Let (a, e) be the common edge of the faces that correspond to the vertices v_{i-1} and v_i of T'. Let the two split graphs with respect to $\{a, e\}$ be Y and Z, where the weak dual T_1 of Y is rooted at v_i. Let z be the leaf of T_1 and let the vertices of Y on the face corresponding to z be a', b', c' in clockwise order such that c' is a vertex of degree two. If z is the left child of its parent then we connect the weak dual of Z rooted at v_{i-1} as a right subtree of z, by merging vertices e, a to the vertices b', c', respectively. Otherwise, we merge the vertices e, a to the vertices a', c', respectively. See Figure 3(c), where $u = v_{i-1}$, $v = v_i$ and z is the right child of its parent.

Lemma 2. *Let G be an outerplanar graph, where the weak dual of G is a path with at least one repeated ancestry. Then there exists an outerplanar graph G' that can be obtained from G by a single flip operation such that $\mathcal{N}(G') \geq \mathcal{N}(G)$.*

We apply flip operations on G repeatedly using Lemma 2. Since at each step the number of repeated ancestries decreases, we finally obtain a ZigZag outerplanar graph. Consequently, we have the following theorem.

Theorem 2. *Let G be a ZigZag outerplanar graph with $n \geq 6$ vertices. Then $\mathcal{N}(G)$ is the maximum among all possible outerplanar graphs with $n \geq 6$ vertices.*

We now prove the correctness of Step 3. Let G be a ZigZag graph with n vertices. See Figure 3(d). We now give a bound on $\mathcal{N}(G)$. For any $n \geq 6$, a ZigZag graph of n vertices has exactly two vertices of degree two and exactly two vertices of degree three; all the other vertices are of degree four. Let a, y and b, x be the vertices of degree two and degree three in G, respectively. Using the zigzag structure of G, it is straightforward to observe that $h(G)_a = h(G)_y$ and $h(G)_b = h(G)_x$, independent of the parity of n. Let $\{h_n\}_i, i \in \{2, 3, 4\}$, be the maximum number of Hamiltonian paths in a ZigZag graph of n vertices starting from any vertex of degree i.

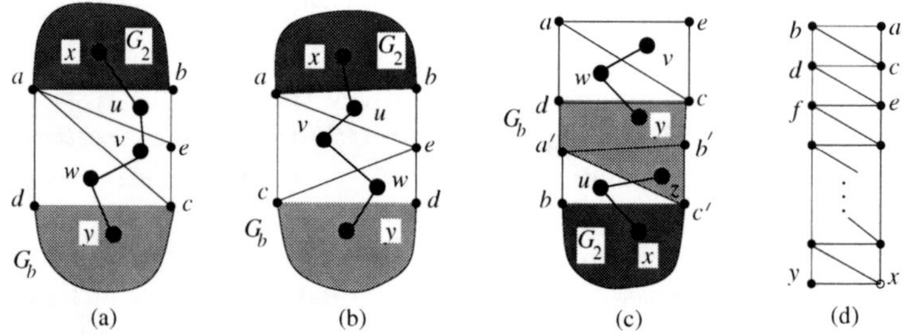

Fig. 3. (a) G'. (b) G'', which is obtained by a child flip on G'. (c) G'', which is obtained by a parent flip on G'. (d) A ZigZag graph.

We first compute $\{h_n\}_2$. Without loss of generality we compute the number of Hamiltonian paths starting at a in G. Any Hamiltonian path that starts from a, chooses either b or c as the next vertex to visit. If the next vertex is b, then the number of such Hamiltonian paths will be equal to the number of Hamiltonian paths in the ZigZag graph $G - \{a\}$ starting at a vertex of degree two, which is $\{h_{n-1}\}_2$. If the next vertex is c, then there are two ways to choose the next vertex. If we visit vertex b and then vertex d, then the number of Hamiltonian paths will be equal to the number of Hamiltonian paths in the ZigZag graph $G - \{a, b, c\}$ starting at a vertex of degree two, which is $\{h_{n-3}\}_2$. Otherwise, we have to visit e after c and then we can complete a Hamiltonian path in only one way, i.e., by visiting the vertices along the outer face. Therefore, $\{h_n\}_2 = \{h_{n-1}\}_2 + \{h_{n-3}\}_2 + 1$. The solution to this recurrence is bounded by $O(\alpha^n)$, where $\alpha \approx 1.46557$ is the unique real root of $\alpha^3 = \alpha^2 + 1$. This recurrence establishes a lower bound of $\Omega(\alpha^n)$ on the maximum number of Hamiltonian paths in an outerplanar graph of n vertices, as follows.

Observe that $\{h_n\}_2 > \{h_{n-1}\}_2 + \{h_{n-3}\}_2$. We claim that $\{h_n\}_2 > \alpha^{n-1}$. The case when $n \in \{6, 7, 8\}$ is straightforward since $\{h_6\}_2 = 9 > \alpha^5$, $\{h_7\}_2 = 14 > \alpha^6$ and $\{h_8\}_2 = 21 > \alpha^7$. Assume that for all k, where $8 < k < n$, $\{h_k\}_2 > \alpha^{k-1}$. Now $\{h_n\}_2 > \{h_{n-1}\}_2 + \{h_{n-3}\}_2 > \alpha^{n-2} + \alpha^{n-4} = \alpha^{n-4}(\alpha^2 + 1) = \alpha^{n-1}$. Consequently, $\{h_n\}_2 \in \Omega(\alpha^n)$.

We compute $\{h_n\}_3$ and $\{h_n\}_4$, and prove that $\{h_n\}_2 > \{h_n\}_3$ and $\{h_n\}_2 > \{h_n\}_4$. We thus have the following theorem.

Theorem 3. *The number of Hamiltonian paths in any outerplanar graph with n vertices is $O(n\alpha^n)$. Furthermore, there exists a maximal outerplanar graph with n vertices that contains $\Omega(\alpha^n)$ Hamiltonian paths.*

4 Hamiltonian Cycles in Planar Graphs

In this section we modify the idea of the proof of Buchin et al. [2] to obtain an improved upper bound on the number of Hamiltonian cycles in planar graphs.

Since the number of simple cycles in a planar graph G is an upper bound on the number of Hamiltonian cycles in G, we first find a recurrence relation for the number of simple cycles in G using a similar argument as in [2, Lemma 1]. We then simplify that recurrence relation to obtain an upper bound on the number of Hamiltonian cycles. Unlike Buchin et al., we impose some restrictions on the cycles that we count, as shown in the following lemma. We will need the concept of *cycle-path*, which is a simple path in G that can be completed to a simple cycle in G.

Lemma 3. *Let $G = (V, E)$ be a maximal plane graph with $n \geq 3$ vertices. For each vertex $v \in V$, partition the edges incident to v into two non-empty sets s_v and s'_v, which are local to v, such that the edges in each set appear consecutively around v in clockwise order. Let $H(G)$ be the number of restricted Hamiltonian cycles in G, where a Hamiltonian cycle h is called restricted if for every vertex v, the two edges that are incident to v in h do not belong to the same set s_v or s'_v. Then $H(G) = O(n2^n)$.*

Proof. A cycle-path P of G is called restricted if for every internal vertex v of P, the two edges that are incident to v in P do not belong to the same set s_v or s'_v. Figure 4(a) illustrates a restricted cycle-path.

Every edge $e \in E$ can have two orientations, which we denote by e' and e''. We first count the number of cycle-paths starting at a fixed edge $e \in E$.

Let P' and P'' be the upper bounds on the total number of restricted cycle-paths starting from e with orientation e' and e'', respectively. Then P' and P'' must be the same by the symmetry of the edge orientations. To simplify the explanation, assume that the total number P of restricted Hamiltonian cycles starting from e is $\min\{P', P''\}$. Without loss of generality we give the starting edge the orientation e'.

We associate restricted cycle-paths with the nodes of a tree. The root of the tree contains the path of length one corresponding to the starting edge. The children of a tree node contain paths starting with the path stored in the predecessor plus an additional edge. Every restricted cycle-path is stored in only one tree node. The children of a tree node ϕ are defined as follows: If the oriented path arrives to a vertex v, and the last edge of the oriented path belongs to s_v (respectively, s'_v), then the children of that tree node consist of the edges in s'_v (respectively, s_v).

No matter which child we choose to continue the path, we will mark all the faces incident to v so that we can avoid reconsidering these faces while continuing the path. Let k_v be the number of unmarked faces that lie among the edges corresponding to the children of ϕ. Then the number of children of ϕ is $k_v + 1$. Figure 4(b) gives an example of ϕ and its children. It is now straightforward to verify that P' can be expressed by the following recurrence:

$$P'(n, f) \leq (k_v + 1) \cdot P'(n - 1, f - k_v) + 1, \tag{1}$$

where f is the number of faces in G and $P'(i, j)$ is the number of cycle-paths in G with i unvisited nodes and j unmarked faces.

Since we want to maximize the number of nodes in the recursion tree, we can assume that the k_vs for all v within a level l of the tree are equal [2, Lemma 1]. Let k_l be the number k_v for the vertices v on level l.

$P' = P'(n - 2, 2n - 6)$ will give us the number of nodes in the tree, where $P'(0, \cdot) = P'(\cdot, 0) = 1$. Observe that for each oriented cycle, we can define another cycle with the opposite orientation. Define a term k'_l analogous to the term k_l for the cycle with opposite direction. Then all k_ls and k'_ls have to be non-negative numbers. Now the k_ls and k'_ls along a cycle have to fulfill the condition $\sum_{l \leq L}(k_l + k'_l) \leq 2n - 6$, where $L \leq n - 1$. We now bound the number of restricted Hamiltonian cycles P as follows:

$$P \leq \min\{P', P''\} \leq \min\left\{\sum_{L=1}^{n-1} \prod_{l \leq L}(k_l + 1), \sum_{L=1}^{n-1} \prod_{l \leq L}(k'_l + 1)\right\}. \tag{2}$$

We are interested in a set k_l which maximizes (1). Due to the convexity of $\prod_{1 \leq l \leq n-1}(k_l + 1)$ (respectively, $\prod_{1 \leq l \leq n-1}(k'_l + 1)$), the maximum will be attained when all k_l (respectively, all k'_l) are equal. To maximize Equation (2), we now need to maximize k_l or k'_l. Since $\sum_{1 \leq l \leq n-1}(k_l + k'_l) \leq 2n - 6$ holds and we are taking $\min\{P', P''\}$, Equation (2) is maximized when k_l and k'_l are equal[1].

[1] Since we are bounding only the number of restricted Hamiltonian cycles, we can safely ignore the effect of restricted cycle-paths that are not Hamiltonian.

Consequently, $k_l = k'_l = \frac{2n-6}{2(n-1)}$ and

$$P \leq \prod_{l=1}^{n-1}(k_l + 1) = \prod_{l=1}^{n-1}(\frac{2n-6}{2(n-1)} + 1). \tag{3}$$

We ignore the summation since P is an upper bound only on the number of restricted Hamiltonian cycles. Therefore, the exponential growth of the maximum number of restricted Hamiltonian cycles that contains the edge e is 2^n. □

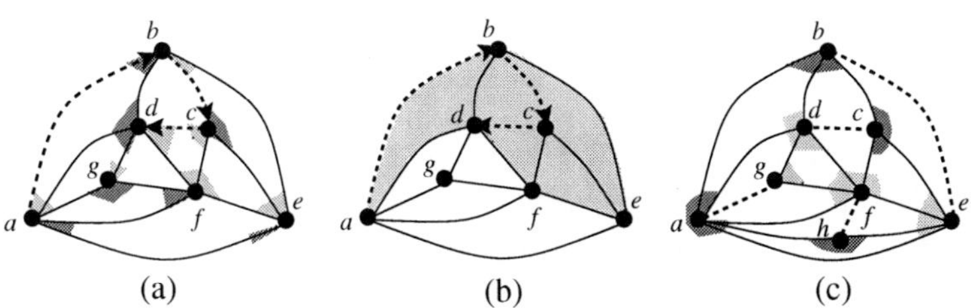

Fig. 4. (a) A restricted cycle-path in G, which is shown in dashed line. For each vertex v, the sets s_v and s'_v are shown in dark-gray and light-gray, respectively. Assume that $\phi = (a, b, c, d)$. Then $(d, b), (d, a)$ and (d, g) are the candidates for the children of ϕ. (b) The faces that are marked are shown in light-gray. Since the faces incident to (d, b) are already marked, we only consider (a, b, c, d, a) and (a, b, c, d, g) as the children of ϕ. Therefore, $k_v + 1 = 2$. (c) Illustration for the proof of Theorem 4, where the edges in M are shown with dashed lines.

Theorem 4. *The exponential growth of the maximum number of Hamiltonian cycles in a planar graph with n vertices is 2.2134^n.*

Proof. First consider the case when n is even. Any Hamiltonian cycle in a maximal planar graph G with an even number of vertices splits into two non-intersecting perfect matchings. We now count the number of ways that a perfect matching M in G can be extended to a Hamiltonian cycle. Let e be an edge of M, where x and y are the end vertices of e. We define two sets s_e and s'_e, which are local to e, such that s_e and s'_e consists of the edges incident to x and y, respectively. We define such pair of sets for every edge in M, as shown in Figure 4(c). Observe that each edge in M plays the role of a single vertex and hence the number of ways that M can be extended to a Hamiltonian cycle is equal to the number of restricted Hamiltonian cycles in G. We count these restricted Hamiltonian cycles in a way similar to Lemma 3 by modifying the parameters n, f, k_l, k'_l as follows.

For each edge e in M we mark the faces adjacent to e. Since for any two edges $\{e_1, e_2\} \subseteq M$, the pair of faces incident to e_1 and the pair of faces incident to e_2 are different, the number of unmarked faces in G is $(2n - 4) - n = n - 4$. Since each edge in M play the role of a single vertex, $k_l = k'_l = \frac{n-4}{2.(n/2)}$. Therefore, the the number of ways that M can be extended to a Hamiltonian cycle is

$$O(n) \cdot \prod_{l=1}^{n/2-1} (k_l + 1) = O(n) \cdot \prod_{l=1}^{n/2-1} (\frac{n-4}{n} + 1) = O(n2^{n/2}). \tag{4}$$

Observe that the upper bound on the number of Hamiltonian cycles in G is the number of perfect matchings $\mathcal{M}$ in G times $O(n2^{n/2})$. Since $\mathcal{M} \leq 6^{n/4}$ [2], the exponential growth of the maximum number of Hamiltonian cycles in a planar graph with n vertices is $6^{n/4} \cdot 2^{n/2} < 2.2134^n$. The case when n is odd can be dealt in a similar way as in [2, Theorem 4]. □

5 Conclusion

In this paper we have given an 2.2134^n upper bound on the number of Hamiltonian cycles in planar graphs. We have also proved an $O(n\alpha^n)$ upper bound and an $\Omega(\alpha^n)$ lower bound on the number of Hamiltonian paths in outerplanar graphs, where $\alpha \approx 1.46557$. It would be interesting to examine whether the techniques of this paper can be extended to establish similar results for planar graphs with bounded treewidth.

We have proved the 2.2134^n upper bound on the number of Hamiltonian cycles in a planar graph under certain assumptions on the recursion tree determined by Equation (1). It would be nice to have an alternative proof that achieves the same upper bound, but does not use any such assumptions.

References

1. Bax, E.T.: Inclusion and exclusion algorithm for the Hamiltonian path problem. Information Processing Letters 47(4), 203–207 (1993)
2. Buchin, K., Knauer, C., Kriegel, K., Schulz, A., Seidel, R.: On the number of cycles in planar graphs. In: Lin, G. (ed.) COCOON 2007. LNCS, vol. 4598, pp. 97–107. Springer, Heidelberg (2007)
3. Collins, K.L., Krompart, L.B.: The number of Hamiltonian paths in a rectangular grid. Discrete Mathematics 169(1-3), 29–38 (1997)
4. Curran, S.J., Gallian, J.A.: Hamiltonian cycles and paths in Cayley graphs and digraphs - A survey. Discrete Mathematics 156(1-3), 1–18 (1996)
5. de Mier, A., Noy, M.: On the maximum number of cycles in outerplanar and series-parallel graphs. Electronic Notes in Discrete Mathematics 34, 489–493 (2009)
6. Eppstein, D.: The traveling salesman problem for cubic graphs. Journal of Graph Algorithms and Applications 11(1), 61–81 (2007)
7. Garey, M., Johnson, D.: Computers and Intractability: A Guide to the Theory of NP-Completeness. W. H. Freeman and Company (1979)
8. Gebauer, H.: Finding and enumerating Hamilton cycles in 4-regular graphs. Theoretical Computer Science 412(35), 4579–4591 (2011)
9. Pichler, R., Rümmele, S., Woltran, S.: Counting and enumeration problems with bounded treewidth. In: Clarke, E.M., Voronkov, A. (eds.) LPAR-16 2010. LNCS (LNAI), vol. 6355, pp. 387–404. Springer, Heidelberg (2010)

Feedback Vertex Sets on Tree Convex Bipartite Graphs

Chaoyi Wang[1], Tian Liu[1,⋆], Wei Jiang[1], and Ke Xu[2,⋆]

[1] Key Laboratory of High Confidence Software Technologies, Ministry of Education, Institute of Software, School of Electronic Engineering and Computer Science, Peking University, Beijing 100871, China
{wchaoyi,lt}@pku.edu.cn

[2] National Lab of Software Development Environment, Beihang University, Beijing 100191, China
kexu@nlsde.buaa.edu.cn

Abstract. A feedback vertex set in a graph is a subset of vertices, such that the complement of this subset induces a forest. Finding a minimum feedback vertex set (FVS) is $\mathcal{NP}$-complete on bipartite graphs, but tractable on chordal bipartite graphs. A bipartite graph is called *tree convex*, if a tree is defined on one part of the vertices, such that for every vertex in the other part, the neighborhood of this vertex induces a subtree. First, we show that chordal bipartite graphs form a proper subset of tree convex bipartite graphs. Second, we show that FVS is $\mathcal{NP}$-complete on the tree convex bipartite graphs where the sum of the degrees of vertices whose degree is at least three on the tree is *unbounded*. Combined with known tractability where this sum is *bounded*, we show a dichotomy of complexity of FVS on tree convex bipartite graphs.

1 Introduction

A *feedback vertex set* in a graph $G = (V, E)$ is a subset D of vertices, such that there is no cycle in the subgraph induced by $V \setminus D$. Finding a minimum feedback vertex set (FVS, in short) is among Karp's twenty one classical $\mathcal{NP}$-complete combinatorial optimization problems [5]. FVS is also $\mathcal{NP}$-complete on bipartite graphs [9], but tractable on convex bipartite graphs [7] and chordal bipartite graphs [6]. A good survey on tractability of FVS in various graph classes as well as on various kind of FVS algorithms before this century is [2].

Recently, the so-called tree convex bipartite graphs were introduced as a generalization to convex bipartite graphs [3,4,8]. A bipartite graph $G = (A, B, E)$ is called *tree convex*, if a tree $T = (A, F)$ is defined, such that for every vertex b in B, the neighborhood of b induces a subtree in T. For T as a path or a star, G is called *convex* or *star convex*, respectively. For T as a triad, which consists

⋆ Corresponding authors. Partially supported by National Natural Science Foundation of China (Grant No. 60973033) and National 973 Program of China (Grant No. 2010CB328103). We are grateful to the unknown reviewers whose comments are helpful to improve our presentations.

G. Lin (Ed.): COCOA 2012, LNCS 7402, pp. 95–102, 2012.

of three paths with a common end, the bipartite graph is called triad convex. It was shown that FVS is $\mathcal{NP}$-complete on star convex bipartite graphs [3], but becomes tractable on triad convex bipartite graphs and even on tree convex bipartite graphs where the sum of the degrees of vertices whose degree is at least three in T is bounded by a constant [4]. These results refined both the known intractability [9] and tractability [7] results of FVS on bipartite graphs. For some applications and other recent progresses on FVS, see introductions in [3,4,8,6].

With these progresses in mind, one naturally asks the following two questions

- *What is the relationship between tree convex bipartite graphs and other well known restricted bipartite graphs, such as the chordal bipartite graphs?*
- *What is the boundary between intractability and tractability of FVS on bipartite graphs?*

In this paper, we give some possible answers to these two questions. We first show that chordal bipartite graphs form a proper subset of tree convex bipartite graphs. Then we show that FVS is $\mathcal{NP}$-complete on the tree convex bipartite graphs where the sum of the degrees of vertices whose degree is at least three on the tree is unbounded. Combined with known tractability results when this sum is bounded [4], this shows a dichotomy of complexity of FVS on tree convex bipartite graphs. Actually, our reduction shows that FVS is $\mathcal{NP}$-complete on tree convex bipartite graphs where the maximum degree of the tree is three.

The remaining part of this paper is organized as follows. The proper containment of the chordal bipartite graphs in the tree convex bipartite graphs is shown in Section 2, and the intractability of FVS on tree convex bipartite graphs is shown in Section 3.

2 Tree Convex Bipartite Graphs versus Chordal Bipartite Graphs

Chordal bipartite graphs have been extensively studied in literature. We refer to [1,6] for different equivalent definitions of the chordal bipartite graphs. A *chord* of a cycle on a graph is an edge between two vertices of the cycle but the edge itself is not a part of the cycle. A graph (not necessarily bipartite) is *chordal* if every cycle of length at least four has a chord. A bipartite graph is *chordal bipartite* if every cycle of length at least six has a chord. Note that in general chordal bipartite graphs are not chordal. Both the tree convex bipartite graphs and the chordal bipartite graphs are generalizations to the convex bipartite graphs.

Theorem 1. *There is a bipartite graph which is tree convex but not chordal bipartite.*

Proof. The bipartite graph G is shown in Figure 1 (left). G is not chordal bipartite, because the cycle $axbzcya$ has no chord. G is tree convex bipartite, because the neighborhoods of a, b, c are $N(a) = \{x, y, w\}$, $N(b) = \{x, z, w\}$ and $N(c) = \{z, y, w\}$, respectively. If we define a tree T on vertices $\{x, y, z, w\}$, which

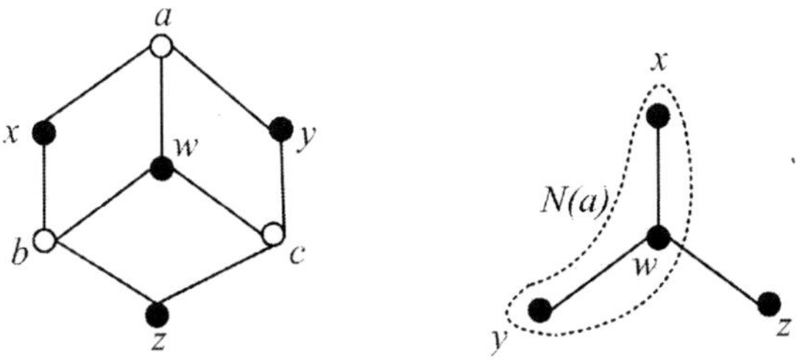

Fig. 1. The bipartite graph G and the tree T

is a star with the center w and three leaves x, y, z, then $N(a) = \{x, y, w\}$ is a subtree of T, as shown in Figure 1 (right). For $N(b)$ and $N(c)$, the situations are similar. □

Theorem 2. *Every chordal bipartite graph is tree convex bipartite.*

Proof. Let $G = (A, B, E)$ be a chordal bipartite graph. From G, we can define a new hypergraph H and a new graph $L(H)$ respectively as follows.

First, we define H. For each vertex b in B, its neighborhood $N(b) = \{a \in A|(a, b) \in E\}$ is a subset of A. The set system $\mathcal{E}$ contains all these neighborhoods of vertices in B, namely $\mathcal{E} = \{N(b)|b \in B\}$. The hypergraph H is defined as $H = (A, \mathcal{E})$. The vertex set of H is A and each $N(b)$ is a hyperedge of H. When G is chordal, we can show that H has the Helly property. A set system $\mathcal{E}$ has the *Helly property*, if for any subsystem $\mathcal{E}'$ of $\mathcal{E}$, whenever the sets in $\mathcal{E}'$ have non-empty pairwise intersections, then all sets in $\mathcal{E}$ have a common element ([1], p.8, Definition 1.3.4). If $\mathcal{E}$ has the Helly property, then $H = (A, \mathcal{E})$ is called a *Helly hypergraph.*

Lemma 1. *If $G = (A, B, E)$ is chordal bipartite, then $H = (A, \mathcal{E})$ is a Helly hypergraph.*

Proof. Let $\mathcal{E}'$ be a subsystem of $\mathcal{E}$. Assume that any two subsets of $\mathcal{E}'$ have a non-empty intersection. We will show that all sets in $\mathcal{E}'$ have a common element. We do induction on the cardinality of $\mathcal{E}'$.

The base step. If $\mathcal{E}'$ has three sets $N(b_1)$, $N(b_2)$, $N(b_3)$, and $a_1 \in N(b_2) \cap N(b_3)$, $a_2 \in N(b_1) \cap N(b_3)$, $a_3 \in N(b_1) \cap N(b_2)$. If any two of a_1, a_2, a_3 are the same, then we have a common element in $N(b_1)$, $N(b_2)$, $N(b_3)$. If a_1, a_2, a_3 are all distinct, then $a_1b_2a_3b_1a_2b_3a_1$ is a cycle of length six in G. Since G is chordal bipartite, the cycle has a chord, say (a_i, b_i) $(1 \leq i \leq 3)$. Then a_i is a common element of sets in $\mathcal{E}'$.

The induction step. Assume that for all $\mathcal{E}'$ of cardinality no more than k $(k \geq 3)$ in which sets are pairwise intersecting, all sets in $\mathcal{E}'$ has a common element. We show that for $\mathcal{E}'$ of cardinality $k+1$ in which sets are pairwise intersecting, all sets in $\mathcal{E}'$ has a common element. Assume that $\mathcal{E}' = \{N(b_i)|i = 1, 2, \ldots, k+1\}$. Let $S = \bigcap\{N(b_i)|i = 3, \ldots, k+1\}$. By the induction hypothesis, $N(b_1) \cap S \neq \emptyset$, $N(b_2) \cap S \neq \emptyset$. By the assumption on $\mathcal{E}'$, $N(b_1) \cap N(b_2) \neq \emptyset$. Similarly, let

$a_1 \in N(b_2) \cap S$, $a_2 \in N(b_1) \cap S$, $a_3 \in N(b_1) \cap N(b_2)$. Suppose that a_1, a_2, a_3 are all distinct, then we have $k-1$ cycles of length 6, that is $a_1b_2a_3b_1a_2b_ia_1$, $i = 3, \ldots, k+1$. If $a_1 \in N(b_1)$ or $a_2 \in N(b_2)$, then $S, N(b_1), N(b_2)$ have a common element. If $a_1 \notin N(b_1)$ and $a_2 \notin N(b_2)$, then all the cycles must have chords a_3b_i. Hence $a_3 \in N(b_i), i = 3, \ldots, k+1$. By definition of S, we obtain $a_3 \in S$. Thus, $S, N(b_1), N(b_2)$ always have a common element, which in turn is a common element of sets in $\mathcal{E}'$. The induction is finished.

Thus H is a Helly hypergraph. □

Next, we define graph $L(H) = (\mathcal{E}, \mathcal{F})$, the line graph or intersection graph of H ([1], p.7-8, Definition 1.3.3), where $\mathcal{F} = \{(N(b_i), N(b_j)) | N(b_i) \cap N(b_j) \neq \emptyset, b_i, b_j \in B\}$. We will show that $L(H)$ is chordal.

Lemma 2. *If $G = (A, B, E)$ is chordal bipartite, then $L(H) = (\mathcal{E}, \mathcal{F})$ is chordal.*

Proof. Assume that $N(b_1)N(b_2) \cdots N(b_k)N(b_1)$ is a cycle of length k ($k \geq 4$) in $L(H)$. By the definition of $L(H)$, there are $a_1, \ldots, a_k$ in A, such that $a_i \in N(b_i) \cap N(b_{i+1})$ for $i = 1, \ldots, k-1$ and $a_k \in N(b_k) \cap N(b_1)$. If all these a_i's are all distinct, then $b_1a_1b_2a_2 \cdots b_ka_kb_1$ is a cycle in G of length at least eight. Since G is chordal bipartite, there is a chord for this cycle of G, say (a_i, b_j) $(1 \leq i, j \leq k)$, then $(N(b_i), N(b_j))$ will be a chord for the cycle of $L(H)$. If some of the a_i's are the same, say $a_i = a_j$, then $(N(b_i), N(b_{j+1}))$ will be a chord for the cycle of $L(H)$. Thus $L(H)$ is chordal. □

A hypergraph H is called a *hypertree*, if there is a tree T with the same vertices of H, such that all hyperedges induce subtrees in T ([1], p.8-9, Definition 1.3.6). A hypergraph H is a hypertree, if and only if H has the Helly property and its line graph $L(H)$ is chordal ([1], p.9, Theorem 1.3.1). Thus, by above two lemmas, the graph H defined from the chordal bipartite G is a hypertree. But H is a hypertree is equivalent to that G is a tree convex bipartite graph, by definitions. □

3 Intractability of FVS on Tree Convex Bipartite Graphs

Theorem 3. *FVS is $\mathcal{NP}$-complete in the tree convex bipartite graphs where the maximum degree of the tree is bounded by three.*

Proof. We reduce from Vertex Cover in general graphs to FVS in the tree convex bipartite graphs where the maximum degree of the tree is bounded by three. Given a graph $G = (V, E)$ with n vertices and m edges, where $V = \{v_1, \ldots, v_n\}$ and $E = \{e_1, \ldots, e_m\}$, without loss of generality, we may assume that G is not a complete graph, that is, there is at least one pair of vertices which are not connected by any edge. We will construct a tree convex bipartite graph $G' = (A, B, E')$ and a tree $T = (A, F)$, such that for every vertex b in B, the neighborhood of b, $N(b) = \{v \in A | (v, b) \in E'\}$, is a subtree on A, and the maximum degree of T is three.

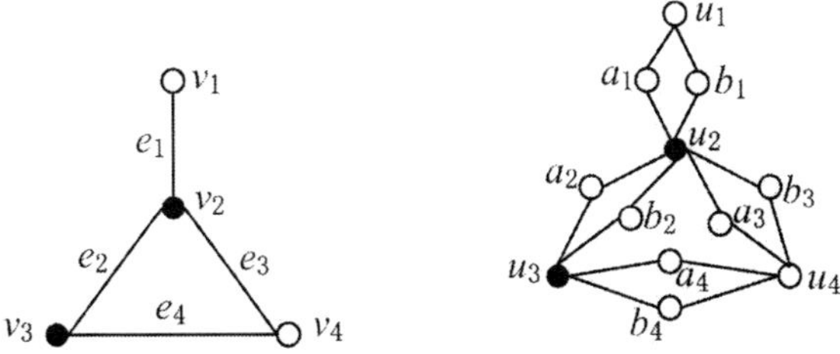

Fig. 2. The input graph G (left) and the constructed graph G' after the first stage (right). The black vertices are solutions to Vertex Cover and FVS, respectively.

The construction will be in four stages. In the first stage, we keep the vertices of graph G unchanged but replace each edges of G with two independent paths both of length two. This stage actually reduces the Vertex Cover to FVS on general bipartite graphs. Figure 2 shows an example of this construction. Formally, after the first stage of the reduction, the constructed graph $G' = (A, B, E')$ has the following form, $A = U = \{u_1, \ldots, u_n\}$, $B = \{a_k, b_k | e_k \in E, k = 1, \ldots, m\}$, and $E' = \{(u_i, a_k), (u_i, b_k), (a_k, u_j), (b_k, u_j) | e_k = (v_i, v_j) \in E, k = 1, \ldots, m\}$. It is easy to see that, for each edge $e_k = (v_i, v_j)$ in G, the corresponding vertices a_k, b_k in G' only appear on a unique cycle $u_i \to a_k \to u_j \to b_k \to u_i$. Then whenever there is a vertex a_k or b_k in a solution to FVS in G', we can safely replace this vertex either by u_i or by u_j without destroying the solution or increasing the size of the solution. Thus, we can assume that the optimal solution of FVS in G' is totally contained in A and has no intersection with B. When the construction is completed in the fourth stage, this property will still hold. Also note that at the moment, the graph G has a minimum vertex cover of size no more than s if and only if G' has a minimum feedback vertex set of size no more than s, in a clear one-to-one correspondence to each other.

In the second stage, we construct a tree T on A whose maximum degree is three and whose vertex set is an extension of U. To this end, we introduce another set of n vertices $W = \{w_1, \ldots, w_n\}$. Now the vertex subset A of G' is changed to $A = U \cup W$ and the other vertex subset B is unchanged. The edges in T are consisted of two parts. The first part of edges connects the vertices in W into a single path of length $n-1$. This part of edges is the set $\{(w_i, w_{i+1}) | i = 1, \ldots, n-1\}$. The second part of edges connects the vertices in U and W in a one-to-one correspondence, namely this part of edges is the set $\{(u_i, w_i) | i = 1, \ldots, n\}$. Formally, the tree $T = (A, F)$, $A = U \cup W = \{u_i, w_i | i = 1, \ldots, n\}$, and $F = \{(w_i, w_{i+1}) | i = 1, \ldots, n-1\} \cup \{(u_i, w_i) | i = 1, \ldots, n\}$. In this way, we not only keep the maximum degree of T no more than three, but also are ready for the next stage to make G' become a tree convex bipartite graph based on the tree T. Figure 3 (left) is an example of the tree T resulted from the input in Figure 2.

In the third stage, we change the graph G' to be a tree convex bipartite graph where the tree T is constructed in last stage. To this end, for each edge $e_k = (v_i, v_j)$ in G ($k = 1, \ldots, m$), without loss of generality, we can assume that

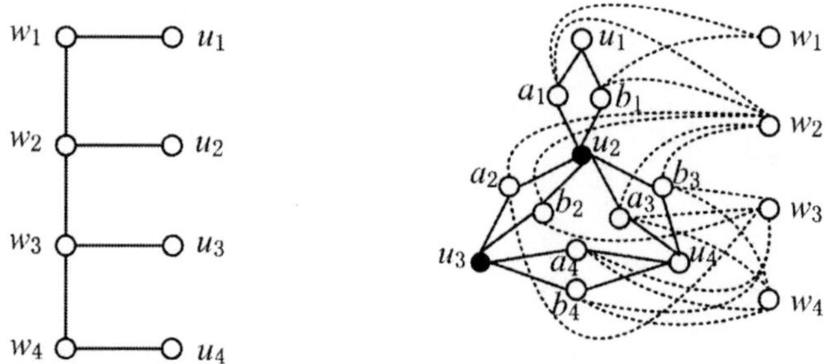

Fig. 3. The tree T (left) and the graph G' after the third stage (right). The edges added in the third stage are shown in dashed lines.

$1 \leq i < j \leq n$, then we add the following set of edges in G', $\{(a_k, w_r), (b_k, w_r) | i \leq r \leq j\}$. This makes G' tree convex bipartite, since for any k $(1 \leq k \leq m)$, if the edge e_k in G has two ends v_i and v_j where we can assume that $i < j$, then the neighborhoods of a_k and b_k in G' are $N(a_k) = N(b_k) = \{u_i, w_i, w_{i+1}, \ldots, w_j, u_j\}$, which induces a path (thus a subtree) on tree T. For example, in Figure 3 (right), the neighborhoods of a_3 and b_3 in G' are $\{u_2, w_2, w_3, w_4, u_4\}$, which are subtrees on T (see Figure 3 (left)).

In the fourth stage, we add more vertices and edges to G' to make sure that the minimum feedback vertex set of G' contains all vertices in W, while keeping G tree convex bipartite. To this end, for each vertex v_i in G $(1 \leq i \leq n)$, we add three new vertices c_i, d_i, f_i to G', and connect each of them to all vertices in $\{u_i\} \cup W$. The added edges form the set $\{(c_i, u_i), (d_i, u_i), (f_i, u_i)\} \cup \{(c_i, w_r), (d_i, w_r), (f_i, w_r) | r = 1, \ldots, n\}$. Figure 4 shows the added vertices and edges for the vertex u_1 only.

Now we have finished the construction of bipartite grpah $G' = (A, B, E')$. The graph G' is still tree convex bipartite, because for any k $(1 \leq k \leq n)$, the neighborhoods of c_k, d_k, f_k are $N(c_k) = N(d_k) = N(f_k) = \{u_k, w_1, w_2, w_3, w_4\}$, which are subtrees on T (see Figure 3 (left)). Moreover, the minimum feedback vertex set of G' contains all vertices in W. This will be shown in the next lemma.

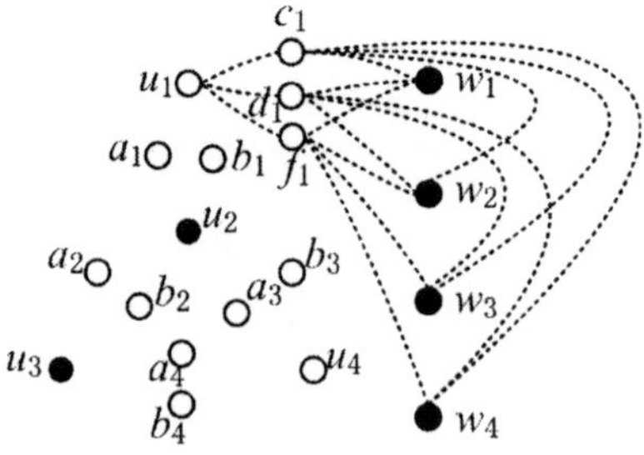

Fig. 4. For vertex u_1, the added three vertices c_1, d_1, f_1 and the added fifteen edges between $\{c_1, d_1, f_1\}$ and $\{u_1, w_1, w_2, w_3, w_4\}$. The edges added are shown in dashed lines. The similar constructions for vertices u_2, u_3, u_4 are not shown.

Lemma 3. *The minimum feedback vertex set of G' contains all vertices in W.*

Proof. Let D be the minimum feedback vertex set of G'. We consider the size of D in the following three cases.

- Case 1: $W \subseteq D$. Since we have assumed that the input graph G is not a complete graph, there is a pair of vertices v_i and v_j which are not connected by any edge. Then $W \cup U \setminus \{u_i, u_j\}$ is a feedback vertex set of G'. This is because when deleting $W \cup U \setminus \{u_i, u_j\}$ from G', all cycles created at the first, third and fourth stages in G' are broken. Since D is the minimum feedback vertex set of G', we have that $|D| \leq |W| + |U| - 2 = 2n - 2$.
- Case 2: $(W \setminus \{w_i\}) \subseteq D$ and $w_i \notin D$ for some $1 \leq i \leq n$. In this case, we must have that $U \subseteq D$. For otherwise, if u_j is not in D for some j $(1 \leq j \leq n)$, then the five vertices u_j, c_j, d_j, f_j, w_i make a complete bipartite graph $K_{2,3}$ by the fourth stage of the construction. Since neither w_i nor u_j is in D, to break the cycles in this small $K_{2,3}$, we have to put at least two of the three vertices c_j, d_j, f_j into D. But this will be a contradiction to the minimum property of D, since instead of at least two of the three vertices c_j, d_j, f_j, we only need to put one vertex u_j into D to broken the cycles in this small $K_{2,3}$. Note that replacing two of the three vertices $\{c_j, d_j, f_j\}$ by u_j will cause no harm to D, since we do not need any vertex in $\{c_j, d_j, f_j\}$ to broken any other cycles. In fact, the other cycles containing $\{c_j, d_j, f_j\}$ have already been broken by $W \setminus \{w_i\}$. Thus, $(U \cup W \setminus \{w_i\}) \subseteq D$ and $|D| \geq 2n - 1$.
- Case 3: $|W \setminus D| \geq 2$. Assume that neither w_i nor w_j is in D for some $1 \leq i \leq j \leq n$. Then for each u_k $(1 \leq k \leq n)$, the five vertices c_k, d_k, f_k, w_i, w_j make a small complete bipartite graph $K_{2,3}$. To broken the cycles in this $K_{2,3}$, D has to contain at least two of the three vertices c_k, d_k, f_k. This holds for each k $(1 \leq k \leq n)$. Thus $|D| \geq 2n$.

Since D is the feedback vertex set in G' with the minimum cardinality, we conclude that the above Case 1 must be true, namely $W \subseteq D$. □

Now the correctness of the reduction is shown by the following two lemmas.

Lemma 4. *If G has a vertex cover of size no more than s, then G' has a feedback vertex set of size no more than $s + n$.*

Proof. Let S be a vertex cover in G of size no more than s, then $\{u_i | v_i \in S\} \cup W$ is feedback vertex set in G' of size no more than $s + n$. Indeed, all the cycles created at the first stage are broken by $\{u_i | v_i \in S\}$ (see the discussions at the end of the description of the first stage), and all the cycles created at the third and fourth stages are broken by W. □

Lemma 5. *If G' has a feedback vertex set of size no more than $s + n$, then G has a vertex cover of size no more than s.*

Proof. Assume that D is the minimum feedback vertex set in G, then $|D| \leq s + n$. By the Lemma 3, we have that $W \subseteq D$. All the cycles created at the third and fourth stages are broken by W. By the discussions in the first stage, all the cycles created at the first stages can be broken by vertices in U, instead of by any vertices in $\{a_k, b_k | k = 1, \ldots, m\}$, without increasing the size of the feedback vertex set. Thus, we can assume that $D \subseteq (U \cup W) = A$. Then the set $S = \{v_i | u_i \in D \setminus W\}$ is a vertex cover in G of size no more than s. For otherwise, there is an edge $e_k = (v_i, v_j)$ in G but neither of v_i and v_j is in S. By the definition of S, we have that neither of u_i and u_j is in D. We have already assumed that neither of a_k and b_k is in D. Thus, the cycle $u_i \to a_k \to u_j \to b_k \to u_i$ in G' is not broken by D, a contradiction to the fact that D is a feedback vertex set. □

The proof of this theorem is finished. □

Note that in above reduction, the sum $t = \sum_{v_i : deg_T(v_i) \geq 3} deg_T(v_i)$ in tree T is unbounded. Thus we have the following intractability result.

Theorem 4. *FVS is $\mathcal{NP}$-complete on tree convex bipartite graphs where the sum $t = \sum_{v_i : deg_T(v_i) \geq 3} deg_T(v_i)$ in tree T is unbounded.*

Combined with the tractability results in [4], we get a dichotomy of complexity of FVS in tree convex bipartite graphs. When t is a constant, the problem is tractable. When t is unbounded, the problem is $\mathcal{NP}$-complete.

A similar dichotomy for Independent Dominating Sets is shown in [8].

References

1. Brandstad, A., Le, V.B., Spinrad, J.P.: Graph Classes - A Survey. Society for Industrial and Applied Mathematics, Philadelphia (1999)
2. Festa, P., Pardalos, P.M., Resende, M.G.C.: Feedback Set Problems. In: Handbook of Combinatorial Optimization (Supplement Volume A), pp. 209–258. Kluwer Academic Publishers (1999)
3. Jiang, W., Liu, T., Ren, T., Xu, K.: Two Hardness Results on Feedback Vertex Sets. In: Atallah, M., Li, X.-Y., Zhu, B. (eds.) FAW-AAIM 2011. LNCS, vol. 6681, pp. 233–243. Springer, Heidelberg (2011)
4. Jiang, W., Liu, T., Xu, K.: Tractable Feedback Vertex Sets in Restricted Bipartite Graphs. In: Wang, W., Zhu, X., Du, D.-Z. (eds.) COCOA 2011. LNCS, vol. 6831, pp. 424–434. Springer, Heidelberg (2011)
5. Karp, R.: Reducibility among combinatorial problems. In: Complexity of Computer Computations, pp. 85–103. Plenum Press, New York (1972)
6. Kloks, T., Liu, C.H., Poon, S.H.: Feedback Vertex Set on Chordal Bipartite Graphs. arXiv:1104.3915v1 (2011)
7. Liang, Y.D., Chang, M.S.: Minimum feedback vertex sets in cocomparability graphs and convex bipartite graphs. Acta Informatica 34, 337–346 (1997)
8. Song, Y., Liu, T., Xu, K.: Independent Domination on Tree Convex Bipartite Graphs. In: Snoeyink, J., Lu, P., Su, K., Wang, L. (eds.) AAIM 2012 and FAW 2012. LNCS, vol. 7285, pp. 129–138. Springer, Heidelberg (2012)
9. Yannakakis, M.: Node-deletion problem on bipartite graphs. SIAM J. Comput. 10, 310–327 (1981)

Crossing Angles of Geometric Graphs*

Karin Arikushi and Csaba D. Tóth

Department of Mathematics and Statistics, University of Calgary, Calgary, AB
{karikush,cdtoth}@ucalgary.ca

Abstract. We study the crossing angles of geometric graphs in the plane. We introduce the *crossing angle number* of a graph G, denoted $\mathrm{can}(G)$, which is the minimum number of angles between crossing edges in a straight line drawing of G. We show that an n-vertex graph G with $\mathrm{can}(G) = O(1)$ has $O(n)$ edges, but there are graphs G with bounded degree and arbitrarily large $\mathrm{can}(G)$. We also initiate studying the *global crossing-angle rigidity* of geometric graphs. We construct bounded degree graphs $G = (V, E)$ such that for any two straight-line drawings of G with the same prescribed crossing angles, there is a subset $V' \subset V$ of $|V'| \geq |V|/2$ vertices that are similar in the two drawings.

1 Introduction

Graphs with n vertices and more than $3n - 6$ edges are not planar and their drawings in the plane have crossing edges. The *crossing number* $\mathrm{cr}(G)$ (resp., *rectilinear crossing number* $\overline{\mathrm{cr}}(G)$) of a graph G is the minimum number of crossings in any drawing (resp., straight-line drawing) of G in the plane. Bienstock and Dean [4] showed that there are families of graphs with unbounded rectilinear crossing number, but crossing number at most k, for any $k \geq 4$. However, $\overline{\mathrm{cr}}(G) = \mathrm{cr}(G)$ if $\mathrm{cr}(G) \leq 3$. Moreover, there are families of bounded degree graphs (even cubic graphs [18]) with arbitrarily large crossing numbers.

Angle conditions on the crossing edges have only been recently considered. The motivation comes from cognitive experiments, which show that having small crossing angles is negatively correlated to path-tracking ability in a graph drawing. Previous research focused on the *crossing resolution* [6] of straight-line (or polyline) drawings, that is, the *minimum* angle at which crossing edges meet. In this paper, we consider the *number* of different angles between crossing edges. A *crossing angle* in a straight-line drawing of a graph is an angle $\alpha, 0 < \alpha \leq \frac{\pi}{2}$, between two crossing edges. The *crossing angle number* of a graph G, denoted $\mathrm{can}(G)$, is the minimum number of crossing angles in a straight line drawing of G. In Section 2, we show that every n-vertex graph G has less than $(6\,\mathrm{can}(G) + 3)n$ edges. We also show that for every $\varepsilon > 0$, there are n-vertex graphs of maximum degree $O(1/\varepsilon)$ such that $\mathrm{can}(G) = \Omega(n^{1/2-\varepsilon})$.

Global Rigidity. A graph $G = (V, E)$ is *globally rigid* in the plane if for every function $\ell : E \to \mathbb{R}^+$, any two straight-line drawings of G in which the Euclidean

* Research supported in part by the NSERC grant RGPIN 35586.

G. Lin (Ed.): COCOA 2012, LNCS 7402, pp. 103–114, 2012.

length of each edge $e \in E$ is $\ell(e)$ are congruent. In other words, the edge lengths determine at most one straight-line drawing up to congruency. For instance, complete graphs are globally rigid. Saxe [19] showed that it is strongly NP-hard to decide whether a graph is globally rigid. Jackson and Jordan [14] gave a simple combinatorial characterization of *generic* global rigidity, where the edge lengths determine at most one straight-line drawing (up to congruency) if the vertices are in general position. They also extended this notion to a so-called *length-direction* rigidity [15], where each edge has either a prescribed length or a prescribed direction vector.

Global Crossing-Angle Rigidity. We adapt the notion of global rigidity to crossing angles. Let $G = (V, E)$ be a graph with a *crossing-angle* function $\alpha : E^2 \to [0, \pi) \cup \{\star\}$. We say that a straight-line drawing of G *complies* with α if for every *crossing* pair of edges (e, f), a counterclockwise rotation through $\alpha(e, f)$ carries the supporting line of e to that of f; and for every *noncrossing* pair of edges, we have $\alpha(e, f) = \star$. In a first approach, we would like to find graphs G where *every* function $\alpha : E^2 \to [0, \pi) \cup \{\star\}$ complies with at most one straight-line drawing up to similarity. This requirement is too strict: we will see that no graph with $n \geq 3$ vertices (not even the complete graph K_n) satisfies this condition. Therefore, we relax this condition, and require α to determine (up to similarity) the locations of at least a constant fraction of the vertices. A graph $G = (V, E)$ is *globally crossing-angle rigid* if for every function $\alpha : E^2 \to [0, \pi) \cup \{\star\}$, and for any two straight-line drawings of G complying with α, there is a vertex set $V'(\alpha) \subset V$ of size $|V'(\alpha)| \geq |V|/2$ such that the two drawings of V' are similar. We prove that K_{24} is globally crossing-angle rigid, and we also construct an infinite family of globally crossing-angle rigid graphs with maximum degree 47 and diameter $O(\log n)$ for $n \geq 24$ vertices.

Extensions and Open Problems. Our result is a first step towards a possible combinatorial characterization of globally crossing-angle rigid graphs. In our definition of globally crossing-angle rigid graphs, α determines at least *half* of the vertices up to similarity, but the choice of the ratio $\frac{1}{2}$ was arbitrary. For every constant $c \in (0, 1)$, there are infinitely many graphs $G = (V, E)$ of maximum degree $\Delta(c)$ where α determines at least $c|V|$ vertices up to similarity. It remains an open problem to find the smallest degree bound $\Delta(c)$ as a function of c. Our crossing angle function $\alpha : E^2 \to [0, \pi) \cup \{\star\}$ encodes the *directed* angle between and *ordered* pair of edges $(e, f) \in E^2$. It would be natural to consider an *undirected* angle crossing function $\beta : \binom{E}{2} \to (0, \pi/2) \cup \{\star\}$ for unordered pairs $\{e, f\} \in \binom{E}{2}$. Our methods can easily be extended to handle this variant of the problem, albeit with higher vertex degrees.

Related Work. Didimo et al. [8] consider graphs that admit straight line drawings where crossing edges meet at a right angle. Such drawings are called *RAC* (right angle crossing) drawings. They prove that graphs with n vertices admitting a RAC drawing have at most $4n - 10$ edges, and this bound is best possible. Argyriou [1] showed that it is NP-hard to decide whether a graph admits a RAC drawing. Refer to [5, 7, 12] for recent results on RAC drawings. Note that if a graph G admits a RAC drawing, then $\mathrm{can}(G) \leq 1$. However, a graph G with $\mathrm{can}(G) = 1$ may have $4.5n - O(\sqrt{n})$ edges. (Let $V(G)$ be a section of a hexagonal lattice, and let $E(G)$ be all hexagon edges and 6 diagonals per hexagon, where diagonals cross at 60°.) Dujmović et al. [10] generalize RAC-drawings and consider so-called α *angle crossing* (αAC) graphs, for $0 < \alpha \leq \frac{\pi}{2}$. These are straight-line drawings where crossing edges meet at an angle *at least* α. They prove that an n-vertex αAC graph has at most $(\pi/\alpha)(3n - 6)$ edges for $0 < \alpha < \pi/2$ and at most $6n - 12$ edges for $2\pi/5 < \alpha < \pi/2$.

The crossing angle number is also related to the *slope number* of a graph G, introduced by Wade and Chu [21]. It is the smallest integer s such that G has a straight line drawing with edges of at most s distinct slopes. Mukkamala and Pálvölgyi [16] show that every cubic graph has slope number at most 4. On the other hand, Pach et al. [17] show that there are graphs of maximum degree $d \geq 5$ with arbitrarily large slope numbers. Dujmović et al. [11] improve the lower bound on slope number for $d \geq 9$ and showed that for every $\varepsilon > 0$, there are Δ-regular graphs with slope number at least $n^{1-(8+\varepsilon)/(\Delta+4)}$. It is unknown whether the slope number is bounded for graphs of maximal degree 4.

2 Graphs with Bounded Crossing Angle Numbers

In a *straight-line drawing* of a graph $G = (V, E)$, the vertices V are represented by distinct points in the plane, and the edges are drawn as line segments between the corresponding vertices that do no pass through any other vertex. A *geometric graph* is a graph $G = (V, E)$ together with a straight-line drawing. The *geometric thickness* [9] of a graph, denoted $\mathrm{gth}(G)$, is the smallest number of layers such that one can draw G in the plane with straight-line edges and assign each edge to a layer so that no two edges on the same layer cross. We establish a relation between the geometric thickness and the crossing angle number.

Theorem 1. *For every graph G, we have* $\mathrm{gth}(G) \leq 2\,\mathrm{can}(G) + 1$.

Proof. Consider a straight-line drawing D of G with $\mathrm{can}(G)$ crossing angles. We begin by decomposing the edges of G into *blocks*. We define a binary relation on the edges of G where two edges are related if and only if they cross in D. The transitive closure of this relation is an equivalence relation. A *block* of G is the union of all edges in an equivalence class.

Let $k = \mathrm{can}(G)$. We partition each block into the union of at most $2k + 1$ subsets, each of which is crossing-free in D. Let $A = \{\alpha_1, \ldots, \alpha_k\}$ denote the set of crossing angles in the drawing D. We construct a (possibly infinite) graph H whose vertices are the elements in $A' = \langle \alpha_1, \ldots, \alpha_k \rangle$, the Abelian group generated by the angles α_i where addition is performed modulo π. Two vertices of H are adjacent if and only if their difference is $\pm\alpha_i$ for some $\alpha_i \in A$.

For a fixed $\alpha \in A$ and $\beta \in A'$, there exists a unique $\beta' \in A'$ such that $\beta - \beta' = \alpha$. Hence, the degree of each $\beta \in H$ is at most $2k$ and there exists a proper coloring of the vertices of H with at most $2k + 1$ colors. Moreover, each color class is an independent set in H. We use the color classes to partition each block of G into planar subgraphs.

Consider a block of G and assume without loss of generality that one edge has slope zero in D. Thus, every edge in the block has a *direction* in $\langle \alpha_1, \ldots, \alpha_k \rangle$. (An edge has *direction* $\alpha \in [0, \pi)$ if it intersects a horizontal line at angle α.) Thus, if β_i and β_j are in a color class of $V(H)$, then edges with direction β_i and β_j do not cross. We may partition the edges in each block independently to obtain a partition of all edges of G into subgraphs, each of which is planar in the drawing D. □

We show that every n-vertex graph with bounded crossing angles number has $O(n)$ edges. Recall that an n-vertex planar graph has at most $3n - 6$ edges.

Corollary 1. *Every n-vertex graph G has at most $(2\,\mathrm{can}(G) + 1)(3n - 6)$ edges.*

Barát et al. [3] proved that for every $\Delta \geq 9$, $\varepsilon > 0$ and $n \in \mathbb{N}$, there is a Δ-regular graph with at least n vertices and geometric thickness $\Omega(\sqrt{\Delta} n^{1/2 - \Delta/4\varepsilon})$.

Corollary 2. *For every $\Delta \geq 9$, $\varepsilon > 0$ and $n \in \mathbb{N}$, there is a Δ-regular graph G with at least n vertices and* $\mathrm{can}(G) = \Omega(\sqrt{\Delta} n^{1/2 - \Delta/4 - \varepsilon})$.

We note here that the crossing angle number of the n-vertex complete graph K_n is less than $\frac{n}{2}$.

Proposition 1. *For $n \geq 2$, we have* $\mathrm{can}(K_n) \leq \lfloor n/2 \rfloor - 2$.

Proof. Consider the straight-line drawing of K_n such that the vertices are represented by the vertices of a regular n-gon with a horizontal side. Let the *direction* of an edge of G be the angle it makes with a horizontal line. The set of edge directions in this drawing is $U = \{i\pi/n : 0 \leq i \leq n/2\}$, and so $|U| = \lfloor n/2 \rfloor$. Hence, the angle between any two edges is in U. However, if the angle is 0, then the two edges are parallel; and if the angle is π/n, then they do not cross. Hence, U yields only $|U| - 2$ crossing angles. □

3 Globally Angle-Rigid Graphs with Bounded Degree

For every $n \geq 24$, we construct a globally crossing-angle rigid graph $F = (V, E)$ with n vertices, bounded vertex degree, and $O(\log n)$ diameter. We start with an auxiliary graph F_0 on a vertex set $V_0 = \{v_1, \ldots, v_{n_0}\}$, for some fixed $n_0 \geq 2$. Let F_0 be a binary tree of diameter $O(\log n_0)$. The graph F is obtained from F_0 by replacing each vertex in V_0 with a clique K_{12}, and replacing each edge of F_0 with a biclique $K_{12,12}$ between the corresponding cliques. The vertex set of F is $V = \bigcup_{i=1}^{n_0} V_i$, where V_i is a set of 12 vertices corresponding to v_i. Hence F has $n = 12n_0$ vertices.

Theorem 2. *Let D_1 be a straight-line drawing of $F = (V, E)$ that complies with a crossing-angle function $\alpha : E^2 \to [0, \pi) \cup \{\star\}$. Then there is a subset $V'(\alpha) \subset V$ of vertices such that (i) $V'(\alpha)$ contains at least 8 vertices from each V_i, $1 \leq i \leq k$, and (ii) for any straight-line drawing D_2 complying with α, a similarity can carry all vertices in $V'(\alpha)$ to the corresponding vertices in D_1.*

For the proof of Theorem 2, we study the crossing-angles of cliques and bicliques in depth. Throughout this section, we assume that we are given a graph $G = (V, E)$ with a crossing-angle function $\alpha : E^2 \to [0, \pi) \cup \{\star\}$, and we consider straight-line drawing complying with α. That is, whenever $\alpha(e, f) = \star$, then the relative interiors of edges e and f are disjoint (although, they may share a common endpoint); and whenever $\alpha(e, f) \in [0, \pi)$, then edges e and f cross at angle $\alpha(e, f)$. We assume that the locations of some vertices are already given and we wish to determine the positions of additional vertices based on the function α. We start with a simple observation.

Proposition 2. *Let $G = (V, E)$ be a graph with edges $pp_1, pp_2, p_1p_2, e_1, e_2 \in E$, and let α be a crossing-angle function such that $\alpha(pp_1, e_1) \neq \star$ and $\alpha(pp_2, e_2) \neq \star$. Then in any straight-line drawing complying with α, the position of p is determined by the slopes of e_1 and e_2, and the positions of p_1 and p_2.*

Proof. The slope of e_1 and the crossing angle $\angle(pp_1, e_1)$ determine the slope of edge pp_1. This slope together with the location of p_1 determines the supporting line of pp_1. Similarly, $\angle(pp_2, e_2)$, the slope of e_2, and the location of p_2 determine the supporting line of pp_2. Note that pp_1 and pp_2 cannot be collinear, otherwise an edge would pass through a vertex in the straight-line drawing of G. Hence, p is the unique intersection point of the supporting lines of pp_1 and pp_2. □

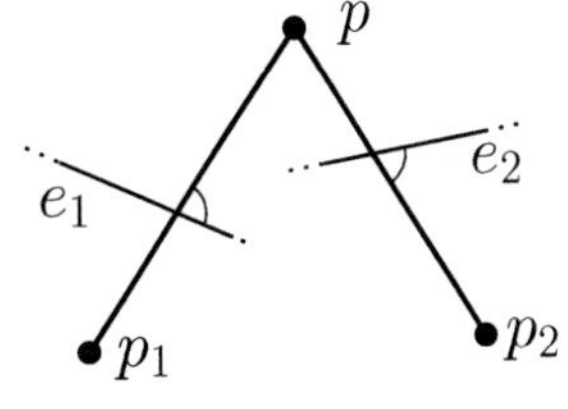

Crossing Free and Almost Crossing Free Vertices. We say that a vertex p of a geometric graph is *crossing free* if no edge incident to p crosses any other edge; and p is *almost crossing free* if exactly one edge incident to p crosses other edges. A point p *convexly avoids* a point set Q if for every $q \in Q$, the segment pq is disjoint from the interior of $\text{conv}(Q)$. Note that if p convexly avoids Q, then it also convexly avoids every subset of Q. Two point sets P and Q are *convexly avoiding* if for every $p \in P$ and $q \in Q$, the segment pq is disjoint from the interior of both convex hulls $\text{conv}(P)$ and $\text{conv}(Q)$. We show that a complete geometric graph has at most one crossing free or almost crossing free vertex.

Proposition 3. *Let $G = (V, E)$ be a complete geometric graph with at least 5 vertices. If $p \in V$ is a crossing free vertex, then it lies on* $\text{conv}(V)$ *and it convexly avoids $V \setminus \{p\}$.*

Proof. Suppose that p lies strictly in the interior of $\text{conv}(G)$, and let q_1, q_2, q_3 be vertices on the boundary of $\text{conv}(G)$ such that $\angle q_1pq_2, \angle q_2pq_3, \angle q_1pq_3 < \pi$. Since $|V| \geq 5$, there is another vertex r of G and w.l.o.g. we may assume that r lies between q_1 and q_2 in the circular order about p. Note that the edge rq_3 must cross either pq_1 or pq_2, otherwise q_1 or q_2 would not lie on the boundary of $\text{conv}(G)$. This contradicts our assumption that p is crossing free.

Now suppose that p does not convexly avoid $V \setminus \{p\}$. Then there are three vertices, say q_1, q_2, and q_3, such that $\text{conv}(p, q_1, q_2, q_3)$ is a convex quadrilateral.

The two diagonals of the quadrilateral cross, hence an edge incident to p crosses another edge of G. This contradicts our assumption that p is crossing free. □

Corollary 3. *Let $G = (V, E)$ be a complete geometric graph with at least 6 vertices. If $p \in V$ is an almost crossing free vertex, then it lies on* conv(V)*, and it convexly avoids a subset of $V \setminus \{p\}$ of size at least $|V| - 2$.* □

Mutually Avoiding Sets. Let P and Q be two point sets in the plane. P *avoids* Q if no supporting line of two points in P intersects conv(Q); and P and Q are *mutually avoiding* if P avoids Q and Q avoids P. Aronov et al. [2] proved that any two point sets, P and Q, of size $|P| = |Q| = n/2$ contain two mutually avoiding subsets $P' \subseteq P$ and $Q' \subseteq Q$ of size $|P'|, |Q'| \geq \sqrt{n/24}$. We strengthen their results when P or Q is in convex position.

Proposition 4. *Let P and Q be disjoint point sets such that every $q \in Q$ convexly avoids a subset of at least $|P| - 1$ elements of P. Then,*

(i) *there is a subset $P' \subseteq P$ of size $|P'| \geq |P| - 3$ such that every $q \in Q$ convexly avoids P', and P' avoids Q;*
(ii) *there are subsets $P'' \subseteq P$ and $Q'' \subseteq Q$, with $|P''| \geq |P| - 1$ and $|Q''| \geq \lceil |Q|/3 \rceil$, such that every $q \in Q''$ convexly avoids P'', and P'' avoids Q''.*

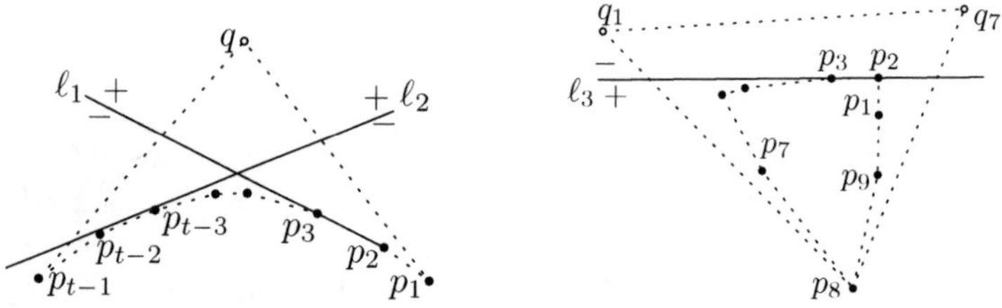

Fig. 1. Illustration for Case 1 (left) and Case 2 (right) in the proof of Proposition 4

Proof. Let $t = |P|$. We distinguish two cases.

Case 1: conv(P) *has $t - 1$ vertices.* In this case, every $q \in Q$ convexly avoids the set of vertices of conv(P). Pick an arbitrary point $q_0 \in Q$. Set conv$(P) = (p_1, \ldots, p_{t-1})$ such that q_0p_1 and q_0p_{t-1} are tangent to conv(P). Let $P' = \{p_2, \ldots, p_{t-2}\}$. It is clear that every $q \in Q$ convexly avoids P'. It remains to show that P' avoids Q. Let ℓ_1 and ℓ_2 be the supporting lines of p_2p_3 and $p_{t-3}p_{t-2}$, resp., (Fig. 1, left). Lines ℓ_1 and ℓ_2 subdivide the plane into 4 wedges. Assume that q lies in the wedge $\ell_1^+ \cap \ell_2^+$. It is clear that every point q' in $\ell_1^+ \cap \ell_2^+$ sees P' in the same order. If $q' \in \ell_1^+ \cap \ell_2^-$ (resp., $q' \in \ell_1^- \cap \ell_2^+$) then $p_{t-2}q'$ (resp., p_1q') intersects the interior of conv(P). For every point $q' \in \ell_1^- \cap \ell_2^-$, segment p_3q' intersects the interior of conv(P). Hence, $Q \subset \ell_1^+ \cap \ell_2^+$, and so P' avoids Q. This proves part (i). For part (ii), notice that every $q \in Q$ sees the vertices of conv(P) in order $(p_1, \ldots, p_{t-1})$, $(p_2 \ldots, p_{t-1}, p_1)$ or $(p_{t-1}, p_1, \ldots, p_{t-2})$, hence the vertices of conv(P) avoid a subset $Q'' \subseteq Q$ of size at least $\lceil |Q|/3 \rceil$.

Case 2: conv(P) has t vertices. Let conv$(P) = (p_1, \ldots, p_t)$ and let B be the set of points $p_i \in P$ such that $p_i q_i$ intersects the interior of conv(P) for some $q_i \in Q$. If $|B| \leq 1$, then *Case 1* applied to $P \setminus B$ completes the proof. If $|B| \geq 2$, then we show that for any two vertices in B, the cyclic distance along conv(P) is at most 2. It follows that $|B| \leq 3$, and we let $P' = P \setminus B$.

Suppose, to the contrary, that $p_1, p_i \in B$, where $4 \leq i \leq t-2$. Then there are $q_1, q_i \in Q$ such that both $p_1 q_1$ and $p_i q_i$ intersect the interior of conv(P). Let ℓ_3 be the supporting line of $p_2 p_3$ such that $p_1 \in \ell_3^+$. Now $q_1, q_i \in \ell_3^-$, since no segment from $\{q_1, q_i\}$ to $\{p_2, p_3\}$ intersects the interior of conv(P) (Fig. 1, right). Let p_j be the furthest point of the convex chain $(p_{i+1}, \ldots, p_t)$ from the line ℓ_3. If $j = t$, then segment $p_{t-1} q_i$ intersects the interior of conv(P). If $j < t$, then $p_t q_1$ intersects the interior of conv(P). Hence, a point in Q convexly avoids a subset of P of size at most $t-2$, contradicting our assumptions. This proves part (i). For part (ii), notice that every $q \in Q$ convexly avoids one of at most three possible subsets of P (namely, subsets $P \setminus \{b\}$ for $b \in B$). Hence some $P'' \subseteq P$ of size $t-1$ avoids a subset $Q'' \subseteq Q$ of size $\lceil |Q|/3 \rceil$. □

Proposition 5. *Let P and Q be disjoint point sets such that every $q \in Q$ convexly avoids P, and P avoids Q. Then, they have subsets $P' \subset P$ and $Q' \subset Q$ of size $|P'| \geq \lceil |P|/2 \rceil$ and $|Q'| \geq \lceil \sqrt{|Q|} \rceil$ such that P' and Q' are mutually avoiding.*

Proof. Let $t = |P|$. Since P and Q are convexly avoiding, the points in $P = \{p_1, \ldots, p_t\}$ are in convex position. Since P avoids Q, every $q \in Q$ sees the same convex arc on the boundary of conv(P), say $(p_1, \ldots, p_t)$ in counterclockwise order. Let ℓ_1 and ℓ_2 be the supporting lines of $p_1 p_2$ and $p_{t-1} p_t$, respectively. The points in Q must lie in the wedge between ℓ_1 and ℓ_2 not containing points of P. Perform an affine transformation on $P \cup Q$ so that ℓ_1 and ℓ_2 are orthogonal and parallel to the coordinate axes. Without loss of generality, assume that P lies in the 1st quadrant and Q lies in the 3rd quadrant.

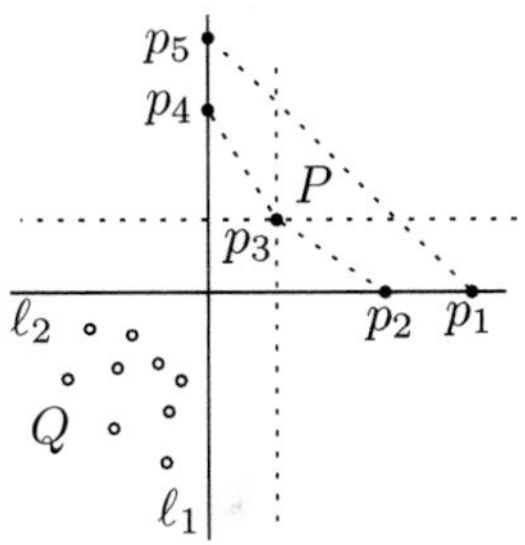

The midpoint m of segment $p_{\lceil t/2 \rceil} p_{\lceil (t+1)/2 \rceil}$ decomposes the convex arc $(p_1, \ldots, p_t)$ into arcs $P_1 = (p_1, \ldots, p_{\lceil t/2 \rceil})$ and $P_2 = (p_{\lceil (t+1)/2 \rceil}, \ldots, p_5)$. Consider the points in Q written in polar coordinates (r, θ) where m is the origin and θ is measured counter-clockwise. First, order $q_i = (r_i, \theta_i) \in Q$ in decreasing distance r_i from p_3. By the Erdős-Szekeres theorem [13], there is a subsequence $Q' = (q_{k_i})$ of length at least $\lceil \sqrt{|Q|} \rceil$ whose angles θ_{k_i} are either increasing or decreasing. Suppose they are increasing. (The argument is analogous if they are decreasing). We show that Q' avoids P_2. If q_{k_i} and q_{k_j} are two points in Q' with $i < j$ then q_{k_j} is to the right of q_{k_i} and below the supporting line of $p q_{k_1}$. Therefore, the supporting line of $q_{k_i} q_{k_j}$ does not intersect conv(P_2). □

Proposition 6. *Let G be the complete graph on the vertex set $V = P \cup Q$, with $|P| \geq 3$ and $|Q| \geq 3$. Let α be a crossing-angle function. Consider a straight-line drawing of G that complies with α such that every $q \in Q$ convexly avoids P, and P and Q are mutually avoiding. Then the positions of the vertices in P uniquely determine the position of a vertex in Q.*

Proof. Let $\{p_1, p_2, p_3\} \subseteq P$ where the vertices are seen in counterclockwise order p_1, p_2, p_3 from any vertex in Q. Similarly, let $\{q_1, q_2, q_3\} \subseteq Q$ where any vertex in P sees q_1, q_2, q_3 in counterclockwise order. The edge p_1q_1 crosses edges q_2p_2 and q_2p_3, and these three edges bound a triangular region in the plane. Since the crossing angles are known, $\angle p_2q_2p_3$ is uniquely determined (Fig. 2(a)). Similarly, using edge p_3q_3, angle $\angle p_1q_2p_2$ is uniquely determined.

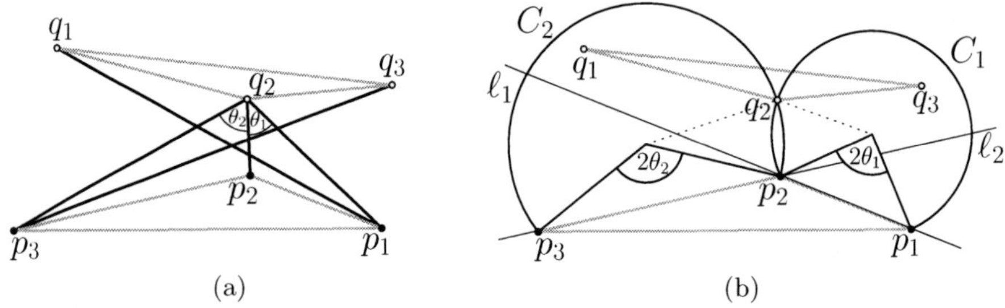

Fig. 2. (a) Determining $\angle p_2q_2p_3$. (b) Inscribed angles.

Let ℓ_1 and ℓ_2 be the supporting lines of p_1p_2 and p_2p_3, respectively. The lines ℓ_1 and ℓ_2 subdivide the plane into 4 wedges, and we may assume w.l.o.g. that Q lies in $\ell_1^+ \cap \ell_2^+$. We use a fact from elementary geometry: given two points a and b and an angle $\theta \in (0, \pi)$, the locus of points c with $\angle abc = \theta$ is the union of two circular arcs. However, $q_2 \in \ell_1^+ \cap \ell_2^+$ so q_2 lies on the intersection of two circles, C_1 defined by $p_1, p_2, \angle p_1q_2p_2$, and C_2 defined by $p_2, p_3, \angle p_2q_2p_3$. Since $p_2 \in C_1 \cap C_2$, the coordinates of q_2 are uniquely determined unless $C_1 = C_2$. If $C_1 = C_2$ then p_1, p_2, p_3, q_2 lie on a common circle and edge p_1q_1 intersects $\text{conv}(P)$, contradicting our assumption that q_1 convexly avoids P. □

Complete Bipartite Graphs. The following lemma states that if we already know the positions of 8 vertices of a complete graph, then the crossing-angle function uniquely determines the positions of all but 4 remaining vertices.

Lemma 1. *Let G be the complete graph with vertices $V = P \cup Q$ such that $|P| \geq 8$ and $|Q| \geq 5$. Let α be a crossing angle function α. In a straight-line drawing complying with α, the positions of the vertices in P uniquely determine the position of a vertex in Q.*

Proof. The vertex set of G is $P \cup Q$, with $|P| \geq 8$ and $|Q| \geq 5$, where the positions of the vertices in P are known. If there is a vertex $q \in Q$ such that edges p_1q, p_2q

cross some edges induced by P for some $p_1, p_2 \in P$, then the position of q is determined by Proposition 2. Otherwise, every vertex $q \in Q$ is crossing free or almost crossing free in the subgraph induced by $P \cup \{q\}$. By Proposition 3, every $q \in Q$ convexly avoids a subset of at least 8 points of P. By Propositions 4(i), there is a subset $P' \subset P$ of size 5 such that every $q \in Q$ convexly avoids P' and P' avoids Q. By Propositions 5 there are mutually avoiding subsets $P'' \subseteq P$ and $Q'' \subseteq Q$, with $|P''| \geq \lceil 5/2 \rceil = 3$ and $|Q''| \geq \lceil \sqrt{5} \rceil = 3$, such that every $q \in Q''$ convexly avoids P''. By Proposition 6, the position of some point $q \in Q''$ is uniquely determined. □

The First 8 Vertices. Lemma 1 states that if we know the positions of 8 vertices, then the position of another vertex is uniquely determined. The positions of the *first* 8 vertices in our construction are determined (up to similarity) by Lemma 2 below, the proof of which relies on properties of points in convex position.

Proposition 7. *Let P be a convex n-gon, and let G be a complete geometric graph on the vertices of P. Then, the slope of one diagonal of P and the directed crossing angles of G determine the slopes of all diagonals of P.*

Proof. Suppose we are given the slope of a diagonal e of P, and all directed crossing angles between diagonals. Let e' be another diagonal of P. If e crosses e', then their crossing angle determines the slope of e'. If e and e' do not cross (Fig. 4(a)), then there is a third diagonal f that crosses both e and e', and the slope of e' is determined by the angles $\angle(e, f)$ and $\angle(f, e')$. □

Proposition 8. *Let P and Q be two convexly avoiding sets of size $|P|, |Q| \geq 3$. Then they have subsets $P' \subseteq P$ and $Q' \subset Q$ such that $|P'| + |Q'| \geq |P| + |Q| - 1$, and P' and Q' are mutually avoiding.*

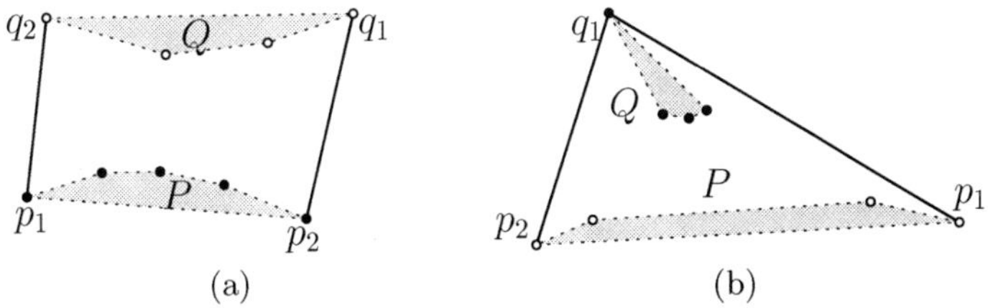

Fig. 3. (a) conv($P \cup Q$) is a quadrilateral. (b) conv($P \cup Q$) is a triangle.

Proof. Notice that the convex hulls conv(P) and conv(Q) are disjoint, so conv(P) and conv(Q) have exactly two common tangents. No three consecutive vertices of conv($P \cup Q$) are in P or Q, otherwise the middle vertex would be incident to an edge $p_i q_i$ that intersects the interior of conv(P) or conv(Q). Therefore, conv($P \cup Q$) is either a triangle or a quadrilateral. If conv($P \cup Q$) is a quadrilateral (Fig. 3(a)), then we show that P and Q are mutually avoiding. Let conv(P) $=$ conv($p_1, \ldots, p_t$) and conv($P \cup Q$) $= (p_1, p_2, q_1, q_2)$ in counterclockwise order.

Suppose that the supporting line of p_ip_j intersects conv(Q). Assume w.l.o.g. that $i < j$ and p_i is closer to the supporting line of p_1p_2 than p_j. Rotate ray $\overrightarrow{p_ip_j}$ about p_i counterclockwise until it reaches a point $q \in Q$. Now p_iq intersects the interior of conv(P).

Suppose that conv($P \cup Q$) is a triangle. Since P avoids Q, we have conv($P \cup Q$) $= (p_1, p_2, q_1)$. In this case, conv($P \cup (Q \setminus \{q_1\})$) is a quadrilateral (Fig. 3(b)). The previous argument readily implies that P and $Q \setminus \{q_1\}$ are mutually avoiding. □

Szekeres and Peters [20] proved, by an exhaustive computer search, that every set of 17 points in the plane, no three of which are collinear, contains 6 points in convex position. We use two convex hexagons in the proof of Lemma 2.

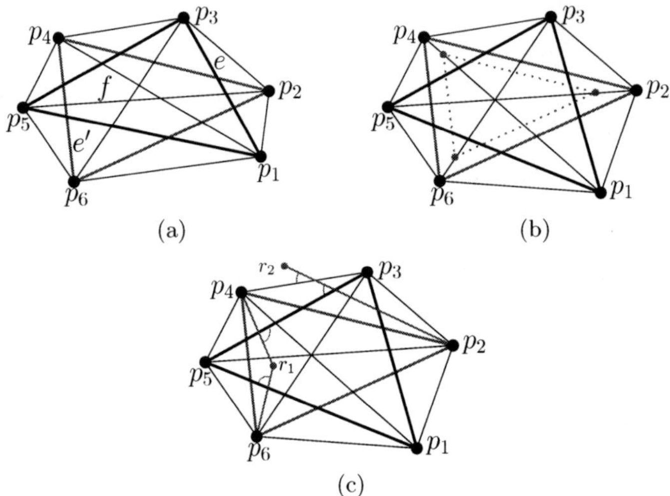

Fig. 4. (a) Diagonals of a convex hexagon. (b) If $\Delta p_1p_3p_5$ is fixed, $\Delta p_2p_4p_6$ has one degree of freedom. (c) A point r_1 in the interior of conv(P), and a point r_2 in the exterior conv(P) that does not convexly avoid P.

Lemma 2. *Let G be a complete graph with 24 vertices and a crossing-angle function α. Then in any straight-line drawing complying with α, the positions of at least 20 vertices of G are uniquely determined up to similarities.*

Proof. Denote by V the vertex set of G. Since $|V| \geq 17+6$ and no three vertices of G are collinear, we can successively choose two set, $P \subset V$ and $Q \subseteq (V \setminus P)$, each consisting of 6 points in convex position. Let conv(P) $= (p_1, \ldots, p_6)$ and conv(Q) $= (q_1, \ldots, q_6)$.

By Proposition 7, the crossing angles determine the slopes of all diagonals of a convex hexagon up to similarities. For instance, if we fix any two vertices in either $\Delta p_1p_3p_5$ or $\Delta p_2p_4p_6$, then the third vertex of the triangle is determined.

However, if we fix $\Delta p_1p_3p_5$, then the vertices of $\Delta p_2p_4p_6$ are not uniquely determined (Fig. 4(b)). We distinguish three cases.

Case 1: There is a vertex $r \in V$ lying in the interior of conv(P) *or* conv(Q). Suppose r lies in the interior of conv(P). Fix the positions of p_1 and p_3. At least two edges in $\{p_1r, p_3r, p_5r\}$ cross some diagonals of conv(P), and by Proposition 2, the position of r is uniquely determined. Then the positions of p_1, p_3, p_5, and r uniquely determine p_2, p_6, and p_6, by repeatedly applying Proposition 2.

Case 2: There is a vertex $r \in V$ lying in the exterior of conv(P) *and* conv(Q) *such that an edge from r to P intersects the interior of* conv(P) *or an edge from r to Q intersects the interior of* conv(Q). Suppose, w.l.o.g., that p_1r crosses the interior of conv(P). Fix the positions of p_1 and p_3. The location of p_i and the crossing angle with a diagonal of conv(P) determine the supporting line of p_1r. The supporting line of p_1r crosses an edge of conv(P), which is incident to a vertex p_j, $j \in \{2, 4, 6\}$. Now the position of p_j is determined by Proposition 2. By repeatedly applying Proposition 2, we determine the positions of p_2, p_4, and p_6.

Case 3: P lies in the exterior of conv(Q), *Q lies in the exterior of* conv(P), *and no edge between P and Q crosses* conv(P) *or* conv(Q). By Proposition 8, P and Q have mutually avoiding subsets of total size 11. W.l.o.g., $\{p_1, \ldots, p_5\}$ and $\{q_1, \ldots, q_6\}$ are mutually avoiding. Fix the positions of p_1, p_3, which immediately determines the position of p_5. By Proposition 6, the positions of $q_2, \ldots, q_5$ are uniquely determined.

In all three cases, the positions of at least 6 vertices are determined up to similarities. Let $A \subset V$ be the set of these vertices. Use Proposition 2 successively to determine as many more vertices as possible. If we know the positions of at most 7 vertices in V, then each of the remaining 17 vertices in V convexly avoids some subset of at least 5 vertices in A. By Proposition 4(ii), there are subsets $A' \subset A$ and $B' \subset (V \setminus A)$, of size $|A'| \geq 5$ and $|B'| \geq 5$, such that every point in B' convexly avoids A', and A' avoids B'. By Propositions 5 and 6, the position of some point in B' is uniquely determined. Finally, if we already know 8 vertices in V, then we can use Lemma 1 to determine the positions of all but at most 4 vertices in V. □

We are now ready to prove Theorem 2.

Proof (of Theorem 2). Applying Lemma 2 for $V_1 \cup V_2$, we find a subset $V_1' \subset V_1$ of size $|V_1'| = 8$ such that the positions of all vertices in V_1' are determined up to similarities by the crossing angles. Pick two arbitrary points $a, b \in V_1'$, and fix the coordinates of a and b. Then the positions of all other points in V_1' are fixed. If we have already chosen a subset $V_i' \subset V_i$, $|V_i'| = 8$, and v_iv_j is an edge of graph F_0, then by Lemma 1 we can choose a set $V_j' \subset V_j$ of 8 vertices (one at a time) such that the positions of the vertices in V_i' and the crossing angles uniquely determine the positions of vertices in V_j'. If we fix the coordinates of a, b and the crossing-angle function α, then the coordinates of all points in $V' = \cup_{i=1}^{n_0} V_i'$ are uniquely determined. □

References

1. Argyriou, E.N., Bekos, M.A., Symvonis, A.: The Straight-Line RAC Drawing Problem Is NP-Hard. In: Černá, I., Gyimóthy, T., Hromkovič, J., Jefferey, K., Králović, R., Vukolić, M., Wolf, S. (eds.) SOFSEM 2011. LNCS, vol. 6543, pp. 74–85. Springer, Heidelberg (2011)
2. Aronov, B., Erdős, P., Goddard, W., Kleitman, D.J., Klugerman, M., Pach, J., Schulman, L.J.: Crossing families. Combinatorica 14(2), 127–134 (1994)
3. Barát, J., Matoušek, J., Wood, D.R.: Bounded-degree graphs have arbitrarily large geometric thickness. Electr. J. Comb. 13(1) (2006)
4. Bienstock, D., Dean, N.: Bounds for rectilinear crossing numbers. J. Graph Theory 17(3), 333–348 (1993)
5. Di Giacomo, E., Didimo, W., Eades, P., Liotta, G.: 2-Layer Right Angle Crossing Drawings. In: Iliopoulos, C.S., Smyth, W.F. (eds.) IWOCA 2011. LNCS, vol. 7056, pp. 156–169. Springer, Heidelberg (2011)
6. Di Giacomo, E., Didimo, W., Liotta, G., Meijer, H.: Area, curve complexity and crossing resolution of non-planar graph drawings. Theory Comput. Syst. 49(3), 565–575 (2011)
7. Didimo, W., Eades, P., Liotta, G.: A characterization of complete bipartite rac graphs. Inf. Process. Lett. 110(16), 687–691 (2010)
8. Didimo, W., Eades, P., Liotta, G.: Drawing graphs with right angle crossings. Theor. Comput. Sci. 412(39), 5156–5166 (2011)
9. Dillencourt, M.B., Eppstein, D., Hirschberg, D.S.: Geometric thickness of complete graphs. J. Graph Alg. & Appl. 4(3), 5–17
10. Dujmović, V., Gudmundsson, J., Morin, P., Wolle, T.: Notes on large angle crossing graphs. Chicago J. Theor. Comput. Sci (2011)
11. Dujmović, V., Suderman, M., Wood, D.R.: Graph drawings with few slopes. Comput. Geom. 38(3), 181–193 (2007)
12. Eades, P., Liotta, G.: Right Angle Crossing Graphs and 1-Planarity. In: Speckmann, B. (ed.) GD 2011. LNCS, vol. 7034, pp. 148–153. Springer, Heidelberg (2011)
13. Erdős, P., Szekeres, G.: A combinatorial problem in geometry. Compositio Math. 2, 463–470 (1935)
14. Jackson, B., Jordán, T.: Connected rigidity matroids and unique realizations of graphs. J. Combin. Theory, Ser. B 94, 1–29 (2005)
15. Jackson, B., Jordán, T.: Globally rigid circuits of the directionlength rigidity matroid. J. Combin. Theory, Ser. B 100, 1–2 (2010)
16. Mukkamala, P., Pálvölgyi, D.: Drawing Cubic Graphs with the Four Basic Slopes. In: Speckmann, B. (ed.) GD 2011. LNCS, vol. 7034, pp. 254–265. Springer, Heidelberg (2011)
17. Pach, J., Pálvölgyi, D.: Bounded-degree graphs can have arbitrarily large slope numbers. Electr. J. Comb. 13(1) (2006)
18. Riskin, A.: The crossing number of a cubic plane polyhedral map plus an edge. Studia Sci. Math. Hungar. 31, 405–413 (1996)
19. Saxe, J.B.: Embeddability of weighted graphs in k-space is strongly NP-hard. In: 17th Allerton Conf. in Communications, Control and Computing, pp. 480–489 (1979)
20. Szekeres, G., Peters, L.: Computer solution to the 17-point Erdős-Szekeres problem. ANZIAM J. 48(2), 151–164 (2006)
21. Wade, G.A., Chu, J.-H.: Drawability of complete graphs using a minimal slope set. Comput. J. 37(2), 139–142 (1994)

Multicut on Graphs of Bounded Clique-Width*

Martin Lackner, Reinhard Pichler,
Stefan Rümmele, and Stefan Woltran

Institute of Information Systems, Vienna University of Technology, Austria
{lackner,pichler,ruemmele,woltran}@dbai.tuwien.ac.at

Abstract. Several variants of Multicut problems arise in applications like circuit and network design. In general, these problems are NP-complete. The goal of our work is to investigate the potential of clique-width for identifying tractable fragments of Multicut. We show for several parameterizations involving clique-width whether they lead to tractability or not. Since bounded tree-width implies bounded clique-width, our tractability results extend previous results via tree-width, in particular to dense graphs.

1 Introduction

Multicut problems are graph problems with many applications to circuit and network design, telecommunication, and recently even databases [15]. An instance of a Multicut problem is given by an undirected graph G and a set H of pairs of so-called *terminal* vertices. The aim is to find a minimum *cut* that separates all terminal pairs. Different kinds of cuts are considered. For the Edge Multicut (EMC) problem, the cut is a set of edges whose removal disconnects each terminal pair. In case of the Restricted (resp. Unrestricted) Vertex Multicut problem (RVMC resp. UVMC), the cut is a set of non-terminal (resp. arbitrary) vertices. All three variants of Multicut problems are intractable, i.e. the corresponding decision problems (asking if a cut of a given cardinality exists) are NP-complete. RVMC and EMC remain NP-hard even on trees [2,7].

An important approach in dealing with intractable problems is to search for *fixed-parameter tractability* in order to confine the combinatorial explosion to certain problem *parameters*. More formally, we say that a problem is in the class FPT with respect to a parameter k, if the problem is solvable in time $f(k) \cdot n^{O(1)}$, where n denotes the size of the input instance. The function f is usually exponential but only depends on k. The related complexity class XP contains the problems solvable in time $\mathcal{O}(n^{g(k)})$ where function g depends only on the parameter k. In general algorithms with FPT runtime are clearly preferable to those with XP runtime. In Table 1 we recall previous complexity results on various parameters of Multicut problems. For several parameters, like the size m of the cut plus cardinality $|H|$ [19] and very recently even the size m alone [1,20],

* This work was supported by the Austrian Science Fund (FWF): P20704-N18.

G. Lin (Ed.): COCOA 2012, LNCS 7402, pp. 115–126, 2012.

Table 1. (Parameterized) Complexity of Multicut problems for various parameters

Graph classes	**UVMC**	**RVMC**	**EMC**
Interval graphs	NP-c [13]	P [13]	NP-c [11]
Trees	P [13]	NP-c [2]	NP-c [11]
Cographs	NP-c (Thm 1)	P (Thm 4)	NP-c [11][3]
Parameters[1]	**UVMC**	**RVMC**	**EMC**
m, $\|H\|$	FPT [19]	FPT [19]	FPT [19]
m	FPT [1,20]	FPT [1,20]	FPT [1,20]
$tw(G)$	NP-c [2]	NP-c [2]	NP-c [11]
$tw(G \cup H)$	FPT [12]	FPT [12]	FPT [12]
$tw(G)$, $\|H\|$	FPT [13]	FPT [13]	FPT [13]
$\|H\|$	NP-c [19]	NP-c [13]	NP-c [7]
$cw(G)$	NP-c (Thm 1)	NP-c [2][4]	NP-c [11][3]
$cw(G \cup H)$	NP-c (Cor 2)	FPT[2] (Thm 9)	NP-c (Thm 3)
$cw(G)$, $\|H\|$	FPT (Thm 7)	FPT (Thm 7)	XP (Section 4.3)

[1] NP-c refers here to NP-completeness even if the parameter value is fixed by a constant.
[2] For graphs without edges between terminal vertices (cf. Section 4.2).
[3] Follows from NP-hardness of EMC on stars (trees of height 1) [11].
[4] Follows from NP-hardness of RVMC on trees [2].

FPT membership could be shown. Also several parameterizations involving tree-width (a parameter which measures the "tree-likeness" of a graph), like the tree-width of the structure representing both G and H (denoted by $tw(G \cup H)$) [12] and the tree-width of G plus $|H|$ [13], lead to FPT membership. In contrast, for some other parameters it was shown that the Multicut problems remain NP-complete even if the parameters are bounded by a constant. This is the case for a parameterization with $tw(G)$ alone [2] and with $|H|$ alone [7,13,19].

We recall that all previous FPT algorithms based on tree-width are only applicable to sparse graphs. To identify FPT fragments of Multicut which additionally apply to dense graphs, we study here the parameter *clique-width*. Clique-width generalizes the class of cographs similarly as tree-width generalizes trees. A formal definition of clique-width is given in Section 2. Many graph classes are known to have bounded clique-width, such as cliques, cographs, trees, tree-cographs, probe cographs, distance-hereditary, P_4-reducible, and series-parallel graphs. Hence, all fixed-parameter tractability results w.r.t. clique-width immediately yield tractability for these graph classes. Moreover, recall the following important connection between clique-width and tree-width (shown in [3], improving a result from [6]). For every graph G, $cw(G) \leq 3 \cdot 2^{tw(G)-1} + 1$ holds. Hence, our tractability results strictly extend previous ones that are based on tree-width. Furthermore, by the relationship between clique-width and rank-width proved in [21], both our NP-completeness results and our tractability results immediately carry over to rank-width.

Our main results, as summarized in the lower part of Table 1, are as follows:

- *NP-completeness results.* The NP-completeness of Multicut parameterized by tree-width $tw(G)$ carries over to parameter $cw(G)$ by the relationship of $tw(G)$ and $cw(G)$ recalled above. However, this leaves a gap for Multicut instances with small clique-width. We fill this gap by extending NP-completeness to these cases, with the notable exception of RVMC, which we show to be tractable on graphs with $cw(G) \leq 2$. Completely new NP-hard cases arise for UVMC and EMC with respect to the parameter $cw(G \cup H)$, as opposed to the parameter $tw(G \cup H)$ for which both problems are in FPT.
- *FPT results.* We prove the FPT membership of UVMC and RVMC with respect to the parameter $cw(G)+|H|$. Under a weak additional assumption, RVMC is also shown to be in FPT with respect to $cw(G \cup H)$. It is noteworthy that these FPT results can also be obtained by exploiting the Monadic Second-Order logic (MSO) characterizations of UVMC and RVMC and by applying the clique-width metatheorem from [5]. However, by designing concrete algorithms we are able to show better upper bounds for the runtime. The presented algorithms have a runtime that is double-exponential in $cw(G)$ but only single-exponential in $|H|$. Furthermore, the runtime depends only linearly on the input size. These algorithms are therefore tailored for large instances with small clique-width and a moderate number of terminal pairs. Note that graphs of small clique-width can have unbounded tree-width. This makes our algorithms more versatile than those based on tree-width.

2 Preliminaries

For a set S and $n \in \mathbb{N}$, we write $S^{[n]}$ to denote the set of all subsets of S with cardinality n, formally $S^{[n]} := \{S' \subseteq S : |S'| = n\}$. For a set of sets S, the *union of* S, denoted $\bigcup S$, is defined as $\{a : \exists B \in S \text{ with } a \in B\}$. We consider here finite simple undirected graphs and refer to edges via two-element subsets of V. An *induced subgraph* (by a vertex set V') is denoted as $G[V']$.

An instance of a Multicut problem is given by a triple (G, H, m), where $G := (V_G, E_G)$ is a graph, $H \subseteq V_G^{[2]}$ is a set of *terminal pairs*, and m is a non-negative integer. We write $V_H := \bigcup H \subseteq V_G$ to denote the *terminal vertices* or *terminals* in such a problem. The Multicut problem asks if a *cut* of size at most m exists, which separates all terminal pairs. In case of the {UVMC, RVMC, EMC} problem, a *cut* is a set of {arbitrary vertices, non-terminal vertices, edges}. A *solution graph* is a valid graph that remains after "cutting".

Clique-width is a graph property introduced in [4]. For a formal definition of clique-width we require *k-expressions.* Each k-expression s has a corresponding (labeled) graph $G(s)$, which is obtained by the following construction ($k \in \mathbb{N}$):

- *Adding a new vertex.* Let v be a vertex and $i \in \{1, ..., k\}$ a label. Then $i(v)$ is a k-expression and $G(i(v))$ consists of the vertex v labeled with i.
- *Renaming labels.* Let $i, j \in \{1, ..., k\}$ be labels with $i \neq j$ and let s be a k-expression. Then $\rho_{j \leftarrow i}(s)$ is a k-expression and $G(\rho_{j \leftarrow i}(s))$ is obtained by re-labeling each i-labeled vertex in $G(s)$ with j.

- *Connecting vertices.* Let $i, j \in \{1, ..., k\}$ be labels with $i \neq j$ and s a k-expression. Then $\eta_{i,j}(s)$ is a k-expression and $G(\eta_{i,j}(s))$ is obtained from $G(s)$ by connecting every i-labeled vertex with every j-labeled vertex.
- *Disjoint union.* Let s, t be k-expressions with no vertices in common. Then $s \oplus t$ is a k-expression and $G(s \oplus t)$ is the disjoint union of $G(s)$ and $G(t)$.

A graph G has *clique-width* k, written $cw(G) = k$, if k is the smallest number such that there is a k-expression s for which the unlabeled version of $G(s)$ is G. In Figure 1 we illustrate these concepts by graph G_{Ex} and the parse tree of a possible 3-expression of G_{Ex}.

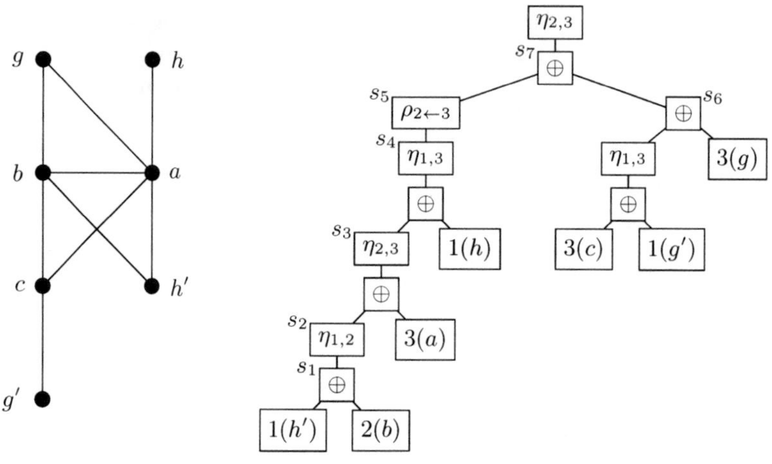

Fig. 1. The graph G_{Ex} and its 3-expression

In general, finding a k-expression is hard: [9] reports NP-completeness for determining whether a graph has clique-width k. However, a (possibly) sub-optimal k-expression can be computed efficiently: [21] contains an FPT algorithm which, for a given k, either concludes that $cw(G) > k$ or outputs a $(2^{3k+2} - 1)$-expression of the graph. It is an open problem whether finding a k-expression is fixed-parameter tractable with respect to k. A more detailed introduction to width parameters and graph decompositions can be found in [18].

In this paper, we want to focus on the complexity of several variants of Multicut rather than the complexity of computing a k-expression. We therefore assume that an appropriate k-expression is given as part of the input. The runtime estimates in this paper are performed under the assumption that basic set operations (union, set difference, intersection) can be computed in linear time with respect to the cardinality of the sets.

3 Complexity Results

In this section we address the complexity of the problems UVMC, RVMC, and EMC with regard to the parameters $cw(G)$ and $cw(G \cup H)$. Let $G \cup H$ denote the

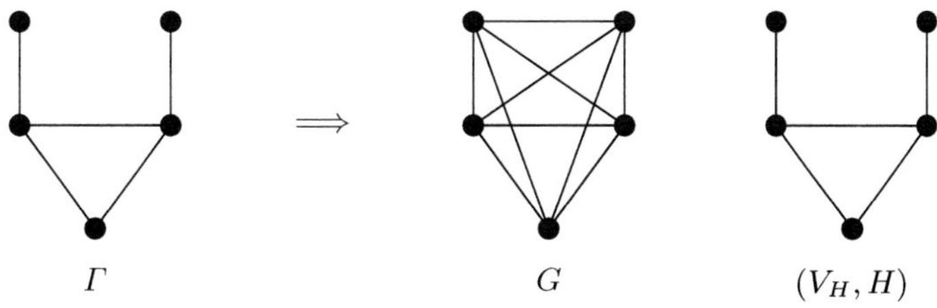

Fig. 2. An example for the reduction in the proof of Theorem 1

graph $(V_G, E_G \cup H)$, i.e. the primal graph of the structure (V_G, E_G, H). Whereas some results for $cw(G)$ follow from previous work, no complexity analysis has been done for $cw(G \cup H)$-bounded graphs yet. Since graphs of clique-width 1 do not contain edges, we do not mention them in this chapter.

UVMC. UVMC is known to be NP-complete on series-parallel graphs [2], i.e. for graphs of clique-width 4. To close the gap for graphs of clique-width 2 and 3, we show that UVMC is NP-complete on cographs (and even on cliques). The analogous result for $cw(G \cup H)$-bounded graphs follows immediately.

Theorem 1. *UVMC remains NP-complete if G is restricted to cliques.*

Proof sketch. We use a reduction from VERTEX COVER. Let $\Gamma := (V_\Gamma, E_\Gamma)$ be an arbitrary graph for which we want to find a vertex cover of size at most k, i.e. a set of vertices that covers every edge. Our input for UVMC is $G := (V_\Gamma, V_\Gamma^{[2]})$, i.e. the complete graph on V_Γ, and $H := E_\Gamma$. An example is given in Figure 2. One can show that a set of vertices is a cut in G iff it is a vertex cover of Γ. □

Corollary 2. *UVMC remains NP-complete if $G \cup H$ is restricted to cliques.*

These results prove NP-hardness for UVMC for a class of input instances with $cw(G) = 2$ (resp. $cw(G \cup H) = 2$) and therefore for input instances where $cw(G)$ (resp. $cw(G \cup H)$) is bounded by any number ≥ 2.

EMC. In [11] it is shown that EMC is NP-complete on trees of depth 1, i.e. stars. We show that EMC also remains NP-complete if $G \cup H$ is restricted to cliques. Since both stars and cliques are cographs and hence their clique-width is 2, it follows that EMC remains NP-complete even if either G or $G \cup H$ is restricted to graphs of clique-width 2.

Theorem 3. *EMC is NP-complete even if $G \cup H$ is a clique.*

Proof sketch. The NP-complete P_3-EDGE DELETION PROBLEM [8] asks if it is possible to obtain a graph containing no induced P_3, i.e. a path on 3 vertices, by deleting a given number of edges. This problem can be reduced to UVMC where $G \cup H$ is a clique. To do so, let G be the given graph and H the edge complement. Now observe that a subgraph of G contains no induced P_3 iff it consists of disjoint cliques, which holds iff no terminal pair is connected. □

RVMC. RVMC for input instances where $cw(G)$ is bounded by a number ≥ 3 is NP-complete. This follows directly from the fact that RVMC is NP-complete on trees [2] and that trees have clique-width 3. However, Theorem 4 shows that RVMC becomes tractable on graphs of clique-width 2. For the parameter $cw(G \cup H)$ a new situation arises, which is analyzed in Section 4.3.

Theorem 4. *RVMC can be solved on cographs in $\mathcal{O}(|H| \cdot |V_G|)$ time.*

Proof sketch. First note that if a cograph contains a path from a to c of length ≥ 2 then either there is a vertex b in this path such that (a, b, c) is itself a path or $\{a, c\}$ is an edge. Now, if there is a connected terminal pair in the cograph, the algorithm rejects the input. Otherwise we construct $C := \{b \in V_G : \exists (a, c) \in H \text{ s.t. } (a, b, c) \text{ is a path}\}$. Because of the aforementioned fact, C contains the vertices that have to be removed. Therefore if C contains a terminal vertex, the algorithm rejects the input. Otherwise C is the minimal cut. □

4 Algorithms

We now present FPT and XP results with several parameterizations related to clique-width. Recall from Section 2 that when dealing with clique-width, we consider a k-expression (referred to as κ) as part of the input. Furthermore, $|\kappa|$ denotes the number of operations in κ.

4.1 Vertex Multicut with $cw(G) + |H|$ as Parameter

We give an FPT algorithm based on dynamic programming for UVMC and RVMC with $cw(G) + |H|$ as parameter. The central idea of our algorithm is to keep track of the connected components of G, while G is built according to its k-expression and potential cuts are performed. Especially the connected components that contain terminal vertices are important. For this purpose, we will use the concept of *connected component sets* (CCSs), which only hold the "relevant" information on each connected component V_C, namely: (i) all terminal vertices occurring in V_C and (ii) all labels of the vertices in V_C. Observe that as long as no terminal pair is in a single connected component, the graph under consideration is a solution graph. Variations of the concept of CCSs will be used later in all following algorithms.

Definition 5. *Let $G = (V, E)$ be a labeled graph, let $V_H \subseteq V$ and let $L(V')$ denote the set of labels of $V' \subseteq V$. The* connected component set (CCS) of the pair (G, H) *is*

$$ccs(G, V_H) := \{(V_C \cap V_H) \cup L(V_C) : V_C \subseteq V \text{ forms a connected component}\}.$$

Observe that in case an element of a CCS contains a terminal vertex, it corresponds to exactly one connected component. However, if an element of a CCS does not contain a terminal, several corresponding components may exist.

In the following we will give a detailed description of the algorithm. We start by describing the *data structure*, which for each subexpression s of the input k-expression κ consists of a set S_s of CCSs and a function $cuts_s$ from S_s to $\mathbb{N}$. The algorithm traverses the parse tree of the k-expression bottom-up manipulating this data structure (details are given below). The main idea of the algorithm is that for each subexpression s of κ,

$$S_s = \{ccs(G(s)[V_G \setminus C], V_H) : C \text{ is a cut w.r.t. } G(s) \wedge |C| \leq m\}.$$

This means that S_s contains all possible CCSs of solution graphs for the UVMC or RVMC problem $(G(s), H, m)$. Furthermore for each CCS $\Delta \in S_s$, $cuts_s(\Delta)$ is the minimum number of cuts required to obtain a graph represented by Δ from the original graph $G(s)$. The function *cuts* is essential to discard CCSs that have size greater than m. Since we are only interested in cuts of size $\leq m$, $cuts_s(\Delta)$ is always $\leq m$. Once the algorithm reaches the root node, which corresponds to the k-expression κ, S_κ contains all possible CCSs of solution graphs for the UVMC or RVMC problem (G, H, m). Hence there is a cut set of size $\leq m$ if and only if S_κ is not empty.

The functions *ren* and *con* will allow us to give a succinct description of the algorithm. *ren* is closely related to the ρ-operation, *con* to the η-operation.

Definition 6. *$ren_{i \leftarrow j}$ and $con_{i,j}$ are functions that map a CCS Δ to another CCS as follows:*

$$ren_{i \leftarrow j}(\Delta) := \{c \cup \{i\} \setminus \{j\} : c \in \Delta \text{ with } j \in c\} \cup \{c : c \in \Delta \text{ with } j \notin c\},$$

$$con_{i,j}(\Delta) := \{c \in \Delta : i \notin c \wedge j \notin c\} \cup \left\{\bigcup \{c \in \Delta : i \in c \vee j \in c\}\right\}.$$

We now describe what the algorithm does in each node of the parse tree of κ. The only distinction between UVMC and RVMC is in the leaves. Below, we let s (and possibly t) be the k-expression of the subtree(s) below the current (internal) node.

$i(v)$**:** If v is a non-terminal vertex, we have two CCSs $S_{i(v)} := \{\emptyset, \{\{i\}\}\}$ with $cuts_{i(v)}(\emptyset) := 1$ and $cuts_{i(v)}(\{\{i\}\}) := 0$. If v is a terminal vertex, we do the same for UVMC but add the vertex to the CCS, i.e. $S_{i(v)} := \{\emptyset, \{\{i, v\}\}\}$, $cuts_{i(v)}(\emptyset) = 1$ and $cuts_{i(v)}(\{\{i, v\}\}) = 0$. For RVMC, we cannot remove v, thus only $S_{i(v)} := \{\{\{i, v\}\}\}$ with $cuts_{i(v)}(\{\{i, v\}\}) = 0$ remains.

$\rho_{i \leftarrow j}(s)$**:** $S_{\rho_{i \leftarrow j}(s)} := \{ren_{i \leftarrow j}(\Delta) : \Delta \in S_s\}$ and for each $\Delta' \in S_{\rho_{i \leftarrow j}(s)}$ we have $cuts_{\rho_{i \leftarrow j}(s)}(\Delta') := min\{cuts_s(\Delta) : ren_{i \leftarrow j}(\Delta) = \Delta'\}$.

$s \oplus t$**:** $S_{s \oplus t} := \{\Delta_s \cup \Delta_t : \Delta_s \in S_s, \Delta_t \in S_t, cuts_s(\Delta_s) + cuts_t(\Delta_t) \leq m\}$ and $cuts_{s \oplus t}(\Delta') := min\{cuts_s(\Delta_s) + cuts_t(\Delta_t) : \Delta_s \cup \Delta_t = \Delta'\}$.

$\eta_{i,j}$**:** For each $\Delta \in S_s$ there are two cases: (1) $\{i, j\} \nsubseteq \bigcup \Delta$, i.e. there is no i-labeled or no j-labeled vertex in the graphs represented by Δ. Therefore no connections are introduced and hence Δ is added to $S_{\eta_{i,j}(s)}$. (2) Otherwise we consider $con_{i,j}(\Delta)$. Here, edges might have been added such that a terminal pair was connected. If this is not the case, i.e. there is no set in $con_{i,j}(\Delta)$

which contains a terminal pair, $con_{i,j}(\Delta)$ is added to $S_{\eta_{i,j}(s)}$. In both cases – (1) and (2) – $cuts_{\eta_{i,j}(s)}(\Delta')$ is the minimum number of cuts of all CCSs in S_s that lead to Δ' if i-labeled and j-labeled vertices are connected.

Theorem 7. *UVMC and RVMC are FPT with respect to the parameters $cw(G)$ and $|V_H|$ and can be solved in time*

$$\mathcal{O}\left(4^{2^{cw(G)}} \cdot (|V_H| + 2^{cw(G)})^{2|V_H|} \cdot (|V_H| \cdot (cw(G)+1) + cw(G) \cdot 2^{cw(G)}) \cdot |\kappa|\right).$$

Proof. The algorithm described above operates on the parse tree of κ. We therefore analyze the cost of each of the four operations. The operation $i(v)$ requires constant time. The operations $\eta_{i,j}(s)$ and $\rho_{i\leftarrow j}(s)$ perform basic set operations on each set in each CCS in S_s. A CCS has size at most $|V_H| \cdot (cw(G)+1) + cw(G) \cdot 2^{cw(G)}$. That is for each element of V_H a set with at most $cw(G)+1$ elements plus at most $2^{cw(G)}$ subsets of $cw(G)$ each having size at most $cw(G)$. In order to bound the size of S_s, we have to bound the number of possible CCSs. Recall that each $h \in V_H$ appears in at most one set in a CCS. Therefore, we can estimate the number of possible CCSs as follows. For the first $h_1 \in V_H$ there are $1+2^{cw(G)}$ many possibilities, since it is either not contained or it occurs together with an arbitrary subset of $cw(G)$. For the second $h_2 \in V_H$ there are at most $2+2^{cw(G)}$ many possibilities, since it is either not contained, or it occurs in the set of h_1, or it occurs together with an arbitrary subset of $cw(G)$. This scheme repeats for the other elements from V_H. After fixing these, we still have $2^{2^{cw(G)}}$ many possibilities for choosing arbitrary subsets of $cw(G)$. This results in less than $2^{2^{cw(G)}} \cdot (|V_H| + 2^{cw(G)})^{|V_H|}$ CCSs in S_s. In total this yields a runtime of at most $\mathcal{O}\left(2^{2^{cw(G)}} \cdot (|V_H| + 2^{cw(G)})^{|V_H|} \cdot (|V_H| \cdot (cw(G)+1) + cw(G) \cdot 2^{cw(G)})\right)$ for η- and ρ-operations. In order to compute $s \oplus t$ we have to consider pairs of CCSs, i.e. at most $4^{2^{cw(G)}} \cdot (|V_H| + 2^{cw(G)})^{2|V_H|}$ combinations. Computing the union is bounded by the maximum size of a CCS. □

Note that although the runtime depends on $|V_H|$, this can also be seen as an FPT algorithm for $cw(G) + |H|$ since $|V_H| \leq 2|H|$. Actually, this FPT result could also be obtained via the metatheorem of [5] and an encoding of the UVMC and RVMC problem by a Monadic Second-Order (MSO) formula in the spirit of [12]. However, a precise upper bound on the runtime in terms of an $\mathcal{O}(\cdot)$-expression would remain obscure if the FPT result were proved via the metatheorem.

Example for RVMC. We apply the above algorithm to the RVMC instance $(G_{Ex}, H_{Ex}, 2)$, where G_{Ex} and a 3-expression of G_{Ex} are shown in Figure 1. Moreover, suppose that the terminal set is $H_{Ex} := \{\{g, g'\}, \{h, h'\}\}$. In the parse tree certain nodes are marked with $s_1, \ldots, s_7$. These denote the subexpressions corresponding to the subtrees rooted at these nodes. For these nodes we give the data structure S_{s_i} and the cuts function in Table 2.

The operation in node s_1 is $\oplus$. Here we take the union of the CCSs from the left and right branch. Observe that h' is present in every CCS, since it is a terminal vertex. The next operations are $\eta_{1,2}$ and $\eta_{2,3}$. Both times the components in the

Table 2. The RVMC algorithm applied to the graph in Figure 1

	$CSSs$	cuts		$CCSs$	cuts
s_1	$\{\{1,h'\}\} \cup \{\{2\}\}$	0	s_5	$\{\{1,2,h'\},\{1,h\}\}$	1
$(\oplus)$	$\{\{1,h'\}\} \cup \{\}$	1	$(\rho_{2\leftarrow 3})$	$\{\{1,h'\},\{1,h\}\}$	2
s_2	$\{\{1,2,h'\}\}$	0	s_6	$\{\{1,3,g'\}\} \cup \{\{3,g\}\}$	0
$(\eta_{1,2})$	$\{\{1,h'\}\}$	1	$(\oplus)$	$\{\{1,g'\}\} \cup \{\{3,g\}\}$	1
s_3	$\{\{1,2,3,h'\}\}$	0	$s7$	$\{\{1,2,h'\},\{1,h\}\} \cup \{\{1,3,g'\},\{3,g\}\}$	1
$(\eta_{2,3})$	$\{\{1,h'\},\{3\}\}$	1	$(\oplus)$	$\{\{1,h'\},\{1,h\}\} \cup \{\{1,3,g'\},\{3,g\}\}$	2
	$\{\{1,2,h'\}\}$	1		$\{\{1,2,h'\},\{1,h\}\} \cup \{\{1,g'\},\{3,g\}\}$	2
	$\{\{1,h'\}\}$	2	root	~~$\{\{1,2,3,g,g',h'\},\{1,h\}\}$~~	1
s_4	$\{\{1,2,h'\},\{1,h\}\}$	1	$(\eta_{2,3})$	$\{\{1,h'\},\{1,h\},\{1,3,g'\},\{3,g\}\}$	2
$(\eta_{1,3})$	$\{\{1,h'\},\{1,h\}\}$	2		$\{\{1,2,3,g,h'\},\{1,h\},\{1,g'\}\}$	2

first CCS are merged, whereas the other CCSs do not change. In s_3 the four rows correspond to (in this order): no vertex cut, vertex b cut, vertex a cut, and both a and b cut. The s_4-operation is $\eta_{1,3}$. Since both h and h' are labeled with 1, CCSs with 3-labeled vertices yield invalid CCSs and are therefore not listed anymore. Subexpression s_6 denotes the right branch of the parse tree. The first CCS required no cuts whereas in the second CCS vertex c has been cut. The root node contains two valid CCSs and therefore there are valid cuts of size 2. The first solution corresponds to the cut set $\{a, b\}$ and the second to the cut set $\{a, c\}$. Note that in general a CCS may correspond to several cuts.

4.2 RVMC with $cw(G \cup H)$ as Parameter

In Section 3 we left open the complexity of RVMC with regard to $cw(G \cup H)$. Now we present an FPT algorithm for that problem on a slightly restricted class of inputs, namely those that do not contain edges (in G) between terminal vertices. If this restriction holds, edges between terminal vertices in $G \cup H$ always correspond to terminal pairs. Note that an edge between two terminal vertices which do not form a terminal pair in H, does not automatically prohibit a cut set. This restriction is not necessary if the following conjecture[1] holds.

Conjecture 8. There is a computable function f such that for every class of graphs $\mathcal{G}$, if the clique-width of $\mathcal{G}$ is bounded by k then the clique-width of $\mathcal{G}'$ is bounded by $f(k)$, where $\mathcal{G}'$ is the class obtained from $\mathcal{G}$ by closing $\mathcal{G}$ under edge contractions.

If this conjecture holds, we can contract all edges in $G \cup H$ between terminal vertices a and b with $\{a, b\} \in E_G$. Then we use the $f(k)$-expression of this graph as input, i.e. the clique-width of the modified graph is still bounded. This modified graph has exactly the same cut sets as the original graph.

The algorithm presented here is based on the one in Section 4.1. Especially the *cuts* function is defined in exactly the same way, but we use a slightly different

[1] A stronger statement, namely that edge contractions do not increase the clique-width at all, is mentioned as an open problem in [14].

form of CCSs. Here, CCSs are subsets of $\mathcal{P}(\{1,\ldots,k,\hat{1},\ldots,\hat{k}\})$. To define such CCSs, $L(V')$ is as before and $\hat{L}(V') := \{\hat{i} : i \in L(V')\}$. Now $\widehat{ccs}(G, V_H) := \{\hat{L}(V_C \cap V_H) \cup L(V_C \setminus V_H) : V_C \subseteq V \text{ forms a conn. comp.}\}$. Let $\hat{K} := \{\hat{1},\ldots,\hat{k}\}$. The data structure consists of a set S of CCSs, *forbidden sets* $N \subseteq \hat{K}^{[1]} \cup \hat{K}^{[2]}$ and a function *cuts* from S to $\mathbb{N}$. The intended meaning is that $\hat{i}$ is contained in an element of a CCS iff this component contains an i-labeled terminal. If $\{\hat{i},\hat{j}\} \in N$, then there must not be an edge introduction between a component containing an i-labeled and one containing a j-labeled terminal. If $\{\hat{i}\} \in N$, then there must not be an edge introduction between a component containing an i-labeled terminal and any other component. The operations are as follows:

$i(v)$**:** If v is a non-terminal vertex, we have two CCSs $S_{i(v)} := \{\emptyset, \{\{i\}\}\}$. If v is a terminal vertex, $S_{i(v)} := \{\{\{\hat{i}\}\}\}$. In both cases $N := \emptyset$.

$\rho_{i\leftarrow j}(s)$**:** All CCSs are relabeled: $j \mapsto i$ and $\hat{j} \mapsto \hat{i}$. N is also changed accordingly; this might lead to N containing sets of size 1.

$s \oplus t$**:** Here $S_{s\oplus t}$ is calculated in exactly the same way as in the algorithm of Section 4.1 and $N_{s\oplus t} := N_s \cup N_t$.

$\eta_{i,j}$**:** If $\hat{i}$ and $\hat{j}$ are present in S_s, $N_{\eta_{i,j}(s)} := N_s \cup \{\hat{i},\hat{j}\}$, otherwise $N_{\eta_{i,j}(s)} := N_s$. The reason for this is that there is no edge between terminal nodes in G. Consequently, there exists a terminal pair in H that has labels i and j. $S_{\eta_{i,j}(s)}$ is calculated similar to the original algorithm. However, checking if a terminal pair is contained in a component works differently. We have to distinguish four cases: both i and j appear in components in the CCS Δ, only i (only j) appears in a component and fourthly neither appears. The reason why the four cases only consider the occurence of i and j but not of $\hat{i}$ and $\hat{j}$ is again the precondition that in G there exists no edge between terminal nodes.

In the first case Δ is not valid iff there is an $h \in N_{\eta_{i,j}(s)}$ with $h \subseteq \bigcup\{c \in \Delta : i \in c \vee j \in c \vee \hat{i} \in c \vee \hat{j} \in c\}$, i.e. if the newly connected component is a superset of a forbidden set in $N_{\eta_{i,j}(s)}$. In the second case we have to check whether $h \subseteq \bigcup\{c \in \Delta : i \in c \vee \hat{j} \in c\}$. If this is the case a terminal pair has been connected and hence the CCS Δ has to be discarded. Case 3 works analogously to Case 2. In Case 4 no new edges are introduced. Nevertheless we have to check if not already an i- and j-labeled terminal pair is contained in a connected component.

Theorem 9. *RVMC is FPT with respect to $cw(G \cup H)$ on graphs that do not contain edges between terminal vertices.*

Runtime estimates can be found similarly to Theorem 7. It can easily be seen that Conjecture 8 holds for cographs. Hence,

Corollary 10. *RVMC is in P if $G \cup H$ is restricted to cographs.*

4.3 Edge Multicut with $cw(G) + |H|$ as Parameter

For EMC, contrary to Vertex Multicut, cuts occur during η-operations, i.e. when new edges are introduced. Therefore it is necessary to store the number of vertices in a component with a certain label. Otherwise we would not be able to calculate the number of cuts required to separate two components. However, adding this information to the data structure yields up to $n^{f(k)}$ branchings in the algorithm, where $f(k)$ is defined as k, the number of labels, times the number of components in a CCS. Clearly such an algorithm is in XP.

5 Conclusion and Future Work

We have pinpointed the parameterized complexity of the UVMC, RVMC and EMC problem for several parameterizations involving clique-width. In the literature, also *weighted versions* of Multicut have been studied where the vertices or the edges are assigned a weight and one seeks to minimize the weight rather than the cardinality of the cut. Our algorithms for UVMC and RVMC can be easily extended so as to also handle weights. In contrast, there is no obvious extension of our EMC algorithm to the weighted version of this problem. For the XP-algorithm for EMC it would be of interest to prove a corresponding W[1]-hardness result which would imply that no FPT algorithm exists.

One drawback of clique-width is that it is in general NP-complete to detect if a given graph has clique-width $\leq k$ [9]. However, there has been recent progress in identifying graph classes for which the computation of k-expression can be done in polynomial time [16,17]. Another approach, by Oum and Seymour [21], is the notion of *rank-width*, which is strongly related to clique-width and which admits good algorithms for finding decompositions. Since $rw(G) \leq cw(G) \leq 2^{1+rw(G)} - 1$ holds [21], all our NP-completeness results as well as all our FPT and XP membership results also hold for rank-width. Nevertheless, it is to be expected that custom-made algorithms for rank-width perform better than algorithms using a “detour” via clique-width. The construction of such algorithms, e.g. building on [10], is a task for future work.

Finally, the question of how much edge contractions can increase the clique-width of graphs (Conjecture 8) seems to be a problem worthwhile to study.

References

1. Bousquet, N., Daligault, J., Thomassé, S.: Multicut is FPT. In: Fortnow, L., Vadhan, S.P. (eds.) Proceedings of the 43rd Annual ACM Symposium on Theory of Computing, STOC 2011, pp. 459–468. ACM, New York (2011)
2. Calinescu, G., Fernandes, C.G., Reed, B.: Multicuts in unweighted graphs and digraphs with bounded degree and bounded tree-width. J. Algorithms 48(2), 333–359 (2003)
3. Corneil, D.G., Rotics, U.: On the relationship between clique-width and treewidth. SIAM J. Comput. 34(4), 825–847 (2005)
4. Courcelle, B., Engelfriet, J., Rozenberg, G.: Handle-rewriting hypergraph grammars. J. Comput. Syst. Sci. 46(2), 218–270 (1993)
5. Courcelle, B., Makowsky, J.A., Rotics, U.: Linear time solvable optimization problems on graphs of bounded clique-width. Theory Comput. Syst. 33(2), 125–150 (2000)

6. Courcelle, B., Olariu, S.: Upper bounds to the clique width of graphs. Discrete Applied Mathematics 101(1-3), 77–114 (2000)
7. Dahlhaus, E., Johnson, D.S., Papadimitriou, C.H., Seymour, P.D., Yannakakis, M.: The complexity of multiterminal cuts. SIAM J. Comput. 23(4), 864–894 (1994)
8. El-Mallah, E.S., Colbourn, C.J.: The complexity of some edge deletion problems. IEEE Transactions on Circuits and Systems (1988)
9. Fellows, M.R., Rosamond, F.A., Rotics, U., Szeider, S.: Clique-width is NP-complete. SIAM J. Discrete Math. 23(2), 909–939 (2009)
10. Ganian, R., Hliněný, P.: On parse trees and Myhill-Nerode-type tools for handling graphs of bounded rank-width. Discrete Applied Mathematics 158(7), 851–867 (2010); Third Workshop on Graph Classes, Optimization and Width Parameters Eugene, Oregon, USA (October 2007)
11. Garg, N., Vazirani, V.V., Yannakakis, M.: Primal-dual approximation algorithms for integral flow and Multicut in trees. Algorithmica 18(1), 3–20 (1997)
12. Gottlob, G., Lee, S.T.: A logical approach to Multicut problems. Information Processing Letters 103(4), 136–141 (2007)
13. Guo, J., Hüffner, F., Kenar, E., Niedermeier, R., Uhlmann, J.: Complexity and Exact Algorithms for Multicut. In: Wiedermann, J., Tel, G., Pokorný, J., Bieliková, M., Štuller, J. (eds.) SOFSEM 2006. LNCS, vol. 3831, pp. 303–312. Springer, Heidelberg (2006)
14. Gurski, F.: Graph operations on clique-width bounded graphs. CoRR abs/cs/0701185 (2007)
15. Gutierrez, C., Hurtado, C., Vaisman, A.: RDFS Update: From Theory to Practice. In: Antoniou, G., Grobelnik, M., Simperl, E., Parsia, B., Plexousakis, D., De Leenheer, P., Pan, J. (eds.) ESWC 2011, Part II. LNCS, vol. 6644, pp. 93–107. Springer, Heidelberg (2011)
16. Heggernes, P., Meister, D., Rotics, U.: Exploiting Restricted Linear Structure to Cope with the Hardness of Clique-Width. In: Kratochvíl, J., Li, A., Fiala, J., Kolman, P. (eds.) TAMC 2010. LNCS, vol. 6108, pp. 284–295. Springer, Heidelberg (2010)
17. Heggernes, P., Meister, D., Rotics, U.: Computing the Clique-Width of Large Path Powers in Linear Time via a New Characterisation of Clique-Width. In: Kulikov, A., Vereshchagin, N. (eds.) CSR 2011. LNCS, vol. 6651, pp. 233–246. Springer, Heidelberg (2011)
18. Hliněný, P., Oum, S., Seese, D., Gottlob, G.: Width parameters beyond tree-width and their applications. The Computer Journal 51(3), 326–362 (2008)
19. Marx, D.: Parameterized graph separation problems. Theor. Comput. Sci. 351(3), 394–406 (2006)
20. Marx, D., Razgon, I.: Fixed-parameter tractability of Multicut parameterized by the size of the cutset. In: Fortnow, L., Vadhan, S.P. (eds.) Proceedings of the 43rd Annual ACM Symposium on Theory of Computing, STOC 2011, pp. 469–478. ACM, New York (2011)
21. Oum, S., Seymour, P.: Approximating clique-width and branch-width. J. Comb. Theory Ser. B 96(4), 514–528 (2006)

Radiation Hybrid Map Construction Problem Parameterized

Chihao Zhang[1], Haitao Jiang[2], and Binhai Zhu[3]

[1] Department of Computer Science, Shanghai Jiao Tong University, Shanghai, 200240, China
chihao.zhang@gmail.com

[2] School of Computer Science and Technology, Shandong University, Jinan, Shandong, 250100, China
htjiang@mail.sdu.edu.cn

[3] Department of Computer Science, Montana State University, Bozeman, MT, 59717-3880, USA
bhz@cs.montana.edu

Abstract. In this paper, we study the Radiation Hybrid Map Construction (RHMC) problem which is about reconstructing a genome from a set of gene clusters. The problem is known to be NP-complete even when all gene clusters are of size two and the corresponding problem (RHMC_2) admits efficient constant-factor approximation algorithms. In this paper, for the first time, we consider the more general case when the gene clusters can have size either two or three (RHMC_3). Let p-RHMC be a parameterized version of RHMC where the parameter is the size of solution. We present a linear kernel for p-RHMC_3 of size $22k$, together with a bounded search tree algorithm, we obtain an FPT algorithm running in $O(6^k k+n)$ time. For p-RHMC_2 we present a bounded search tree algorithm which runs in $O^*(2.45^k)$ time, greatly improving the previous bound using weak kernels.

1 Introduction

Radiation hybrid (Rh) mapping is a popular and powerful technique for mapping unique DNA sequences onto chromosomes and whole genomes. The achieved map of these DNA sequences provide a basis for association studies in modern genetics. The technique has been used since 1990 for construction maps of small chromosomal regions for human and several other mammals [5,10,11].

In Rh mapping experiments, chromosomes of the target organism are randomly broken into small DNA fragments through gamma radiation. The underlying mechanism is that, when two markers are physically close to each other on the chromosome, the probability that these two markers are broken down by the gamma radiation is low, and so with a high probability they are either co-present in or co-absent from a DNA fragment. The *radiation hybrid map construction* (RHMC) problem is to determine the most likely linear order of the markers using the observed co-occurrences. We will formally define this problem in the next section.

G. Lin (Ed.): COCOA 2012, LNCS 7402, pp. 127–137, 2012.

Traditional Rh map construction methods are mostly heuristics, and often they are only able to produce framework maps on a small portion of all the markers [6]. Slonim et al. proposed a hidden Markov model on the Rh mapping data and used a maximum-likelihood approach to compute the map [11]; Givry et al. proposed to take advantage of known sequence information for target chromosomes for building more robust maps [3]. In [2], the RHMC_2 problem (i.e., each cluster contains two genes) was shown to be NP-hard and a 2-approximation algorithm was presented. The approximation ratio was then improved to 10/7 in [1].

In this paper, we study the problem under the framework of parameterized complexity. We restrict our attention on the case when the size of each cluster is at most 3. We show that this problem is Fixed-Parameter Tractable (FPT) by presenting a $22k$ kernel for it. Moreover, on top of the kernel, we present a bounded search tree algorithm which runs in $O(6^k k + n)$ time. Furthermore, in the case when the size of each cluster is at most 2, we give an improved FPT algorithm which runs in $O^*(2.45^k)$ time.

This paper is organized as follows. In Section 2, we give some necessary definitions regarding the problem as well as FPT algorithms. In Section 3, we present the linear kernel for RHMC_3, where each cluster has at most three genes, and then give an FPT algorithm based on the kernel. In Section 4, we present an improved FPT algorithm for RHMC_2. Finally in Section 5, we conclude the paper with several open questions.

2 Preliminaries

Radiation Hybrid Map Construction Problem. Let Σ be a set of markers, and $\mathcal{C} = \{C_i \subseteq \Sigma : 1 \leq i \leq n\}$ be a set of clusters, with $|C_i| \leq d$. The problem is to decide whether after deleting some k clusters, there is a total order $\leq$ on Σ under which for every remaining cluster C, the markers in C are consecutive.

For example, take $\Sigma = \{a, b, c, d\}$ and $C_1 = \{b, c, d\}, C_2 = \{a, c, d\}, C_3 = \{a, b\}$. If none of C_1, C_2, C_3 is deleted, there is no total order on Σ satisfying our requirement. However, if we delete C_1, the order $c \leq d \leq a \leq b$ satisfies the condition since both $C_2 = \{a, c, d\}$ and $C_3 = \{a, b\}$ appear consecutively.

We denote this problem by RHMC. For fixed d, we call it RHMC_d. It is easy to see that RHMC_2 is equivalent to the so-called Minimum co-Path problem: Given a simple undirected graph, decide whether one can delete some k edges, resulting a graph which is a disjoint union of paths. Minimum co-Path problem can be viewed as the complement of the Hamiltonian path problem, hence it is NP-hard [2].

Fixed-Parameter Tractable Algorithm. Let (I, k) be an instance of parameterized problem. An FPT algorithm decides (I, k) in time $O(f(k) \cdot n^c)$, where f is an arbitrary computable function that only depends on k and c is a constant. We often use the notation $O^*(f(k))$ to suppress the polynomial term. A basic approach towards FPT algorithm is to consider the *problem kernel*. Formally, a polynomial time algorithm K is a *kernelization* if it reduces the instance (I, k)

to another instance (I', k') such that (1) (I, k) is a YES instance if and only if (I', k') is a YES instance, and (2) there is a computable function h such that $|I'| \leq h(k)$. The reduced instance (I', k') is called the *kernel*, and if h is a linear function, we say that the kernel is linear. It is well known that a parameterized problem is FPT if and only if it has a kernel. In many cases, the kernelization K consists of many reduction rules which reduce the size of the input instance. For more information on parameterized complexity and algorithms, one can refer to [4,7,9]. The parameterized version of $\mathsf{RHMC_d}$ is defined as follows:

p-$\mathsf{RHMC_d}$ PROBLEM

Input: A set of clusters $\mathcal{C} = \{C_i \subseteq \Sigma : 1 \leq i \leq n\}$ with $|C_i| \leq d$, and an integer $k \in \mathbb{N}$.

Parameter: k.

Problem: Decide whether after deleting some k clusters, there is a total order $\leq$ on Σ under which for every remaining cluster C, the markers in C are consecutive.

3 A Linear Kernel for p-$\mathsf{RHMC_3}$

Let Σ be a set of markers, and $I = (\mathcal{C} = \{C_i \subset \Sigma : 1 \leq i \leq n\}, k)$ be an instance of p-$\mathsf{RHMC_3}$ such that $|C_i| \leq 3$ for all $1 \leq i \leq n$. We can rephrase p-$\mathsf{RHMC_3}$ as a problem on graphs:

Let $G(I) = (V(I), E(I))$ be an undirected graph. We consider each marker $c \in \Sigma$ as a vertex in $G(I)$. We add edge $\{u, v\}$ to $E(I)$ if $u, v \in C_i$ for some i and $u \neq v$. Since each cluster contains at most 3 markers, it can be viewed as a subgraph (a vertex, or two vertices on an edge, or a K_3) in the $G(I)$. We call a subgraph of $G(I)$ *legal* if it corresponds to some cluster. We say two clusters C_i, C_j are neighbors if $C_i \cap C_j \neq \varnothing$, i.e. their corresponding subgraphs share at least one vertex.

Then $\mathsf{RHMC_3}$ is equivalent to deciding whether one can remove a set $S(I)$ of k clusters such that there exists a set of disjoint paths $T(I)$ in $G(I)$ and for each remaining legal subgraph: (1) if it is a triangle, then it contains exactly two edges covered by some path in $T(I)$; (2) if it is an edge or a vertex, it belongs to some path in $T(I)$. We call $T(I)$ the *valid* set.

In Section 3.1 we will define the notion of *good* pattern and present a linear kernel in Section 3.2.

3.1 Good Patterns

To ease the presentation, we first consider the case that all clusters are of size 3, i.e. all the legal subgraphs are triangles. We denote this problem by p-$\mathsf{RHMC_3}^*$. At the end of Section 3.2, we will consider the general case of p-$\mathsf{RHMC_3}$.

For a fixed cluster, consider all its neighbors, only 10 patterns as illustrated in Figure 1 are allowed to draw a valid path. (The fixed cluster is the one shaded and the thick line denote the path.)

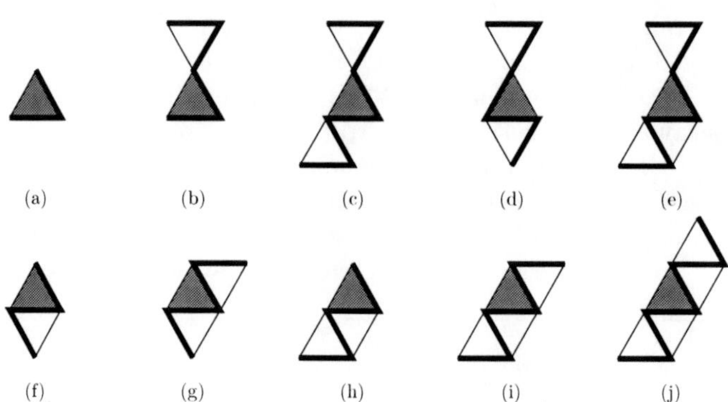

Fig. 1. 10 good patterns of shaded cluster

A cluster C is *good* if its neighbors and itself form a pattern in Figure 1 where C is the shaded one; a good cluster is *free* if all of its neighbors are good. If a cluster is not good, we call it *bad.*

Given an instance I, if all clusters are good, then it is easy to solve this instance. If $G(I)$ contains more than one component, then we can consider each component independently. Hence in the following we assume that at least one cluster is bad and $G(I)$ is connected.

We immediately have the following lemma:

Lemma 1. *Let* $I = (\{C_i : 1 \leq i \leq n\}, 0)$ *be an instance of* $\mathsf{RHMC_3}^*$*, then* I *is a YES instance if and only if (1) all the clusters are good (hence free) and (2)* $G(I)$ *contains no cycle.*

Proof. The "only if" direction is straightforward.

If all clusters are good, by recognizing all patterns in Figure 1, it is easy to verify that we can also connect two valid paths when attaching two good clusters. □

This lemma directly implies a bounded search tree algorithm for p-$\mathsf{RHMC_3}^*$: For a bad cluster, consider (at most) 5 of its neighbors and itself, and branch on all possible ways to form a good pattern (in Figure 1) by removing at least one of the 6 clusters. The worst case occurs around case (j) in Figure 1, where we have a band of 6 triangles forming a cycle — with each triangle sharing exactly one edge with each of its two neighbors. In this case, any of the 6 clusters can be deleted to obtain the good pattern (j) and we need to branch on each of these 6 cases. Hence this bounded search tree algorithm runs in $O^*(6^k)$ time. After all the bad patterns are destroyed, we need to in addition break all the cycles (which can be done in an arbitrary fashion as these cycles must be all disjoint).

Theorem 1. p-$\mathsf{RHMC_3}^*$ *can be solved in* $O(6^k n)$ *time.*

3.2 Kernelization Algorithm

In this section, we present a kernelization algorithm for p-RHMC$_3$*. Let I be an instance. We first define an operation on good patterns called *contraction.* Let C be a good cluster of size 3, if its pattern is among (a)(b)(f)(g)(h)(i)(j) in Figure 1, then contracting C means removing C from I. Otherwise, if C's pattern is among (c)(d)(e), then by contraction we mean removing C from I and identifying two vertices of its neighbors, as depicted in Figure 2.

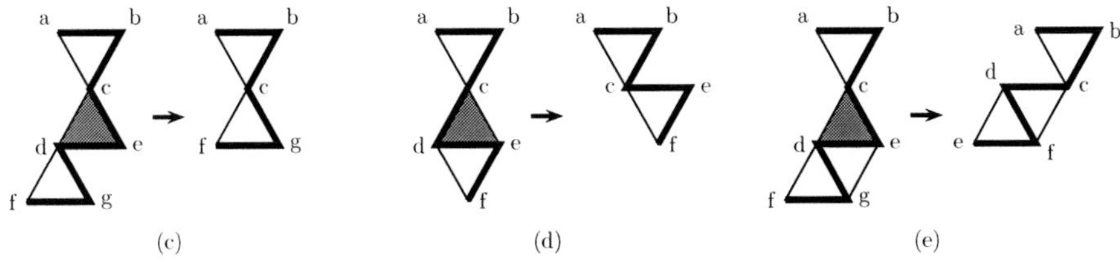

Fig. 2. Contraction of patterns (c)(d) and (e)

Our kernelization algorithm is exhaustively applying the following rule:

Rule: If there is a free cluster C in G, then contract it.

Lemma 2. *Let $I = (\{C_i : 1 \leq i \leq n\}, k)$ be an instance of p-RHMC$_3$*, C be a free cluster and I' be the instance obtained from I by contracting C. Then I is a YES instance if and only if I' is a YES instance.*

Proof. We first prove the "only if" direction: Assume that I is a YES instance, if C is of pattern (a)(b)(f)(g)(h)(i)(j), contracting C is equivalent to deleting C, hence a solution set of I is also a solution set of I'. Otherwise, let $S(I)$ and $T(I)$ be the solution set and valid set of I respectively. If $C \notin S(I)$, then $S(I)$ is also a solution set of I'. If $C \in S(I)$, we distinguish between three cases:

(1) If C's pattern is (c), since $C \in S(I)$, $T(I)$ must contain edges ac, cb, fd, dg. In this case, we know $S(I)\backslash\{C\} \cup \{C'\}$ is a solution set of I', where C' is the cluster $\{a, b, c\}$.
(2) If C's pattern is (d), first let $T(I') = T(I)$. If it is the case that $T(I)$ contains edges ab, bc, de, df, then we can safely change df to ef in $T(I')$. Thus $S(I)$ is also a solution set of I'. The other cases are symmetric.
(3) If C' pattern is (e), again let $T(I') = T(I)$. It must be the case that $T(I)$ contains edges ac, bc, and de (or eg), then we can remove de (or eg) from $T(I')$ after contacting C. Thus $S(I)$ is also a solution set of I'.

To prove the "if" part we need more effort. Assume I' is a YES instance and $S(I')$ be one of its solution of size at most k, we show that $S(I')$ is also a solution set of I.

We know that after removing $S(I')$ in $G(I')$, we can find a valid set $T(I')$. We claim that after restoring C in $G(I')$, either

(1) there exists a path in C that connects two paths in $T(I')$, or
(2) C already has two edges covered by some path in $T(I')$.

We first prove the claim. Since C is free, all its neighbors are good, and hence it is among the 10 patterns illustrated in Figure 1. Let C' be one of its neighbors, we justify the claim by examing the pattern of C'.

The pattern illustrated in Figure 3 encompasses all situations, where C' is cluster 2(the one shaded). We analyse this pattern in detail and other patterns can be checked similarly.

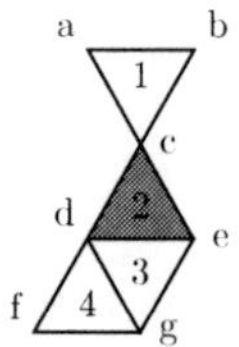

Fig. 3. One good pattern of C'

a. If C is cluster 1, after contracting C, the path in $T(I')$ must be $P = cedgf$. thus after restoring C, we can extend P with path cba or cab (depending on which one of a and b is identified with c while contracting).
b. If C is cluster 3, after contracting C, the path in $T(I')$ across this pattern may be $P_1 = abcedgf$, $P_2 = bacedgf$, $P_3 = abcedfg$ or $P_4 = bacedfg$. Restoring C does not affect P_1 and P_2. If it is the case of P_3 (resp. P_4), we can safely replace the path by $abcedgf$ (resp. $bacedgf$). This is because we only have two ways to force the existence of edge df in some path:
   (1) We have some cluster $\{d, f, x\}$ (x is not among $\{a, b, c, d, e, f, g\}$, but this cluster shares a vertex with cluster 2 and hence it is impossible in this pattern.
   (2) We have some cluster $\{g, x, y\}$ (x, y are not among $\{a, b, c, d, e, f, g\}$), but in this situation, cluster 3 is no longer good.
c. If C is cluster 4, after contracting C, the path in $T(I')$ across this pattern may be $P_1 = abcedg$, $P_2 = bacedg$, $P_3 = abcdeg$ or $P_4 = bacdeg$. Restoring C does not affect P_1 and P_2. If it is the case of P_3 (resp. P_3), we can safely replace the path by $abcedg$ (resp. $bacedg$). The reason is similar to case b above.

By the claim, if (1) happens, restoring C only extends some path in $T(I')$; if (2) happens, restoring C does not affect $T(I')$. Then we know $S(I')$ is also a solution set of I. □

Now we come to the general case of p-RHMC$_3$, i.e. some clusters may be of size 2. We first generalize the operation of contraction in the following way: Let I be an instance and C a 2-sized cluster. If there is another 3-sized cluster C' such that $C \subset C'$, then contracting C is equivalent to removing C. Similarly, for a 3-sized cluster C, if there is another 2-sized cluster C' such that $C' \subset C$, then contracting C is equivalent to removing C. Otherwise, for a 2-sized cluster C, the operation is equivalent to contracting the edge in $G(I)$ and for a 3-sized cluster, the definition of contraction is the same as the case in p-$\mathsf{RHMC_3}^*$. Secondly, in $\mathsf{RHMC_3}$, some new patterns are introduced, both for 2-sized clusters and 3-sized clusters. Figure 4 illustrates good patterns of a 2-sized cluster whose neighbors are triangles.

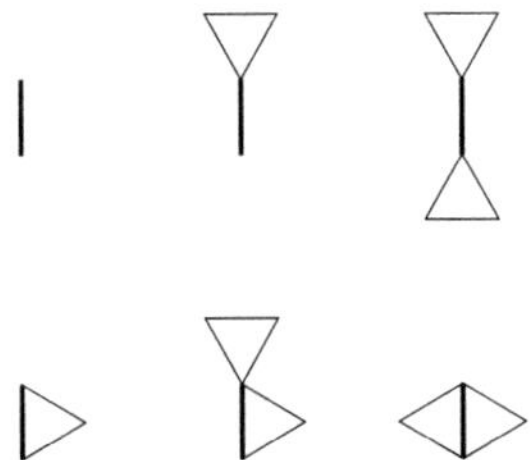

Fig. 4. Good patterns of 2-sized clusters such that all the neighbors are triangles

The analysis of free 2-sized clusters is similar to 3-sized ones. Now we consider the influence of 2-sized clusters to proof of Lemma 2, i.e. the analysis of 3-sized cluster. It introduces a new way to force the existence of some edge in case (b) and (c). Taking (b) for instance, if edge *df* is some 2-sized cluster, then we know that we must keep it in the valid set. However, it is easy to check that, in this case, cluster *df* is not good as we cannot draw a path across its pattern. The similar argument holds for (c).

Therefore both Theorem 1 and Lemma 2 can be generalized to p-$\mathsf{RHMC_3}$.

Theorem 2. *p-$\mathsf{RHMC_3}$ can be solved in $O(6^k n)$ time.*

Lemma 3. *Let $I = (\{C_i : 1 \leq i \leq n\}, k)$ be an instance of p-$\mathsf{RHMC_3}$ and C a free cluster, I' be the instance obtained from I by contracting C. Then I is a YES instance if and only if I' is a YES instance.*

Lemma 4. *Let $I' = (\{C_i : 1 \leq i \leq n'\}, k)$ be the reduced instance of p-$\mathsf{RHMC_3}$ after exhaustively applying the reduction rule. If I' is a YES instance, then $n' \leq 22k$.*

Proof. Since I' is a YES instance, there exists a set of clusters $S(I')$ with $|S(I')| \leq k$ such that after removing $S(I')$, all the clusters left are free. Let $\mathcal{L}$ be these free clusters left and $C \in S(I')$ be a cluster. Consider the set $\mathcal{L} \cup \{C\}$, some free cluster may become non-free. But this number is bounded by some constant because C can only touch a constant number of clusters which are of good pattern, and since these clusters are good before adding C, they can only have a constant number of neighbors. After an exhaustive seach for all the patterns, we conclude that adding C to $\mathcal{L}$ can change at most 21 clusters from free to non-free ones, the extreme case is shown in Figure 5. Therefore, the total number of clusters in I' does not exceed $22k$. □

Fig. 5. The shaded cluster affects 21 clusters

Theorem 3. *p-RHMC$_3$ has a kernel of size* $22k$.

Remark 1. The $22k$ kernel directly implies a 22-approximation algorithm for RHMC$_3$, i.e. choose all the clusters in the kernel as solution.

Combining Theorem 2 and Theorem 3, we have

Corollary 1. *p-RHMC$_3$ can be solved in time* $O(k \cdot 6^k + n)$.

4 An FPT Algorithm for p-RHMC$_2$

In [8], a $5k$ weak kernel (loosely speaking, parameterized search space) is constructed for p-RHMC$_2$. That immediately implies an FPT algorithm which runs in $O^*(\binom{5k}{k}) = O^*(2^{3.61k})$ time. Here we present an FPT algorithm which runs in $O^*(2.45^k)$ time, using the well-known bounded search tree algorithm.

First notice that RHMC$_2$ problem is equivalent to Minimum co-Path Set problem.

Given a simple undirected graph G, a *co-path set* is a set S of edges in G whose removal leaves a graph in which every connected component is a path. And the problem is to decide whether there exists a co-path set of size k.

For an instance of RHMC$_2$, we let $V(G)$ be the set of markers Σ, and $\{u, v\} \in E(G)$ if there is some cluster $C_i = \{u, v\}$. It is then easy to verify these two problems are equivalent.

Hence in the following, we describe the algorithm in term of Minimum co-Path Set problem.

We start with a simple lemma.

Lemma 5. *If there are two edges $e_1 = (u, v)$ and $e_2 = (v, w)$ in G, where $d(u) = d(w) = 3$ and $d(v) = 2$, then there exists an optimal solution for* RHMC_2 *which does not delete e_1 and e_2.*

Proof. Let $N(u) = \{v, u', u''\}$ and $N(w) = \{v, w', w''\}$. It suffices to prove that if e_1 or e_2 (or both) is deleted in some optimal solution S, then we can replace e_1 with one edge from (u, u') and (u, u'') or replace e_2 with one of (w, w') and (w, w'') (or both) to obtain another optimal solution which is at least as good as S. Suppose that e_1 is deleted in some optimal solution S for RHMC_2, then after all the edges in S are deleted, v is the end of some path P. We consider three cases.

(1) If $u' \in P$ and $u'' \notin P$ (or vice versa), which means (u, u') (resp. (u, u'')) is also deleted for S, then replace e_1 with this deleted edge (u, u') (resp. (u, u'')). Clearly, P is replaced by a new path of the same length.

(2) If $u' \in P$ and $u'' \in P$, then we can assume that $u \in P$ (otherwise, we can add e_1 back to P to obtain a longer path whose endpoint is u). Consequently, the path is either in the form $\langle v, \cdots, u', u, u'', \cdots \rangle$ or $\langle v, \cdots, u'', u, u', \cdots, \rangle$. We can replace e_1 with (u, u') in the former case or replace e_1 with (u, u'') in the latter case, to obtain a new path with the same size as P.

(3) If $u' \notin P$ and $u'' \notin P$, then we can replace e_1 with either (u, u') or (u, u'') to have a new solution of the same size as that of S. The argument for e_2 is similar hence omitted. □

Corollary 2. *If there is a path $P = \langle u, v_1, \cdots, v_k, w \rangle$ in G with $d(u) = d(w) = 3$, and $d(v_i) = 2$ for all $1 \leq i \leq k$, then there is an optimal solution for* RHMC_2 *in which all the edges along the path P are reserved (i.e., not deleted).*

Lemma 6. *Given a vertex v, if $N(v) = \{v_1, v_2, v_3\}$ and $d(v_1) = d(v_2) = d(v_3) = 2$, then there exists an optimal solution for* RHMC_2 *which deletes either (v, v_1) or (v, v_2).*

Proof. Assume to the contrary that the optimal solution does not delete (v, v_1) and (v, v_2), instead it deletes (v, v_3). Then, due to the fact that $d(v_3) = 2$ before the deletion, v_3 is the endpoint of some path P. We have two cases. (1) If $v_1 \in P$ and $v_2 \in P$, then the path is either in the form $\langle v_3, \cdots, v_1, v, v_2, \cdots, \rangle$ or in the form $\langle v_3, \cdots, v_2, v, v_1, \cdots, \rangle$, and we can replace (v, v_3) with (v, v_1) in the former case or replace (v, v_3) with (v, v_2) in the latter case. (2) If $v_1 \notin P$ and $v_2 \notin P$, we can replace (v, v_3) with either (v, v_1) or (v, v_2). In both cases we obtain an alternative optimal solution for RHMC_2. □

Algorithm *Bounded Search Co-path Set*
Input: *Graph G, integer k*
Output: *A Co-path set S of size k*
1 If some component of G contains at most 10 vertices, then use brute-force to solve that component optimally.
2 While there exists a vertex v such that $d = d(v) \geq 4$, choose all but two of its incident edges and put them in S.
3 While there exists a vertex v satisfying Lemma 6, choose one of its two incident edges and put it in S following Lemma 6.
4 While there exists vertices u, v, w satisfying Lemma 5, reserve edges (u, v) and (v, w) following Lemma 5.
5 For a path $P = \langle u, v, w \rangle$, where $N(v) = \{u, w, a\}$, $N(u) = \{v, b, c\}$, and $N(w) = \{v, g, h\}$
  5.1 add (u, v) and (w, g) to S;
  5.2 or add (u, v) and (w, h) to S;
  5.3 or add (u, b) and (v, w) to S;
  5.4 or add (u, b) and (v, a) to S;
  5.5 or add (u, c) and (v, w) to S;
  5.6 or add (u, c) and (v, a) to S.
6 Repeat Steps 3,4,5 until every vertex has degree less than 3.
7 Choose an arbitrary edge from each cycle and put it in S.
8 Return S.

Theorem 4. *Algorithm* Bounded Search Co-path Set *computes a co-path set in $O^*(6^{k/2}) \approx O^*(2.45^k)$ time.*

Proof. Step 1 uses constant time.
Step 2 has a recurrence relation

$$f(k) = \binom{d}{d-2} f(k - (d - 2)),\ d \geq 4.$$

Step 3 has a recurrence relation

$$f(k) = 2f(k - 1).$$

Step 4 has a recurrence relation

$$f(k) = 4f(k - 2).$$

Step 5 branches on whether (u, v) is deleted or not. If (u, v) is deleted, then (v, w) is reserved and either (w, g) or (w, h) is also deleted; if (u, v) is reserved then at least one of (u, b) and (u, c), as well as at least one of (v, w) and (v, a) are deleted. So step 5 has a recurrence relation

$$f(k) = 6f(k - 2).$$

$f(k)$ achieves its maximum value when $d = 4$ or $f(k) = 6f(k - 2)$, so we have $f(k) \leq O^*((6)^{k/2}) \approx O^*(2.45^k)$. □

5 Conclusing Remarks

In this paper, we studied the Radiation Hybrid Map Construction problem using parameterized algorithms. For p-RHMC$_3$, where each gene cluster contains at most three genes, we showed an FPT algorithm based on a linear kernel of it. For p-RHMC$_2$, we presented a bounded search tree algorithm which runs in $O^*(2.45^k)$ time, greatly improving the previous bound using weak kernels. An important open question is whether one can extend these methods to handle p-RHMC$_d$, where each gene cluster contains at most d genes. Furthermore, does the generalized version p-RHMC remain FPT?

Acknowledgments. This research is partially supported by NSF of China under project 60928006 and 60970011, and by the Open Fund of Top Key Discipline of Computer Software and Theory in Zhejiang Provincial Colleges at Zhejiang Normal University.

References

1. Chen, Z.-Z., Lin, G., Wang, L.: An approximation algorithm for the minimum co-path set problem. Algorithmica 60(4), 969–986 (2011)
2. Cheng, Y., Cai, Z., Goebel, R., Lin, G., Zhu, B.: The radiation hybrid map construction problem: recognition, hardness, and approximation algorithms (2008) (unpublished manuscript)
3. De Givry, S., Bouchez, M., Chabrier, P., Milan, D., Schiex, T.: Carh ta Gene: multipopulation integrated genetic and radiation hybrid mapping. Bioinformatics 21(8), 1703 (2005)
4. Downey, R.G., Fellows, M.R.: Parameterized complexity. Springer, New York (1999)
5. Cox, D.R., Burmeister, M., Price, E.R., Kim, S., Myers, R.M.: Radiation hybrid mapping: a somatic cell genetic method for constructing high resolution maps of mammalian chromosomes. Science 250, 245–250 (1990)
6. Faraut, T., De Givry, S., Chabrier, P., Derrien, T., Galibert, F., Hitte, C., Schiex, T.: A comparative genome approach to marker ordering. Bioinformatics 23(2), e50 (2007)
7. Flum, J., Grohe, M.: Parameterized complexity theory. Springer, New York (2006)
8. Jiang, H., Zhang, C., Zhu, B.: Weak kernels. Electronic Colloquium on Computational Complexity, ECCC Report TR10-005 (October 2010)
9. Niedermeier, R.: Invitation to fixed-parameter algorithms. Oxford University Press, USA (2006)
10. Richard, C.W., et al.: A radiation hybrid map of the proximal long arm of human chromosome 11 containing the multiple endocrine neoplasia type 1 (MEN-1) and bcl-1 disease loci. American Journal of Human Genetics 49(6), 1189 (1991)
11. Slonim, D., Kruglyak, L., Stein, L., Lander, E.: Building human genome maps with radiation hybrids. Journal of Computational Biology 4(4), 487–504 (1997)

On the Central Path Problem*

Yongding Zhu and Jinhui Xu

Department of Computer Science and Engineering
State University of New York at Buffalo
Buffalo, NY 14260, USA
{yzhu3,jinhui}@buffalo.edu

Abstract. In this paper we consider the following Central Path Problem (CPP): Given a set of m arbitrary (i.e., non-simple) polygonal curves $Q = \{P_1, P_2, \ldots, P_m\}$ in 2D space, find a curve P, called *central path*, that best represents all curves in Q. In order for P to best represent Q, P is required to minimize the maximum distance (measured by the directed Hausdorff distance) to all curves in Q and is the locus of the center of minimal spanning disk of Q. For the CPP problem, a direct approach is to first construct the farthest-path Voronoi diagram $FPVD(Q)$ of Q and then derive the central path from it, which could be rather costly. In this paper, we present a novel approach which computes the central path in an "output-sensitive" fashion. Our approach sweeps a minimal spanning disk through Q and computes only a partial structure of the $FPVD(Q)$ directly related to P. The running time of our approach is thus $O((H + mk + n + s) \log m \log^2 n)$ and the worst case running time is $O(n^2 2^{\alpha(n)} \log n)$, where n is the size of Q, s is the total number of self-intersecting points of each individual curve in Q, k is the size of the visited portion of $FPVD(Q)$ by the central path algorithm, and H is the number of intersections between the visited portion of $FPVD(Q)$ and $VD(P_i)(i = 1, 2, \ldots, m)$.

1 Introduction

In this paper, we consider the following *Central Path Problem (CPP)*: Given a set of 2D polygonal curves (also called paths) $Q = \{P_1, P_2, \ldots, P_m\}$ with a total complexity of n, find a curve P (called *central path*) which minimizes the maximum distance to Q and best represents Q, where the distance from P to any curve $P_i \in Q$ is measured by the directed Hausdorff distance (DHD) $\overline{\delta_H}(P, P_i)$. To ensure that the computed central path P best represents Q, we require that P be formed by centers of minimal spanning disks (MSD) of Q. In CPP, the set of input curves may be non-simple and could entwine each other, making the central path computation quite expensive. In the worst case, the central path has size $\Omega(n^2)$.

The CPP studied in this paper is motivated by applications in several areas, such as segmentation, medical imaging, data mining and pattern recognition. In segmentation, the accurate boundary of an object in noisy images often needs to first generate a set of candidate segmentations using algorithms following different optimization criteria and then derive the most likely solution from them, which can be formulated as a CPP.

* This research was partially supported by NSF through a CAREER award CCF-0546509 and grants IIS-0713489 and IIS-1115220.

G. Lin (Ed.): COCOA 2012, LNCS 7402, pp. 138–150, 2012.

In data mining or pattern recognition, CPP can be used to model the problem of finding moving patterns of vehicles or other objects in data collected from sensor or other networks. In such applications, the trajectory of each moving object can be viewed as an arbitrary polygonal curve and the common pattern of a set of moving objects can be captured by the median (such as central path) of the set of trajectories [4,8].

In this paper, we use directed Hausdorff distance (DHD) and minimal spanning disks (MSD) to compute P. DHD ensures that the computed path P has the smallest distance to Q, and MSD allows P to preserve the "shape" of Q. The rationale behind using the direct Hausdorff distance from P to Q, instead of undirected Hausdorff distance, is that the directed Hausdorff distance from Q to P could lead to a 0-distance path which is meaningless to CPP as shown in Figure 2. More importantly, undirected Hausdorff distance could make the central path lose the critical "center" property. In [4], Buchin *et al.* suggested some criteria for median trajectory. For example, they require that the median trajectory should be part of the input and be conceptually in the "middle" of the input trajectories defined by the level of the arrangement. Similar to their criteria, we also require that our central path be conceptually at the "center" of input curves. Roughly speaking, the central path should be within the region bounded by the upper and lower envelopes of the input curves. Furthermore, our central path preserves the shapes of input curves (e.g., P_1 and P_2 in Figure 3) and is capable of reducing possible spike noises associated with the input curves in applications like segmentation. It also worth mentioning that Fréchet distance has exactly the same weakness as undirected Hausdorff distance.

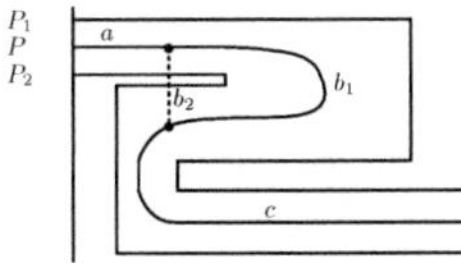

Fig. 1. Central paths of $\{P_1, P_2\}$, $\overline{ab_1c}$: using DHD and MSD, and $\overline{ab_2c}$: using DHD only

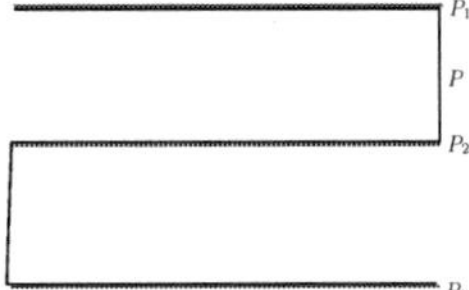

Fig. 2. 0-distance central path for CCP under $\overline{\delta_H}(Q, P)$

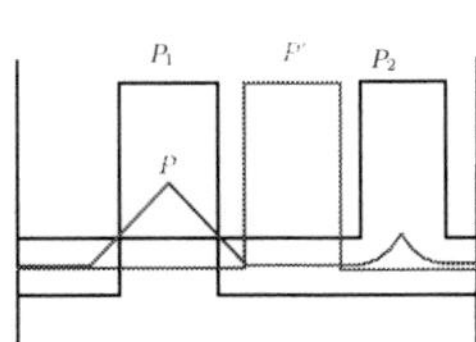

Fig. 3. The central paths of P_1 and P_2 constructed by using undirected Hausdorff distance (green path) and our definition (red one)

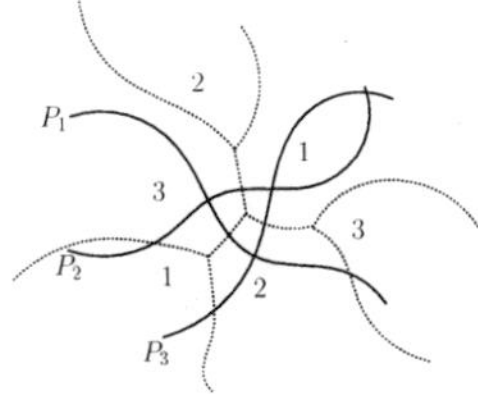

Fig. 4. FPVD of P_1, P_2 and P_3. The plane is subdivided into farthest-path Voronoi cell of P_i (marked by i, $i = 1, 2, 3$)

Although techniques exist for computing median of trajectories [4], their objectives are quite different (e.g., the median trajectory is part of the input trajectories). To our best knowledge, no existing algorithm directly solves the CPP problem. One possible solution is to first construct the farthest-path Voronoi diagram $FPVD(Q)$ [9,10] of Q (see Figure 4), and then find the central path P from it (since P is a substructure of $FPVD(Q)$). Thus the time complexity of such an approach is at least as large as that of constructing $FPVD(Q)$ (whose complexity can be easily quadratic). Available efficient algorithms for farthest Voronoi diagram are only for points, disjoint segments, and disjoint polygons [2,3,6,5,10]. For arbitrary non-disjoint polygonal curves, no efficient and output sensitive algorithm is previously known, and existing solutions have a complexity which is at least as large as the number of intersections of the set of curves (i.e., $\Omega(n^2)$ in the worst case), and could be much higher than the actual size of the $FPVD(Q)$. (Note that the size of $FPVD(Q)$ could be $o(n)$ even if the number of intersections of Q is quadratic.) Another possible solution to the CPP problem is to first determine all critical minimal spanning disks of Q and then connect (part of) them to form the central path. However, as shown in article [1], determining the minimum spanning disks takes $O(m^3 n \log n)$ time, which is much higher than the actual complexity of the central path.

To obtain a more efficient solution, in this paper we present a novel algorithm for the CPP problem. Our algorithm is based on the observation that the central path P does not need to have the whole $FPVD(Q)$. Thus it constructs only the portion of $FPVD(Q)$ directly related to the central path. Consequently, it significantly improves the aforementioned possible solutions. More specifically, it first sweeps a minimal spanning disk through the set of curves to find a backbone path of the central path and then search around the backbone to obtain the central path. Our algorithm finds the central path in $O((H + mk + n + s) \log m \log^2 n)$ time which is bounded by $O(n^2 2^{\alpha(n)} \log n)$ in the worst case, where s is the total number of self-intersections (e.g. intersections of the same curve) in Q, k is the size of the portion of $FPVD(Q)$ the algorithm visited, and H is the number of intersections between the visited portion of $FPVD(Q)$ and the Voronoi diagram of each P_i. Our result is based on a number of interesting geometric observations and can be easily implemented for practical purpose. Furthermore, our algorithm can be naturally extended to find FPVD or colored farthest Voronoi diagrams in an output-sensitive fashion.

2 Preliminaries

Let $Q = \{P_1, P_2, \cdots, P_m\}$ be the set of input polygonal curves. Each curve in Q is viewed as a set of vertices and open segments, both called sites. For the simplicity of our description, we make the following assumptions on Q: (1) General position assumption: Sites in Q are in general positions (i.e., no 4 sites are co-circular). In particular, no 3 segments intersect at one common point. (2) Non-overlapping assumption: No two segments partially or fully overlap, and no vertex of one input curve lies on another curve. (3) Common starting and ending assumption: All curves start at the same vertical starting line, end at the same vertical ending line, and lie completely between them.

The assumptions are reasonable and can be easily satisfied. For the first two assumptions, we can slightly perturb the vertices of the curves. For the third assumption, we extend each violating curve in the way as shown in Figure 5.

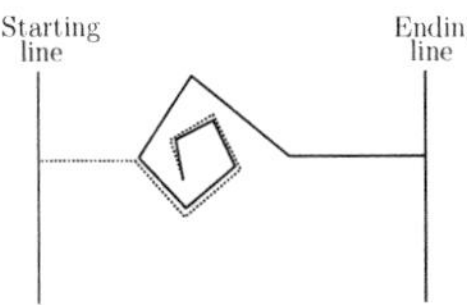

Fig. 5. Extend the starting point of a curve to the common starting line by using infinitesimally close edges

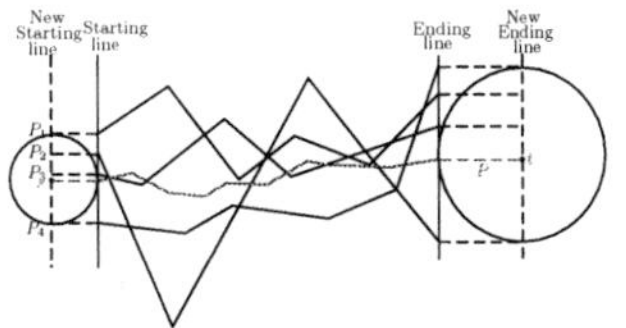

Fig. 6. Dummy extension of Q; the red line is the central path; the dotted lines segments are parallel

A (closed) disk D is called a spanning disk of Q if D intersects every curve in Q. D is a minimal spanning disk (MSD) if any disk shrinking from D (a disk D' is shrunk from D if $D' \subset D$) is no longer a spanning disk. D is a critical minimal spanning disk of Q if there are three curves in Q intersecting D only at its boundary circle. The center of D is called a *center point* and the three curves are called *touching curves* of D. Clearly each non-critical MSD D has two touching curves and each critical MSD has three touching curves. No MSD touches more than three curves due to the above assumptions.

In this paper, we use two types of Voronoi diagrams, the Voronoi diagram $VD(P_i)$ of each input curve P_i and the farthest-path Voronoi diagram $FPVD(Q)$ of Q. Each $VD(P_i)$ is defined as the Voronoi diagram of the set of sites of P_i. $VD(P_i)$ has a size of $O(n_i + s_i)$ and can be computed in $O((n_i + s_i)\log(n_i + s_i))$ time [7], where n_i is the complexity of P_i and s_i is the number of self-intersections in P_i.

$FPVD(Q)$ is a partition of the 2D space into cells such that each cell is the union of points which have the farthest distance to one input curve than to any other curve in Q (see Figure 4). Note that the farthest Voronoi region of an input curve could be empty or consist of a number of disjoint cells. Each edge in $FPVD(Q)$ is a line segment or parabolic curve. The size of $FPVD(Q)$ varies from $O(1)$ to $\Omega(n^2)$. As mentioned earlier, the central path can be found by solving a shortest path problem on $FPVD(Q)$ as shown in the following lemma.

Lemma 1. *CCP can be solved in* $O(n^2 2^{\alpha(n)} \log n)$ *time, where n is the size of Q.*

3 Algorithms for the Central Path Problem

In this section, we present our algorithms for computing the central path P.

Lemma 2. *The central path P lies on the farthest-path Voronoi diagram* $FPVD(Q)$.

From the above lemma, we know that one way to compute the central path P is to construct $FPVD(Q)$ first and then find P from it by some shortest path algorithm. Since constructing $FPVD(Q)$ takes at least quadratic time in the worst case, the time complexity for computing P is bottlenecked by the computation of $FPVD(Q)$. To obtain a better solution, we have to avoid explicitly constructing the whole $FPVD(Q)$. For this purpose, we first perform a *dummy extension* to all input curves as shown in Figure 6. More specifically, we extend the starting and ending vertices of each input curve horizontally by a large enough distance (e.g., $\frac{d}{2}$, where d is the vertical distance between the top and bottom curves along the starting or ending line) so that the disks centered at the points (denoted as **s** and **t** respectively) on the vertical starting and ending lines and tangent to the top and bottom curves are minimal spanning disks of Q. With the dummy extension, our main idea is thus, starting from **s**, to construct on the fly a partial structure of $FPVD(Q)$ which is barely enough for determining P. More specifically, our approach sweeps a minimal spanning disk D from **s** to **t**, and determines the moving direction and radius of D using local information of Q and $FPVD(Q)$.

To implement this idea is actually quite challenging. This is because the local structure of $FPVD(Q)$ close to D could be rather complicated and its construction depends on curves far away from D. This seemingly suggests that we have to maintain a substantial amount of information of $FPVD(Q)$ and Q in order to sweep the minimal spanning disk D. To overcome this difficulty, our main idea is to make use of the continuity property of each input curve. For this reason, we view each input curve as a directed graph. Let $V_{P_i} = \{v_1, v_2, \ldots, v_{n_i+1}\}$ and $E_{P_i} = \{\overrightarrow{e_1}, \overrightarrow{e_2}, \ldots, \overrightarrow{e_{n_i}}\}$ be the sets of vertices and directed edges of P_i, where $P_i \in Q, i = 1, 2, \ldots, m$. Each edge is oriented such that P_i starts at the starting line and ends at the ending line.

Since the input curves may not be simple, it is hard to maintain the sweeping disk in a consistent manner. To resolve this issue, we first preprocess them.

Lemma 3. *Q can be converted into a set of simple curves in $O(s + n\log n)$ time without changing the structures of $VD(P_i)$, $1 \le i \le m$, and $FPVD(Q)$, where $s = \sum_{i=1}^{m} s_i$ and s_i is the number of self-intersecting points of P_i.*

With the modified input curves, we now consider how to sweep D. First we notice that at each center point, D faces two possible directions while sweeping along the bisector of two touching curves, as D touches 3 curves at such a location. To determine the correct moving direction, we define a direction for each point of the bisector.

Definition 1. *Let $\overline{ab}$ be an edge (i.e., a segment or a parabolic curve) in a bisector of two sites in Q with D sweeping from a to b. Let $f(t)$, $t \in [0, 1]$, be a parametrization of $\overline{ab}$ such that $f(0) = a$ and $f(1) = b$. Then the* forward direction *of any interior point p of $\overline{ab}$ (with respect to D) is $f'(t_p)$, where $f(t_p) = p$. If p is the left endpoint a, the* forward direction *is the right derivative of the parametrization function, i.e., $f'(x^+)$.*

From the above definition, we know that any point other than the center points of $FPVD(Q)$ can have at most one forward direction depending on how D sweeps. At a center point, we have two candidate forward directions. If D sweeps toward both directions, the sweeping will lead to a full construction of $FPVD(Q)$. However, for the CPP, this could be too costly. Therefore it is crucial to determine which forward direction to sweep when D reaches a center point.

With the orientations of input curves and bisectors, we now consider how to sweep an MSD D starting from **s**. From the definition of **s**, it is easy to see that **s** is an interior point of a bisector, say $\overline{sa}$, for the the top and bottom sites of Q at the starting line, and has a unique forward direction (i.e., toward the ending line). Thus, D can be moved along $\overline{sa}$ until a center point is encountered.

To determine the next center point, our idea is to observe the change of the touching curves of the sweeping disk D. The next definition and lemma give some property of any new touching curve.

Definition 2. *Let D be the sweeping disk touching two input curves P_1 and P_2 at p_1 and p_2 respectively. p_1 and p_2 divide the bounding circle of D into two portions. The portion intersected by the ray (or half line) originated from the center of D in the forward direction is the* front portion *and the other is* rear portion.

Lemma 4. *While moving the sweeping disk D in a given forward direction, D touches a new input curve only at its rear portion.*

As mentioned earlier, the sweeping disk D keeps moving along the bisector of its two touching curves until it reaches the next center point. Each point on the bisector is then associated with a forward direction by Definition 2. A bisector oriented in such a way is called *directed bisector*. At a center point, D picks only one of the two directed bisectors to continue the sweeping. Our algorithm chooses the one such that the orientations of the directed bisector and the two touching sites are *coherent*. As we will show later in Lemma 9, the choice uniquely exists.

Definition 3. *Let $\overline{ab}$ be an edge of a directed bisector of two sites S_1 and S_2 in Q, $c \neq b$ be a point in $\overline{ab}$, and $\overline{\eta}_c$ be the forward direction of c. $\overline{\eta}_c$ is* coherent *with S_1 (similarly defined for S_2), if one of the following conditions is satisfied:*

1. *S_1 is a segment; the angle between $\overline{\eta}_c$ and the associated direction of S_1 is acute;*
2. *S_1 is a vertex, say p, incident to segments $\overline{p_1p}$ and $\overline{pp_2}$, and c' is any point in $\overline{cb}$. The angle from $\overline{pc}$ to $\overline{pp_2}$ is larger than that from $\overline{pc'}$ to $\overline{pp_2}$.*

Definition 4. *A directed bisector $\overline{ab}$ of S_1 and S_2 is* coherent *with S_1 if for any point of $\overline{ab}$ (other than b), its forward direction is coherent with S_1. If $\overline{ab}$ is coherent with both S_1 and S_2, $\overline{ab}$ is a* coherent bisector.

Intuitively, the coherence of a bisector and one of its bisecting sites means that they point to the "same" direction. Note that if a directed bisector is coherent with neither sites that it bisects, then the bisector associated with the opposite forward direction is coherent with both sites that it bisects.

Lemma 5. *Let L_1 be a closed simple curve on a 2D plane $\mathcal{P}$ and L_2 be a continuous curve on $\mathcal{P}$ with endpoints a and b, which touches, but not intersects, L_1 at two points α and β. Let R be the closed region bounded by L_1 and L_2 and with α and β on its boundary. If both a and b are located outside of R and there exists a point $\gamma \in L_2$ in the interior of R, then L_2 is not simple.*

Lemma 6. *During the sweeping process, a coherent bisector b keeps its coherence until the center of the sweeping disk D reaches the next center point.*

Definition 5. *Let b and b' be two directed bisectors incident to a center point O. b and b' are consecutive if the forward direction of b points toward O and the forward direction of b' points away from O, or vice versa.*

Lemma 7. *Let b and b' be two consecutive bisectors incident to a center point O with b being the bisector of curves P_i and P_j and b' being the bisector of P_i and P_k. Then if b is coherent with P_i, b' is coherent with P_i.*

Let S_i and S_j be two sites of an input curve with S_i appearing before S_j in the curve order. We say that D retrospectively visits S_i if it touches S_j before S_i.

Lemma 8. *Let $B = \{b_1, b_2, \ldots, b_m\}$ be a sequence of consecutive and coherent bisectors (i.e., b_i and b_{i+1} are consecutive bisectors for $i = 1, 2, \ldots, m-1$), where the MSD D centered at any point on any bisector of B touches an input curve P_r. Let $S = \{S_1, S_2, \ldots, S_l\}$ be the set of sites of P_r touched by their corresponding bisectors in B. Then the sites can only be retrospectively visited at most once by B.*

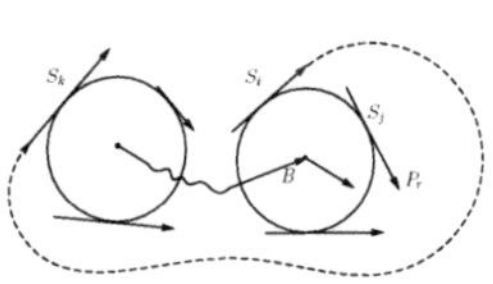

Fig. 7. Illustration of Lemma 8

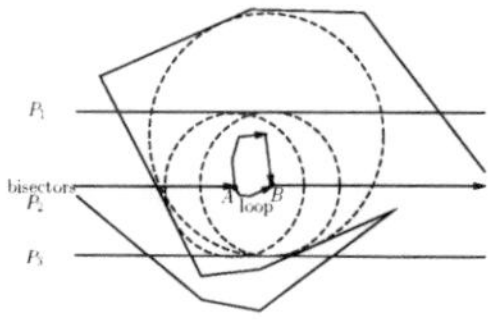

Fig. 8. A closed cell in *FPVD(Q)*

The following lemma shows that if D is swept along a coherent bisector and reaches a center point, then we can find one and only one coherent and consecutive bisector for D to sweep away from the center point. Thus if D always sweeps along coherent bisectors, it will never be stalled unless it encounters $\mathbf{t}$. This means that the sweeping disk will generate a simple path from $\mathbf{s}$ to $\mathbf{t}$ such that all constituent bisectors are coherent.

Lemma 9. *At any center point O, if there exists an incident bisector which can be oriented to be coherent, then exactly one of the other incident bisectors can be oriented to be coherent. Furthermore, the two coherent bisectors are consecutive.*

The above lemmas give properties of coherent bisectors. Next, we show that coherent bisectors can be used to identify a *backbone path* for the central path.

Definition 6. *A* backbone *is a path on $FPVD(Q)$ starting from the* **s** *and ending at* **t** *with each of its segments being a coherent bisector.*

Lemma 10. *For any input set of simple polygonal curves Q, there exists a unique and simple backbone.*

Below we discuss our idea for finding the next center point on the backbone.

Let v be the current center point on the backbone. To find the next center point, we sweep a minimal spanning disk D along the unvisited coherent bisector $b(v)$ (by Lemma 9) until a third input curve P_r becomes a touching curve. Let P_i and P_j be the two touching curves bisected by $b(v)$. By Lemma 6, we know that the coherence of $b(v)$ will be preserved until the center O of D reaches the center point u defined by the three touching curves P_i, P_j and P_r. Clearly, the main difficulty is how to efficiently find P_r. By Lemma 4, we know that P_r touches D at its rear portion.

To facilitate the search of P_r, we maintain a status array *stA* for the sweeping disk D. In this array, we store the current center point v, the next coherent bisector $b(v)$, and the closest site $s_{i\mu_i}$, $i = 1, 2, \ldots, m$, from the center O of D to the input curve P_i. The set of

$s_{i\mu_i}$ is maintained for determining the third touching curve P_r. Clearly, when D moves along $b(v)$, the set of closest sites changes, and thus needs to be updated.

To keep track of $s_{i\mu_i}$, we construct Voronoi diagram $VD(P_i)$ for each input curve P_i, and for each cell C of $VD(P_i)$, build a binary search tree $T(C)$ for all vertices of C according to their circular orders along the boundary of C. With $T(C)$, we can search the intersection of $b(v)$ and the boundary of C in $O(\log |C|)$ time, if C is the Voronoi cell of $s_{i\mu_i}$ [10]. This means that when sweeping D along $b(v)$, we can efficiently determine when and where O leaves C and enters into a new Voronoi cell C', and accordingly replace $s_{i\mu_i}$ with the site associated with C' in *stA*.

To determine the new center point u, we maintain a priority queue *priQ* for candidate vertices, as well as the intersections of $b(v)$ and the boundary of the Voronoi cells $VDC(s_{l\mu_l})$ $(l = 1, 2, \ldots, m)$ as follows. For each site $s_{l\mu_l}$ $(1 \le l \le m, l \ne i, j)$ in the status array *stA*, we compute the center o of the circle that touches $s_{l\mu_l}$, $s_{i\mu_i}$ and $s_{j\mu_j}$ (in case that there are 2 such circles, one at each side of $s_{l\mu_l}$, the one that does not contain v is considered). There are 3 cases to consider. (1) If the center o is on $b(v)$ and within $VDC(s_{l\mu_l})$, add it to *priQ* according to its distance to v. (2) If the center o is on $b(v)$ but located outside of $VDC(s_{l\mu_l})$, add the intersection point between $b(v)$ and $VDC(s_{l\mu_l})$ to *priQ* according to its distance to v. Note that if there are 2 intersection points due to the existence of parabolic arc in either $b(v)$ or some $VDC(s_{l\mu_l})$ edge, we always use the first one along $b(v)$ from v. (3)If the center o is not on $b(v)$, add ∞ to *priQ*.

With the above data structures, we can determine u by keeping extracting event points from *priQ*, and process them accordingly. Depending on the type of extracted event, we have 3 cases to consider.

(a) If the event is a center point, say c. Assume that the new touching site is $s_{l\mu_l}$, $l \ne i, j$. We first add the bisector portion $\overline{vc} \subset b(v)$ as part of the backbone, and replace v by c (i.e., c becomes the new current center point). Then compute the bisectors between $s_{l\mu_l}$ and $s_{i\mu_i}$ and between $s_{l\mu_l}$ and $s_{j\mu_j}$ (both starting from c), replace $b(v)$ by the new coherent bisector (the existence of a new coherent bisector is ensured by Lemma 9), and update *stA* accordingly. Finally, clear *priQ* and recompute *priQ* according to the new *stA*.

(b) If the event is an intersection point of $b(v)$ and $VDC(s_{l\mu_l})$ for some l, replace $s_{l\mu_l}$ by the site associated with the neighboring Voronoi cell, compute the new center of the circle that touches $s_{l\mu'_l}$, $s_{i\mu_i}$ and $s_{j\mu_j}$, and update *priQ*.

(c) If the event is ∞, find the third touching site of the sweeping disk D when it is centered at the other endpoint v' of $b(v)$. The third site must belong to either P_i or P_j. In this case, we replace either $s_{i\mu_i}$ or $s_{j\mu_j}$ by the new touching site, compute the new bisector $b(v')$, and update *stA* and *priQ* accordingly.

Backbone Algorithm

1. Compute the Voronoi diagram $VD(P_i)$ for each input curve P_i and build binary search tree $T(C)$ for each Voronoi cell C;
2. Construct an empty priority queue *priQ* and an empty array *stA*;
3. Let v be the current center point (initially, $v \leftarrow \mathbf{s}$); compute *stA* and *priQ*;
4. Extract an event from *priQ* and update *stA* accordingly;
5. Repeat step 4 until D reaches $\mathbf{t}$.

Lemma 11. *The total number of intersections between the backbone B and each Voronoi diagram $VD(P_i)(i = 1, 2, \ldots, m)$, denoted by H, is at most $O((n + s)k)$, where n is the total complexity of Q, s is the total number of self-intersecting points in Q, and k is the complexity of backbone B.*

Theorem 1. *The above algorithm correctly computes the backbone in $O((H + mk + n + s) \log m \log n)$ time, where k is the size of backbone B, m is the number of input curves in Q, H is the total number intersection points between the backbone path and each Voronoi Diagram, n is the total complexity of Q and s is the number of self-intersecting points in Q.*

With the backbone B, one might tempt to think that B is the central path. Unfortunately, this is not always true. When all cells of $FPVD(Q)$ are unbounded, B is indeed the central path. However, when $FPVD(Q)$ contains some closed cells, the backbone may not always be the central path. For example, in Figure 8, the backbone passes the two points A and B through the upper curve of the loop, whereas the central path uses the lower curve. In the presence of closed cells in $FPVD(Q)$, the central path P may not faithfully follow the backbone. This means that P may go through some incoherent bisectors in $E[FPVD(Q)]$. Since each center point is adjacent to at least one incoherent bisector, this seemingly suggests that we may need to examine an exponential number of paths. To efficiently determine the central path, our idea is to augment Q without changing $FPVD(Q)$ so that the central path follows the coherent bisectors.

Definition 7. *Given an input path P_i $(i = 1, 2, \ldots, m)$, an* aperture *is a point p on an edge of $VD(P_i)$ not bisecting two adjacent edges of P_i, such that p has the local minimum distance to P_i (see Figure 9). The distance from p to P_i is the* size *of the aperture. Let p_1 and p_2 be the two points in P_i closest to p (there are exactly two such points to be shown in Lemma 12). Then the line segment connecting p_1 and p_2 is called the* bridge. *All bridges and edges in P_i form a set of polygons called* pockets. *The radius of the maximum inscribed circle of a pocket is called the* radius *of the pocket (Fig. 9).*

Lemma 12. *Any aperture p is on an open edge of $VD(P_i)$. The total number of apertures in Q is $O(n + s)$.*

Lemma 13. *Let d_1 be the smallest size of apertures in Q, and d_2 be the minimum directed Hausdorff distance from backbone B to input curves in Q. If $d_1 > d_2$, then B is an optimal central path.*

Let $MS(P_i, d)$ be the Minkowski sum of P_i and a disk centered at the origin and with radius d.

Corollary 1. *Let d be the minimum directed Hausdorff distance from backbone B to input curves in Q. If there are no holes in $\cap_{i=1}^{m} MS(P_i, d)$ and $\cap_{i=1}^{m} MS(P_i, d)$ contains* **s** *and* **t**, *then B is an optimal central path.*

Lemma 14. *Let $d > 0$ and $r > 0$ be the maximum aperture size and minimum pocket radius in Q respectively, and d_1 and d_2 be the minimum directed Hausdorff distances from backbone B and central path to input curves in Q respectively. Then $d_1 \leq d$, if and only if $d_2 \leq d$; $d_1 \geq r$, if and only if $d_2 \geq r$.*

Lemma 13 and Corollary 1 imply that holes in $\cap_{i=1}^{m} MS(P_i, d)$ could cause the central path and backbone behave differently, where d is the minimum directed Hausdorff distance from backbone B to Q. Although detecting holes in $\cap_{i=1}^{m} MS(P_i, d)$ could be quite costly, discovering holes in each $MS(P_i, d)$ is relatively easy, since it is closely related to the sizes of apertures in P_i. A hole in $MS(P_i, d)$ indicates that the central path may have to go through the aperture or its surroundings instead of following the forward bisectors. To make use of this observation, our idea for computing central path is to modify Q so that the backbone will ultimately become the central path.

Definition 8. *Given an input curve P_i, another curve $\tilde{P}_i$ starting from the common starting line and ending at the ending line is a* twin curve, *if it satisfies the following conditions:*

1. *P_i and $\tilde{P}$ have the same complexity;*
2. *given an infinitesimally small number ϵ, for any point p_1 on P_i, there exist a point p_2 on $\tilde{P}$ such that the distance between p_1 and p_2 is less than ϵ, and vice versa.*

Furthermore, $\tilde{P}_i$ is the left twin curve, *denoted by P_i^l, if it is on the left side of P_i, and the* right twin curve, *denoted by P_i^r, if it is on the right side of P_i.*

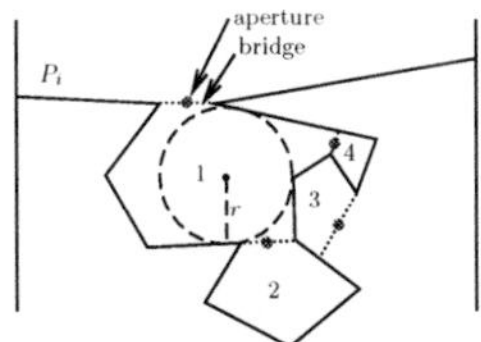

Fig. 9. P_i and its apertures (red dots), bridges (dotted segments), and 4 pockets. r is the radius of pocket 1

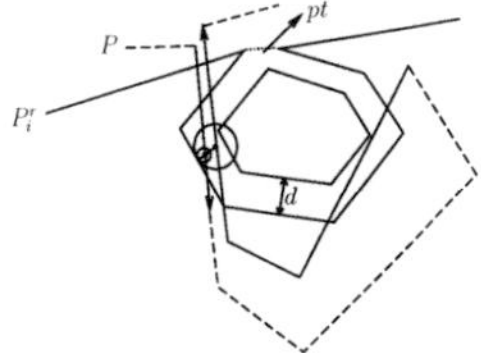

Fig. 10. A simple curve P coherent with some portion of a pocket pt of P_i^r but not coherent with other portions

For any $P_i \in Q$, we can construct a pair of twin curves P_i^l and p_i^r in $O(n_i + s_i)$ time. If we replace P_i by P_i^l and p_i^r in Q for all $i = 1, 2, \ldots, m$, the size of Q is still $O(n + s)$. After the replacement, the MSD D originally touching P_i on its left-hand (or right-hand) side now can be considered touching P_i^r (or P_i^l). This means that no MSD touches P_i^l on its left-hand side and touches P_i^r on its right-hand side. That also indicates that each pocket, aperture, or bridge is only associated with one of the two twin paths, not both. Since P_i^l (or P_i^r) and P_i are infinitesimally close to each other, the replacement does not change the structure of $FPVD(Q)$. Let Q' be the new set of input curves. It is sufficient for us to find the central path in Q'.

Definition 9. *A* virtual edge *is an open edge added to a curve P_i that is not considered as part of it while computing bisectors.*

Intuitively, a virtual edge is added to ensure the continuity of an input curve during pocket orientation reversal (described below). It is not a real edge of the curve. If a virtual edge connects two vertices v_1 and v_2 on an input curve, the next site to be visited after v_1 is not the virtual edge but v_2 during the computation of backbone or central

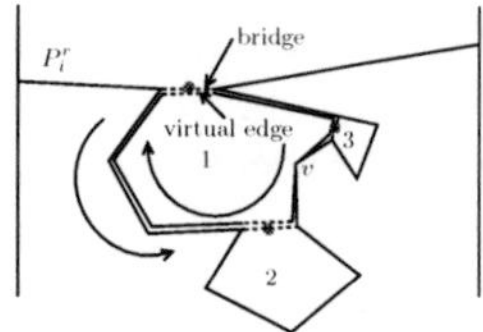

Fig. 11. Orientation reversal for pocket 1 of P_i^r

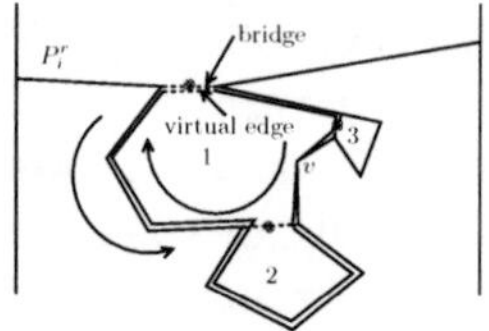

Fig. 12. Orientation reversal for pocket 2 of P_i^r after reversing pocket 1

path. Since virtual edges have no effect on the Voronoi diagram of the corresponding input curves, *Backbone Algorithm* still works for input curves with virtual edges.

With twin curves and virtual edges, we now modify the structure of pockets so that holes in $MS(P_i, d)$ can be easily detected. We first define an operation called *Pocket Orientation Reversal (POR)*. Given a pocket of P_i^r (P_i^l can be handled similarly) in Q', pick an arbitrary point v of P_i^r which is not an endpoint of any bridge. Let v' and v'' be two points on P_i^r infinitesimally close to v. Without loss of generality, we assume v', v, v'' are ordered so that $v' \rightarrow v$ and $v \rightarrow v''$ are forward directions of P_i^r. For each vertex and endpoint of bridges (denoted by u) of the pocket, generate an infinitesimally close point u' on the internal angular bisector of the two adjacent edges or bridges of u. For each bridge of the pocket, let v_1 and v_2 be the two endpoints and v_1' and v_2' be the generated infinitesimally close points. If $v_1'v_2'$ is not a virtual edge of P_i^r, then add $v_1'v_2'$ to P_i^r as a virtual edge (see Figure 11). Otherwise, delete it from P_i^r (see Figure 12). Consider a curve $v_1 v_2 \cdots v_\eta$ of the pocket that satisfies the following conditions: (1) For v_i $(i \neq 1, \eta)$, $v_i \neq v', v''$, and v_i is not an endpoint of some bridge; (2) v_1 is either endpoint of a bridge or v''; (3) If v_η is either endpoint of a bridge or v'. Then we connect $v_1', v_2', \ldots, v_\eta'$(if $v_1 = v''$, then $v_1' = v_1$; similarly, if $v_\eta = v'$, then $v_\eta' = v_\eta$) and add the new curve $v_1'v_2' \cdots v_\eta'$ to P_i^r. Obviously, the two curves do not intersect but do share one endpoint if v' or v'' is involved. Thus, after adding all possible new curves to P_i^r, P_i^r is still simple.

To make use of the POR operation, we first consider the following decision problem Backbone-POR. Given a number $d > 0$ and a set of twin curves Q', determine whether there exists a backbone $B(d)$ with $\overline{\delta_H}(B(d), Q') \leq d$ by allowing POR operations in Q'. To find the backbone $B(d)$ with POR, we sweep a MSD D from **s** to **t** as in *Backbone Algorithm*. In order to efficiently perform POR operations, we use a data structure *PKT* for pocket. *PKT* maintains the following information for each pocket *pt*: 1) The bounding edges and vertices; 2) a flag indicating whether the pocket has been touched by the sweeping disk D; 3) the center point (called the first center point of *pt*) when D first touches one site of the pocket; 4) a flag indicating whether the pocket orientation has been reversed. We also maintain a stack called *pocket stack* for all the pockets that have been touched by the MSD D and whose orientations have not been reversed. For a newly touched (by D) pocket *pt*, we can first check its second field of *PKT*. If its status is "not touched", we can change its status and push it into the pocket stack. While sweeping D, if the radius of D is larger than d and the pocket stack is not empty, then pop a pocket off the pocket stack and reverse its orientation. Then continue sweeping D from the first center point of the popped pocket. If, on the other hand, the pocket stack is empty, then no such backbone is found. Below are the main steps of the algorithm.

BACKBONE-POR (d) Algorithm

1. Generate twin curves to form Q';
2. Compute the Voronoi diagrams $VD(P_i^l)$ and $VD(P_i^r)$ for $i = 1, 2, \ldots, m$;
3. Initialize B as the backbone, and PS as the pocket stack;
4. Sweep D from **s** to **t** as in Backbone Algorithm; let v be the current center point;
5. Let f be the latest site touched by D, and pt is the pocket associated with f (if there exists a pocket); Mark pt as touched and push pt to PS;
6. If the radius of D is larger than d and PS is not empty, pop a pocket pt from PS and reverse its orientation. Move D to the first center point O of pt, discard the computed portion of B between O and v and continue sweeping D;
7. If the radius of D is larger than d and PS is empty, there does not exist a backbone B such that $\overline{\delta_H}(B, Q') \leq d$; Return FALSE;
8. If the radius of D is smaller than or equal to d, continue sweeping D to find the next center point v and repeat step $5-7$ until $v = \mathbf{t}$;
9. Return TRUE and output B as the backbone.

Lemma 15. *Let P be a simple path on $FPVD(Q')$ from* **s** *to* **t**. *If the maximum directed Hausdorff distance from P to Q' is no larger than d, then for any pocket pt in Q' with radius larger than d, P is coherent with either pt or the reversed pt.*

Lemma 16. *If there exists a path in $FPVD(Q')$ from* **s** *to* **t** *whose directed Hausdorff distance to Q' is at most d, then the backbone algorithm with POR (w.r.t. d) correctly generates the backbone of Q'.*

Lemma 17. *The running time of BACKBONE-POR(Q,d) is $O((H{+}mk{+}n{+}s)\log m \log n)$, where k is the complexity of the portion of $FPVD(Q')$ the center of D traversed and H is the number of intersections between the traversed portion of $FPVD(Q')$ and Voronoi diagrams $VD(P_i)$ $(i = 1, 2, \ldots, k)$.*

Now we are ready to present our central path algorithm. The idea is to perform a binary search on the radii of all pockets in Q' to find a backbone with POR.

CENTRAL-PATH Algorithm

1. Preprocess Q to make every input curve simple;
2. Perform dummy extension on Q, and determine **s** and **t**;
3. Construct twin curves to form Q';
4. Compute the Voronoi diagrams $VD(P_i^l)$ and $VD(P_i^r)$ for $i = 1, 2, \ldots, m$;
5. Compute all pockets $PT = \{pt_1, pt_2, \ldots, pt_\beta\}$ of Q' and sort them by radii;
6. Binary search on PT; let pt be the current pocket and d be its radius; call BACKBONE-POR Algorithm with Q and d as input and maintain a counter to record the number of steps elapsed (initially it is 0); if the counter is larger than $\Theta(n^2 2^{\alpha(n)} \log n)$, suspend the current algorithm and call the basic algorithm in Lemma 1 instead; return the final backbone path B.

Theorem 2. *The Central Path Algorithm returns an optimal central path in* $O((H + mk + n + s)\log m \log^2 n)$ *time bounded by* $O(n^2 2^{\alpha(n)} \log n)$*, where* k *is size of the visited portion of* $FPVD(Q)$*,* H *is the number of intersections between the visited portion of* $FPVD(Q)$ *and* $VD(P_i)(i = 1, 2, \ldots, k)$*,* m *is the number of input curves in* Q*,* n *is the total complexity of* Q *and* s *is the number of self-intersecting points in* Q*.*

Acknowledgment. The authors would like to thank Professor Joseph S.B. Mitchell, State University of New York at Stony Brook, for helpful suggestions and discussions.

References

1. Abellanas, M., Hurtado, F., Icking, C., Klein, R., Langetepe, E., Ma, L., Sacristán, V.: The farthest color voronoi diagram and related problems (extended abstract). In: 17th European Workshop Computational Geometry, pp. 113–116 (2001)
2. Aurenhammer, F.: Voronoi diagrams—a survey of a fundamental geometric data structure. ACM Computing Surveys 23(3), 345–405 (1991)
3. Aurenhammer, F., Drysdale, R.L.S., Krasser, H.: Farthest line segment voronoi diagrams. Information Processing Letters 100(6), 220–225 (2006)
4. Buchin, K., Buchin, M., Van Kreveld, M., Löffler, M., Silveira, R.I., Wenk, C., Wiratma, L.: Median trajectories. In: Proceedings of the 18th Annual European Conference on Algorithms: Part I, pp. 463–474 (2010)
5. Cheong, O., Everett, H., Glisse, M., Gudmundsson, J., Hornus, S., Lazard, S., Lee, M., Na, H.S.: Farthest-polygon voronoi diagrams. In: Proceedings of the 15th Annual European Symposium on Algorithms, pp. 407–418 (2007)
6. de Berg, M., Van Kreveld, M., Overmars, M., Schwarzkopf, O.: Computational Geometry: Algorithms and Applications, 2nd edn. (2000)
7. Fortune, S.: A sweepline algorithm for voronoi diagrams. In: Proceedings of the 2nd Annual Symposium on Computational Geometry, pp. 313–322 (1986)
8. Har-Peled, S., Raichel, B.: The frechet distance revisited and extended. In: Proc. of the 27th Annual ACM Symposium on Computational Geometry, SoCG 2011, pp. 448–457 (2011)
9. Huttenlocher, D.P., Kedem, K., Sharir, M.: The upper envelope of voronoi surfaces and its applications. In: Proceedings of the 7th Annual Symposium on Computational Geometry, pp. 194–203 (1991)
10. Zhu, Y., Xu, J.: Improved Algorithms for Farthest Colored Voronoi Diagram of Segments. In: Wang, W., Zhu, X., Du, D.-Z. (eds.) COCOA 2011. LNCS, vol. 6831, pp. 372–386. Springer, Heidelberg (2011)

On the Generalized Multiway Cut in Trees Problem

Hong Liu* and Peng Zhang **

School of Computer Science and Technology,
Shandong University, Jinan 250101, China
{hong-liu,algzhang}@sdu.edu.cn

Abstract. Given a tree $T = (V, E)$ with n vertices and a collection of terminal sets $D = \{S_1, S_2, \ldots, S_c\}$, where each S_i is a subset of V and c is a constant, the generalized Multiway Cut in trees problem (GMWC(T)) asks to find a minimum size edge subset $E' \subseteq E$ such that its removal from the tree separates all terminals in S_i from each other for each terminal set S_i. The GMWC(T) problem is a natural generalization of the classical Multiway Cut in trees problem, and has an implicit relation to the Densest k-Subgraph problem. In this paper, we show that the GMWC(T) problem is fixed-parameter tractable by giving an $O(n^2 + 2^k)$ time algorithm, where k is the size of an optimal solution, and the GMWC(T) problem is polynomial time solvable when the problem is restricted in paths. We also discuss some heuristics for the GMWC(T) problem.

1 Introduction

The cut problem is a classical and long-standing active topic in combinatorial optimization. In this paper, we study a generalized version of the Multiway Cut in trees problem. The related definitions are given below.

Given a graph $G = (V, E)$ and a vertex subset $S \subseteq V$, we say that the removal of an edge subset $E' \subseteq E$ *cuts* S if every vertex pair coming from S is disconnected when E' is removed from G. In this case we say that E' is a (multiway) cut for S.

The Multiway Cut Problem. Given a graph $G = (V, E)$ and a terminal set $S \subseteq V$, the Multiway Cut Problem asks to find a minimum size edge subset $E' \subseteq E$ whose removal cuts S.

The Generalized Multiway Cut Problem (GMWC). Given a graph $G = (V, E)$ and a collection of terminal sets $D = \{S_1, S_2, \ldots, S_c\}$, where c is a constant, the problem asks to find a minimum size edge subset $E' \subseteq E$ whose removal cuts all terminal sets in D.

* The author is supported by the Independent Innovation Foundation of Shandong University (2012TS071).

** Corresponding author. The author is supported by the National Natural Science Foundation of China (60970003), China Postdoctoral Science Foundation (200902562), the Special Foundation of Shandong Province Postdoctoral Innovation Project (200901010), and the Independent Innovation Foundation of Shandong University (2012TS072).

G. Lin (Ed.): COCOA 2012, LNCS 7402, pp. 151–162, 2012.

The GMWC problem generalizes the Multiway Cut problem in the sense that the number of terminal sets is generalized from one to a constant c. Note that the terminal set number c, which is a constant, is *not* a part of the GMWC instance. We denote by GMWC(T) the GMWC problem when the input graph is a tree. Actually, we have one specified problem corresponding to each distinct value of c. We thus denote these problems by 1-MWC(T), 2-MWC(T), and so on. Note that 1-MWC(T) is just the traditional Multiway Cut in trees problem.

Throughout the paper, we use n to denote the number of vertices in an input graph, $OPT(\mathcal{I})$ (or OPT) to denote the value of the optimal solution to an instance $\mathcal{I}$. We also use OPT to denote the optimal solution itself by abusing notations slightly.

1.1 Motivation

Zhang [16,17] studied the generalized k-Multicut in trees problem (k-GMC(T)), which is a generalization of the k-Multicut in trees problem [13].

The Multicut Problem. Given a graph $G = (V, E)$ and a collection of terminal pairs $D = \{(s_1, t_1), (s_2, t_2), \cdots, (s_q, t_q)\}$, the Multicut problem asks to find a minimum size edge subset $E' \subseteq E$ such that its removal cuts all terminal pairs in D.

The k-Multicut Problem. Given a graph $G = (V, E)$, a collection of terminal pairs $D = \{(s_1, t_1), (s_2, t_2), \cdots, (s_q, t_q)\}$, and an integer $k > 0$, the k-Multicut problem asks to find a minimum size edge subset $E' \subseteq E$ such that its removal cuts at least k terminal pairs in D.

The Generalized k-Multicut Problem (k-GMC). Given a graph $G = (V, E)$, a collection of terminal sets $D = \{S_1, S_2, \cdots, S_q\}$, and an integer $k > 0$, the k-GMC problem asks to find a minimum size edge subset $E' \subseteq E$ such that its removal cuts at least k terminal sets in D.

The k-GMC problem generalizes the k-Multicut problem in the sense that a terminal pair (s, t) is generalized to a terminal set S.

The generalized k-Multicut problem restricted in trees (i.e., k-GMC(T)), is a particularly interesting problem, since there is an approximation preserving reduction from the famous Densest k-Subgraph problem (DkS) to k-GMC(T), which was given in [16]. The reduction says that an $f(n)$-approximation algorithm for k-GMC(T) will lead to a $2f^2(n)$-approximation algorithm for DkS. A remarkable feature of the k-GMC(T) instance used in the reduction [16] is that the input tree is only a star and each terminal set is of size exactly 3.

Therefore, the k-GMC(T) problem is almost as hard as the DkS problem, although k-GMC(T) is defined in trees. After a series of approximation results for the DkS problem, an improved ratio $O(n^{\frac{1}{4}+\epsilon})$ for the problem just appeared [1] recently, where $\epsilon > 0$ is any constant. By contrast, the k-Multicut in trees problem admits $2 + \epsilon$-approximation [13]. This means that for the k-GMC(T) problem, simply augmenting the size of each terminal set from 2 to 3 significantly increase the difficulty of approximating it.

Very recently, using the linear programming technique, Zhang and Zhu et al. [17] designed an $O(\sqrt{q})$-approximation algorithm for the k-GMC(T) problem and an $O(\sqrt{q \log n})$-approximation algorithm for the k-GMC problem in general graphs. Then we would like to consider some special cases of the k-GMC(T) problem. An interesting problem arises: How about the k-GMC(T) problem when k is a constant? This problem can be solved by enumerating all $\binom{q}{k}$ sub-collections of D and solving the resulting problem corresponding to each enumeration, that is, to cut constant number terminal sets in a tree, which is precisely the GMWC(T) problem we study in this paper.

1.2 Related Work

The Multiway Cut in trees problem (that is, 1-MWC(T)) is proved to be polynomial time solvable [5,6]. Chopra and Rao [5] gave a polynomial-time greedy algorithm for 1-MWC(T) and proved that the solution found by their algorithm is optimal via the optimal criteria in linear programming. Costa and Billionnet [6] proved that 1-MWC(T) can even be solved in linear time by dynamic programming.

The Multiway Cut in graphs problem was proved to be NP-hard and APX-hard by Dahlhaus et al. [7]. The current best approximation ratio for Multiway Cut in graphs is 1.3438, according to Karger et al. [10]. The Multicut in trees problem is already NP-hard and APX-hard, and its current best approximation ratio is 2, according to Garg et al. [8]. From the negative side, if the Unique Games Conjecture is true, then the Multiway Cut in graphs problem can not be approximated within a factor equal to the integrality gap of the so-called earth-mover linear program for the problem [12], and the Multicut in trees problem can not be approximated within $2-\epsilon$ [11].

From the aspect of parameterized algorithms, Xiao [15] gave an $O(2^k rT)$ time parameterized algorithm for the Multiway Cut in graphs problem, where r is the number of terminals and T is the time of finding a minimum s-t cut in a graph. Chen et al. [4] gave an $O(k4^k n^3)$ time parameterized algorithm for the node version of the Multiway Cut problem. Guo and Niedermeier [9] gave a parameterized algorithm for the Multicut in trees problem, whose time complexity is $O(k^{3k} + q^3 n + n^3 q)$. Very recently, Bousquet et al. [3] solved a long-standing open problem, showing that the Multicut in graphs problem is fixed-parameter tractable.

1.3 Our Results

In this paper, we investigate several solving approaches for the GMWC(T) problem. The precise complexity of GMWC(T) is still unknown at the current time.

We design an $O(n^2 + 2^k)$ time algorithm for the GMWC(T) problem, where the parameter k is the size of an optimal cut for the problem. This means that GMWC(T) is fixed-parameter tractable. First, we give three simple data reduction rules which are used to reduce the size of the input instance, obtaining the so-called problem kernel. Then, by a careful classification of the vertices in the kernel, we prove that the kernel is of a small size. Finally, we design a parameterized $O(n^2 + 2^k)$ time algorithm based on the kernel for the GMWC(T) problem, by using the basic depth-bounded search tree technique and the interleaving technique [14].

Note that GMWC(T) can be solved by converting to the Multicut in trees problem and hence admits a parameterized algorithm by Guo and Niedermeier [9], since there are at most $O(n^2)$ terminal pairs for all the c terminal sets in D. However, the running time of our algorithm for GMWC(T) is much faster than that of the above trivial conversion approach.

We prove that when restricted in paths, the GMWC(T) problem is polynomial time solvable by a dynamic programming approach. We also give counterexamples for several heuristics for GMWC(T) based on dynamic programming and greedy approach.

2 An FPT Algorithm

2.1 Data Reduction Rules

A trivial solution to the GMWC(T) problem is to convert the problem to the Multicut in trees problem by enumerating all possible terminal pairs from the terminal set collection D. This will lead to $O(n^2)$ terminal pairs in the worst case. Thus, the time complexity would be $O(k^{3k}+n^7)$ by the parameterized algorithm in [9]. In this section and the subsequent section, we shall give a parameterized algorithm for GMWC(T) with significantly improved time complexity.

Recall that the input graph of the GMWC(T) problem is a tree T. At the first step, we root the tree T at any non-leaf vertex.

(**Remove nonterminal leaves.**) If a leaf is not a terminal in any terminal set, then remove the leaf from the tree.

(**Remove nonterminal degree-2 vertices.**) If a degree-2 vertex is not a terminal, then merge its two incident edges into a single edge and delete this degree-2 vertex from the tree.

The correctness of the removing nonterminal leaves rule and the removing nonterminal degree-2 vertices rule is straightforward.

(**Remove terminal leaves conditionally.**) If a terminal leaf v appears only once in terminal set collection D, and has its parent u not being a terminal of any terminal set, then replace v by u in terminal set collection D and remove vertex v from the tree.

Lemma 1. *The optimal value does not change after applying the conditionally removing terminal leaves rule.*

Proof. Let $\mathcal{I}$ be the instance before applying the conditionally removing terminal leaves rule, and $\mathcal{I}'$ be the new instance after applying the rule. As the optimal solution to instance $\mathcal{I}'$ is a feasible solution to instance $\mathcal{I}$, obviously we have $OPT(\mathcal{I}) \leq OPT(\mathcal{I}')$.

For the opposite direction, let $OPT(\mathcal{I})$ be an optimal solution to instance $\mathcal{I}$. Suppose the edge (u, v) removed by the rule is in $OPT(\mathcal{I})$. (Otherwise we have $OPT(\mathcal{I}') \leq OPT(\mathcal{I})$ and are done.) After removing all edges in $OPT(\mathcal{I})$

from the tree, all terminal sets are cut. If we put edge (u, v) back to the tree again, then by the optimality of $OPT(\mathcal{I})$, there is a unique vertex w on the tree which is in the same terminal set as that of v, such that v and w are connected. Since u is not a terminal, u itself can not be vertex w. So, there is a path P between u and w and we can replace edge (u, v) in $OPT(\mathcal{I})$ by any edge on path P, without changing the feasibility (w.r.t. instance $\mathcal{I}$) of $OPT(\mathcal{I})$. This suggests that $OPT(\mathcal{I})$ is also a feasible solution to instance $\mathcal{I}'$, resulting in $OPT(\mathcal{I}') \leq OPT(\mathcal{I})$. This concludes the lemma. □

2.2 Problem Kernel and the Algorithm

To show that we can get a small kernel for the GMWC(T) problem by repeatedly applying the data reduction rules given in Section 2.1 until none of them can apply further, we need the following classification for vertices of the input tree.

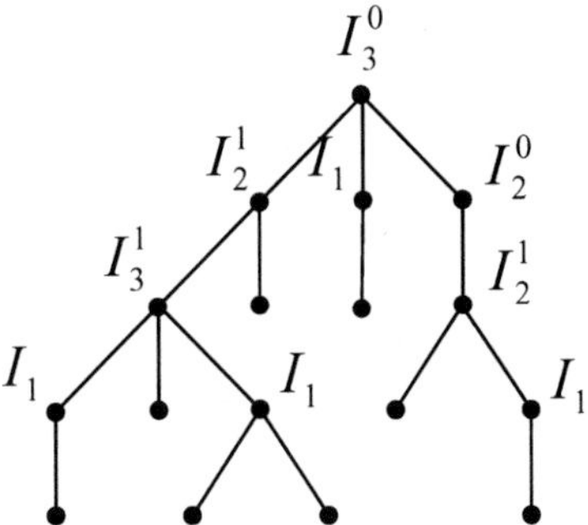

Fig. 1. Classification of internal vertices

First, let I be the set of all internal vertices, and L be the set of all leaves. Then, all the internal vertices are classified into four categories according to the adjacency to other internal vertices. Given a vertex v in the tree, let $N(v)$ be the neighborhood of v, i.e., the set of all vertices in the tree adjacent to v.

- $I_0 = \{v \in I \mid N(v) \cap I = 0\}$. That is, each internal vertex in I_0 has no any adjacent internal vertex.
- $I_1 = \{v \in I \mid N(v) \cap I = 1\}$. Each internal vertex in I_1 has just one adjacent internal vertex.
- $I_2 = \{v \in I \mid N(v) \cap I = 2\}$. Each internal vertex in I_2 has exactly two adjacent internal vertices.
- $I_3 = \{v \in I \mid N(v) \cap I \geq 3\}$. Each internal vertex in I_3 has at least three adjacent internal vertices.

Finally, vertices in I_2 are further classified into two categories. Suppose u is a vertex in the tree. Let $p(u)$ be the parent vertex of u.

- $I_2^1 = \{v \in I_2 \mid \exists u \in L, p(u) = v\}$. That is, I_2^1 is the set of all I_2-vertices that have at least one leaf child.
- $I_2^0 = I_2 \setminus I_2^1$ is the set of all I_2-vertices that have no leaf child.

Similarly, we have two categories I_3^1 and I_3^0.

- $I_3^1 = \{v \in I_3 \mid \exists u \in L, p(u) = v\}$.
- $I_3^0 = I_3 \setminus I_3^1$.

For a vertex v in subset I_j, we also say that v is of type I_j and call v an I_j-vertex. We note here that I_0 is a very special internal vertex set, since it can contain only one element and this will happen only if the input tree is a star. For a tree that is not a star, there is no any I_0-vertex at all.

We bound the size of the problem kernel by parameter k, which is the size of an optimal multiway cut, i.e., $OPT(\mathcal{I})$.

Lemma 2. *If the input tree is a star, then* $|I_0| + |L| \leq c + k$.

Proof. By the removing nonterminal leaves rule, each leaf is a terminal.

If the unique I_0-vertex, i.e., the star center, is not a terminal, then by the conditionally removing terminal leaves rule, there must be at least one leaf that appears at least twice in terminal set collection D. We thus can charge one of the terminal appearances to the star center.

So, we may assume that all vertices in the star are terminals.

When all edges in OPT are removed from the star, the star is broken into $k + 1$ connected components. In the component which the star center belongs to, by the definition of the generalized multiway cut, there are exactly c distinct terminals, while all the remaining components are of size one. This gives the lemma. □

Lemma 3. *If the input tree is not a star, then* $|I_1| + |I_2| + |I_3^1| + |L| \leq c(k+1)$.

Proof. By the removing nonterminal degree-2 vertices rule, each I_2^0-vertex is a terminal. By the removing nonterminal leaves rule, each leaf is also a terminal.

If a I_1-vertex, to say, u is not a terminal, then by the conditionally removing terminal leaves rule, u must have at least one leaf v as its child such that v appears at least twice in terminal set collection D. So we can charge one terminal appearance on v to the I_1-vertex u. The same argument applies to each I_2^1-vertex and each I_3^1-vertex.

Therefore, the sum $|I_1| + |I_2| + |I_3^1| + |L|$ is bounded from above by the total number of terminal appearances.

When all edges in OPT are removed from the input tree, the tree is broken into $k + 1$ connected components. In each component, by the definition of the generalized multiway cut, there are at most c distinct terminals. Thus there are at most $c(k + 1)$ terminal appearances in total. The lemma follows. □

Lemma 4. *If the input tree is not a star, then* $|I_3^1| + |I_3^0| \leq |I_1| - 1$.

Proof. By removing all leaves from the input tree T, we get a tree T' whose vertices are all I-vertices of T. Denote by n'_i the number of vertices in T' each of which has exactly i child, for $i = 0, 1$, and by n'_2 the number of vertices in T' that has at least 2 children. Then we have $n'_0 = |I_1|$, $n'_1 = |I_2|$ and $n'_2 = |I_3|$. It is well-known that $n'_2 = n'_0 - 1$ in a binary tree. For a tree such as T' in which internal vertices may have more than two children, we have $n'_2 \leq n'_0 - 1$. □

Theorem 1. *The GMWC(T) problem admits a problem kernel of size at most* $2c(k+1)$.

Proof. By Lemmas 2, 3 and 4. □

Lemma 5. *The problem kernel of GMWC(T) can be obtained in* $O(n)$ *time.*

Proof. By scanning each vertex in the input tree T, all nonterminal leaves (corresponding to the removing nonterminal leaves rule) and nonterminal degree-2 vertices (corresponding to the removing nonterminal degree-2 vertices rule) can be removed in $O(n)$ time.

Noticing that the terminal set collection D contains only c terminal sets, which is a constant number, the conditionally removing terminal leaves rule can also be performed in $O(n)$ time by scanning each terminal in D. □

Before giving the the main theorem of this section, we first show that there is a so-called *interleaving technique* which can be used to accelerate a class of FPT algorithms. Let $\mathcal{I}$ be an instance of a problem Π and $(\mathcal{I}, k)$ be the input to an FPT algorithm $\mathcal{A}$ for Π. Suppose algorithm $\mathcal{A}$ works in two stages that the first stage is a reduction to problem kernel and the second stage is a bounded search tree of size $O(\xi^k)$, where ξ is a constant. Reduction to problem kernel takes $P(|\mathcal{I}|)$ steps and results in an instance of size at most $q(k)$. The expansion of a node in the search tree takes $R(|\mathcal{I}|)$ steps. All of P, q and R are polynomially bounded. The overall time complexity of algorithm $\mathcal{A}$ is then $O(P(|\mathcal{I}|) + R(q(k))\xi^k)$. The following theorem shows that algorithm $\mathcal{A}$ can be accelerated.

Theorem 2 ([14]). *Suppose algorithm $\mathcal{A}$ is a two-stage FPT algorithm for problem Π running in $O(P(|\mathcal{I}|)+R(q(k))\xi^k)$ time as described above. Then there is an FPT algorithm for problem Π with improved time complexity $O(P(|\mathcal{I}|) + \xi^k)$.*

The parameterized algorithm solving the GWMC(T) problem is given in the proof of the following theorem.

Theorem 3. *The Generalized Multiway Cut in trees problem can be solved in* $O(n^2 + 2^k)$ *time.*

Proof. The basic strategy of solving GMWC(T) is to use a bounded depth search tree. First we obtain the problem kernel of size at most $2c(k+1)$. By Theorem 1 and Lemma 5, this can be done in $O(n)$ time.

Next we convert terminal set collection D into a terminal pair set by generating all possible $\binom{|S|}{2}$ terminal pairs for each terminal set $S \in D$. Let $LIST$ be the resulting set of terminal pair. The size of $LIST$ is at most $\binom{2c(k+1)}{2} = O(k^2)$. For each pair in $LIST$, we record its least common ancestor in the kernel. By [2], this will consume a preprocessing time linear in the size of the kernel and a constant query time for each terminal pair. Thus the total time of computing least common ancestor for all terminal pairs in $LIST$ is $O(k^2)$.

Then, root the kernel at any vertex. This is the only one node of the initial search tree at the current time. In the bottom-up manner, we scan each internal vertex of the kernel. Suppose the current scanned vertex u is the least common ancestor for some pair (s, t) in $LIST$. If it is not the case, then we scan the next vertex.

Let P be the unique path between s and t in the kernel. If u is s or t, then remove the edge on P incident to u from the kernel, and remove all pairs disconnected by such an edge removal from $LIST$, resulting in a new node of the search tree which is the unique child of the current node. We get in the new node and scan the next vertex of the kernel. Otherwise u is an intermediate vertex on P. Let (l, u) and (u, r) be the two edges on P incident to u. The search tree splits into two branches at the current node, which corresponds to each of the two edges. In the either new node, we remove the corresponding edge from the kernel, and remove all pairs thus disconnected from $LIST$. Then we get in each of the two nodes and scan the next vertex of the kernel recursively.

Since $OPT(\mathcal{I})$ is equal to k, the depth of the search tree is bounded by k. The work at each node of the search tree (including determining on u, removing edge from the kernel, and removing disconnected pairs) can be done in $O(k^3)$ time. Therefore, the total time of computing the desired optimal solution is $O(n + k^2 + k^3 2^k) = O(n^2 + k^3 2^k)$.

Finally, using the interleaving technique introduced in [14], the above parameterized algorithm can be accelerated to run in $O(n^2 + 2^k)$ time. The proof is finished. □

3 Some Observations

3.1 About the Complexity of GMWC(T)

It is well-known that the Multiway Cut in trees problem (i.e., 1-MWC(T)) is polynomial-time solvable [5,6], while the Multicut in trees problem is NP-hard [8], even the problem is restricted in stars. From the viewpoint of problem goal, the GMWC(T) problem lies exactly between the Multiway Cut in trees problem and the Multicut in trees problem. To determine the precise complexity of GMWC(T) thus becomes an intriguing and subtle problem.

In the Multicut in trees problem, we have to disconnect q terminal pairs $\{(s_i, t_i)\}$, where q is a part of the problem input. In the GMWC(T) problem, we have to disconnect c terminal sets $\{S_i\}$, where c is independent of the input of the problem. If there is a polynomial time algorithm for GMWC(T) whose time complexity has nothing to do with the constant c, then we can solve the Multicut in trees problem in polynomial time, since Multicut in trees is nothing more than q-MWC(T). However, this would be absurd, since the Multicut in trees problem is NP-hard and P is not equal to NP, as believed.

Theorem 4. *If the GMWC(T) problem can be solved in polynomial time, the time complexity of the supposed algorithm must depend on constant c, assuming $P \neq NP$.* □

3.2 A Special Case of GMWC(T)

Suppose we are given a GMWC(T) instance (T, D). Let us consider the weighted version of GMWC(T) and thus $\{w(e)\}$ be the weights defined on edges in T. For each terminal set $S \in D$, let T_S be the subtree of T obtained by uniting the unique paths between all the $\binom{|S|}{2}$ possible terminal pairs coming from S.

Consider an optimal solution $E^* \subseteq E(T)$ to the instance (T, D). E^* must contain at least one edge for the subtree T_S corresponding to terminal set S. This observation suggests we can express the optimal value as

$$OPT = \min_{E' \subseteq E(T):\, |E'| \leq c} \Big\{ w(E') + \sum_{T_j} OPT(T_j) \Big\}, \tag{1}$$

where T_j stands for a tree in the forest led by removing E' from T, and $OPT(T_j)$ is the optimal value of the instance (T, D) restricted on subtree T_j.

Equation (1) may suggest a dynamic programming algorithm for GMWC(T): Just compute the value of the GMWC(T) instance on each subtree of T in a systematic way. Unfortunately, the number of subtrees of a tree is unpractically large and consequently we can not get a polynomial time algorithm for GMWC(T) by this way. For example, a star with h leaves may have $\Omega(2^h)$ subtrees.

When the input tree is simply a path, the number of subtrees is $\binom{n}{2}$, a polynomially bounded quantity. The above dynamic programming algorithm will work in polynomial time for this special case. So we have

Theorem 5. *If the input tree is a path, then the GMWC(T) problem can be solved in polynomial time.* □

3.3 Counterexamples for Some Heuristics

In this section we show the counterexamples for some simple heuristics for GMWC(T). First we define two terminologies for the problem. Two terminals are *homologous* if they come from the same terminal set. Given a tree, a terminal in the tree is called an *orphan* if there is no any other terminal in the tree that is homologous with this terminal.

Heuristics Based on Dynamic Programming. Since the GMWC(T) problem is defined in trees, the heuristic that is easily thought may be the dynamic programming approach. We root the input tree at an arbitrary vertex, and then scan each vertex v in the tree from the bottom to the top one by one. For the subtree $T(v)$ rooted at v, we find by enumerating the minimum cost edge subset $E' \subseteq E(T(v))$ such that the removal of E' can disconnect all homologous terminal pairs. Note that this dynamic programming approach is different to the one in Theorem 5.

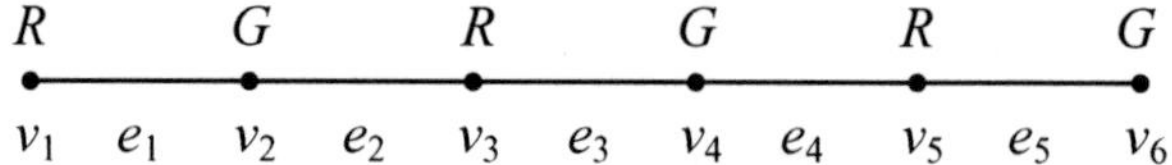

Fig. 2. Instance A of GMWC(T)

However, the instance A shown in Figure 2 is a counterexample to the above dynamic programming approach. In Instance A, there are two (that is, $c = 2$) terminal sets with $R = \{v_1, v_3, v_5\}$ and $G = \{v_2, v_4, v_6\}$. When the tree is rooted at v_6, the dynamic programming approach will finally output $\{e_1, e_3, e_5\}$ as a solution. On the other hand, the optimal solution is $\{e_2, e_4\}$. (Note that Instance A can be optimally solved by the dynamic programming approach in Theorem 5.) Some other variants of this dynamic programming approach also do not work. We omit their descriptions here due to the limitation of space.

Heuristics Based on Greedy Approach. Based on the observations in Theorems 4 and 5, one can suggest a framework of greedy approach for GMWC(T) as follows. (i) Find by enumerating an edge subset E' with $|E'| \leq c'$ in the current tree T' (which is the input tree T initially) such that a certain criterion is met, where c' is the number of terminal sets (of size at least 2) in T'. (ii) Remove E' from T'. This will break T' into several subtrees. (iii) Recursively repeat the above two steps on each resulting subtree, until all terminal sets are disconnected. (iv) Remove the redundant edges (with respect to constituting a feasible cut) from the edge set finally obtained.

We discuss several criteria that can be used in step (i).

Firstly, the criterion could be a minimum weight edge subset E' of size at most c' such that the removal of E' splits every terminal set. In this case, Instance A in Figure 2 acts as a counterexample. The algorithm will pick edge e_3 in its first iteration, resulting in two subtrees. It will pick edge e_1 and then pick edge e_5 in the two resulting subtrees. Of course, this is not an optimal solution.

Secondly, the criterion could be an edge subset E' of size at most c' such that the ratio of the total weight $w(E')$ to pairs(E') is minimized, where pairs(E') is the number of homologous terminal pairs disconnected by removing E'. In this case, Instance A in Figure 2 also acts as a counterexample. The algorithm will pick edge e_3 (with ratio of $\frac{1}{4}$) in its first iteration, resulting in two subtrees. It will pick edge e_1 and then pick edge e_5 in the two resulting subtrees (with ratio of $\frac{1}{1}$ in both of the two subcases).

Thirdly, the criterion could be an edge subset E' of size at most c' such that the ratio of the total weight $w(E')$ to orphans(E') is minimized, where orphans(E') is the number of orphans generated by removing E'. This greedy approach will compute the optimal solution to Instance A.

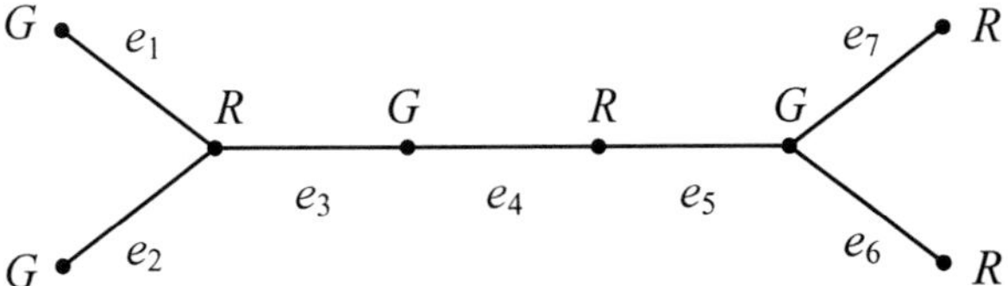

Fig. 3. Instance B of GMWC(T)

However, the instance B in Figure 3 is a counterexample. In the first iteration, the algorithm will remove e_4 with corresponding ratio being $\frac{1}{2}$ (one edge and two resulting orphans). In its subsequent four iterations, the algorithm will pick e_1, e_2, e_6 and e_7 successively. However, an optimal solution to Instance B is $\{e_2, e_3, e_5, e_6\}$.

Finally, let us consider as the criterion an edge subset E' of size at most c' such that orphans(E') is maximized. (In case of ties, a minimum weight edge subset is picked.) For this greedy approach, Instances A and B are no longer counterexamples. However, there is still a counterexample for this approach, namely, Instance C as shown in Figure 4.

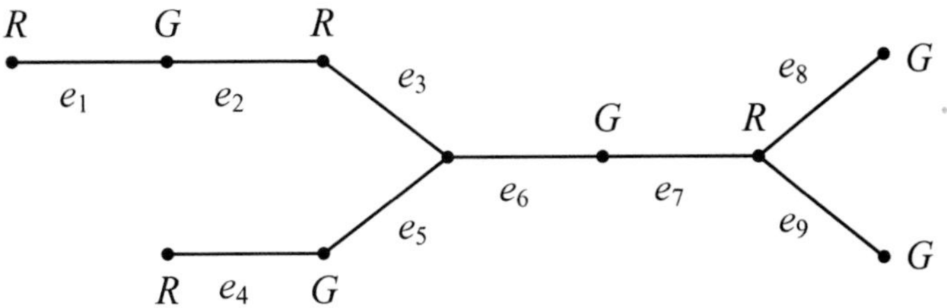

Fig. 4. Instance C of GMWC(T)

When running on Instance C, the algorithm will pick $\{e_3, e_5\}$ in its first iteration, generating four orphans. In its subsequent four iterations, the algorithm will pick edges e_1, e_4, e_8 and e_9 successively. However, an optimal solution to Instance C is $\{e_2, e_5, e_7, e_9\}$.

4 Conclusions

In this paper, we propose the GMWC(T) problem. We show the GMWC(T) problem is fixed-parameter tractable and is polynomial time solvable when restricted in paths. At the current time, we have neither an algorithm to show the GMWC(T) problem is in P, nor a proof to show the problem is NP-hard. To determine the complexity of GMWC(T) remains an interesting open problem.

References

1. Bhaskara, A., Charikar, M., Chlamtac, E., Feige, U., Vijayaraghavan, A.: Detecting high log-densities – an $O(n^{1/4})$ approximation for densest k-subgraph. In: Schulman, L. (ed.) Proceedings of the 42nd Annual ACM Symposium on Theory of Computing, STOC, pp. 201–210. ACM (2010)
2. Bender, M.A., Farach-Colton, M.: The LCA Problem Revisited. In: Gonnet, G.H., Viola, A. (eds.) LATIN 2000. LNCS, vol. 1776, pp. 88–94. Springer, Heidelberg (2000)
3. Bousquet, N., Daligault, J., Thomassé, S.: Multicut is FPT. In: Fortnow, L., Vadhan, S. (eds.) Proceedings of the 43rd Annual ACM Symposium on Theory of Computing, STOC, pp. 459–468. ACM (2011)
4. Chen, J.-E., Liu, Y., Lu, S.-J.: An improved parameterized algorithm for the minimum node multiway cut problem. Algorithmica 55, 1–13 (2009)
5. Chopra, S., Rao, M.: On the Multiway Cut Polyhedron. Networks 21, 51–89 (1991)
6. Costa, M.-C., Billionnet, A.: Multiway cut and integer flow problems in trees. Electronic Notes in Discrete Mathematics 17, 105–109 (2004)
7. Dahlhaus, E., Johnson, D., Papadimitriou, C., Seymour, P., Yannakakis, M.: The complexity of multiterminal cuts. SIAM Journal on Computing 23, 864–894 (1994)
8. Garg, N., Vazirani, V., Yannakakis, M.: Primal-dual approximation algorithm for integral flow and multicut in trees. Algorithmica 18, 3–20 (1997)
9. Guo, J., Niedermeier, R.: Fixed-parameter tractability and data reduction for multicut in trees. Networks 46(3), 124–135 (2005)
10. Karger, D., Klein, P., Stein, C., Thorup, M., Young, N.: Rounding algorithms for a geometric embedding of minimum multiway cut. Mathematics of Operations Research 29(3), 436–461 (2004)
11. Khot, S., Regev, O.: Vertex cover might be hard to approximate to within $2 - \epsilon$. Journal of Computer and System Sciences 74(3), 335–349 (2008)
12. Manokaran, R., Naor, J., Raghavendra, P., Schwartz, R.: SDP gaps and UGC hardness for multiway cut, 0-extension, and metric labeling. In: Dwork, C. (ed.) Proceedings of the 40th Annual ACM Symposium on Theory of Computing, STOC, pp. 11–20. ACM (2008)
13. Mestre, J.: Lagrangian relaxation and partial cover (extended abstract). In: Albers, S., Weil, P. (eds.) Proceedings of the 25th International Symposium on Theoretical Aspects of Computer Science (STACS), pp. 539–550. LIPIcs 1 Schloss Dagstuhl - Leibniz-Zentrum fuer Informatik, Germany (2008)
14. Niedermeier, R., Rossmanith, P.: A general method to speed up fixed-parameter-tractable algorithms. Information Processing Letters 73, 125–129 (2000)
15. Xiao, M.-Y.: Simple and improved parameterized algorithms for multiterminal cuts. Theory of Computing Systems 46, 723–736 (2010)
16. Zhang, P.: Approximating Generalized Multicut on Trees. In: Cooper, S.B., Löwe, B., Sorbi, A. (eds.) CiE 2007. LNCS, vol. 4497, pp. 799–808. Springer, Heidelberg (2007)
17. Zhang, P., Zhu, D.-M., Luan, J.-F.: An approximation algorithm for the generalized k-Multicut problem. Discrete Applied Mathematics 160(7-8), 1240–1247 (2012)

Algorithms for Forest Local Similarity

Zhewei Liang and Kaizhong Zhang

Department of Computer Science, The University of Western Ontario,
London, Ontario, Canada, N6A 5B7
{zliang32,kzhang}@csd.uwo.ca

Abstract. An ordered labelled tree is a tree where the left-to-right order among siblings is significant. Ordered labelled forests are sequences of ordered labelled trees. Given two ordered labelled forests F and G, the *local forest similarity* is to find two sub-forests F' and G' of F and G respectively such that they are the most similar over all possible F' and G'. In this paper, we present efficient algorithms for the local forest similarity problem for two types of sub-forests: sibling subforests and closed subforests. Our algorithms can be used to locate the structurally similar regions in RNA secondary structures since RNA molecules' secondary structures could be represented as ordered labelled forests.

Keywords: local forest similarity, closed subforests, sibling subforests, forest removing similarity, RNA secondary structure comparison.

1 Introduction

Ordered labelled trees are trees where each node has a label and the left-to-right order among siblings is significant. An ordered labelled forest is a sequence of ordered labelled trees. Ordered labelled trees and forests are very useful data structures for hierarchical data representation such as RNA secondary structures [11,7,14,4] and XML documents [1]. Fig.1 [10] shows an example of the RNA GI:2347024 structure. Algorithms for the edit distance between two forests (trees) [15,3] could be used to measure the global similarity of forests (trees). Recently the *Forest (tree) Pattern Matching* (FPM) problem and the *Local Forest (tree) Similarity* (LFS) problem became interesting and attracted some attention [13,4,10,5,6,16].

In this paper, the FPM problem is defined as the following: Given a target forest F and a pattern forest G, find a sub-forest F' of F which is the most similar to G over all possible F'. And the LFS problem is defined as the following: Given two forests F and G, find two sub-forests F' and G' of F and G respectively, such that they are the most similar over all possible F' and G'. There are various ways to define the term "sub-forest". Here we will consider two types of "sub-forests": sibling subforests and closed subforests.

For sibling subforests and closed subforests, we present two efficient algorithms for solving the LFS problem. Our algorithms can be used to locate the structurally similar regions in RNA secondary structures since RNA molecules' secondary structures could be represented as ordered labelled forests.

G. Lin (Ed.): COCOA 2012, LNCS 7402, pp. 163–175, 2012.

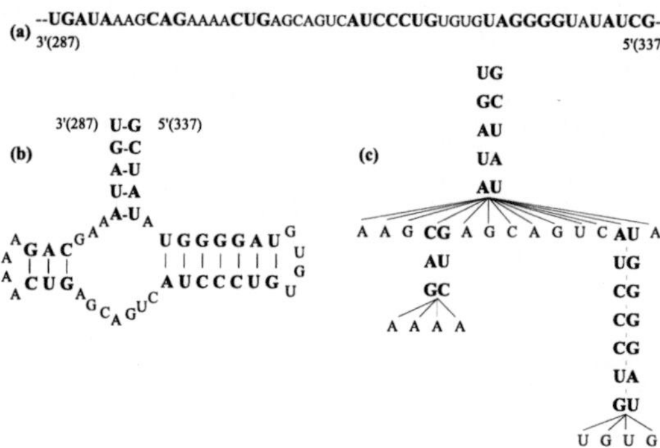

Fig. 1. RNA structures and forest representation. (a) A segment of the RNA GI: 2347024 primary structure [8], (b) its secondary structure, (c) its forest representation.

2 Preliminaries

We shall use the following definitions and notations in this paper.

Let F be any given ordered forest, and the nodes in F be labelled with a left-to-right postorder numbering. Given any two nodes, they are either in ancestor-descendant relationship, or in left-right relationship. Let $|F|$ be the number of nodes in F. A subtree of F is any connected sub-graph of F, and a subforest of F is an ordered sequence of subtrees of F. A complete-subtree of F is a subtree consisting of a root and all of its descendants in F. A complete-subforest of F is an ordered sequence of complete-subtrees of F. $F[i..j]$ is generally an ordered subforest of F in the postorder numbering, induced by the nodes numbered from i to j inclusively. $f[i]$ denotes the label of ith node in F. We also use $f[i]$ to represent the ith node if there is no confusion. $F[i]$ denotes the complete-subtree rooted at $f[i]$. Let $l(i)$ be the postorder number of the leftmost leaf descendant of $f[i]$. If $f[i_1]$ and $f[i_2]$ (or just i_1 and i_2) have the same parent, they are siblings. D_F and L_F denote the depth and the number of leaves of F, respectively. We assume that the forest F has an imaginary root node, denoted by $p(F)$. Let the key roots of F be the set $K(F) = \{p(F)\} \cup \{i \mid i \in F \text{ and } i \text{ has a left sibling}\}$ and from [15] we have $|K(F)| \leq L_F$.

2.1 Forest Edit Similarity

Similarity and distance measures are used in sequence alignment [12] and forest edit distance (similarity) [15] problems. Local alignment is normally based on similarity measures. Our algorithms are based on forest edit similarity metric.

The forest edit similarity uses a similarity measure, $s(a, b)$, between labels of forest nodes or λ. If both a and b are labels, then $s(a, b)$ is a score for match or mismatch; if b is λ, then $s(a, b)$ is a score for deletion; and if a is λ, then $s(a, b)$ is a score for insertion. The forest edit similarity also uses a forest edit mapping [15]. An forest edit mapping between F and G is defined as (M, F, G), where M

is a set of integer pairs (i, j) satisfying: (1) $1 \le i \le |F|$, $1 \le j \le |G|$; (2) for any pair (i_1, j_1) and (i_2, j_2) in M, (a) $i_1 = i_2$ if and only if $j_1 = j_2$ (one-to-one condition); (b) $f(i_1)$ is an ancestor of $f(i_2)$ if and only if $g(j_1)$ is an ancestor of $g(j_2)$ (ancestor condition); (c) $f(i_1)$ is to the left of $f(i_2)$ if and only if $g(j_1)$ is to the left of $g(j_2)$ (sibling condition). The similarity score of an edit mapping M is given by:

$$s(M) = \sum_{(i,j)\in M} s(f[i], g[j]) + \sum_{i \notin M} s(f[i], \lambda) + \sum_{j \notin M} s(\lambda, g[j]).$$

And the forest similarity between F and G is defined as

$$\phi(F, G) = \max\{s(M) \mid M \text{ is an edit mapping between } F \text{ and } G\}.$$

If $s(a, b)$ is a similarity metric, then $\phi(F, G)$ is also a similarity metric [2].

2.2 Sub-Forests Definitions

In this paper, we consider local forest similarity problem for two types of sub-forests of forest F: (1) a *sibling subforest*: a sequence of subtrees of F such that their roots are siblings. (2) a *closed subforest*: a sequence of subtrees of F such that their roots are consecutive siblings. Examples of these two types of sub-forests are shown in Fig. 2 as the forests enclosed by dashed lines.

Restricting subtrees to complete-subtrees, we can have another two types of sub-forests: (1′) a *sibling complete-subforest*: a sequence of complete-subtrees of F such that their roots are siblings. (2′) a *closed complete-subforest*: a sequence of complete-subtrees of F such that their roots are consecutive siblings.

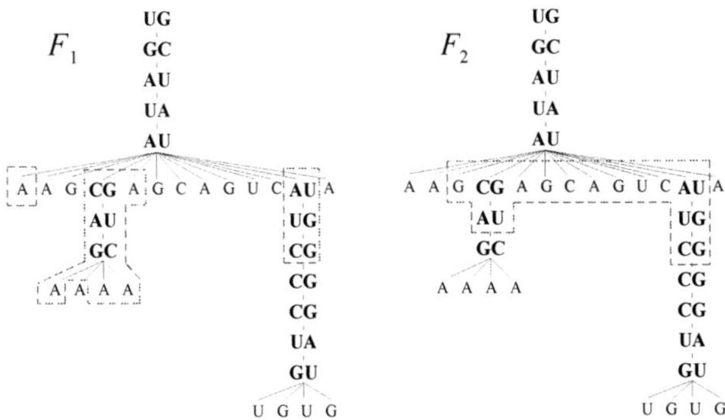

Fig. 2. Examples of two types of sub-forests of the forest in Fig. 1.(c). F_1: a sibling subforest, F_2: a closed subforest.

FPM and LFS problems for the complete-subtree and complete-subforest were considered though the subtree and subforest were used to represent the complete-subtree and complete-subforest in [10,6,16]. While in this paper we use subtree, subforest, and closed subforest to represent a more general situation.

2.3 Previous Work

For the local forest similarity problem for sibling subforests and closed subforests, we present two efficient algorithms. To the best of our knowledge, these are the first algorithms for this problem.

In a related work [10], Peng gave an algorithm for LFS on "closed complete-subforests" using distance metrics that runs in $O(|F| \cdot |G| \cdot L_F \cdot L_G)$ time and $O(|F| \cdot |G|)$ space. Note that under distance metrics, the goal is to find the minimum score, so the local similarity may end up with two identical leaves matched that produced an optimal distance score zero. Therefore for local forest similarity, we will have to use similarity metrics [14,2].

The problem considered by Jansson and Peng [6] and Zhang and Zhu [16] is the forest pattern matching (FPM) problem which is different from local forest similarity (LFS) problem considered in this paper.

2.4 Our Results

In this paper, we show how to solve the LFS problem on "sibling subforests" and "closed subforests" efficiently. The time and space complexities are summarized in Table 1.

Table 1. Our algorithms for LFS problem for sibling subforests and closed subforests

LFS	Time	Space	Section
Sibling Subforests	$O(\|F\| \cdot \|G\| \cdot \min\{D_F, L_F\} \cdot \min\{D_G, L_G\})$	$O(\|F\| \cdot \|G\|)$	3.1
Closed Subforests	$O(\|F\| \cdot \|G\| \cdot \min\{D_F, L_F\} \cdot \min\{D_G, L_G\})$	$O(\|F\| \cdot \|G\|)$	3.2

Our first algorithm modifies the global forest distance algorithm [15]. Our second algorithm combines the idea of [15] and the method of Smith-Waterman for local alignment between two sequences [12].

3 Algorithms for the Local Forest Similarity Problem

In this section, we present efficient algorithms for the LFS problem for two types of sub-forests, sibling subforests and closed subforests, respectively. We also refer the LFS problem as the problem of *finding two most similar sub-forests.*

3.1 An Algorithm for Finding Two Most Similar Sibling Subforests

Now we consider the problem of finding two most similar sibling subforests, which is also called *forest removing similarity* problem.

Let the degree of node i be d_i and its children be $i_1, i_2, \ldots, i_{d_i}$, let the degree of node j be d_j and its children be $j_1, j_2, \ldots, j_{d_j}$. Let $subf(F)$ be the set of complete-subforests of F. Let $F \setminus f$ represent the subforest resulting from the deletion of complete-subforest f from F.

Given two forests F and G, the forest removing similarity (FRS) between F and G, $\Phi_{rr}(,)$, is defined as follows where $f \in subf(F)$ and $g \in subf(G)$:

$$\Phi_{rr}(F, G) = \max\{\phi(F \setminus f, G \setminus g)\}.$$

Fig. 3 shows the situation of finding two most similar sibling subforests between F and G.

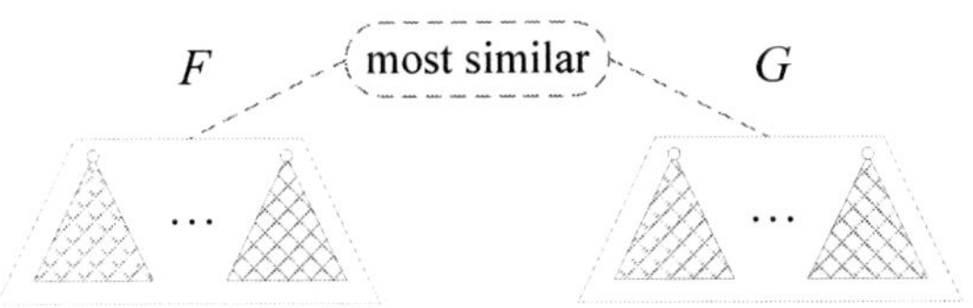

Fig. 3. Find two most similar sibling subforests between F and G

From this definition and the algorithm in [15,16], we have the following lemma for $\Phi_{rr}(F[l(i)..i'], G[l(j)..j'])$ when consider two cases: (1) $F[l(i)..i']$ and $G[l(j)..j']$ are both trees; (2) $F[l(i)..i']$ or $G[l(j)..j']$ is a forest.

Lemma 1. *Let i, j, F and G be defined as above, where $l(i) \leq i' \leq i$ and $l(j) \leq j' \leq j$, then*
(1) If $l(i') = l(i)$ and $l(j') = l(j)$,

$$\Phi_{rr}(F[l(i)..i'], G[l(j)..j']) = \max \begin{cases} 0, \\ \Phi_{rr}(F[l(i)..i'-1], G[l(j)..j']) + s(f[i'], \lambda), \\ \Phi_{rr}(F[l(i)..i'], G[l(j)..j'-1]) + s(\lambda, g[j']), \\ \Phi_{rr}(F[l(i)..i'-1], G[l(j)..j'-1]) + s(f[i'], g[j']). \end{cases}$$

(2) If $l(i') \neq l(i)$ or $l(j') \neq l(j)$,

$$\Phi_{rr}(F[l(i)..i'], G[l(j)..j']) = \max \begin{cases} \Phi_{rr}(F[l(i)..l(i')-1], G[l(j)..j']), \\ \Phi_{rr}(F[l(i)..i'], G[l(j)..l(j')-1]), \\ \Phi_{rr}(F[l(i)..i'-1], G[l(j)..j']) + s(f[i'], \lambda), \\ \Phi_{rr}(F[l(i)..i'], G[l(j)..j'-1]) + s(\lambda, g[j']), \\ \Phi_{rr}(F[l(i)..l(i')-1], G[l(j)..l(j')-1]) \\ \qquad + \Phi_{rr}(F[i'], G[j']). \end{cases}$$

Proof. Due to the page limitation, we omit the proof. Please refer to [9] for detail.

Our algorithm for forest removing similarity problem is based on Lemma 1.

Theorem 1. *Our algorithm for FRS problem can be implemented to run in $O(|F| \cdot |G| \cdot min\{D_F, L_F\} \cdot min\{D_G, L_G\})$ time and $O(|F| \cdot |G|)$ space.*

Proof. This is same to Theorem 2 in [15].

3.2 An Algorithm for Finding Two Most Similar Closed Subforests

We first examine Local Sequence Similarity (LSS) method and then give a natural extension from LSS to LFS for closed subforests.

Given two sequences $A[1..m]$ and $B[1..n]$, the LSS problem [12] is to find a pair of subsequences that satisfies $\max\{\phi(F[k..l], G[q..r])| \ 1 \leq k \leq l \leq m, \ 1 \leq q \leq r \leq n\}$. Smith and Waterman [12] gave an efficient algorithm for this problem using dynamic programming. The idea is that when calculating a similarity score for $F[1..i](1 \leq i \leq m)$ and $G[1..j](1 \leq j \leq n)$, any prefix of $F[1..i]$ or $G[1..j]$ could be deleted without any penalty. In other words, the maximum similarity score is $\max\{\phi(F[i'..i], G[j'..j]) \mid 1 \leq i' \leq i+1, \ 1 \leq j' \leq j+1\}$.

In order to solve the LFS problem for closed subforests, we need to use the removing similarity $\Phi_{rr}(,)$ described in the last section. We also need another removing similarity: *the maximum similarity between a closed subforest and a sibling subforest*, which we shall introduce first.

Let i, j, F, and G be defined as above, now we consider finding the maximum similarity between a *closed subforest* under one node i of forest F and a *sibling subforest* under one node j of forest G.

Let m be a node which satisfies $i_1 \leq m < i$, let n be a node which satisfies $j_1 \leq n < j$, we define $r_1(m)$ and $r_2(n)$ as follows:

$$r_1(m) = \max\{k_1 \mid 1 \leq k_1 \leq d_i, \ i_{k_1} \leq m, \ k_1 \in \mathbb{Z}\},$$
$$r_2(n) = \max\{k_2 \mid 1 \leq k_2 \leq d_j, \ j_{k_2} \leq n, \ k_2 \in \mathbb{Z}\}.$$

Fig. 4 shows the definition of $r_1(m)$.

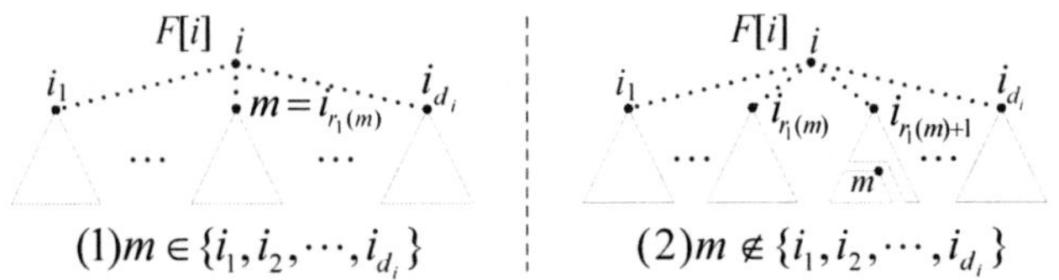

Fig. 4. Definition of $r_1(m)$

Let $subf(F, node\_set)$ be the set of complete-subforests of F such that nodes in $node\_set$ are not in any of the complete-subforests. We define another removing edit similarity $\Phi_{Rr}(,)$ where $f \in subf(F[l(i_1)..m], \{i_1, \ldots i_{r_1(m)}\})$ and $g \in subf(G[l(j_1)..n])$:

$$\Phi_{Rr}(F[l(i_1)..m], G[l(j_1)..n]) = \max\{\phi(F[l(i_1)..m] \setminus f, G[l(j_1)..n] \setminus g])\}.$$

Now we can define the score $\Theta_1(F[l(i_1)..m], G[l(j_1)..n])$ where $i_1 \leq m \leq i_{d_i}$ and $j_1 \leq n \leq j_{d_j}$, using $\Phi_{Rr}(,)$ and $r_1(m)$ as follows:

$$\max\{\Phi_{Rr}(F[l(i_u)..m], G[l(j_1)..n]) \mid 1 \leq u \leq r_1(m) + 1\}.$$

Notice that if $m = i_{k_1}$ where $1 \le k_1 \le d_i$, the value of $r_1(m)$ is therefore k_1. Under this situation, for convenience, when $u = k_1+1$, we let $F[l(i_{k_1+1})..i_{k_1}] = \emptyset$. Therefore, if $m \in \{i_1, \ldots, i_{d_i}\}$, then $\{F[l(i_u)..m] \mid 1 \le u \le r_1(m) + 1\}$ can be written as $\{F[l(i_1)..m], \ldots, F[l(m)..m], \emptyset\}$; otherwise it can be written as $\{F[l(i_1)..m], \ldots, F[l(i_{r_1(m)+1})..m]\}$. Fig. 5 shows these two different cases.

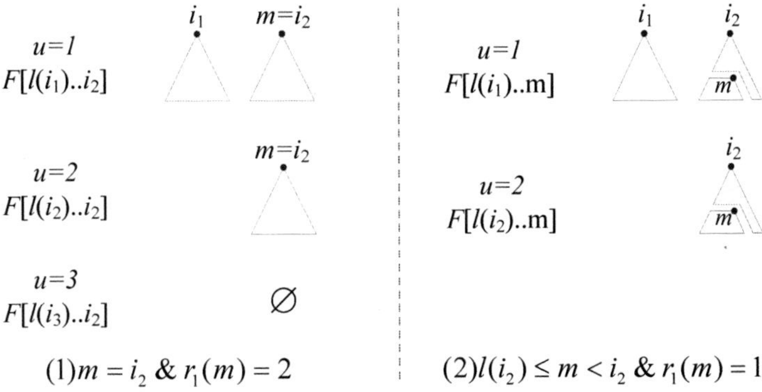

Fig. 5. Two cases of $F[l(i_u)..m]$ where $l(i_2) \le m \le i_2$ and $1 \le u \le r_1(m) + 1$

With these definitions, what we want to compute for $F[i]$ and $G[j]$ is $\max\{\Theta_1(F[l(i_1)..i_u], G[l(j_1)..j_{d_j}]) \mid 1 \le u \le d_i\}$. Fig. 6 shows this situation.

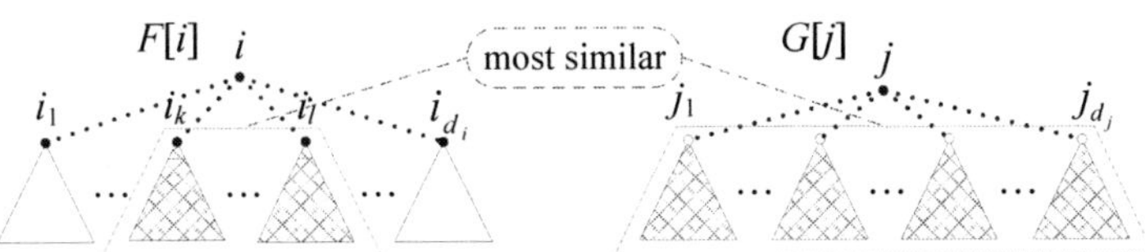

Fig. 6. Find the maximum similarity between a closed subforest of $F[i]$ and a sibling subforest of $G[j]$

To calculate $\Theta_1(F[l(i_1)..m], G[l(j_1)..n])$, we have the following lemmas.

Lemma 2. *Let i, j, F, and G be defined as above, where $l(i_u) \le m \le i_{d_i}$, $j_1 \le n \le j_{d_j}$, and $1 \le u \le r_1(m)$, then*

$$\Phi_{Rr}(F[l(i_u)..m], G[l(j_1)..n]) = \max \begin{cases} \Phi_{Rr}(F[l(i_u)..l(m)-1], G[l(j_1)..n]), \ if\ m \notin \{i_1, i_2, \ldots, i_{d_i}\} \\ \Phi_{Rr}(F[l(i_u)..m], G[l(j_1)..l(n)-1]), \\ \Phi_{Rr}(F[l(i_u)..m-1], G[l(j_1)..n]) + s(f[m], \lambda), \\ \Phi_{Rr}(F[l(i_u)..m], G[l(j_1)..n-1]) + s(\lambda, g[n]), \\ \Phi_{Rr}(F[l(i_u)..l(m)-1], G[l(j_1)..l(n)-1]) \\ \quad + \Phi_{rr}(F[l(m)..m-1], G[l(n)..n-1]) + s(f[m], g[n]). \end{cases}$$

Proof. Please refer to [9] for detail.

Lemma 3. *Let i, j, F, and G be defined as above, then*
(1) $\Theta_1(F[i_1], G[j_1]) = \Phi_{rr}(F[i_1], G[j_1])$.
(2) $\Theta_1(F[i_1], G[l(j_1)..n]) = \Phi_{rr}(F[i_1], G[l(j_1)..n])$.
(3) If $i_1 < m \leq i_{d_i}$ *and* $n = j_1$*, then*

$$\Theta_1(F[l(i_1)..m], G[j_1]) = \max \begin{cases} \Theta_1(F[l(i_1)..l(m)-1], G[j_1]), \ if\ m \notin \{i_2, \ldots, i_{d_i}\} \\ \Theta_1(F[l(i_1)..m-1], G[j_1]) + s(f[m], \lambda), \\ \Theta_1(F[l(i_1)..m], G[l(j_1)..j_1-1]) + s(\lambda, g[j_1]), \\ \Phi_{rr}(F[m], G[j_1]). \end{cases}$$

(4) If $i_1 < m \leq i_{d_i}$ *and* $j_1 < n \leq j_{d_j}$*, then*

$$\Theta_1(F[l(i_1)..m], G[l(j_1)..n]) = \max \begin{cases} \Theta_1(F[l(i_1)..l(m)-1], G[l(j_1)..n]), \ if\ m \notin \{i_2, \ldots, i_{d_i}\} \\ \Theta_1(F[l(i_1)..m], G[l(j_1)..l(n)-1]), \\ \Theta_1(F[l(i_1)..m-1], G[l(j_1)..n]) + s(f[m], \lambda), \\ \Theta_1(F[l(i_1)..m], G[l(j_1)..n-1]) + s(\lambda, g[n]), \\ \Theta_1(F[l(i_1)..l(m)-1], G[l(j_1)..l(n)-1]) \\ \quad + \Phi_{rr}(F[l(m)..m-1], G[l(n)..n-1]) + s(f[m], g[n]). \end{cases}$$

Proof. Due to the page limitation, here we only prove case (3).
(1) For $m \in \{i_2, \ldots, i_{d_i}\}$, we know that $m = i_{r_1(m)}$. From the definition of $\Theta_1(,)$, we have

$$\max \begin{cases} \Theta_1(F[l(i_1)..m-1], G[j_1]) + s(f[m], \lambda), \\ \Theta_1(F[l(i_1)..m], G[l(j_1)..j_1-1]) + s(\lambda, g[j_1]), \\ \Phi_{rr}(F[m], G[j_1]). \end{cases}$$

$$= \max \begin{cases} \max\{\Phi_{Rr}(F[l(i_u)..i_{r_1(m)}-1], G[j_1]) \mid 1 \leq u \leq r_1(m-1)+1\} + s(f[i_{r_1(m)}], \lambda), \\ \max\{\Phi_{Rr}(F[l(i_u)..i_{r_1(m)}], G[l(j_1)..j_1-1]) \mid 1 \leq u \leq r_1(m)+1\} + s(\lambda, g[j_1]), \\ \Phi_{rr}(F[m], G[j_1]). \end{cases}$$

According to Lemma 2, we have

$$\max\{\Phi_{Rr}(F[l(i_u)..i_{r_1(m)}], G[j_1]) \mid 1 \leq u \leq r_1(m)+1\}$$
$$= \max \left\{ \max \begin{cases} \Phi_{Rr}(F[l(i_u)..i_{r_1(m)}], G[l(j_1)..l(j_1)-1]), \\ \Phi_{Rr}(F[l(i_u)..i_{r_1(m)}-1], G[j_1]) + s(f[i_{r_1(m)}], \lambda), \\ \Phi_{Rr}(F[l(i_u)..i_{r_1(m)}], G[l(j_1)..j_1-1]) + s(\lambda, g[j_1]), \\ \Phi_{Rr}(F[l(i_u)..l(i_{r_1(m)})-1], G[l(j_1)..l(j_1)-1]) \\ \quad + \Phi_{rr}(F[l(i_{r_1(m)})..i_{r_1(m)}-1], G[l(j_1)..j_1-1]) \\ \quad + s(f[i_{r_1(m)}], g[j_1]). \end{cases} \right\} 1 \leq u \leq r_1(m)+1$$

$$= \max \begin{cases} \max\{\Phi_{Rr}(F[l(i_u)..i_{r_1(m)}], \emptyset) \mid 1 \leq u \leq r_1(m)+1\}, \\ \max\{\Phi_{Rr}(F[l(i_u)..i_{r_1(m)}-1], G[j_1]) \mid 1 \leq u \leq r_1(m-1)+1\} + s(f[i_{r_1(m)}], \lambda), \\ \max\{\Phi_{Rr}(F[l(i_u)..i_{r_1(m)}], G[l(j_1)..j_1-1]) \mid 1 \leq u \leq r_1(m)+1\} + s(\lambda, g[j_1]), \\ \max\{\Phi_{Rr}(F[l(i_u)..l(i_{r_1(m)})-1], \emptyset) \mid 1 \leq u \leq r_1(l(m)-1)+1\} \\ \quad + \Phi_{rr}(F[l(i_{r_1(m)})..i_{r_1(m)}-1], G[l(j_1)..j_1-1]) + s(f[i_{r_1(m)}], g[j_1]). \end{cases}$$

$$= \max \begin{cases} 0, \\ \max\{\Phi_{Rr}(F[l(i_u)..i_{r_1(m)}-1], G[j_1]) \mid 1 \le u \le r_1(m-1)+1\} + s(f[i_{r_1(m)}], \lambda), \\ \max\{\Phi_{Rr}(F[l(i_u)..i_{r_1(m)}], G[l(j_1)..j_1-1]) \mid 1 \le u \le r_1(m)+1\} + s(\lambda, g[j_1]), \\ \Phi_{rr}(F[l(i_{r_1(m)})..i_{r_1(m)}-1], G[l(j_1)..j_1-1]) + s(f[i_{r_1(m)}], g[j_1]). \end{cases}$$

And we know that $\Phi_{Rr}(F[i_{r_1(m)}], G[j_1]) \ge \Phi_{rr}(F[l(i_{r_1(m)})..i_{r_1(m)} - 1], G[l(j_1)..j_1 - 1]) + s(f[i_{r_1(m)}], g[j_1])$ since the latter formula represents a particular mapping of $F[i_{r_1(m)}]$ to $G[j_1]$, so we can use $\Phi_{Rr}(F[i_{r_1(m)}], G[j_1])$ to substitute $\Phi_{rr}(F[l(i_{r_1(m)})..i_{r_1(m)} - 1], G[l(j_1)..j_1 - 1]) + s(f[i_{r_1(m)}], g[j_1])$ here because $\Phi_{Rr}(F[i_{r_1(m)}], G[j_1])$ is a possible mapping of $\max\{\Phi_{Rr}(F[l(i_u)..i_{r_1(m)}], G[j_1]) \mid 1 \le u \le r_1(m) + 1\}$. From the definition of $\Phi_{Rr}(,)$ and $\Phi_{rr}(,)$, we have $\max\{0, \Phi_{Rr}(F[m], G[j_1])\} = \Phi_{rr}(F[m], G[j_1])$. Therefore

$$\max \begin{cases} \Theta_1(F[l(i_1)..m-1], G[j_1]) + s(f[m], \lambda), \\ \Theta_1(F[l(i_1)..m], G[l(j_1)..j_1-1]) + s(\lambda, g[j_1]), \\ \Phi_{rr}(F[m], G[j_1]). \end{cases}$$

$$= \max\{\Phi_{Rr}(F[l(i_u)..i_{r_1(m)}], G[j_1]) \mid 1 \le u \le r_1(m)+1\}$$
$$= \max\{\Phi_{Rr}(F[l(i_u)..i_{r_1(m)}], G[j_1]) \mid 1 \le u \le r_1(m),\ 0\}.$$

From the definition of $\Theta_1(F[l(i_1)..m], G[j_1])$, we have $\Theta_1(F[l(i_1)..m], G[j_1]) = \max\{\Phi_{Rr}(F[l(i_u)..i_{r_1(m)}], G[j_1]) \mid 1 \le u \le r_1(m),\ 0\}$. Therefore,

$$\Theta_1(F[l(i_1)..m], G[j_1]) = \max \begin{cases} \Theta_1(F[l(i_1)..m-1], G[j_1]) + s(f[m], -), \\ \Theta_1(F[l(i_1)..m], G[l(j_1)..j_1-1]) + s(-, g[j_1]), \\ \Phi_{rr}(F[m], G[j_1]). \end{cases}$$

(2) For $m \notin \{i_2, \ldots, i_{d_i}\}$, it is similar to above (1).

To find the maximum similarity between a sibling subforest and a closed subforest, we can also define the third removing edit similarity $\Phi_{rR}(,)$ and the score $\Theta_2(,)$ using $\Phi_{rR}(,)$ and $r_2(n)$. They are symmetric to $\Phi_{Rr}(,)$ and $\Theta_1(,)$, we do not describe them again.

Finally, we are ready to find the maximum similarity between two closed subforests. Fig. 7 shows this situation under two nodes and our goal is to find the maximum score between all of the nodes of F and G, respectively.

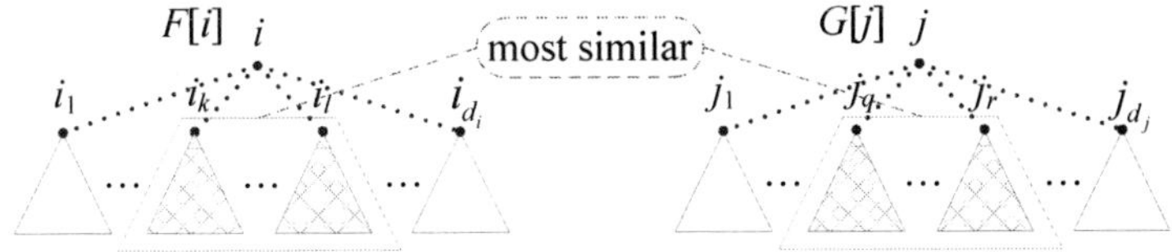

Fig. 7. Find two most similar closed subforests between $F[i]$ and $G[j]$

We give the definition for the fourth removing edit similarity $\Phi_{RR}(,)$ where $f \in subf(F[l(i_1)..m], \{i_1, \ldots i_{r_1(m)}\})$ and $g \in subf(G[l(j_1)..n], \{j_1, \ldots j_{r_2(n)}\})$:

$$\Phi_{RR}(F[l(i_1)..m], G[l(j_1)..n]) = \max\{\phi(F[l(i_1)..m] \setminus f, G[l(j_1)..n] \setminus g])\}.$$

Now we can define the score $\Psi(F[l(i_1)..m], G[l(j_1)..n])$ for $F[l(i_1)..m]$ and $G[l(j_1)..n]$ using $\Phi_{RR}(,)$ for closed subforests as follows:

$$\max\{\Phi_{RR}(F[l(i_u)..m], G[l(j_v)..n]) \mid 1 \leq u \leq r_1(m) + 1,\ 1 \leq v \leq r_2(n) + 1\}.$$

With these definitions, what we want to compute for node i, j is $\max\{\Psi(F[l(i_1)..i_u], G[l(j_1)..j_v]) \mid 1 \leq u \leq d_i,\ 1 \leq v \leq d_j\}$. For the calculation of $\Psi(F[l(i_1)..m], G[l(j_1)..n])$, we have the following lemma.

Lemma 4. *Let i, j, F, and G be defined as above, then*
(1) $\Psi(F[i_1], G[j_1]) = \Phi_{rr}(F[i_1], G[j_1])$.
(2) If $i_1 < m \leq i_{d_i}$ and $n = j_1$, then

$$\Psi(F[l(i_1)..m], G[j_1]) = \max \begin{cases} \Psi(F[l(i_1)..l(m) - 1], G[j_1]), & if\ m \notin \{i_2, \ldots, i_{d_i}\} \\ \Psi(F[l(i_1)..m - 1], G[j_1]) + s(f[m], \lambda), \\ \Theta_1(F[l(i_1)..m], G[l(j_1)..j_1 - 1]) + s(\lambda, g[j_1]), \\ \Phi_{rr}(F[m], G[j_1]). \end{cases}$$

(3) If $m = i_1$ and $j_1 < n \leq j_{d_j}$, then

$$\Psi(F[i_1], G[l(j_1)..n]) = \max \begin{cases} \Psi(F[i_1], G[l(j_1)..l(n) - 1]), & if\ n \notin \{j_2, \ldots, j_{d_j}\} \\ \Theta_2(F[l(i_1)..i_1 - 1], G[l(j_1)..n]) + s(f[i_1], \lambda), \\ \Psi(F[i_1], G[l(j_1)..n - 1]) + s(\lambda, g[n]), \\ \Phi_{rr}(F[i_1], G[n]). \end{cases}$$

(4) If $i_1 < m \leq i_{d_i}$ and $j_1 < n \leq j_{d_j}$, then

$$\Psi(F[l(i_1)..m], G[l(j_1)..n]) = \max \begin{cases} 0, \\ \Psi(F[l(i_1)..l(m) - 1], G[l(j_1)..n]), & if\ m \notin \{i_2, \ldots, i_{d_i}\} \\ \Psi(F[l(i_1)..m], G[l(j_1)..l(n) - 1]), & if\ n \notin \{j_2, \ldots, j_{d_j}\} \\ \Psi(F[l(i_1)..m - 1], G[l(j_1)..n]) + s(f[m], \lambda), \\ \Psi(F[l(i_1)..m], G[l(j_1)..n - 1]) + s(\lambda, g[n]), \\ \Psi(F[l(i_1)..l(m) - 1], G[l(j_1)..l(n) - 1]) \\ \quad + \Phi_{rr}(F[l(m)..m - 1], G[l(n)..n - 1]) + s(f[m], g[n]). \end{cases}$$

Proof. Please refer to [9] for detail.

With above lemmas, we can calculate $\max\{\Psi(F[l(i_1)..i_u], G[l(j_1)..j_v]) \mid 1 \leq u \leq d_i,\ 1 \leq v \leq d_j\}$ using dynamic programming. However, we have to do this for every node i of F and every node j of G, respectively. Because for each child subtree of $F[i]$ and $G[j]$, the calculation starts at i_1 (the leftmost child of i) and j_1 (the leftmost child of j) instead of $l(i_1)$ and $l(j_1)$ respectively, and $\Phi_{rr}(F[i_1], G[j_1])$, $\Theta_1(F[l(i_1)..m], G[j_1])$ and $\Theta_2(F[i_1], G[l(j_1)..n])$ are needed in the calculation. It is better to do the calculations for all the nodes on the paths of F and G from the leaves to their nearest ancestor key roots together. In this way, we do the computation layer by layer on both F and G.

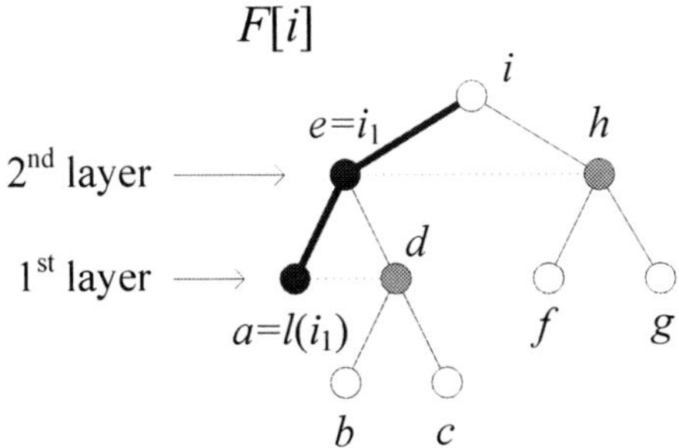

Fig. 8. Layer and path representation

We shall need the following definitions: $lp(i)$: a set which contains the nodes on the leftmost path of $F[i]$ except the root i; $layer(i)$: a set which contains all of the sibling nodes of nodes in $lp(i)$ including $lp(i)$. In the Fig. 8, the bold line is the leftmost path of $F[i]$, the black nodes $\{a, e\}$ belong to $lp(i)$ and the black and gray nodes $\{a, d, e, h\}$ belong to $layer(i)$. With these definitions, we can extend Lemma 3 - 4 from two nodes to two leftmost paths of two key roots of F and G, respectively.

Our algorithm for the LFS problem is a dynamic programming algorithm.

In the first stage of our algorithm, we call FRS algorithm $RemovingSimilarity(F, G)$ in Section 3.1 for F and G to get $\Phi_{rr}(F[i'], G[j'])$ and $\Phi_{rr}(F[l(i')..i'-1], G[l(j')..j'-1])$ ($1 \leq i' \leq p(F)$, $1 \leq j' \leq p(G)$) needed in $ForestRemovingSimi(,)$, $Theta_1(,)$, $Theta_2(,)$ and $Psi(,)$ in the second stage.

In the second stage, the key roots of F and G are sorted in the increasing order and put in two arrays K_F and K_G, respectively. For any key root i of F and any key root j of G, we first call $LeftmostPath(i, j)$ to get $lp(i)$ and $lp(j)$ needed in $Theta_1(,)$, $Theta_2(,)$ and $Psi(,)$; we second call $ForestRemovingSimi(,)$ for $F[i]$ and $G[j]$ to get $\Phi_{rr}(,)$ needed in $Theta_1(,)$, $Theta_2(,)$ and $Psi(,)$; then we call $Theta_1(,)$ and $Theta_2(,)$ for $F[i]$ and $G[j]$ to get $\Theta_1(,)$ and $\Theta_2(,)$ needed in $Psi(,)$; and at last we call $Psi(,)$ for $F[i]$ and $G[j]$.

We can now show our algorithm for finding two most closed subforests between F and G.

Theorem 2. *Algorithm LFS correctly computes the cost of an optimal solution.*

Proof. In Algorithm LFS, because of step 1, all the $\Phi_{rr}(i', j')$ used in step 7, 8, 9, 10 are available. At the same time, all the $\Phi_{rr}(F[l(i')..i'-1], G[l(j')..j'-1])$ used in step 8, 9, 10 are available.

Input: Two forests F and G.
Output: $\max\{\Psi(F[l(i_1)..i_l], G[l(j_1)..j_r]) \mid i_1$ is i_l's leftmost sibling, j_1 is j_r's leftmost sibling$\}$.

1 Call *RemovingSimilarity*(F, G) according to Lemma 1
2 **for** $i' := 1$ **to** $|K_F|$ **do**
3   **for** $j' := 1$ **to** $|K_G|$ **do**
4         $i := K_F[i']$
5         $j := K_G[j']$
6   Call $LeftmostPath(i, j)$
7   Call $ForestRemovingSimi(i, j)$ according to Lemma 1
8   Call $Theta_1(i, j)$ according to Lemma 3
9   Call $Theta_2(i, j)$ according to Lemma for $\Theta_2(i, j)$
10   Call $Psi(i, j)$ according to Lemma 4
11   **end**
12 **end**

Algorithm LFS**:** Find two most closed subforests between F and G

Because of step 6, all the $lp(i)$ and $lp(j)$ used in step 8, 9, 10 are available.
Because of step 7, all the $\Phi_{rr}(,)$ used in step 8,9 are available.
Because of step 8, all the $\Theta_1(,)$ used in step 10 are available.
Because of step 9, all the $\Theta_2(,)$ used in step 10 are available.

Theorem 3. *Our algorithm for LFS problem can be implemented to run in $O(|F| \cdot |G| \cdot min\{D_F, L_F\} \cdot min\{D_G, L_G\})$ time and $O(|F| \cdot |G|)$ space.*

Proof. The time and space complexity of the computation for the removing similarity of all subtree pairs of F and G are $O(|F| \cdot |G| \cdot \min\{D_F, L_F\} \cdot \min\{D_G, L_G\})$ and $O(|F| \cdot |G|)$ respectively, due to Theorem 1.

For one key root i of F and one key root j of G, the time and space complexity for $LeftmostPath(i, j)$, $ForestRemovingSimi(i, j)$, $Theta_1(i, j)$, $Theta_2(i, j)$, and $Psi(i, j)$ are the same: $O(|F[i]| \cdot |G[j]|)$. And for all key roots of F and G, the time complexity is $O(\min\{D_F, L_F\} \cdot \min\{D_G, L_G\})$ and the space complexity is $O(|F| \cdot |G|)$ due to Lemma 7 in [15]. Therefore, the total time and space complexity of our algorithm are $O(|F| \cdot |G| \cdot \min\{D_F, L_F\} \cdot \min\{D_G, L_G\})$ and $O(|F| \cdot |G|)$, respectively.

4 Conclusion

In this paper, we studied the local forest similarity problem. We presented two efficient algorithms for this problem when sub-forests are sibling subforests and closed subforests. These algorithms could be used to locate structurally similar regions in RNA secondary structures or in other hierarchical data.

When the inputs are two sequences represented as linear forests (trees), our second algorithm can be reduced to the sequence local alignment algorithm [12].

References

1. Bray, T., Paoli, J., Sperberg-McQueen, C.M., Maler, E., Yergeau, F.: Extensible markup language (XML) 1.0. W3C Recommendation 6 (2000)
2. Chen, S., Ma, B., Zhang, K.: On the similarity metric and the distance metric. Theor. Comput. Sci. 410(24-25), 2365–2376 (2009)
3. Demaine, E.D., Mozes, S., Rossman, B., Weimann, O.: An Optimal Decomposition Algorithm for Tree Edit Distance. In: Arge, L., Cachin, C., Jurdziński, T., Tarlecki, A. (eds.) ICALP 2007. LNCS, vol. 4596, pp. 146–157. Springer, Heidelberg (2007)
4. Höchsmann, M., Töller, T., Giegerich, R., Kurtz, S.: Local similarity in RNA secondary structures. In: Proceedings of the IEEE Computational Systems Bioinformatics Conference, pp. 159–168 (2003)
5. Jansson, J., Hieu, N.T., Sung, W.K.: Local Gapped Subforest Alignment and Its Application in Finding RNA Structural Motifs. Journal of Computational Biology 13(3), 702–718 (2006)
6. Jansson, J., Peng, Z.: Algorithms for Finding a Most Similar Subforest. In: Lewenstein, M., Valiente, G. (eds.) CPM 2006. LNCS, vol. 4009, pp. 377–388. Springer, Heidelberg (2006)
7. Jiang, T., Wang, L., Zhang, K.: Alignment of trees - an alternative to tree edit. Theoretical Computer Science 143, 137–148 (1995)
8. Motifs database, `http://subviral.med.uottawa.ca/cgi-bin/motifs.cgi`
9. Liang, Z.: Efficient Algorithms for Local Forest Similarity. Thesis(M.Sc), School of Graduate and Postdoctoral Studies, University of Western Ontario, London, Ontario, Canada (2011)
10. Peng, Z.: Algorithms for Local Forest Similarity. In: Deng, X., Du, D.-Z. (eds.) ISAAC 2005. LNCS, vol. 3827, pp. 704–713. Springer, Heidelberg (2005)
11. Shapiro, B.A., Zhang, K.: Comparing multiple RNA secondary structures using tree comparisons. Computer Applications in the Biosciences 6(4), 309–318 (1990)
12. Smith, T.F., Waterman, M.S.: Identification of common molecular subsequences. Journal of Molecular Biology 147(1), 195–197 (1981)
13. Wang, J., Shapiro, B.A., Shasha, D., Zhang, K., Currey, K.M.: An algorithm for finding the largest approximately common substructures of two trees. IEEE Trans. Pattern Anal. Mach. Intell. 20(8), 889–895 (1998)
14. Zhang, K.: Computing similarity between RNA secondary structures. In: Proceedings of IEEE International Joint Symposia on Intelligence and Systems, Rockville, Maryland, pp. 126–132 (May 1998)
15. Zhang, K., Shasha, D.: Simple fast algorithms for the editing distance between trees and related problems. SIAM Journal on Computing 18(6), 1245–1262 (1989)
16. Zhang, K., Zhu, Y.: Algorithms for Forest Pattern Matching. In: Amir, A., Parida, L. (eds.) CPM 2010. LNCS, vol. 6129, pp. 1–12. Springer, Heidelberg (2010)

Speedup of RNA Pseudoknotted Secondary Structure Recurrence Computation with the Four-Russians Method

Yelena Frid and Dan Gusfield

Department of Computer Science, U.C. Davis

Abstract. While secondary pseudoknotted structure prediction is computationally challenging, such structures appear to play biologically important roles in both cells and viral RNA [1]. Restricting the class of possible structures and then finding the optimal structure for that restricted class is a common method employed to deal with the computational complexity.

We derive a practical and worst-case speedup algorithm using the Four-Russians method for the $O(n^6)$ time *Rivas&Eddy Algorithm* [2] describing the broadest set of structures. *Fast R&E* algorithm finds the optimal Rivas&Eddy fold in $O(n^6/q)$-time, where $q \geq log(n)$.

Because the solution matrix produced by *Fast R&E* algorithm is identical to the one produced by the original Rivas&Eddy algorithm, the contribution of the algorithm lies not only in its stand alone practicality but also in its ability to be implemented alongside heuristic speedups, leading to even greater reductions in time. Our approach is the first to achieve a $\Omega(log(n))$ time speedup without reducing the set of possible Rivas&Eddy pseudoknotted structures. The analysis presented here of the original algorithm could be used to improve other pseudoknot algorithms with similar recurrences.

1 Introduction

The algorithmic goal of structure prediction is motivated by the understanding that RNA structure helps to determine function. It has been particularly observed that in eukaryotic genomes ncRNA (Non-coding RNA) function is seen more clearly from structure [3–5]. Pseudoknot structures, specifically, play an important role in transcription regulation, as well as RNA splicing and catalysis [1, 6]. While algorithms that compute the optimal pseudoknot free fold for RNA[1] [7] are solved in polynomial time $O(n^3)$ [7–9] or $O(n^3/log(n))$ [10] depending on problem formulation, the optimal secondary structure including pseudoknots computation is NP-hard[11]. However, there are available dynamic programs that find the optimal secondary structure of RNA for a subclass of pseudoknotted structures [2, 12–18]. These algorithms range from $O(n^4)$ to $O(n^6)$ asymptotic computation time

[1] A fold does not contain a pseudoknot if for $seq(1..n)$ an RNA sequence of size n the folding set for seq does not contain both (i,j) and (i',j') if $i < i' < j < j'$.

G. Lin (Ed.): COCOA 2012, LNCS 7402, pp. 176–187, 2012.

for a sequence of size n. The classification of RNA structure prediction algorithms based on the size of possible structures has been examined by Condon et al.[19] and Saule et al. [20]. The Four-Russians technique, which has been used to speedup many dynamic programs has not yet been applied to the pseudoknotted secondary structure problem.

Secondary structure algorithms for RNA do not easily lend themselves to the traditional Four-Russians technique of performing some preprocessing for a subset of all possible inputs and then computing using that preprocessing. The Four-Russians speedup technique for non-pseudoknotted RNA secondary structure, discussed in Frid and Gusfield [10], employed simultaneous and interleaved computation and preprocessing. Unfortunately, pseudoknotted secondary structure nesting is not guaranteed, thus requiring more analysis in order to execute preprocessing, and computation. Through the analysis presented here featuring *encoding* some optimal structures, preprocessing and sped up computation becomes possible. *Rivas&Eddy Algorithm* describes the broadest set of structures through its recurrences and has an asymptotic time bound of $O(n^6)$ for an RNA sequence of size n. We present *Fast R&E* algorithm which applies the Four-Russians technique to a modified maximum matching *Rivas&Eddy Algorithm.* Our approach is the first to achieve a $\Omega(log(n))$ speedup without reducing the set of possible structures.

Since the solution set of *Fast R&E* is identical to the solution set of the original *Rivas&Eddy Algorithm*, both algorithms contain equivalent limitations: the set of possible structures is restricted and the optimal structure chosen is dependent on the scoring scheme. Therefore, like in the original algorithm, there is a need to apply scoring schemes that lead to prediction of biologically accurate structures. Our approach is also compatible with Mohl et al. [21] speedup, which based on some simple pruning, in practice achieves a linear speedup of the Reeder and Giegerich Algorithm [15].

The *Fast R&E* algorithm computes pseudoknotted RNA secondary structure in $O(n^6/log(n))$ time for the standard Four-Russians speedup and in $O(n^6/log^2(n))$ time based on Pinhas et al.[22] and Williams et al. [23]. The ideas that allow for both a $log(n)$ and a $log^2(n)$ speedup are examined in the following sections.

2 The Basic Optimal Folding Problem

Let *seq* be an RNA sequence over the four-letter alphabet {A,U,C,G}, where each letter in the alphabet represents an RNA *nucleotide*. Let nucleotides x, y at position i and j in the sequence be a ***permitted pair*** of nucleotides if (x, y) or $(y, x) \in$ {(A,U), (C,G), (G,U)}. For a given sequence *seq* we define the ***folding set*** M as a set containing disjoint permitted pairs of sites in sequence *seq*. Let β be a scoring scheme such that $\beta(i, j)$ returns the contribution of pairing nucleotide x at site i with the nucleotide y at site j. The basic scoring scheme sets $\beta(i, j)$ equal to '1' if (x, y) is a permitted pair with $|j - i| > d$ and set $\beta(i, j)$ to '0' otherwise. Richer scoring schemes β allow more biologically significant information to be captured by the algorithm. Let $foldScore$ be the score associated with a folding set M where $foldScore = \sum_{(i,j)\in M} \beta(i,j)$.

The *optimal folding problem*: Find the set M for which $foldScore$ is maximum under some constraints. Unconstrained, there is an exponential number of possible sets M.

3 Rivas&Eddy Algorithm

We are interested in the optimal folding problem under the constraints of the Rivas&Eddy recurrence relations. Rivas&Eddy recurrences examine the largest subset of pseudoknot structures for which an optimal solution can be found in polynomial time. We will make use of a maximization of base pairs version of the *Rivas&Eddy Algorithm* for folding, as described by Mohl et al. [21] . That version maximizes the pair contributions to the fold instead of minimizing the energy of a fold.

3.1 Rivas&Eddy Recurrences

Let $\mathbf{S[i,j;k,l]}$, where $i < j < k < l \leq N$, contain the *foldscore* for optimal Rivas&Eddy fold for the subsequence $seq(i..l)^2$, where each nucleotide in position r such that $r \in j+1..k-1$ is unpaired. Clearly, S is a four dimensional matrix.

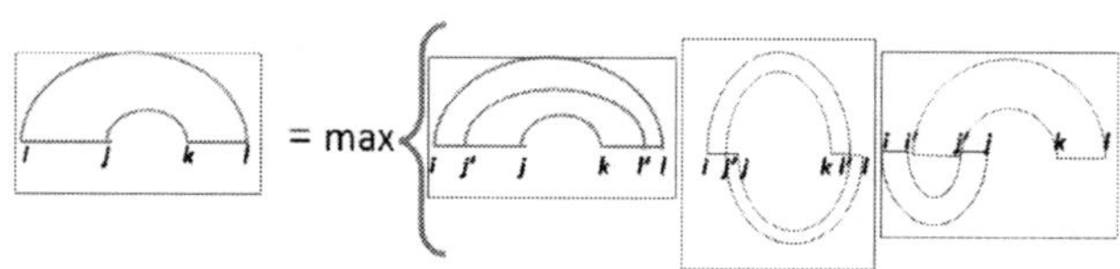

Fig. 1. Simplified S matrix recurrence as seen in Lyngso et. al. [11]

Let W[i,j] be the optimal Rivas&Eddy foldScore for subsequence $seq(i,j)$. We make use of the recurrences describing the different possible fold options, as seen below:

$$W[i,j] = \max \begin{cases} \text{Rule a.} & W[i,j-1] \\ \text{Rule b.} & W[i+1,j-1] + \beta(i,j) \\ \text{Rule c.} & \max_{k\in\{i+1,\ldots,j-1\}} \underline{\mathtt{W[i,k]}} + \mathtt{W[k+1,j]} \\ \text{Rule d.} & \max_{j',k',l'\in\{i+1,\ldots,j-1\}\wedge\{j'<k'<l'\}} \underline{\mathtt{S[i,j';k',l']}} + \mathtt{S[j'+1,k'-1;l'+1,j]} \end{cases} \tag{1}$$

[2] Notational note: All subsequences will be represented as seq(a..b) where a is the starting index of the subsequence and b is the index of the final character in that subsequence.

$$S[i,j;k,l] = \max \begin{cases} \text{Rule A.} & \max\{S[i+1,j;k,l], S[i,j-1;k,l], S[i,j;k+1,l], S[i,j;k,l-1]\} \\ \text{Rule B.} & \max_{j' \in \{i+1,\dots,j-1\}} \underline{W[i,j']} + S[j'+1,j;k,l] \\ \text{Rule C.} & \max_{j' \in \{i+1,\dots,j-1\}} \underline{W[j',j]} + S[i,j'-1;k,l] \\ \text{Rule D.} & \max_{l' \in \{k+1,\dots,l-1\}} \underline{W[k,l']} + S[i,j;l'+1,l] \\ \text{Rule E.} & \max_{l' \in \{k+1,\dots,l-1\}} \underline{W[l',l]} + S[i,j;k,l'-1] \\ \text{Rule F.} & \max_{j' \in \{i+1,\dots,j-1\} \wedge k' \in \{k+1,\dots,l-1\}\}} \underline{S[i,j';k'+1,l]} + S[j'+1,j;k,k'] \\ \text{Rule G.} & \max_{j' \in \{i+1,\dots,j-1\} \wedge k' \in \{k+1,\dots,l-1\}\}} \underline{S[i,j';k,k'-1]} + S[j'+1,j;k',l] \\ \text{Rule H.} & \max_{\{\{i',j'\} \in \{i+1,\dots,j-1\} \wedge \{i'<j'\}} \underline{S[i,i';j'+1,j]} + S[i'+1,j';k,l] \\ \text{Rule I.} & \max_{\{\{k',l'\} \in \{k+1,\dots,l-1\} \wedge \{k'<l'\}} \underline{S[k,k';l',l]} + S[i,j;k'+1,l'-1] \end{cases} \tag{2}$$

Rules c, d for recurrence of matrix W and Rules B to I for the recurrence of matrix S maximize the sum of the foldScores for two sections or the RNA string. Let the section described first in each rule be called the ***head*** of the *split* and the second score corresponding to an optimal foldScore for a subsequence be called the ***tail***. For example in Rule I $\underbrace{S[k,k';l',l]}_{\textbf{head}} + \underbrace{S[i,j;k'+1,l'-1']}_{\textbf{tail}}$. We could interpret each rule as finding the optimal head and tail sum under certain conditions as seen in Figure 1. In the above recurrences Rules c,d, and Rules B to I the underlined part of each rule is the *head*, and the non-underlined is the *tail*.

3.2 Rivas&Eddy Algorithm

While many orders of evaluation are possible for these recurrences, we explicitly describe an order of evaluation that will make the Four-Russians speedup possible. Any evaluation not shown below can be done in any logical order.

The W and S matrices can be computed by an algorithm that goes through all the possible subsequences of *seq* finding the optimal Rivas&Eddy fold for each. As shown in *Rivas&Eddy Algorithm* below, the recurrences are evaluated in increasing order of the right endpoint of *seq*.

Rivas&Eddy Algorithm *Compute Matrix W::*

Initialization: **W**=0; $S(i,i;k,k)=\beta(i,k)$ or $S(i,j;k,l) = 0$
for j=1 to N **do**
  for i=$j-1$ to 1 **do**
    Compute Matrix S given (i,j) (Rules A to I)
    branch_max= **compute** Rule_c(i,j) (Rule c)
    max_Rule_d = **compute** Rule_d(i,j) (Rule d)
    W[i,j]=max(*Rule a, Rule b, branch_max, max_Rule_d*)

Compute Matrix S given (i,l)::
for k=$l-1$ to $i+1$ **do**
  for j=$i+1$ to $k-1$ **do**
    S[i,j;k,l]=max(Rules A to I);

The score corresponding to the optimal Rivas&Eddy fold is found in $W[1,n]$. The overall asymptotic time to compute *Rivas&Eddy Algorithm* is equal to the total time taken to compute Rules a to d for the W matrix and Rules A to I for the S matrix. The total asymptotic time results in { $O(n^2 * O(Rules\, a$ to $d))$ + $O(n^4 * O(RulesAtoI))$ } or $O(n^6)$. *Compute Matrix S give(i,j)* call is the bottleneck of the matrix W computation. For a specific i, j Compute Matrix S takes $O(n^4)$ time. Within *Compute Matrix* S the computations of Rules F-I take $O(n^2)$ time and are the bottlenecks. If you we could reduce the computation time of Rules F-I from $O(n^2)$ to $O(n^2/q)$, the overall asymptotic time of the entire algorithm would improve to $O(n^6/q)$.

We present the *Fast R&E* algorithm that, not only reduces the computation time for the bottleneck Rules F to I, but for all the Rules B to I, as well as *Rule* c and *Rule* d by a factor of q. This will lead to an overall speedup of $O(n^6/q)$ where q is $log(n)$ for the standard Four-Russians speedup. A further improvement of $log^2(n)$ maybe achieved, for the Four-Russians implementation of Pinhas et al. [22].

4 Conceptual Speedup of Rivas&Eddy

For simplicity of exposition we develop the speedup for Rule I. Rule I computes in $0(n^2)$ time and is one of the bottleneck computations. The analysis presented below leading to the speedup is then generalized to all the Rules. Given some i, j, k, l, Rule I can be computed by the following two loops:

compute Rule I for i, j, k, l*::*
for $l' = l - 1$to $k + 1$ **do**
  for $k' = k + 1$ to $l' - 1$ **do**
    $rule\_I\_max = \mathtt{max}\ (rule\_I\_max,\ S[k, k'; l', l] + S[i, j; k' + 1, l' - 1])$

We conceptually divide all the possible index points of seq into sets of size q called **Mgroups**. Let $M_{g=0}$ be the first such group that contains the possible index points $\{0, 1, ...q - 1\}$, let $M_{g=1}$ contain $\{q, ...2q - 1\}$ and so on ... the last group of which is $M_{g=n/q} = \{n - q, ...n - 1\}$. In general:

$$M_g = \{g \cdot q, g \cdot q + 1, ...g \cdot q + q - 1\}$$

Lets define the function $\boldsymbol{K^*(RuleX, \mathbb{Y}, M_g)}$ where $Rule\, X$ ∈{Rule b - d, Rule B - I} and $\mathbb{Y}$ is a set of indexes referencing seq that correspond to the particular Rule X. Let $K^*(Rule\, X, \mathbb{Y}, M_g)$ return some index z, where $z \in M_g$ and maximizes Rule X under the $\mathbb{Y}$ constraint. For example:

$\mathbf{K}^*$(Rule $I, \{i, j, k, l, l'\}, \boldsymbol{M_g}$)::
for $k' \in M_g$
  return $\boldsymbol{k'}$ such that $\mathtt{max}\{S[k, k'; l', l] + S[i, j; k' + 1, l' - 1]\}$

Incorporating the $\boldsymbol{K^*}$-function into the computation of each Rule does not change the number of head and tail combinations examined. For example, the computation of Rule I:

```
compute Rule I for i, j, k, l::
    for l' = l − 1to k + 1 do
        for g = (k+1)/q to (l'−1)/q do
            k' = K*( Rule I, {i, j, k, l, l'}, M_g)
            rule_I_max=max (rule_I_max, S[k, k'; l', l]+S[i, j; k'+1, l'−1])
```

The $\boldsymbol{K}^*$-function examines overall $O(q)$ head and tail combinations, returning the index in Mgroup M_g that results in the maximum sum. Therefore, *Rule I* compares $O(\frac{n^2}{q} \cdot O(K^*))$ head and tail combinations, totaling in $O(n^2)$ comparisons.

4.1 Breaking Up S and W into q Size Vectors

Let us conceptually break the solution matrices **S**, and **W** into vectors of size q such that for each Rule there is a vector that corresponds to the set of tails for the indexes in M_g. For Rule I let V_g be a q size vector that contains the possible tails for the indexes k' in M_g. For a particular i, j, l' and M_g, the tails examined by the $\boldsymbol{K}^*$-function are $\{ S[i, j; gq + 1, l' - 1]) \ldots S[i, j; gq + q, l' - 1] \}$.

Hence, $V_g = \{V_g(1) = S[i, j; \boldsymbol{gq + 1}, l' - 1], \ldots V_g(q) = S[i, j; \boldsymbol{gq + q}, l' - 1]\}$.

More precisely for *Rule I*: $V_g(m + 1 - gq) = S[i, j; m + 1, l' - 1]$.

In general for some i, j, k, l, g the vector V_g of size q indexed on $x \in \{1, ..., q\}$ is defined for each rule as follows:

For Rule D and I: $V_g(x) = S[i, j; gq + x, l]$;
For Rule B, F, G and H: $V_g(x) = S[gq + x; j; k, l]$;
For Rule C: $V_g(x) = S[i, gq + x - 2; k, l]$;
For Rule E: $V_g(x) = S[i, j; k, gq + x - 2]$.

We can then rewrite the computation of $\boldsymbol{K}^*$-function replacing the input set of indexes M_g with the vector V_g of values for the tails. We will also add the value g to the inputs, which references which Mgroup we are maximizing.

```
K*(Rule I, {[i,j], k, l, l'}, g, V_g)::
for x = 1 to q
    k' = gq + x − 1
    return k' such that max {S[k, k'; l', l]+ V_g(x)}
```

Note that index $\boldsymbol{i, j}$ are no longer input to the $\boldsymbol{K}^*$-function for *RuleI*. as the values of the scores reference by these indexes are stored in V_g

Fact 1. If x is the index point that leads to the maximum of the sum of $S[k, k'; l', l] + V_g(x)$ where $k' = gq + x - 1$ then k' is also the index point that leads to the maximum sum of $S[k, k'; l', l] + S[i, j; \boldsymbol{k' + 1}, l' - 1]$ where $k' \in M_g$.

4.2 Encoding

Optimal Rivas&Eddy scores stored in $S[i,j;\boldsymbol{k},l]$ and $S[i,j;\boldsymbol{k+1},l]$ can differ by the effect of only one more nucleotide i.e. $seq[k+1]$. Therefore, we can observe that for the scoring scheme and the recurrences of the Rivas&Eddy Algorithm, $|S[i,j;k,l] - S[i,j;k+1,l]|$ belongs to a finite set of differences $\mathbb{D}$, where $\mathbb{D}$ is the set of scores created as the result of the Scoring function β. The cardinality or size of $|\mathbb{D}|$ is $O(1)$ as a function of n. For the simple β scoring function of $+1$ for every permitted pair and 0 otherwise, the $\mathbb{D}$ set is equal to $\{0,1\}$ and therefor $|\mathbb{D}| = 2$.

For the V_g vectors of Rule I the ***base*** or smallest element of the vector is $S[i,j;gq+q,l]$. Let E_g be a q size vector of differences from the *base* of V_g. For Rule I, we define $\mathbf{E_g(x)} = (S[i,j;a+x,l] - \underbrace{S[i,j;a+q,l]}_{base})$, where $a = gq$.

For a particular i,j,k,l we can create and store all the E_g vectors as soon as the corresponding values in the S matrix are computed. Once computed, retrieval of any desired E_g clearly takes $O(1)$ time. The overall overhead for encoding the S matrix into a set of E matrices for the entire algorithm requires an addition of $O(n^4)$ time.

Fact 2 If x is the index point that leads to the maximum of the sum of $S[k,k';l',l]+E_g(x)$ where $k' = gq+x-1$ then x is also the index point that leads to the maximum sum of $S[k,k';l',l]+V_g(x)$. Based on Fact 1, $K^*(RuleI, \{k,l,l'\}, M_g)$ will therefore return k'.

```
K*(Rule I, {k, l, l'}, g, E_g)::
for x = 1 to q
   k' = gq + x − 1
   return k' such that max {S[k, k'; l', l]+ E_g(x)}
```

We can now incorporate encoded E_g vectors in the *compute Rules* functions. For example:

```
compute RuleI for i, j, k, l::
   for l' = l − 1to k + 1 do
      for g = (k+1)/q to (l'−1)/q do
         retrieve E_g
         k' = K*(Rule I, {k, l, l', }, g, E_g )
         rule_I_max=max (rule_I_max, S[k, k'; l', l]+S[i, j; k' + 1, l' − 1])
```

For Rule I if K*()-function could be computed in $O(1)$ time, the asymptotic run-time of each rule would be reduced to $O(n^2/q)$.

Introducing table R. Let $\boldsymbol{R}$ be a table such that $R[X,\mathbb{Y},g,E]$ contains the output to the $K^*(X,\mathbb{Y},g,E_g)$-function. For example, $\mathbf{R}[I,\{k,l,l'\},g,\boldsymbol{E_g}] = k'$ where k' is the index point that leads to the maximum of the sum of $S[k,k';l',l]+E_g(x)$, where $x = k' - gq + 1$. Clearly,

$$\mathbf{K}^*(Rule\,I,\{k,l,l'\},g,E_g) = \mathbf{R}[I,\{k,l,l'\},g,E_g].$$

We, therefore, can replace all calls to function $\boldsymbol{K}^*$ with references into table $\boldsymbol{R}$.

The precomputation of table R, developed below, will allow to achieve the $\Omega(q)$ time speedup.

5 Precomputing the $\boldsymbol{R}$ Table

Assume the function call *Compute Matrix S given(1,n)* has been made, leading the variables i,l to equal: $i = 1$; $l = n$. During the $k = n-1$ iteration of the outer loop, the inner loop computes optimal *foldscore* for $S[\mathtt{1},\boldsymbol{j}\,;\mathtt{n}-\mathtt{1},\mathtt{n}]$ where $j = 2,3...$ and so on. Assume that the inner loop has completed the $j = q-1$ iteration. Hence, we have an optimal solution for $S[1,\mathbf{1};n-1,n]$ to $S[1,\boldsymbol{q}-\mathbf{1};n-1,n]$. These values correspond to the *heads* of $g = 0$ for Rules d,F,G,H,I. Therefore the following algorithm precomputes $K^*(Rule\ X,\{1,n-1,n,\},g=0,E_{\boldsymbol{g=0}})$ for all Rules X ,for all possible E_g vectors, where $X \in \{d,F,G,H,I\}$ and stores the result in $\boldsymbol{R}$-table.

for each difference vector v of size $q-1$ such that $\forall_{1<x\leq q-1} v[x] \in \mathbb{D}$ **do**
  `compute` (an encode vector $\boldsymbol{E}$ from v) [3]
  for x=1 to q **do**
    t=$g \cdot q + x - 1$
    let max_i be the index t that makes $S[1,t;n-1,n] + \boldsymbol{E}(x)$ is maximum.
  `set` for all $X \in \{d,F,G,H,I\}$ $\boldsymbol{R}(X,\{1,n-1,n\},g,E)$=$max\_i$

The algorithm above makes use of the fact that you can enumerate all possible sets of differences $E_{g=0}$ in $|\mathbb{D}|^q$-time.

We can generalize this algorithm for any g, as well as any i,k,l by creating an ***S_update_table*** function. Assume we have completed some M_g iteration of j (i.e. we have completed j from gq to $gq+q-1$). We therefore have the scores for all the heads of Rules d, F-I for Mgroup M_g. Then an ***S_update_table*** function can precompute which the index in M_g would give the maximum score for every possible variation of vector E.

---

[3] set $\boldsymbol{E}(q) = 0$ and $\forall_{1<x\leqq -1}\boldsymbol{E}(x) = \sum_{i=q-1}^{x} v[i]$

```
S_update_table function (i,k,l,g)::
for each difference vector v size q−1 such ∀_{1<x≤q−1} v[x] ∈ 𝔻 do
  compute ( an encode vector E from v)
  for x=1 to q do
    t=g·q+x−1
    max_i=t if S[i,t;k,l] + E(x) is max.
  for each Rule X ∈ {d,F,G,H,I} do
    set R(X,{i,k,l},g,E) = max_i
```

```
Compute fast Matrix S for(i,l)::
for k=l−1 to i+1 do
  for g=(i+1)/q to (k−1)/q do
    for j = g·q to gq+q−1 ∧ j < k do
      S[i,j;k,l]=max(Rules A-I);
    call S_update_table(i,k,l,g)
```

S_update_table function would be called by the *Compute fast Matrix S for (i,l)* algorithm $O(n \cdot n/q)$ times and each call would take $O(|\mathbb{D}|^q \cdot q)$ time to compute. The *Compute* ***fast*** *Matrix S for(i,l)* algorithm is presented above. While an extra loop iterated over g was added, it is clear that the call to maximize Rules A to I is made still only $O(n^2)$ times.

We would need to create a similar *W1_update_table* function Rules c, B, and D and *W2_update_table* function for Rules C, E. For example Rules d, F,G,H the precomputation occurs when the *heads* for each *Mgroup* have their corresponding optimal solutions.

```
W1_update_table function(g) ::
for each vector v size q−1 such ∀_{1<x≤q−1} v[x] ∈ 𝔻 compute E from v do
  for i = 0 to gq−1
    for x = 1 to q
      max_i=t=gq+x−1 if W[i,t] + E(x) is max
    for each Rule X∈{c,B,D} set R(X,{i},g,E) = max_i
```

The total asymptotic time for a single call to the *W1_update_table function(g)* function is $O(|\mathbb{D}|^q \cdot n \cdot q)$.

```
W2_update_table function(j,g) ::
for each vector v size q−1 such ∀_{1<x≤q−1} v[x] ∈ 𝔻 compute E from v do
  for x = 1 to q
    max_i=t=gq+x−1 if W[t,j] + E(x) is max
  for each Rule X∈{C,E} set R(X,{i},g,E) = max_i
```

The total asymptotic time for a single call to the *W2_update_table function(g)* function is $O(|\mathbb{D}|^q \cdot q)$.

6 Fast R&E Algorithm

```
Compute Matrix W::
for g = 0 to N/q do
    for j=gq to gq + q − 1 do
        for i=j − 1 to 1 do
            Compute Matrix S for (i, j)
            branch_max= compute Rule_c(i, j)
            max_Rule_d = Rule_d(i, j)
            W[i,j]=max(Rule a, Rule b, branch_max, max_Rule_d)
            if (i%q == 0) W2_update_table(j,g) (Updating Table R for Rules C, E)
    W1_update_table(g) (Updating Table R for Rules c, B, D)
```

6.1 Asymptotic Analysis of Fast R&E Algorithm

For a particular g the *Fast R&E* calls the *W1_update_function* O(1) times. During one iteration of g the *W2_update_function* is called O(n) times. The total overhead for precomputing Rules c, B, D and Rules C, E is $0((n/q) \cdot [n|\mathbb{D}|^q q + |\mathbb{D}|^q nq])$ time, or simplified $O(n^2 * |\mathbb{D}|^q)$ time. For a particular i, j algorithm computes Rules A to I with function *Compute fast Matrix* S in $O(n^4/q)$ time and calls the *S_update_function* to precompute Rules d, F, G, H and I $O(n^2/q)$ times. In total, asymptotic run-time for all calls made to *Compute Matrix* S is $O(n^6/q) + O(n^4 * |\mathbb{D}|^q)$.

Finally for a particular i, j $Rule\_c(i, j)$ is computed in $O(n/q)$ time, $Rule\_d$ is computed in $O(n^3/q)$ time. The total asymptotic time for the entire *Fast R&E Algorithm* is $\underbrace{O(n^6/q + n^3/q + n^5/q))}_{computation} + \underbrace{O(n^2|\mathbb{D}|^q) + O(n^4|\mathbb{D}|^q)}_{preprocessing}$

If q=$\log_b(n)$ where the log base b is constrained by $|\mathbb{D}| < b < N$ then the asymptotic run-time is $O(n^6/log(n))$.

Memory: The original *Rivas&Eddy Algorithm* requires $O(n^4)$-space. For simplicity of exposition we chose to describe the speedup using the preprocessing of all possible tails and that requires a factor of $O(|\mathbb{D}|^q)$-space for table R. When preprocessing both heads and tails table R requires $O((|\mathbb{D}|^{2q}))$- space in total. If preprocessing all possible heads for a specific tail there is $O(|\mathbb{D}|^q)$-space requirement for table R.

Empirical Results: We ran empirical tests comparing *Fast R&E Algorithm* to the *Rivas&Eddy Algorithm* for sequences ranging in length from 100-225 nucleotides. We used the 0,1 scoring scheme setting $|\mathbb{D}|$=2. The average times for 15 sequences of each length are reported below. The standard deviation for all tests is within 5 seconds. The theoretical speedup for a sequence of 100 nucleotides is 6.64 ($log_{2.001}(100) = 6.64$) we achieved 2.33 time improvement. For a sequence size 225 nucleotides we achieved a 2.28 improvement compared to the theoretical 7.96 potential speedup.

size(n)	q	Rivas&Eddy run-time(seconds)	Fast R&E (seconds)	ratio
100	3	776.0	333.30	2.32
150	3	3,915.0	2,178.1	1.79
200	3	20,544.0	9,493.33	2.16
225	3	40,687.31	17,782.04	2.28

Conclusion and Future Work. We presented the Four-Russians speedup of $\Omega(log(n))$ for the algorithm that examines broadest set of polynomial time computed pseudoknotted secondary structures - the *Rivas&Edddy Algorithm*. The analysis explored here could be used on other pseudoknotted algorithms [12–18]. Because the solution matrices of *Fast R&E* is the same as the solution matrices of the original *Rivas&Eddy* this algorithm could be used in conjunction with other heuristic speedups [21]. It is also interesting to note that the preprocessing done here takes at most $O(n^5/log(n))$ leading to the question of whether through further preprocessing an even great speedup could be achieved.

Acknowledgments. This research was partially supported by grant IIS-0803564 from the National Science Foundation.

References

1. Condon, A., Jabbari, H.: Computational prediction of nucleic acid secondary structure: Methods, applications, and challenges. Theor. Comput. Sci. 410(4-5), 294–301 (2009)
2. Rivas, E., Eddy, S.R.: A dynamic programming algorithm for RNA structure prediction including pseudoknots. Journal of Molecular Biology 285(5), 2053–2068 (1999)
3. Torarinsson, E., Havgaard, J.H., Gorodkin, J.: Multiple structural alignment and clustering of RNA sequences. Bioinformatics 23(8), 926–932 (2007)
4. Rose, D., Hackermuller, J., Washietl, S., Reiche, K., Hertel, J., FindeiSZ, S., Stadler, P., Prohaska, S.: Computational rnomics of drosophilids. BMC Genomics 8(1), 406 (2007)
5. Torarinsson, E., Yao, Z., Wiklund, E.D., Bramsen, J.B., Hansen, C., Kjems, J., Tommerup, N., Ruzzo, W.L., Gorodkin, J.: Comparative genomics beyond sequence-based alignments: RNA structures in the encode regions. Genome Res. 18(2), 242–251 (2008)
6. Liu, C., Song, Y., Shapiro, L.: RNA Folding Including Pseudoknots: A New Parameterized Algorithm and Improved Upper Bound. In: Giancarlo, R., Hannenhalli, S. (eds.) WABI 2007. LNCS (LNBI), vol. 4645, pp. 310–322. Springer, Heidelberg (2007)
7. Nussinov, R., Pieczenik, G., Griggs, J.R., Kleitman, D.J.: Algorithms for loop matchings. SIAM Journal on Applied Mathematics 35(1), 68–82 (1978)
8. Zuker, M., Sankoff, D.: RNA secondary structures and their prediction. Bulletin of Mathematical Biology 46(4), 591–621 (1984)
9. Waterman, M.S., Smith, T.F.: RNA secondary structure: A complete mathematical analysis. Math. Biosc. 42, 257–266 (1978)

10. Frid, Y., Gusfield, D.: A Simple, Practical and Complete $O(\frac{n^3}{\log n})$-Time Algorithm for RNA Folding Using the *Four-Russians* Speedup. In: Salzberg, S.L., Warnow, T. (eds.) WABI 2009. LNCS, vol. 5724, pp. 97–107. Springer, Heidelberg (2009)
11. Lyngsø, R.B., Pedersen, C.N.S.: RNA pseudoknot prediction in energy-based models. Journal of Computational Biology 7(3-4), 409–427 (2000)
12. Akutsu, T.: Dynamic programming algorithms for RNA secondary structure prediction with pseudoknots. Discrete Applied Mathematics 104(1-3), 45–62 (2000)
13. Dirks, R.M., Pierce, N.A.: A partition function algorithm for nucleic acid secondary structure including pseudoknots. Journal of Computational Chemistry 24(13), 1664–1677 (2003)
14. Mathews, D.H., Turner, D.H.: Prediction of RNA secondary structure by free energy minimization. Current Opinion in Structural Biology 16(3), 270–278 (2006); Nucleic acids/Sequences and topology - Anna Marie Pyle and Jonathan Widom/Nick V Grishin and Sarah A Teichmann
15. Reeder, J., Giegerich, R.: Design, implementation and evaluation of a practical pseudoknot folding algorithm based on thermodynamics. BMC Bioinformatics 5(1), 104 (2004)
16. Uemura, Y., Hasegawa, A., Kobayashi, S., Yokomori, T.: Tree adjoining grammars for RNA structure prediction. Theoretical Computer Science 210(2), 277–303 (1999)
17. Deogun, J.S., Donts, R., Komina, O., Ma, F.: RNA secondary structure prediction with simple pseudoknots. In: Chen, Y.-P.P. (ed.) APBC. CRPIT, vol. 29, pp. 239–246. Australian Computer Society (2004)
18. Cao, S., Chen, S.-J.: Predicting structures and stabilities for h-type pseudoknots with interhelix loops. RNA 15(4), 696–706 (2009)
19. Condon, A., Davy, B., Rastegari, B., Zhao, S., Tarrant, F.: Classifying RNA pseudoknotted structures. Theoretical Computer Science 320(1), 35–50 (2004)
20. Saule, C., Régnier, J.-M.S.M., Denise, A.: Counting RNA pseudoknotted structures. Journal of Computational Biology 18(10), 1339–1351 (2011)
21. Möhl, M., Salari, R., Will, S., Backofen, R., Sahinalp, S.C.: Sparsification of RNA Structure Prediction Including Pseudoknots. In: Moulton, V., Singh, M. (eds.) WABI 2010. LNCS, vol. 6293, pp. 40–51. Springer, Heidelberg (2010)
22. Pinhas, T., Tsur, D., Zakov, S., Ziv-Ukelson, M.: Edit Distance with Duplications and Contractions Revisited. In: Giancarlo, R., Manzini, G. (eds.) CPM 2011. LNCS, vol. 6661, pp. 441–454. Springer, Heidelberg (2011)
23. Williams, R.: Matrix-vector multiplication in sub-quadratic time (some preprocessing required). In: Bansal, N., Pruhs, K., Stein, C. (eds.) SODA, pp. 995–1001. SIAM (2007)

An Improved Approximation Algorithm for the Bandpass-2 Problem

Zhi-Zhong Chen[1] and Lusheng Wang[2]

[1] Division of Information System Design, Tokyo Denki University, Hatoyama, Saitama 350-0394, Japan
zzchen@mail.dendai.ac.jp

[2] Department of Computer Science, City University of Hong Kong, Tat Chee Avenue, Kowloon, Hong Kong SAR
lwang@cs.cityu.edu.hk

Abstract. The bandpass-2 problem (Bandpass-2, for short) is a generalization of the maximum traveling salesman problem (Max TSP, for short). Of particular interest is the difference between the two problems, where the edge weights in Bandpass-2 are dynamic rather than given at the front. A trivial approximation algorithm for Bandpass-2 can achieve a ratio of 0.5. Recently, Tong *et al.* [19] have presented a nontrivial approximation algorithm for Bandpass-2 that achieves a ratio of $\frac{21}{40}$. In this paper, we present a new approximation algorithm that achieves a ratio of 0.5318.

Keywords: Bandpass-2 problem, approximation algorithm, maximum weight matching, maximum weight 2-matching, worst-case performance ratio.

1 Introduction

Let M be a binary matrix and suppose that each column of M has a nonnegative weight. A *bandpass* B in M consists of two consecutive 1's in the same column of M. The *weight* of B is the same as that of the column in which B resides. Two bandpasses B_1 and B_2 are *disjoint* if they contain no common entry of M. The *weight* of a set S of pairwise-disjoint bandpasses in M is the total weight of bandpasses in S. The *weight* of M is the maximum weight of a set of pairwise-disjoint bandpasses in M. In the bandpass-2 problem (Bandpass-2, for short), we are given a binary matrix M and want to permute the rows of M so that the weight of the permuted matrix is maximized. The unweighted version of Bandpass-2 has been studied in [1, 3, 15, 19], and it has applications in optical communication networks [1].

Let M be the incidence matrix of an edge-weighted complete undirected graph G. Clearly, solving Bandpass-2 for M is equivalent to finding a maximum-weight Hamiltonian path in G. In this sense, Bandpass-2 is a generalization of Max TSP and is hence NP-hard and APX-hard because so is Max TSP [2]. Thus, we want to design efficient approximation algorithms for Bandpass-2, just like that many

G. Lin (Ed.): COCOA 2012, LNCS 7402, pp. 188–199, 2012.

approximation algorithms for Max TSP have been designed in the literature [4–6, 9–14, 16, 17].

A trivial approximation algorithm for Bandpass-2 proceeds as follows. Given a column-weighted binary matrix M with n rows, the algorithm first constructs a complete undirected graph G from M where the vertex set of G is $\{1, 2, \ldots, n\}$ and the weight of each edge $\{i, j\}$ in G is the total weight of bandpasses that can be formed between the i-th and the j-th rows of M. It then computes a maximum-weight matching N and further permutes the rows of M so that the two rows corresponding to the endpoints of each edge in N appear consecutively in the permuted matrix. It is easy to see that this algorithm achieves a ratio of 0.5.

Recently, Tong *et al.* [19] have presented a refinement on the trivial algorithm and their algorithm achieves a ratio of $\frac{21}{40}$. Their idea can be sketched as follows. First, we use M and N to modify G into another graph G' (by deleting the edges of N from G and decreasing the weights of some of the remaining edges). We then compute a maximum-weight matching N' in G'. The subgraph of G formed by the edges in $N \cup N'$ is a collection of vertex-disjoint paths or even cycles and hence can be transformed into a collection $\mathcal{P}$ of vertex-disjoint paths by removing one edge of the smallest weight from each cycle. Finally, we use $\mathcal{P}$ to permute the rows of M so that for each path Q in $\mathcal{P}$, the rows corresponding to the vertices of Q appear consecutively in the permuted matrix.

In one word, Tong *et al.*'s algorithm is a *matching-based* algorithm. In contrast, all the known approximation algorithms for Max TSP are *cycle-cover-based* algorithms [4–6, 9–14, 16, 17]. Intuitively speaking, computing a cycle cover C in G is not useful for the following reason: For every two adjacent edges $\{i, j\}$ and $\{j, k\}$ of C, even if we permute the rows of M so that the i-th, j-th, and k-th rows appear consecutively (in this order), it is possible that the total weight of edges $\{i, j\}$ and $\{j, k\}$ in G is twice the total weight of bandpasses that can be formed between the i-th and the j-th rows or between the j-th and the k-th rows. It is also not clear how to use cycle covers to improve the second step of Tong *et al.*'s algorithm because N can be assumed to be a perfect matching of G (without loss of generality) and in turn only a matching can be added to N in order to expand N to a collection of vertex-disjoint paths. Indeed, Tong *et al.* [19] ask if we can use cycle covers to obtain a better approximation algorithm for Bandpass-2.

In this paper, we show that cycle covers can be used to design a better approximation algorithm for Bandpass-2 than Tong *et al.*'s algorithm. More precisely, we show that we can use cycle covers to improve the second step of their algorithm. One of our key lemmas says that if N is a matching of a graph H and $\mathcal{P}$ is a collection of vertex-disjoint paths of H such that no edge of N is also an edge of a path in $\mathcal{P}$, then we can partition the edge set of $\mathcal{P}$ into three matchings X_0, X_1, and X_2 such that for all $j \in \{0, 1, 2\}$, the edges in $N \cup X_j$ form a collection of vertex-disjoint paths. This lemma may be of independent interest and can be applied to other problems. Besides this lemma, the design of our algorithm is based on a deeper understanding of the structure of Bandpass-2 than that of Tong *et al.*'s algorithm.

The remainder of this paper is organized as follows. In Section 2, we prove two general lemmas that are independent of the problem studied in this paper (namely, Bandpass-2). In Section 3, we present our new approximation algorithm for Bandpass-2. In Section 4, we use the two lemmas to prove that our algorithm achieves a ratio of 0.5318.

2 Two Useful Lemmas

Throughout this section, fix an edge-weighted complete undirected graph $G = (V, E)$ and a matching N in G. Let n be the number of vertices in G. We assume that $n \geq 3$. For a set F of edges in G, we use $w(F)$ to denote the total weight of edges in F, and use $G[F]$ to denote the graph (V, F). Similarly, for a subgraph H of G, we use $w(H)$ to denote the total weight of edges in H.

A *2-matching* of G is a subgraph of G in which the degree of each vertex is at most 2. Note that a 2-matching is a collection of vertex-disjoint cycles or paths. A 2-matching of G is *acyclic* if it does not contain any cycle (i.e., it is a collection of vertex-disjoint paths). A 2-matching of G is a *cycle cover* of G if it is a collection of vertex-disjoint cycles and contains all vertices of G.

Lemma 1. *Let $\mathcal{C}$ be a 2-matching of G such that no edge of N is also an edge of $\mathcal{C}$. Then, we can partition the edge set of $\mathcal{C}$ into four matchings $X_0, \ldots, X_3$ such that $G[N \cup X_j]$ is an acyclic 2-matching for all $j \in \{0, \ldots, 3\}$. Moreover, the partitioning takes $O(n\alpha(n))$ time, where α is the inverse Ackermann function.*

In general, Lemma 1 cannot be improved by partitioning the edge set of $\mathcal{C}$ into three matchings instead of four matchings. To see this, it suffices to consider a concrete example, where $\mathcal{C}$ is just a cyle of length 4 and $\mathcal{N}$ consists of the two edges connecting nonadjacent vertices in $\mathcal{C}$. The next lemma says that Lemma 1 can be improved if $\mathcal{C}$ is acyclic.

Lemma 2. *Let $\mathcal{P}$ be an acyclic 2-matching of G such that no edge of N is also an edge of $\mathcal{P}$. Then, we can partition the edge set of $\mathcal{P}$ into three matchings Y_0, Y_1, and Y_2 such that $G[N \cup Y_j]$ is an acyclic 2-matching for all $j \in \{0, 1, 2\}$. Moreover, the partitioning takes $O(n\alpha(n))$ time.*

In general, Lemma 2 cannot be improved by partitioning the edge set of $\mathcal{P}$ into two matchings instead of three matchings. To see this, it suffices to consider a concrete example, where $\mathcal{P}$ is just a path with edges $\{v_1, v_2\}$, $\{v_2, v_3\}$, $\{v_3, v_4\}$ and $\mathcal{N}$ consists of edges $\{v_1, v_3\}$ and $\{v_2, v_4\}$.

3 The New Algorithm

Throughout this section, fix a column-weighted matrix M. Let n (respectively, m) be the number of rows (respectively, columns) of M. We next detail how to solve Bandpass-2 for M.

As in the trivial algorithm, we first construct an edge-weighted complete undirected graph G from M, where the vertex set of G is $\{1, 2, \ldots, n\}$ and the weight of each edge $\{i, j\}$ in G is the total weight of bandpasses that can be formed between the i-th and the j-th rows of M. Note that the construction of G takes $O(n^2 m)$ time.

Again as in the trivial algorithm, we then compute a maximum-weight matching N in G. This takes $O(n^3)$ time [7]. Let M' be the matrix obtained by modifying M as follows:

- For each $\{i, j\} \in N$ and for each column k of M, if the (i, k) and the (j, k) entries of M are both a 1, then change both of them to a 0.

Similar to the construction of G from M, we construct an edge-weighted complete undirected graph G' from M' as follows. The vertex set of G' is $\{1, 2, \ldots, n\}$ and the weight of each edge $\{i, j\}$ in G' is the total weight of bandpasses that can be formed between the i-th and the j-th rows of M'. Note that N is also a matching of G' but the weight of each edge of N in G' is 0.

From here on, our algorithm differs from Tong *et al.*'s algorithm [19]. More specifically, our algorithm proceeds as follows.

1. Compute a maximum-weight cycle cover $\mathcal{C}$ in G'. (*Comment:* This step can be done in $O(n^3)$ time [8]. Moreover, $\mathcal{C}$ is also a cycle cover in G.)
2. Modify $\mathcal{C}$ by remove those edges e with $e \in N$.
3. Use Lemma 1 to partition the edge set of $\mathcal{C}$ into four matchings $X_0, \ldots, X_3$ such tha $G[N \cup X_j]$ is an acyclic 2-matching of G for all $j \in \{0, \ldots, 3\}$.
4. Compute a Hamiltonian path $\mathcal{P}$ in G' whose weight is at least $\frac{7}{9}$ times the maximum weight of a Hamiltonian path in G'. (*Comment:* Paluch *et al.* [16] have shown that this step can be done in $O(n^3)$ time. Moreover, $\mathcal{P}$ is also an acyclic 2-matching in G.)
5. Modify $\mathcal{P}$ by remove those edges e with $e \in N$.
6. Use Lemma 2 to partition the edge set of $\mathcal{P}$ into three matchings Y_0, Y_1, and Y_2 such tha $G[N \cup Y_j]$ is an acyclic 2-matching of G for all $j \in \{0, 1, 2\}$.
7. Let Z be the heaviest matching among $X_0, \ldots, X_3, Y_0, \ldots, Y_2$. (*Comment:* $G[N \cup Z]$ is an aylic 2-matching of G and hence is a collection of vertex-disjoint paths.)
8. Permute the rows of M so that for each path Q in $G[N \cup Z]$, the rows corresponding to the vertices of Q appear consecutively in the permuted matrix.
9. Output the permuted matrix M''.

Obviously, we have the next lemma:

Lemma 3. *The above algorithm takes* $O\left(n^2(n+m)\right)$ *time.*

We note that both the trivial algorithm and Tong *et al.*'s algorithm [19] takes $O\left(n^2(n+m)\right)$ time.

4 Performance Analysis

We inherit the notations from Section 3. If n is odd, then we can add a new row consisting of 0's only without changing the problem. So, without loss of generality, we may assume that n is even. By this assumption, we may further assume that N is a perfect matching of G. Then, we can view N as a bijection from $\{1, 2, \ldots, n\}$ to itself by defining $N(i) = j$ and $N(j) = i$ for each edge $\{i, j\} \in N$.

The weight of an edge $\{i, j\}$ in G is denoted by $w(i, j)$ and the total weight of edges in a subgraph H of G is denoted by $w(H)$. Similarly, the weight of an edge $\{i, j\}$ in G' is denoted by $w'(i, j)$ and the total weight of edges in a subgraph H of G' is denoted by $w'(H)$.

Several more definitions are in order. For each integer $c \in \{1, 2, \ldots, m\}$, we use w_c to denote the weight given to the c-th column of M. A *permutation* of $\{1, 2, \ldots, n\}$ is a bijection from $\{1, 2, \ldots, n\}$ to itself. Let π be a permutation of $\{1, 2, \ldots, n\}$. We use M_π to denote the matrix whose i-th row is the $\pi(i)$-th row of M for each $i \in \{1, 2, \ldots n\}$. A *strip* of M_π is a maximal segment of consecutive 1's in a column of M_π. The *length* of a strip is the number of 1's therein. For each integer $c \in \{1, 2, \ldots, m\}$ and each integer $\ell \geq 2$, we use $S_{c,\ell}(\pi)$ to denote the set of length-ℓ strips in the c-th column of M_π, and define $s_{c,\ell}(\pi) = |S_{c,\ell}(\pi)|$. We use $W(\pi)$ to denote the weight of M_π. Since each strip in $S_{c,\ell}(\pi)$ contains a set of $\lfloor \frac{\ell}{2} \rfloor$ pairwise-disjoint bandpasses, we have

$$W(\pi) = \sum_{c=1}^{m} \sum_{\ell=2}^{n} w_c s_{c,\ell}(\pi) \lfloor \frac{\ell}{2} \rfloor = \sum_{c=1}^{m} w_c s_{c,2}(\pi) + \sum_{c=1}^{m} \sum_{\ell=3}^{n} w_c s_{c,\ell}(\pi) \lfloor \frac{\ell}{2} \rfloor. \quad (1)$$

We say that π is *optimal* if $W(\pi)$ is maximized over all permutations of $\{1, 2, \ldots, n\}$.

In the remainder of this section, fix an optimal permutation π^* of $\{1, 2, \ldots, n\}$ and let π^o be the permutation of $\{1, 2, \ldots, n\}$ with $M_{\pi^o} = M''$, where M'' is the output of our algorithm on input M. By Equation 1, we have

$$W(\pi^*) = \sum_{c=1}^{m} w_c s_{c,2}(\pi^*) + \sum_{c=1}^{m} \sum_{\ell=3}^{n} w_c s_{c,\ell}(\pi^*) \lfloor \frac{\ell}{2} \rfloor. \quad (2)$$

Obviously, G contains a Hamiltonian path P^* in which $\pi^*(1)$, $\pi^*(2)$, ..., $\pi^*(n)$ appear (as the vertices) in this order. Obviously, we have the following equation:

$$w(P^*) = \sum_{c=1}^{m} w_c s_{c,2}(\pi^*) + \sum_{c=1}^{m} \sum_{\ell=3}^{n} w_c s_{c,\ell}(\pi^*)(\ell - 1). \quad (3)$$

Since the edge set of P^* can be partitioned into two matchings of G and N is a maximum-weight matching of G, we have:

$$\begin{aligned} w(N) \geq \frac{1}{2} w(P^*) &= \frac{1}{2} \sum_{c=1}^{m} w_c s_{c,2}(\pi^*) + \frac{1}{2} \sum_{c=1}^{m} \sum_{\ell=3}^{n} w_c s_{c,\ell}(\pi^*)(\ell - 1) \\ &\geq \frac{1}{2} \sum_{c=1}^{m} w_c s_{c,2}(\pi^*) + \frac{3}{4} \sum_{c=1}^{m} \sum_{\ell=3}^{n} w_c s_{c,\ell}(\pi^*) \lfloor \frac{\ell}{2} \rfloor, \end{aligned} \quad (4)$$

where the last inequality holds because $\ell - 1 \geq \frac{3}{2}\lfloor \frac{\ell}{2} \rfloor$ for all integers $\ell \geq 3$.

By Equation 2, Inequality 4, and the trivial fact that $W(\pi^o) \geq w(N)$, we know that if $\sum_{c=1}^{m} w_c s_{c,2}(\pi^*)$ is significantly smaller than $W(\pi^*)$, then $W(\pi^o)$ is already significantly larger than $\frac{1}{2}W(\pi^*)$. So, we can concentrate on the case where $\sum_{c=1}^{m} w_c s_{c,2}(\pi^*)$ is not significantly smaller than $W(\pi^*)$. To take care of this case, we want to partition $S_{c,2}(\pi^*)$ into four subsets $S_{c,2}^1(\pi^*)$, ..., $S_{c,2}^4(\pi^*)$ below.

First, we need several definitions. Let B and B' be two disjoint bandpasses in $S_{c,2}(\pi^*)$. Suppose that i (respectively, i') is the integer in $\{1, 2, \ldots, n-1\}$ such that B (respectively, B') is formed between the i-th and the $(i+1)$-th (respectively, the i'-th and the $(i'+1)$-th) rows of M_{π^*}. Since B and B' are disjoint and reside in the same column of M_{π^*}, $\{i, i+1\} \cap \{i', i'+1\} = \emptyset$. Consider the four edges of G:

$$\{\pi^*(i),\ \pi^*(i')\},\ \{\pi^*(i), \pi^*(i'+1)\},\ \{\pi^*(i+1), \pi^*(i')\},\ \{\pi^*(i+1), \pi^*(i'+1)\}.$$

Since N is a matching, N contains at most two of the four edges. If N contains at least one of the four, then B is *linked* to B' by N. In particular, if N contains exactly one (respectively, two) of the four, then B is *singly* (respectively, *doubly*) *linked* to B' by N. If B is singly linked to B' by N, then the edge among the four that is contained in N is referred to as the *edge of* N *supporting the linkage* between B and B'.

Obviously, B can be linked to at most two bandpasses in $S_{c,2}(\pi^*)$ by N. Indeed, if B is linked to two bandpasses in $S_{c,2}(\pi^*)$ by N, then B is singly linked to both of them by N.

B is *kept* by N if $\{\pi^*(i), \pi^*(i+1)\} \in N$. Note that if B is kept by N, then B is also a bandpass in M_{π^o}. B is *completely isolated* by N if both the $(N(\pi^*(i)), c)$ and the $(N(\pi^*(i+1)), c)$ entries of M are a 0. Similarly, B is *partially isolated* by N if exactly one of the $(N(\pi^*(i)), c)$ and the $(N(\pi^*(i+1)), c)$ entries of M is a 0. Obviously, if B is kept by N, then it is neither partially nor completely isolated by N. Moreover, if B is completely isolated by N, then B is linked to no bandpass in $S_{c,2}(\pi^*)$ by N. Furthermore, if B is partially isolated by N, then B is linked to at most one bandpass in $S_{c,2}(\pi^*)$ by N. In particular, if B is partially isolated by N and is also linked to a bandpass B' in $S_{c,2}(\pi^*)$ by N, then B is singly linked to B' by N.

Now, we are ready to partition $S_{c,2}(\pi^*)$ into four subsets $S_{c,2}^1(\pi^*), \ldots, S_{c,2}^4(\pi^*)$ as follows. For each bandpass $B \in S_{c,2}(\pi^*)$,

- if B is kept by N, then B belongs to $S_{c,2}^1(\pi^*)$;
- if B is completely isolated by N, then B belongs to $S_{c,2}^2(\pi^*)$;
- if B is partially isolated by N, B is linked to one bandpass B' in $S_{c,2}(\pi^*)$ by N, and B' is also partially isolated by N, then B belongs to $S_{c,2}^3(\pi^*)$;
- otherwise, B belongs to $S_{c,2}^4(\pi^*)$.

Obviously, we have the following equation:

$$s_{c,2}(\pi^*) = |S_{c,2}^1(\pi^*)| + |S_{c,2}^2(\pi^*)| + |S_{c,2}^3(\pi^*)| + |S_{c,2}^4(\pi^*)|. \tag{5}$$

So, by Inequality 4, we have:

$$w(N) \geq \frac{1}{2}\sum_{c=1}^{m} w_c \cdot \left(|S_{c,2}^{1}(\pi^*)| + |S_{c,2}^{2}(\pi^*)| + |S_{c,2}^{3}(\pi^*)| + |S_{c,2}^{4}(\pi^*)|\right) + \frac{3}{4}\sum_{c=1}^{m}\sum_{\ell=3}^{n} w_c s_{c,\ell}(\pi^*)\lfloor\frac{\ell}{2}\rfloor \quad (6)$$

Essentially, the next lemma has been shown by Tong *et al.* [19].

Lemma 4. $w(N) \geq \sum_{c=1}^{m} w_c \cdot \left(|S_{c,2}^{1}(\pi^*)| + \frac{1}{2}|S_{c,2}^{3}(\pi^*)| + \frac{2}{3}|S_{c,2}^{4}(\pi^*)|\right)$.

Lemma 5. *The maximum weight of a Hamiltonian path in G' is not smaller than $\sum_{c=1}^{m} w_c|S_{c,2}^{2}(\pi^*)|$. Consequently, $w'(Z) \geq \frac{7}{27}\sum_{c=1}^{m} w_c|S_{c,2}^{2}(\pi^*)|$.*

Proof. Obviously, $\mathcal{P}^*$ is a Hamiltonian path in G'. For each bandpass $B \in S_{c,2}^{2}(\pi^*)$, both entries of B remain to be a 1 in M' because B is completely isolated by N. Consequently, B contributes a weight of w_c to $w'(P^*)$.

Finally, $w'(Z) \geq \max\{w'(Y_0), \ldots, w'(Y_2)\} \geq \frac{1}{3}\sum_{i=0}^{2} w'(Y_i) = \frac{1}{3}w'(\mathcal{P}) \geq \frac{7}{27}w'(\mathcal{P}^*) \geq \frac{7}{27}\sum_{c=1}^{m} |S_{c,2}^{2}(\pi^*)|w_c$. Q.E.D.

The next lemma is a key in the analysis of our algorithm.

Lemma 6. $w'(\mathcal{C}) \geq \frac{1}{4}\sum_{c=1}^{m} w_c|S_{c,2}^{3}(\pi^*)|$. *Consequently,* $w'(Z) \geq \frac{1}{16}\sum_{c=1}^{m} w_c|S_{c,2}^{3}(\pi^*)|$.

Proof. Consider an arbitrary column $c \in \{1, 2, \ldots, m\}$. The bandpasses in $S_{c,2}^{3}(\pi^*)$ can be paired up so that the two bandpasses in each pair are singly linked to each other by N. In the remainder of this proof, we assume that the bandpasses in $S_{c,2}^{3}(\pi^*)$ have been paired up in this way. For each bandpass $B \in S_{c,2}^{3}(\pi^*)$, we refer to the bandpass $B' \in S_{c,2}^{3}(\pi^*)$ paired up with B as the *mate* of B. We also call $\{B, B'\}$ a *couple* in column c.

We construct an edge-weighted multigraph H_c (possibly with parallel edges and self-loops) as follows. The vertices of H_c are $\{1, 2, \ldots, n\}$ and each of them is initially incident to no edge in H_c. Then, for each edge $\{\pi^*(i), \pi^*(i')\} \in N$ (say, with $i < i'$), we perform the following steps:

S1. If (1) the $(i-1, c)$-th and the (i, c)-th entries of M^* exist and form a bandpass B in $S_{c,2}^{3}(\pi^*)$ and (2) the $(i'-1, c)$-th and the (i', c)-th entries of M^* exist and form a bandpass B' in $S_{c,2}^{3}(\pi^*)$, then add an edge between vertices $i-1$ and $i'-1$ in H_c.

S2. If (1) the $(i-1, c)$-th and the (i, c)-th entries of M^* exist and form a bandpass B in $S_{c,2}^{3}(\pi^*)$ and (2) the $(i'+1, c)$-th and the (i', c)-th entries of M^* exist and form a bandpass B' in $S_{c,2}^{3}(\pi^*)$, then add an edge between vertices $i-1$ and $i'+1$ in H_c.

S3. If (1) the $(i+1,c)$-th and the (i,c)-th entries of M^* exist and form a bandpass B in $S^3_{c,2}(\pi^*)$ and (2) the $(i'-1,c)$-th and the (i',c)-th entries of M^* exist and form a bandpass B' in $S^3_{c,2}(\pi^*)$, then add an edge between vertices $i+1$ and $i'-1$ in H_c.

S4. If (1) the $(i+1,c)$-th and the (i,c)-th entries of M^* exist and form a bandpass B in $S^3_{c,2}(\pi^*)$ and (2) the $(i'+1,c)$-th and the (i',c)-th entries of M^* exist and form a bandpass B' in $S^3_{c,2}(\pi^*)$, then add an edge between vertices $i+1$ and $i'+1$ in H_c.

Furthermore, we assign a weight of w_c to each edge in H_c.

Note that only one "if"-condition in Steps S1 through S4 can be true. Moreover, if the "if"-condition in a step is true, then the bandpasses B and B' in that step form a couple in column c. Conversely, if $\{B, B'\}$ is a couple in column c, then for the edge $\{\pi^*(i), \pi^*(i')\}$ of N supporting the linkage between B and B', the "if" condition in one of Steps S1 through S4 holds. Thus, H_c has as many edges as couples in column c. In other words, H_c has $\frac{1}{2}|S^3_{c,2}(\pi^*)|$ edges. So, the total weight of edges in H_c is $\frac{1}{2}|S^3_{c,2}(\pi^*)|w_c$.

We claim that H_c contains no self-loops. To see this, first note that a self-loop can be added to H_c only in Step S3 and only when $i' = i+2$, because we assume $i < i'$. Moreover, when $i' = i+2$, we have $i+1 = i'-1$ and so the "if"-condition in Step S3 implies that the $(i+1,c)$-th and the (i,c)-th entries of M^* form a bandpass B in $S^3_{c,2}(\pi^*)$ and the $(i+1,c)$-th and the (i',c)-th entries of M^* form a bandpass B' in $S^3_{c,2}(\pi^*)$. However, since the bandpasses in $S^3_{c,2}(\pi^*)$ are disjoint, B and B' cannot coexist. Thus, H_c contains no self-loops.

We further claim that H_c does not contain parallel edges, either. To see this, let j and j' be two (distinct) adjacent vertices in H_c. Then, by Steps S1 through S4, at least one of $\{\pi^*(j-1), \pi^*(j'-1)\}$, $\{\pi^*(j-1), \pi^*(j'+1)\}$, $\{\pi^*(j+1), \pi^*(j'-1)\}$, and $\{\pi^*(j+1), \pi^*(j'+1)\}$ belongs to N. Obviously, if only one of the four edges belongs to N, then there is at most one edge between vertices j and j' in H_c. So, we assume that at least two of the four edges belong to N. Then, since N is a matching, N contains either both $\{\pi^*(j-1), \pi^*(j'-1)\}$ and $\{\pi^*(j+1), \pi^*(j'+1)\}$, or both $\{\pi^*(j-1), \pi^*(j'+1)\}$ and $\{\pi^*(j+1), \pi^*(j'-1)\}$. We assume that N contains both $\{\pi^*(j-1), \pi^*(j'-1)\}$ and $\{\pi^*(j+1), \pi^*(j'+1)\}$; the other case is similar. Clearly, if performing Steps S1 through S4 for the edge $\{\pi^*(j-1), \pi^*(j'-1)\} \in N$ results in the addition of an edge to H_c between vertices j and j', then the $(j-1,c)$ and the (j,c) entries of M^* form a bandpass B_1 in $S^3_{c,2}$. Similarly, if performing Steps S1 through S4 for the edge $\{\pi^*(j+1), \pi^*(j'+1)\} \in N$ results in the addition of an edge to H_c between vertices j and j', then the (j,c) and the $(j+1,c)$ entries of M^* form a bandpass B_2 in $S^3_{c,2}$. Now, since the bandpasses in $S^3_{c,2}$ are disjoint, B_1 and B_2 cannot coexist in $S^3_{c,2}$ and in turn we know that at most one edge can exist between vertices j and j' in H_c even if N contains both $\{\pi^*(j-1), \pi^*(j'-1)\}$ and $\{\pi^*(j+1), \pi^*(j'+1)\}$. Hence, H_c does not contain parallel edges. Consequently, H_c is a (simple) graph by the first claim in the proof.

We next construct an edge-weighted multigraph H (possibly with parallel edges but without self-loops) as follows. The vertices of H are $\{1, 2, \ldots, n\}$. For every column $c \in \{1, 2, \ldots, m\}$, every edge of H_c is also an edge of H with weight w_c. So, the total weight of edges in H is $\frac{1}{2}\sum_{c=1}^{m} |S_{c,2}^{3}(\pi^*)|w_c$.

We further construct an edge-weighted graph H' by modifying H as follows. For every $i \in \{1, 2, \ldots, n\}$ and every $j \in \{1, 2, \ldots, n\}$ such that H contains at least one edge between i and j, we merge the edges between i and j into a single edge whose weight is the total weight of edges between i and j in H. Note that the weight of each edge between two distinct vertices in H' does not exceed the weight of the same edge in G'. This follows from the simplicity of graph H_c and the fact that for each edge $\{k, \ell\}$ added to H_c in Steps S1 through S4, a bandpass can be formed in column c between the $\pi^*(k)$-th and the $\pi^*(\ell)$-th rows of matrix M'.

Consider an arbitrary $j \in \{1, 2, \ldots, n\}$. By Steps S1 through S4, we know that for each neighbor j' of vertex j in H', at least one of $\{\pi^*(j-1), \pi^*(j'-1)\}$, $\{\pi^*(j-1), \pi^*(j'+1)\}$, $\{\pi^*(j+1), \pi^*(j'-1)\}$, and $\{\pi^*(j+1), \pi^*(j'+1)\}$ belongs to N. So, the following claim holds:

C1. Let j be an arbitrary integer in $\{1, 2, \ldots, n\}$. Then, only $k-1$, $k+1$, $\ell-1$, and $\ell+1$ can be neighbors of vertex j in H', where k and ℓ are the integers in $\{1, 2, \ldots, n\}$ with $\{\pi^*(j-1), \pi^*(k)\} \in N$ and $\{\pi^*(j+1), \pi^*(\ell)\} \in N$.

We finally claim that the edge set of H' can be partitioned into two subsets E_1 and E_2 such that both $H'[E_1]$ and $H'[E_2]$ are 2-matchings of H'. To see this, it suffices to show that we can color the edges of H' *red* or *blue* so that for each vertex j of H', no three edges incident to j get the same color. To find such a coloring, consider an arbitrary edge $\{\pi^*(i), \pi^*(i')\}$ in N with $1 \le i < i' \le n$. When we perform Steps S1 through S4 for this edge (in any column $c \in \{1, 2, \ldots, m\}$), we add to H_c zero or more edges among $\{i-1, i'-1\}$, $\{i-1, i'+1\}$, $\{i+1, i'-1\}$, and $\{i+1, i'+1\}$. In other words, the edge $\{\pi^*(i), \pi^*(i')\} \in N$ contributes at most four edges (namely, $\{i-1, i'-1\}$, $\{i-1, i'+1\}$, $\{i+1, i'-1\}$, and $\{i+1, i'+1\}$) to H'. In the following case-analysis, we describe how to color these possible edges in H'.

Case 1: $i' > i+2$. In this case, we color $\{i-1, i'-1\}$ and $\{i+1, i'+1\}$ *red* and color $\{i-1, i'+1\}$ and $\{i+1, i'-1\}$ *blue*. Note that no two adjacent edges receive the same color here.

Case 2: $i' = i+2$. In this case, $\{i-1, i'-1\}$, $\{i-1, i'+1\}$, $\{i+1, i'-1\}$, and $\{i+1, i'+1\}$ become $\{i-1, i+1\}$, $\{i-1, i+3\}$, $\{i+1, i+1\}$, and $\{i+1, i+3\}$, respectively. So, only $\{i-1, i+1\}$, $\{i-1, i+3\}$, and $\{i+1, i+3\}$ can be possible edges in H'. We color $\{i-1, i+1\}$ and $\{i+1, i+3\}$ *red* and color $\{i-1, i+3\}$ *blue*. Note that no two edges incident to vertex $i-1$ (respectively, $i+3$) receive the same color here. However, two edges incident to vertex $i+1$ receive the same color here. Fortunately, by Claim C1, these two edges are the only possible edges incident to vertex $i+1$ in H'. For convenience, we refer to this kind of vertex $i+1$ as a *done vertex*.

Case 3: $i' = i+1$. In this case, $\{i-1, i'-1\}$, $\{i-1, i'+1\}$, $\{i+1, i'-1\}$, and $\{i+1, i'+1\}$ become $\{i-1, i\}$, $\{i-1, i+2\}$, $\{i+1, i\}$, and $\{i+1, i+2\}$, respectively. We color $\{i-1, i\}$ and $\{i+1, i+2\}$ *red* and color $\{i-1, i+2\}$ and $\{i+1, i\}$ *blue*. Note that no two adjacent edges receive the same color here.

We want to show that for each undone vertex j in H', no three edges incident to j in H' get the same color. To this end, let j, k, and ℓ be as in Claim C1. By Claim C1, the possible edges incident to j in H' are $\{j, k-1\}$, $\{j, k+1\}$, $\{j, \ell-1\}$, and $\{j, \ell+1\}$. Obviously, the first two possible edges receive different colors when we consider the above cases (namely, Cases 1 through 3) for the edge $\{\pi^*(j-1), \pi^*(k)\} \in N$, while the last two receive different colors when we consider the above cases for the edge $\{\pi^*(j+1), \pi^*(\ell)\} \in N$. Therefore, no three edges incident to j in H' get the same color.

Recall that the total weight of edges in H' is the same as that in H, namely, $\frac{1}{2}\sum_{c=1}^{m} |S_{c,2}^3(\pi^*)|w_c$. So, by the final claim, there is an $i \in \{1, 2\}$ such that the total weight of edges in $H'[E_i]$ is at least $\frac{1}{4}\sum_{c=1}^{m} |S_{c,2}^3(\pi^*)|w_c$. Also recall that the weight of each edge between two distinct vertices in H' does not exceed the weight of the same edge in G'. Therefore, $w'(G'[E_i]) \geq \frac{1}{4}\sum_{c=1}^{m} |S_{c,2}^3(\pi^*)|w_c$. Since $G'[E_i]$ is isomorphic to $H'[E_i]$, $G'[E_i]$ is also a 2-matching of G'. Hence, $w'(G'[E_i]) \leq w'(\mathcal{C})$ because $\mathcal{C}$ is a maximum-weight matching in G'. At last, we have $w'(\mathcal{C}) \geq \frac{1}{4}\sum_{c=1}^{m} |S_{c,2}^3(\pi^*)|w_c$.

Finally, $w'(Z) \geq \max\{w'(X_0), \ldots, w'(X_3)\} \geq \frac{1}{4}\sum_{i=0}^{3} w'(X_i) = \frac{1}{4}w'(\mathcal{C}) \geq \frac{1}{16}\sum_{c=1}^{m} |S_{c,2}^3(\pi^*)|w_c$. Q.E.D.

We are now ready to show the following lemma:

Lemma 7. *The algorithm achieves a ratio of* 0.5318.

Proof. Let x be a real number with $0 \leq x \leq 1$. Then, we can obtain the following inequality by (1) multiplying both sides of the second inequality in Lemma 6 by x, (2) multiplying both sides of the inequality in Lemma 5 by $1-x$, and (3) adding the left-hand sides and the right-hand sides of the resulting two inequalities, respectively.

$$w'(Z) \geq \frac{x}{16}\sum_{c=1}^{m} w_c|S_{c,2}^3(\pi^*)| + \frac{7(1-x)}{27}\sum_{c=1}^{m} w_c|S_{c,2}^2(\pi^*)|. \tag{7}$$

Let y be a real number with $0 \leq y \leq 1$. Then, we can obtain the following inequality by (1) multiplying both sides of Inequality 6 by y, (2) multiplying both sides of the inequality in Lemma 4 by $1-y$, and (3) adding the left-hand sides and the right-hand sides of the resulting two inequalities, respectively.

$$\begin{aligned} w(N) \geq{} & \frac{2-y}{2}\sum_{c=1}^{m} w_c|S_{c,2}^1(\pi^*)| + \frac{y}{2}\sum_{c=1}^{m} w_c|S_{c,2}^2(\pi^*)| + \frac{1}{2}\sum_{c=1}^{m} w_c|S_{c,2}^3(\pi^*)| \\ & + \frac{4-y}{6}\sum_{c=1}^{m} w_c|S_{c,2}^4(\pi^*)| + \frac{3y}{4}\sum_{c=1}^{m}\sum_{\ell=3}^{n} w_c s_{c,\ell}(\pi^*)\lfloor\frac{\ell}{2}\rfloor. \end{aligned} \tag{8}$$

By the algorithm, $W(\pi^o) \geq w(N) + w'(Z)$. So, by Inequalities 8 and 7, we have:

$$W(\pi^o) \geq \frac{2-y}{2}\sum_{c=1}^{m} w_c|S_{c,2}^1(\pi^*)| + \frac{14-14x+27y}{54}\sum_{c=1}^{m} w_c|S_{c,2}^2(\pi^*)|$$

$$+\frac{8+x}{16}\sum_{c=1}^{m} w_c|S^3_{c,2}(\pi^*)| + \frac{4-y}{6}\sum_{c=1}^{m} w_c|S^4_{c,2}(\pi^*)|$$
$$+\frac{3y}{4}\sum_{c=1}^{m}\sum_{\ell=3}^{n} w_c s_{c,\ell}(\pi^*)\lfloor\frac{\ell}{2}\rfloor. \quad (9)$$

Consider the following linear programming (LP) problem:

$$\begin{aligned}
&\text{Maximize } z \\
&\text{Subject to: } z \le \frac{2-y}{2} \\
&z \le \frac{14-14x+27y}{54} \\
&z \le \frac{8+x}{16} \\
&z \le \frac{4-y}{6} \\
&z \le \frac{3y}{4} \\
&0 \le x,\ y,\ z \le 1.
\end{aligned}$$

Solving the above LP, we obtain $z = 0.531818\ldots$ So, by Inequality 9 and Equation 5, we have:

$$W(\pi^o) \ge 0.5318\left(\sum_{c=1}^{m} w_c s_{c,2}(\pi^*) + \sum_{c=1}^{m}\sum_{\ell=3}^{n} w_c s_{c,\ell}(\pi^*)\lfloor\frac{\ell}{2}\rfloor\right).$$

Now, by Equation 2, we have $W(\pi^o) \ge 0.5318 \cdot W(\pi^*)$. This completes the proof. Q.E.D.

By Lemmas 3 and 7, we finally have the following theorem:

Theorem 1. *There is an $O\big(n^2(n+m)$-time approximation algorithm for Bandpass-2 which achieves a ratio of* 0.5318.

References

1. Babayev, D.A., Bell, G.I., Nuriyev, U.G.: The Bandpass Problem: Combinatorial Optimization and Library of Problems. Journal of Combinatorial Optimization 18, 151–172 (2009)
2. Barvinok, A., Johnson, D.S., Woeginger, G.J., Woodroofe, R.: The Maximum Traveling Salesman Problem Under Polyhedral Norms. In: Bixby, R.E., Boyd, E.A., Ríos-Mercado, R.Z. (eds.) IPCO 1998. LNCS, vol. 1412, pp. 195–201. Springer, Heidelberg (1998)
3. Bell, G.I., Babayev, D.A.: Bandpass Problem. In: Annual INFORMS Meeting, Denver, CO, USA (2004)
4. Chen, Z.-Z., Nagoya, T.: Improved Approximation Algorithms for Metric Max TSP. Journal of Combinatorial Optimization 13, 321–336 (2007)

5. Chen, Z.-Z., Okamoto, Y., Wang, L.: Improved deterministic approximation algorithms for Max TSP. Information Processing Letters 95, 333–342 (2005)
6. Chen, Z.-Z., Wang, L.: An Improved Randomized Approximation Algorithm for Max TSP. Journal of Combinatorial Optimization 9, 401–432 (2005)
7. Gabow, H.: Implementation of Algorithms for Maximum Matching on Nonbipartite Graphs. Ph.D. Thesis, Department of Computer Science, Stanford University, Stanford, California (1973)
8. Gabow, H.: An Efficient Reduction Technique for Degree-Constrained Subgraph and Bidirected Network Flow Problems. In: Proceedings of the 15th Annual ACM Symposium on Theory of Computing, STOC 1983, pp. 448–456. ACM (1983)
9. Hassin, R., Rubinstein, S.: An Approximation Algorithm for the Maximum Traveling Salesman Problem. Information Processing Letters 67, 125–130 (1998)
10. Hassin, R., Rubinstein, S.: Better Approximations for Max TSP. Information Processing Letters 75, 181–186 (2000)
11. Hassin, R., Rubinstein, S.: A 7/8-Approximation Approximations for Metric Max TSP. Information Processing Letters 81, 247–251 (2002)
12. Kaplan, H., Lewenstein, M., Shafrir, N., Sviridenko, M.: Approximation Algorithms for Asymmetric TSP by Decomposing Directed Regular Multigraphs. Journal of the ACM 52, 602–626 (2005)
13. Kostochka, A.V., Serdyukov, A.I.: Polynomial Algorithms with the Estimates $\frac{3}{4}$ and $\frac{5}{6}$ for the Traveling Salesman Problem of Maximum. Upravlyaemye Sistemy 26, 55–59 (1985) (in Russian)
14. Kowalik, L., Mucha, M.: Deterministic 7/8-Approximation for the Metric Maximum TSP. Theor. Comput. Sci. 410, 5000–5009 (2009)
15. Lin, G.: On the Bandpass Problem. Journal of Combinatorial Optimization 22, 71–77 (2011)
16. Paluch, K., Mucha, M., Mądry, A.: A 7/9 - Approximation Algorithm for the Maximum Traveling Salesman Problem. In: Dinur, I., Jansen, K., Naor, J., Rolim, J. (eds.) APPROX 2009. LNCS, vol. 5687, pp. 298–311. Springer, Heidelberg (2009)
17. Serdyukov, A.I.: An Algorithm with an Estimate for the Traveling Salesman Problem of Maximum. Upravlyaemye Sistemy 25, 80–86 (1984) (in Russian)
18. Tarjan, R.E.: Efficiency of a Good But Not Linear Set Union Algorithm. Journal of the ACM 225, 215–225 (1975)
19. Tong, W., Goebel, R., Ding, W., Lin, G.: An Improved Approximation Algorithm for the Bandpass Problem. In: Snoeyink, J., Lu, P., Su, K., Wang, L. (eds.) AAIM 2012 and FAW 2012. LNCS, vol. 7285, pp. 351–358. Springer, Heidelberg (2012)

The b-Matching Problem in Hypergraphs: Hardness and Approximability

Mourad El Ouali[1] and Gerold Jäger[2]

[1] Computer Science Institute
Christian-Albrechts-University of Kiel
Christian-Albrechts-Platz 4
D-24098 Kiel, Germany
meo@informatik.uni-kiel.de

[2] Department of Mathematics and Mathematical Statistics
University of Umeå
SE-901-87 Umeå, Sweden
gerold.jaeger@math.umu.se

Abstract. In this paper we analyze the maximum cardinality b-matching problem in l-uniform hypergraphs with respect to the complexity class MAX-SNP, where b-matching is defined as follows: for given $b \in \mathbb{N}$ and a hypergraph $\mathcal{H} = (V, \mathcal{E})$ a subset $M_b \subseteq \mathcal{E}$ with maximum cardinality is sought so that no vertex is contained in more than b hyperedges of M_b. We show that if the maximum degree of the vertices is bounded by a constant $B \in \mathbb{N}$, this problem has no approximation scheme, unless $\mathcal{P} = \mathcal{NP}$. This result generalizes a result of Kann from $b = 1$ to the case that $b \in \mathbb{N}$ with $0 < b \leq \frac{B}{3}$. Furthermore, we extend a result of Srivastav and Stangier, who gave an approximation algorithm for the unweighted b-matching problem.

Keywords: Hypergraphs, matching, L-reduction, Boolean satisfiability, randomized rounding, MAX-SNP-hardness.

1 Introduction

Packing problems are widely explored in discrete optimization. Classical problems in this field include the b-matching problem which is a generalization of the set packing problem. An instance of the b-matching problem provides a hypergraph $\mathcal{H} = (V, \mathcal{E})$, $|V| = n$, $\mathcal{E} \subseteq 2^V$ and $b \in \mathbb{N}$ and asks for a subset $M_b \subseteq \mathcal{E}$ with maximum cardinality (or, in a weighted version, with maximum weight) such that no vertex is contained in more than b sets. One of our goals is to prove that the restriction of the b-matching problem to a certain class of hypergraphs has no polynomial time approximation scheme PTAS, unless $\mathcal{P} = \mathcal{NP}$. We utilize the concept of L-reduction (see [7]) instead of directly using a PTAS-reduction, because it is often easier to show that a reduction is an L-reduction than a PTAS-reduction. As mentioned above, the main part of this work deals with the b-matching problem on the subclass of l-uniform hypergraphs with vertex degree at most a constant B.

G. Lin (Ed.): COCOA 2012, LNCS 7402, pp. 200–211, 2012.

Several positive results (i.e., approximation ratios) were achieved under various aspects for the b-matching problem in hypergraphs [5, 8–10, 12]. However, concerning its algorithmic complexity, the problem has still not been investigated extensively, which motivated us to study the problem with respect to this aspect. The recent known inapproximability result was found by El Ouali et al. [2]. They showed that for $1 \leq b \leq \frac{l}{\ln l}$ the b-matching problem in l-uniform hypergraphs cannot be efficiently approximated with a factor of $O(\frac{l}{b \ln l})$, unless $\mathcal{P} = \mathcal{NP}$. To our knowledge, the only known result of the problem related to the class MAX-SNP is due to Kann [3], who proved that there is no approximation scheme for the case $b = 1$, unless $\mathcal{P} = \mathcal{NP}$. No similar result is known for $b > 1$.

There are several applications for the b-matching problem in medicine, computational geometry, and combinatorial auctions (CA). Here we briefly explain an example for combinatorial auctions (see [4]); the reader may consult the references for more details.

An auctioneer wants to sell n kinds of goods (vertices) to m bidders. Every good $v \in V$ is available only b times. Assume that each bidder $j \in \{1, \ldots, m\}$ can valuate a subset (hyperedge) $E \in \{E_1, \ldots, E_m\} \subseteq 2^V$ of goods. In the weighted case of the problem, a valuation $w_j(E) \in \mathbb{R}_{\geq 0}$ is the maximum amount of money that the bidder j is ready to pay for the subset E. The objective of the auctioneer is to get the maximum profit, i.e., $\max \sum_{i=1}^{m} w_j$.

In this work, we generalize Kann's result and show that for an arbitrary $b \leq \frac{B}{3}$, the b-matching problem in l-uniform hypergraphs with vertex degree at most a constant B has no PTAS, unless $\mathcal{P} = \mathcal{NP}$. Further we will analyze the approximability of this problem and present a $(1 - \epsilon)$-approximation algorithm for $\epsilon \in (0, 1)$ and $b \geq 2\epsilon^{-1}\sqrt{\Delta \ln(\sqrt{n})}$, where Δ is the maximum vertex degree in the hypergraph. This improves the result of [12] for unweighted hypergraphs.

2 Definitions and Preliminaries

A hypergraph $\mathcal{H}$ is an ordered pair $(V, \mathcal{E})$ where V is the set of vertices and $\mathcal{E}$ is the set of (hyper-)edges (i.e., a collection of distinct non-empty subsets of V). As usually, the **degree** of a vertex $v \in V$ is the number of hyperedges it appears in (notation: $\deg(v)$). $\mathcal{H}$ is called of **bounded degree** B if the degree of each vertex is at most B. Furthermore if the degree of every vertex is exactly B, then the hypergraph is called B-**regular**. We call the number of vertices of a hyperedge as its size. If the size of each hyperedge is exactly l, i.e., $\forall E \in \mathcal{E} : |E| = l$, then $\mathcal{H}$ is called l-**uniform**.

Formulation of the Problems

b-matching Problem: For given $b \in \mathbb{N}$ we call a set $M_b \subseteq \mathcal{E}$ a b-**matching** if no vertex is contained in more than b edges of M_b. The b-matching problem is the problem of finding a b-matching with maximum cardinality. We denote the b-matching problem in hypergraphs with bounded degree B by (b, B)-matching.

l-dimensional b-matching Problem: This problem is a variant of b-matching in l-uniform hypergraphs, where the vertices of the input hypergraph are a union of l disjoint sets, $V = V_1 \dot{\cup} V_2 \dot{\cup} \ldots \dot{\cup} V_l$, and each hyperedge contains exactly one vertex from each set such that we have $E \subseteq V_1 \times V_2 \times \cdots \times V_l$. We will denote this problem in hypergraphs with bounded degree B by l-dimensional (b, B)-matching problem

The following reduction was introduced by Papadimitriou and Yannakakis [7].

Definition 1. *Let $P = (X, (S_x)_{x \in X}, w, goal)$ and $P' = (X', (S'_x)_{x \in X}, w', goal')$ be two optimization problems with non-negative weights. An L-***reduction** *from P to P' is a pair of functions f and g, both computable in polynomial time, and two constants $\alpha, \beta \in \mathbb{R}^+$ such that it holds for any instance x of P:*

1. *$f(x)$ is an instance of P' with $Opt(f(x)) \leq \alpha \cdot Opt(x)$,*
2. *For any feasible solution y' of $f(x)$, $g(x, y')$ is a feasible solution of x such that $|w(g(x, y') - Opt(x)| \leq \beta \cdot |w'(y') - Opt(f(x)|$.*

We denote by $Opt(x)$ the optimum value of instance x.
We say that P is L-reducible to P' if there is an L-reduction from P to P'.

Note that "L" stands for "linear". The L-reduction can be composed as follows:

Proposition 1 (Composition of two L-Reductions).
Let P_1, P_2, P_3 be optimization problems with non-negative weights. If $(f_1, g_1, \alpha_1, \beta_1)$ is an L-reduction from P_1 to P_2 and $(f_2, g_2, \alpha_2, \beta_2)$ is an L-reduction from P_2 to P_3, then their composition $(f_3, g_3, \alpha_1\alpha_2, \beta_1\beta_2)$ is an L-reduction from P_1 to P_3, where $f_3(x) = f_2(f_1(x))$ and $g_3(x, y_3) = g_1(x, g_2(f_1(x), y_3))$.

To introduce the MAX-SNP-hardness of an optimization problem we have to define the maximum 3-satisfiability problem:

Definition 2. *Consider a set $X = \{x_1, \ldots, x_n\}$ of variables and a family $\mathcal{C} = \{c_1, \ldots, c_m\}$ of different clauses, each involving exactly three literals (a variable or a negated variable) over X.*

Maximum 3-satisfiability (in short MAX-3-SAT**)** *is the problem of finding an assignment A of X so that the number of clauses in $\mathcal{C}$ satisfied by A is maximum. The variant of the problem, where the number of occurrences of each variable is bounded by a constant $B \in \mathbb{R}^+$ will be denoted by* MAX-3-SAT-B.

Definition 3. *An optimization problem P with non-negative weights is called* MAX-SNP*-hard if* MAX-3-SAT *is L-reducible to P.*

Corollary 1. *No approximation scheme for any* MAX-SNP*-hard problem exists, unless $\mathcal{P} = \mathcal{NP}$.*

For the approximation result we need the following theorem, which is well-known as the independent bounded differences inequality theorem:

Theorem 1. *[1] Let $X = (X_1, \ldots, X_n)$ be a family of independent random variables with X_k taking values in a set A_k for each k. Suppose that the real valued function f defined on $\prod_{k=1}^{n} A_k$ satisfies $|f(x) - f(x')| \leq c_k$, where the vector x and x' differ only in in the k-th coordinate. Let $\mathbb{E}(X)$ be the expected value of the random variable $f(X)$. Then for any $t \geq 0$,*

$$\Pr[f(X) \geq \mathbb{E}(f(X)) + t] \leq \exp\left(\frac{-2t^2}{\sum_{k=1}^{n} c_k^2}\right).$$

3 Main Result

Theorem 2. *For every $b \leq \frac{B}{3}$, the 3-dimensional (b, B)-matching problem is* MAX-SNP*-hard.*

Proof. To prove Theorem 2, we will describe an L-reduction (f, g, α, β) from a MAX-3-SAT-B instance to 3-dimensional (b, B)-matching instance. This is an extension of the reduction used to prove that 3-dimensional $(1, B)$-matching problem is MAX-SNP-hard in [3]. The reduction consists of two steps. We consider an instance I of MAX-3-SAT-B with a set of variables $X = \{x_1, \ldots, x_n\}$ and a set of clauses $\mathcal{C} = \{c_1, \ldots, c_m\}$.

First Step: We make $b - 1$ copies of $\mathcal{C}$ in a way to obtain b identical sets of clauses $\mathcal{C}^1, \ldots, \mathcal{C}^b$. For every set of clauses $\mathcal{C}^j$, $j \in [b]$, this part of the construction remains the same as in [3] and can be described as follows:

We consider a set of clauses $\mathcal{C}^j = \{c_1^j, \ldots, c_m^j\}$. We assume that every variable $x_i \in X$ appears $c(x_i)$ times in $\mathcal{C}^j$ (either as x_i or as $\bar{x}_i$). Moreover, let $K = 2^{\lfloor \log \frac{3}{2} b^2 B + 1 \rfloor}$ be the largest power of 2 such that $K \leq \frac{3}{2} b^2 B + 1$.

For every variable x_i we construct K identical rings of triples (in a triangle form) that we denote by $R(x_i, k, j)$, $k \in [K]$. Each ring contains $2c(x_i)$ triples. The free vertices in the ring triples (the apices of the triangles) are denoted by $x(\rho, \lambda, j)$ and $\bar{x}(\rho, \lambda, j)$, where $(\rho, \lambda, j) \in [c(x_i)] \times [K] \times [b]$ (see Figure 1a). The K rings $R(x_i, 1, j), \ldots, R(x_i, K, j)$ are connected by tree triples in $2c(x_i)$ binary trees, denoted by $T_{x_i(\rho,j)}$ and $T_{\bar{x}_i(\rho,j)}$ ($\rho \in [c(x_i)]$) in such a way that $\bar{x}_i(\rho, 1, j), \bar{x}_i(\rho, 2, j), \ldots, \bar{x}_i(\rho, k, j)$ are leaves in the tree $T_{x_i(\rho,j)}$ and $\bar{x}_i(\rho, 1, j), \bar{x}_i(\rho, 2, j), \ldots, \bar{x}_i(\rho, k, j)$ are leaves in the tree $T_{\bar{x}_i(\rho,j)}$ (see Figure 1b). The root of $T_{x_i(\rho,j)}$ is $x_i(\rho, j)$ and the root of $T_{\bar{x}_i(\rho,j)}$ is $\bar{x}_i(\rho, j)$. We denote by $\hat{T}_{x_i(\rho,j)}$ and $\hat{T}_{\bar{x}_i(\rho,j)}$ the binary tree consisting of $T_{x_i(\rho,j)}$ and $T_{\bar{x}_i(\rho,j)}$, respectively, and the K ring triples with the nodes $x_i(\rho, 1, j), x_i(\rho, 2, j), \ldots, x_i(\rho, k, j)$ and $\bar{x}_i(\rho, 1, j), \bar{x}_i(\rho, 2, j), \ldots, \bar{x}_i(\rho, k, j)$.

Finally, the clause triples connect some of the roots. For each clause c_l^j, $l \in [m]$, we introduce two new elements $s_1(l, j)$ and $s_2(l, j)$. If the variable x_i occurs in this clause and this is its ρ-th occurrence in $\mathcal{C}^j$, then the root element in the triple is $x_i(\rho, j)$ or $\bar{x}_i(\rho, j)$, depending on whether the occurrence is x_i or $\bar{x}_i$ (see Figure 2).

Second Step: The construction in the first step holds for every set of clauses $\mathcal{C}^j$ for all $j \in [b]$. Consider two constructed binary trees $\hat{T}_{x_i(\rho,j)}$ and $\hat{T}_{x_i(\rho,j')}$

corresponding to the same variable $x_i \in X$ and the same occurrence number $\rho \in [c(x_i)]$, but belonging to two different sets of clauses $\mathcal{C}^j$ and $\mathcal{C}^{j'}$ analogously for trees $\hat{T}_{\bar{x}_i(\rho,j)}$ and $\hat{T}_{\bar{x}_i(\rho,j')}$.

Furthermore let L be the number of levels of $\hat{T}_{x_i(\rho,j)}$ and $\hat{T}_{x_i(\rho,j')}$. We consider a triple (triangle) T in $\hat{T}_{x_i(\rho,j)}$. Then the apex of T belongs to level r for an $r \in \{0, \ldots, L-1\}$ and the two vertices from the base belong to level $r+1$. In the tree $\hat{T}_{x_i(\rho,j')}$ we have exactly one triangle T' between levels r and $r+1$ corresponding to T. We connect the apex of T with two vertices of T' in level $r+1$ and the apex of T' with two vertices of T in level $r+1$ to obtain two new triangles. This is done for every pair of corresponding triangles of $\hat{T}_{x_i(\rho,j)}$ and $\hat{T}_{x_i(\rho,j')}$. The same will be done for the clause triples $(s_1(l,j), s_2(l,j), x_i(\rho,j))$ and $(s_1(l,j'), s_2(l,j'), x_i(\rho,j'))$ in order to obtain two new clause triples $(s_1(l,j), s_2(l,j), x_i(\rho,j'))$ and $(s_1(l,j'), s_2(l,j'), x_i(\rho,j))$. This process is applied to every pair of symmetric triangles for every pair of symmetric trees $\hat{T}_{x_i(\rho,j)}$ and $\hat{T}_{x_i(\rho,j')}$. Finally, we obtain two new binary trees $\hat{T}_{x_i(\rho,j\to j')}$ and $\hat{T}_{x_i(\rho,j'\to j)}$ between each pair of binary trees such that the set of their vertices alternates between the sets of vertices of the trees $\hat{T}_{x_i(\rho,j)}$ and $\hat{T}_{x_i(\rho,j')}$ (or $\hat{T}_{\bar{x}_i(\rho,j)}$ and $\hat{T}_{\bar{x}_i(\rho,j')}$) (see Figure 3). The constructed instance contains $b^2\sum_{x\in X} c(x)$ clause triples, $2Kb^2\sum_{x\in X} c(x)$ ring triples and $2b^2(K-1)\sum_{x\in X} c(x)$ tree triples. To achieve the final instance, namely a 3-dimensional hypergraph, we label the vertices of the constructed instance above as follows: For the ring, tree and clause triples, we define a 3-dimensional hypergraph $\mathcal{H}$ with $E(\mathcal{H}) \subseteq Z \times U \times Y$, and label the elements of $\mathcal{H}$ with Z, U or Y. Consider the set $\mathcal{C}^j$, and label all trees in $\mathcal{C}^j$ identically as follows. Start with the root and label it with Z and label the elements in every tree triple Z, U and Y anti-clockwise. The elements in ring triples are labelled anti-clockwise in x_i-trees and clockwise in $\bar{x}_i$-trees. The elements $s_1(l,j)$ and $s_2(l,j)$ are labelled with U and Y, respectively (see Figure 4). We obtain an instance $\mathcal{H}$ for which the set of hyperedges has cardinality at most $9b^2m + 6b^4B^2n$:

$$
\begin{aligned}
|E(\mathcal{H})| &= b^2\sum_{x\in X} c(x) + 2Kb^2\sum_{x\in X} c(x) + 2b^2(K-1)\sum_{x\in X} c(x)\\
&\leq 3b^2m + b^2\sum_{x\in X} 2c(x)(K+K-1)\\
&\leq 3b^2m + b^2\sum_{x\in X} 2c(x)(3b^2B+1)\\
&\overset{(\sum_{x\in X} c(x)\leq 3m)}{\leq} 3b^2m + b^2\sum_{x\in X} 6b^2B^2 + 6b^2m\\
&= b^2(9m + 6b^2B^2n).
\end{aligned}
$$

We conclude that f can be computed in time polynomial in m and n. Moreover, every element in a ring triangle or tree triangle occurs exactly $2b$ times in the edges of $\mathcal{H}$. The half of the root elements occurs b times and the remaining elements are $s_1(l,j)$ and $s_2(l,j)$ in the clause triples. They occur at most $3b$

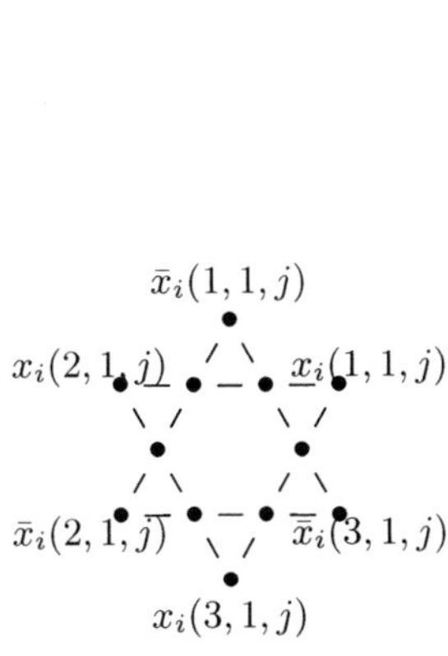

(a) The ring $R(x_i, 1, j)$, the first of K rings for the set of clauses $\mathcal{C}^j$ for a variable x_i with $c(x_i) = 3$ occurrences.

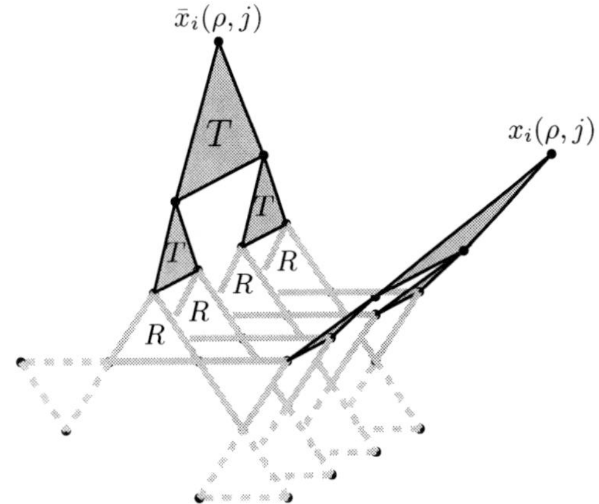

(b) An example of binary trees of triples with $c(x_i) = 3$ and $K = 4$.

Fig. 1.

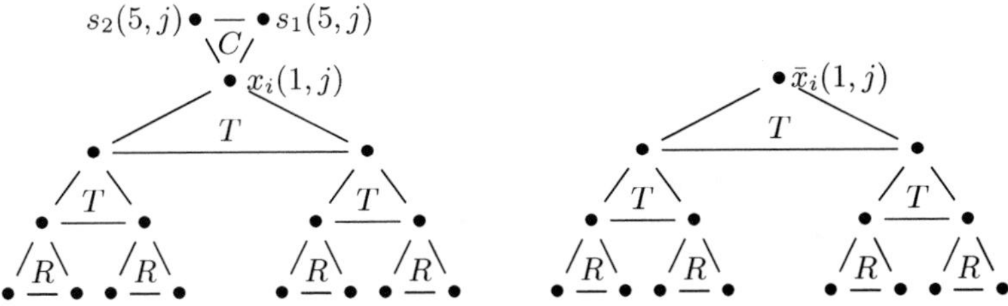

Fig. 2. An example of binary trees for x_i and the adjacent clause triple and ring triples, where the first occurrence of x_i in $\mathcal{C}^j$ is in the 5-th clause. The triples are marked with R, T and C for ring, tree and clause, respectively.

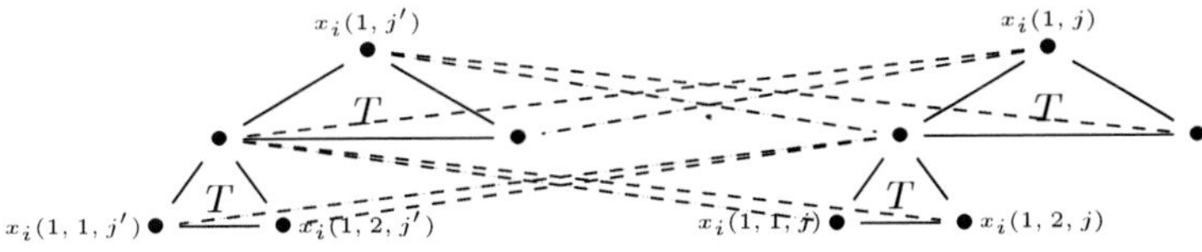

Fig. 3. An example of connecting the elements between two symmetric trees, namely $\hat{T}_{x_i(\rho,j)}$ and $\hat{T}_{x_i(\rho,j')}$ in two different sets of clauses $\mathcal{C}^j$ and $\mathcal{C}^{j'}$

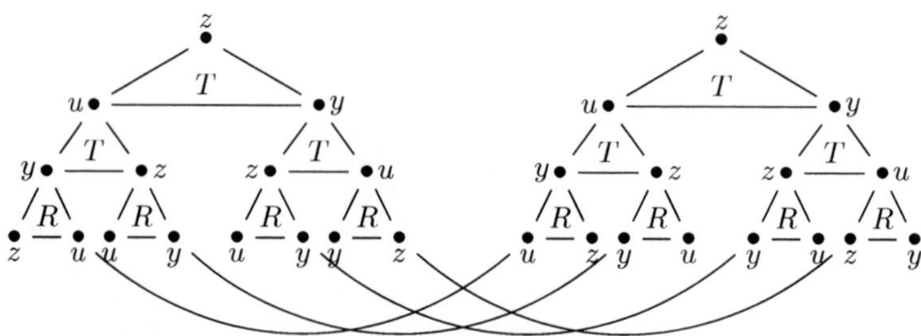

Fig. 4. An example of element labeling in a tree with two levels. Dots representing identical elements are connected with arcs.

times each, because a clause contains at most three literals and is connected with b root elements in the b sets of clauses. Thus, $f(I)$ is an instance of the Max 3-dimensional (b, B)-matching problem for $B \geq 3b$.

Definition 4. *A b-matching M in f(I) is called* **standard b-matching** *if its ring triples are matched in an even distance. We denote it by stand(M). Moreover, for every $i \in [n]$ an optimal standard b-matching contains either all ring triples corresponding to x_i or all corresponding to $\bar{x}_i$.*

Let denote by $\mathcal{R}_{x,j}$ resp. $\mathcal{T}_{x,j}$ the set of ring triples resp. the set of tree triples corresponding to the variable x and a set of clauses $\mathcal{C}^j$ in the final instance. By $\mathcal{R}_{x,j}(k)$ we denote the set $\mathcal{R}_{x,j}$ restricted to the k-th set of ring triples. The following assertions are useful to understand the subsequent analysis.

Assertion 1. In an optimal b-matching of set $\bigcup_{j\in[b]} \mathcal{R}_{x,j}$, for every $j \in [b]$ all sets of ring triples are matched in the same way.

Assertion 2. For our choice of K, any maximum b-matching of the whole problem contains an maximum b-matching of the subset $\hat{\mathcal{T}}_{x_i,j} := \mathcal{R}_{x,j} \cup \mathcal{T}_{x,j}$ (i.e., $f(I)$ without clause triples).

Assertion 3. To every maximum b-matching M we can construct a corresponding standard b-matching.

Proof. *Assertion 1:* Let M be a given 3-dimensional b-matching in $f(I)$. For every variable x_i we consider the sets $M \cap \mathcal{R}_{x_i,j}(1)$ and $M \cap \mathcal{R}_{x_i,j}(2)$. Suppose that the first set has cardinality t_1 and the second set has cardinality t_2 and w.l.o.g., $t_1 \geq t_2$. Since the two sets of ring triples have an empty cut ($\mathcal{R}_{x_i,j}(1) \cap \mathcal{R}_{x_i,j}(2) = \emptyset$), we are able to construct on $\mathcal{R}_{x_i,j}(1) \cup \mathcal{R}_{x_i,j}(2)$ a new b-matching that is larger than the original one by matching the triples in $\mathcal{R}_{x_i,j}(2)$ like $\mathcal{R}_{x_i,j}(1)$. This increases its cardinality to t_1 and also increases the b-matching.

Now we consider two adjacent pairs of sets of rings $(M \cap \mathcal{R}_{x_i,j}(1), M \cap \mathcal{R}_{x_i,j}(2))$ and $(M \cap \mathcal{R}_{x_i,j}(3), M \cap \mathcal{R}_{x_i,j}(4))$ and its connecting tree triples situated in the lowest level of $\mathcal{T}_{x_i,j}$. We suppose as above that the cardinality of the first pair and its connecting tree triples is t_1 and of the second pair and its connecting tree triples is t_2 with $t_1 \geq t_2$. In this way along the K rings, we can gain a

b-matching from the sets of rings and its connecting tree triples that is larger than the original one. □

Assertion 2: After finishing the b-matching of the set of ring triples of the instance, we continue on $\hat{\mathcal{T}}_{x_i,j}$ with the same procedure as described above. In each iteration we go one level up and continue in $\log_2 K - 1$ steps in the same manner.

Now we are done with constructing a new b-matching M' from the given b-matching M on the structure of ring triples and tree triples $\hat{\mathcal{T}}_{x_i,j}$, and we want to extend the constructed new b-matching over the entire instance. Therefore, we have to carefully include some triples from the clause triples. We suppose that we cannot include all the p clause triples belonging to the b-matching M in the new b-matching M'. This case occurs if in M' there are some roots $x_i(\rho', j)$ for $\rho' \in [c(x_i)]$ with b incident tree triples.

The easiest way to include all the p clause triples is by sacrificing some other tree or ring triples. W.l.o.g., sacrifice δ ring triples to include the p clauses triples in the b-matching.

Moreover, the cardinality of the b-matching on the structure of ring and tree triples $\hat{\mathcal{T}}_{x_i,j}$ is the cardinality of the matched ring triples plus the cardinality of the tree triples in the tree connected to the matched ring plus the cardinality of the matched tree triples in trees connected to non-matched ring triples. In the case of an odd number of levels this cardinality is smaller than

$$((bc(x_i) - \delta) \cdot K + ((bc(x_i) - \delta) \cdot \frac{K-2}{3} + (bc(x_i) + \delta) \cdot \frac{2K-1}{3}$$
$$= bc(x_i)(2K - 1) - \delta \cdot \frac{2K-1}{3}.$$

The above equality shows how many triples compared with a maximum b-matching we can lose if we sacrifice δ ring triples. If we don't sacrifice any ring triples, i.e., $\delta = 0$, we obtain a maximum b-matching of cardinality $bc(x_i) \cdot (2K - 1)$.

To extend a given maximum b-matching of $\hat{\mathcal{T}}_{x_i,j}$ over the set of clause triples in order to gain a large b-matching over the whole instance, we have to include as many of the clause triples as possible. For every clause in $\mathcal{C}^j$ at most b triples can be included. Depending on how the substructure $\hat{\mathcal{T}}_{x_i(\rho,j)}$ is matched, we include $b - r(\rho)$ with $r(\rho) \leq b$ clause triples if $r(\rho)$ triples of $\hat{\mathcal{T}}_{x_i(\rho,j)}$ are included that contain the root $x_i(\rho, j)$.

A further important property of b-matchings in the constructed instance is that every ring triple included in the b-matching is more valuable than the inclusion of some clause triples. We will verify this in the following paragraph.

We still concentrate on a variable x_i and consider $\hat{\mathcal{T}}_{x_i,j}$. Assume that we include only $bc(x_i) - \delta$ ring triples in order to include p clause triples in the b-matching. From the included p clause triples we can always choose $\lceil \frac{p}{2} \rceil$ ones which contain either $x(\rho, j)$ or $\bar{x}(\rho, j)$ for $\rho \in [c(x_i)]$. Therefore, we can always obtain a b-matching of $bc(x_i)(2K - 1) + \gamma \lceil \frac{p'}{2} \rceil$ triples for $\gamma = \frac{p}{p'}$ without sacrificing any ring triple. We show that since $K > \frac{3}{2}\gamma \lfloor \frac{p'}{2} \rfloor + \frac{1}{2}$, this is greater than any

b-matching of $bc(x_i)(2K-1) - \delta(\frac{2K-1}{3}) + p$ triples, where some ring triples are sacrificed. Then we have

$$K = 2^{\lfloor \log \frac{3}{2} bB^2 + 1 \rfloor} \geq 2^{\lfloor \log \frac{3}{2} b(c(x_i))^2 + 1 \rfloor} > 2^{\log(\frac{3}{4} b(c(x_i))^2 + \frac{1}{2})}$$
$$= \frac{3}{4} \cdot b(c(x_i))^2 + \frac{1}{2} \geq \frac{3}{2} \cdot \gamma \left\lfloor \frac{c(x_i)}{2} \right\rfloor + \frac{1}{2}.$$

Since $p' \leq c(x_i)$, it follows

$$\frac{2}{3} \cdot \left(K - \frac{1}{2}\right) > \gamma \left\lfloor \frac{p'}{2} \right\rfloor = \gamma \cdot \left(p' - \left\lceil \frac{p'}{2} \right\rceil\right)$$

and since $\delta \geq 1$,

$$bc(x_i) \cdot (2K-1) + \gamma \cdot \left\lceil \frac{p'}{2} \right\rceil > bc(x_i) \cdot (2K-1) - b\delta \cdot \left(\frac{2K-1}{3}\right) + p$$

□

Assertion 3: A standard matching of cardinality $bc(x_i)(2K-1)$ on the substructure $\hat{\mathcal{T}}_{x_i,j}$ is easy to construct, depending on the structure of the given b-matching and can be described as follows:

Let M be a b-matching in $f(I)$. For every $x_i \in X$ we consider the restriction of M on the ring triples. If $|M \cap \mathcal{R}_{x_i,j}| \geq |M \cap \mathcal{R}_{\bar{x}_i,j}|$. Then we include for every vertex $x_i(\rho, \lambda, j)$ for $(\rho, \lambda, j) \in [c(x_i)] \times [K] \times [b]$, b incident triples from the sets of ring triples $\mathcal{R}_{x_i,j}$ and delete all ring triples in $M \cap \mathcal{R}_{\bar{x}_i,j}$. Otherwise the same is applied to $\bar{x}_i$. Assume that we deal with the first case and suppose that the set of tree triples $\mathcal{T}_{x_i,j}$ (i.e., binary tree corresponding to x_i without ring and clause triples), and $\mathcal{T}_{\bar{x}_i,j}$, respectively, has L levels. Then we include from $\mathcal{T}_{x_i,j}$ all triples in level $L-1$ and from $\mathcal{T}_{\bar{x}_i,j}$ all triples in level L. This procedure is applied along both sets of trees by going two levels up in each step. As K and the occurrence for every x_i are constants ($K, c(x_i) = O(1)$), the construction of *stand*(M) can be done in deterministic polynomial time. As a property of the standard b-matching is that both ways of including triples (whether x_i or $\bar{x}_i$) correspond to the truth values of x_i, we are done. □

Let M be a b-matching in $f(I)$ and consider the set $M_{|\hat{\mathcal{T}}_{x_i,j}}$ which is the restriction of M on the set $\hat{\mathcal{T}}_{x_i,j}$ for all $i \in [n]$ (i.e., M without the clause triples). As mentioned in the above assertions, it is easy to find a maximum b-matching for $M_{|\hat{\mathcal{T}}_{x_i,j}}$ and therewith a corresponding standard matching *stand*$(M_{|\hat{\mathcal{T}}_{x_i,j}})$ such that both b-matchings have the same cardinality. As described in the appendix, part B, there are two manners to construct such a standard b-matching *stand*$(M_{|\hat{\mathcal{T}}_{x_i,j}})$. These two ways conform with the truth values of x_i. Furthermore we know that every maximum b-matching of $f(I)$ contains a maximum b-matching of the subset $\hat{\mathcal{T}}_{x_i,j}$. In order to obtain the optimal b-matching on $f(I)$, we have to include as many of the clause triples as possible and this depends on how the ring triples in $\hat{\mathcal{T}}_{x_i,j}$ were chosen in the b-matching.

Hence, we can claim that solving a Max-3-Sat-B problem is equivalent to solving the b-matching problem in $f(I)$.

Lemma 1. *The transformation* f : MAX-3-SAT-B $\longrightarrow$ MAX-3-*dimensional-*$(b,\ B)$*–matching, gives an* L*-reduction, with* $\alpha = b^2(18b^2B + 7)$ *and* $\beta = 1$.

Proof: Let I be an instance of MAX-3-SAT-B. Then the following holds

$$Opt(f(I)) \leq b^2 \sum_{i=1}^{n} (c(x_i)(2K-1)) + b^2 Opt(I) = b^2(2K-1) \sum_{i=1}^{n} c(x_i) + b^2 Opt(I)$$
$$\leq b^2 \left(3m \left(2\frac{3}{2} b^2 B + 1 \right) + Opt(I) \right) \leq b^2(18b^2B + 7) Opt(I).$$

because $Opt(I) \geq m/2$. Therefore, $\alpha = b^2(18b^2B+7)$ satisfies the first constraint of an L-reduction.

Furthermore, for every b-matching M of cardinality c_2 we can construct in polynomial time a solution of I with c_1 satisfied clauses and $Opt(f(I)) - c_2 \geq \beta^{-1}(Opt(I) - c_1)$, where $\beta^{-1} = 1$. As explained above, if a given b-matching M restricted on $\hat{\mathcal{T}}_{x_i,j}$ is not optimal, we can make it optimal on this substructure. We presume that the b-matching $M_{|\hat{\mathcal{T}}_{x_i,j}}$ is optimal. We construct a standard optimal b-matching $Stand(M_{|\hat{\mathcal{T}}_{x_i,j}})$ over $\hat{\mathcal{T}}_{x_i,j}$ corresponding to M. Thus, it follows $Opt(f(I)) - |M_{|\hat{\mathcal{T}}_{x_i,j}}| = Opt(f(I)) - |Stand(M_{|\hat{\mathcal{T}}_{x_i,j}})|$. We set the variables of I, as the b-matching $Stand(M_{|\hat{\mathcal{T}}_{x_i,j}})$ indicates. By looking at the ring triples in the b-matching, we obtain an approximate solution to I that satisfies c_1 clauses and $Opt(f(I)) - c_2 \geq (Opt(I) - c_1)$. □

From the construction and Lemma 1, Theorem 2, it follows.

Theorem 3. *There exists an* L*-Reduction from the* 3*-dimensional* (b, B)*-matching problem to the* (b, B)*-matching problem in* l*-uniform hypergraphs with* $\alpha = \beta = 1$.

Proof: It is easy to transform a 3-dimensional (b, B)-matching instance to an instance of b-matching in a 3-uniform hypergraph, where the set of vertices is the union of the three sets of the partition, and the set of hyperedges is still the same. This transformation is an L-reduction with $\alpha = 1$ and $\beta = 1$. If we compose the above two transformations, we obtain an L-reduction f' from the bounded MAX-3-SAT-B to the bounded (b, B)-matching restricted to 3-uniform hypergraphs with the same α and β as given in the first reduction. So we conclude that b-matching restricted to 3-uniform hypergraphs is also MAX-SNP-hard.

For $l > 3$, by extending all hyperedges of the constructed 3-uniform hypergraph to l elements and by introducing extra dummy elements, we obtain also an L-reduction with $\alpha = 1$ and $\beta = 1$. With the same argument, by composing this transformation and f', we obtain a new L-reduction with α and β like for f. □

The following corollary is an easy consequence of Theorem 2.

Corollary 2. *The* (b, B)*-matching problem in* l*-uniform hypergraphs is* MAX-SNP*-hard.*

Proof. By Theorem 2, Theorem 3 and Proposition 1 we get an L-reduction from MAX-3-SAT-B to the (b, B)-matching problem in l-uniform hypergraphs with $\alpha = b^2(18b^2B + 7)$ and $\beta = 1$. Hence, the (b, B)-matching problem in l-uniform hypergraphs is MAX-SNP-hard. □

4 Approximation Result

Let $\mathcal{H}$ be a hypergraph with maximum vertex degree Δ. The following result is an improvement of a result presented by Srivastav and Stangier [12] for the unweighted b-matching problem. We reach the same performance as in [12], but by a deeper analysis we improve the value of b from $b \geq 24\epsilon^{-2}\ln n$ to $b \geq 2\epsilon^{-1}\sqrt{\Delta\ln(\sqrt{n})}$.

Theorem 4. *Let $\epsilon \in (0,1)$ and $b \geq 2\epsilon^{-1}\sqrt{\Delta\ln(\sqrt{n})}$. Then an approximation algorithm exists which returns a b-matching M with $|M| \geq (1-\epsilon)Opt$ with probability at least 0.5, where Opt is the cardinality of a maximum b-matching.*

Let $b \in \mathbb{N}$ and $A = (a_{ij})_{i\in[n],\, j\in[m]} \in \{0,1\}^{n\times m}$ be the vertex-edge incidence matrix. An integer linear programming formulation of b-matching (ILP-b-matching) in $\mathcal{H}$ follows:

$$\max \sum_{j=1}^{m} x_j \,:\, Ax \leq b,\; x \in \{0,1\}^m \tag{ILP}$$

Its linear programming relaxation, denoted by LP-b-matching, is given by relaxing the integrality constraints to $x_j \in [0,1]\ \forall j \in [m]$. Let Opt^* be the value of an optimal solution to LP-b-matching and $(x_j^*)_{j\in[m]}$ be an optimal solution of LP-b-matching. The basic randomized algorithm for b-matching was introduced by Raghavan and Thomson in [8] and later by Srivastav and Stangier [12] and consists of two steps: randomized rounding and scaling down the probability of setting the variables x_i to 1 by a factor of $1 - \frac{\epsilon}{2}$ for all $j = 1, \ldots, m$.

We may assume that $|\mathcal{E}| \geq b$ and $|V| \geq 4$ (otherwise the problem can be solved by enumeration). Let us denote by M the b-matching set returned by the algorithm, i.e., $M := \{E_j \in \mathcal{E};\ X_j = 1\}$.

Claim 1: $\Pr[X \text{ is not feasible}] \leq \frac{1}{4}$.

Proof of Claim 1: For $i \in [n]$ let $(AX)_i = \sum_{j=1}^m a_{ij}X_j$ be a random variable. We have to prove: $\Pr[(AX)_i > b] \leq \frac{1}{4n}$ which implies

$$\Pr[\exists i \in [n] : (AX)_i > b] \leq \sum_{i=1}^{n} \Pr[(AX)_i > b] \leq n \cdot \frac{1}{4n} = \frac{1}{4}$$

First it is easy to check that $\mathbb{E}[(AX)_i] \leq (1 - \frac{\epsilon}{2})b$.

Let $\delta := (1 - \frac{\epsilon}{2})b - \mathbb{E}[(AX)_i] \geq 0$ and $\mathcal{Y} := (AX)_i + \delta$. It follows that $\mathbb{E}[\mathcal{Y}] = \left(1 - \frac{\epsilon}{2}\right) b$. On the other hand we have:

$$\Pr[(AX)_i > b] \leq \Pr[(AX)_i + \delta > b] \;=\; \Pr\left[\mathcal{Y} > \left(1 - \frac{\epsilon}{2}\right) b + \frac{\epsilon}{2} b\right]$$

For each $i \in [n]$ we consider the function $f_i : \{0,1\}^m \to \mathbb{N}$ with $f_i(X_1, \ldots, X_m) := Y = \sum_{j=1}^m a_{ij}X_j$. For a coordinate j we get: According to Theorem 1 we get:

$$\sum_{j=1}^{m} c_j^2 = \sum_{j=1}^{m} a_{ij}^2 = \sum_{j=1}^{m} a_{ij} = \deg(i) \leq \Delta \tag{1}$$

Let $b \geq 2\frac{\sqrt{\Delta \ln(\sqrt{4n})}}{\epsilon}$, applying Theorem 1 to the function f_i, it follows:
$\Pr\left[Y > \left(1-\frac{\epsilon}{2}\right)b + \frac{\epsilon}{2}b\right] = \Pr\left[Y > \mathbb{E}(Y) + \frac{\epsilon}{2}b\right] \leq \Pr\left[Y > \mathbb{E}(Y) + \sqrt{\Delta \ln(\sqrt{n})}\right] \leq$ $\exp\left(\frac{-2(\sqrt{\Delta \ln(\sqrt{4n})})^2}{\Delta}\right)$.
Hence, Claim 1 holds.

Claim 2: $\Pr[\sum_{j=1}^{m} X_j < (1-\epsilon) \cdot Opt^*] \leq \frac{1}{4}$.

Proof of Claim 2: (See [12].)

$$\Pr\left[Ax \leq b \wedge \sum_{j=1}^{m} X_j \geq (1-\epsilon)Opt^*\right] = 1 - \Pr\left[AX \not\leq b \vee \sum_{j=1}^{m} X_j < (1-\epsilon)Opt^*\right]$$
$$\geq 1 - \frac{1}{4} - \frac{1}{4} = \frac{1}{2}$$

As $Opt \leq Opt^*$, Theorem 4 follows. □

References

1. Devroye, L.: Non-Uniform Random Variate Generation. Springer, New York (1986)
2. El Ouali, M., Fretwurst, A., Srivastav, A.: Inapproximability of b-Matching in k-Uniform Hypergraphs. In: Katoh, N., Kumar, A. (eds.) WALCOM 2011. LNCS, vol. 6552, pp. 57–69. Springer, Heidelberg (2011)
3. Kann, V.: Maximum bounded 3-dimensional matching is Max-Snp-complete. Inf. Process. Lett. 37(1), 27–35 (1991)
4. Krysta, P.: Greedy Approximation via Duality for Packing, Combinatorial Auctions and Routing. In: Jedrzejowicz, J., Szepietowski, A. (eds.) MFCS 2005. LNCS, vol. 3618, pp. 615–627. Springer, Heidelberg (2005)
5. Lovász, L.: On the ratio of optimal integral and fractional covers. Discrete Math. 13, 383–390 (1975)
6. McDiarmid, C.: On the method of bounded differences. Surveys in Combinatorics. In: Siemons, J. (ed.) London Math. Soc., vol. 141, pp. 148–188. Cambridge University Press (1989)
7. Papadimitriou, C.H., Yannakakis, M.: Optimization, Approximation, and Complexity Classes. In: Proc. 20th STOC, pp. 229–234 (1988)
8. Raghavan, P., Thompson, C.D.: Randomized rounding: a technique for provably good algorithms and algorithmic proofs. Combinatorica 7, 365–374 (1987)
9. Srinivasan, A.: Improved approximation guarantees for packing and covering integer programs. SIAM J. Comput. 29(2), 648–670 (1999)
10. Srinivasan, A.: An extension of the Lovász Local Lemma and its applications to integer programming. SIAM J. Comput. 36, 609–634 (2006)
11. Srinivasan, A.: New approaches to covering and packing problems. In: Kosaraju, S.R. (ed.) Proc. 12th SODA, pp. 567–576 (2001)
12. Srivastav, A., Stangier, P.: Weighted fractional and integral k-matching in hypergraphs. Disc. Appl. Math. 57, 255–269 (1995)

Resource Scheduling with Supply Constraint and Linear Cost

Qiang Zhang[1], Weiwei Wu[2], and Minming Li[1]

[1] Department of Computer Science
City University of Hong Kong, Hong Kong SAR, China
qianzhang8@student.cityu.edu.hk, minming.li@cityu.edu.hk
[2] Division of Mathematical Sciences
Nanyang Technological University, Singapore
wweiwei2@gmail.com

Abstract. We consider the following resource scheduling problem to minimize the total weighted completion time. There are m resources available at each time unit, and n jobs, each requiring an arbitrary number s_i of resources. Each resource can only be assigned to one job. The objective is to find a schedule that minimizes $\sum w_i c_i$, where w_i is the weight/importance of job J_i and c_i is the time that job J_i receives all resources it requires. We show this problem is NP-hard when m is the input of the problem. We then give a simple greedy algorithm with 2-approximation ratio. Finally, we present a polynomial time algorithm with complexity $O(n^{d+1})$ to solve this problem when the number of different resources requirements that are not multiples of m is at most d.

Keywords: Algorithms, Machine scheduling, Parallel tasks, Supply allocation.

1 Introduction

Resource scheduling has a long history in operations research, economics and computer science literature. In this paper, we consider the following resource scheduling problem. There are a fixed number of resources available at each time unit and a lot of projects or jobs. Each of the projects or jobs needs a number of resources. They value their cost based on the time they receive all the resources. The fundamental problem here is how to allocate these resources to the projects or jobs such that the social cost is minimized. This problem can be easily found in the real life. For example, a steel factory has a fixed amount of steel production each week. The factory receives orders from a lot of customers. Each of them needs a specific amount of steel. The question left to the factory is how to allocate its steel to customers in order to minimize the social cost under the supply constraint.

This problem also falls in the field of machine scheduling in computer science. In the literature of machine scheduling, a machine scheduling problem is denoted by a 3-tuple $\alpha|\beta|\gamma$ introduced in [10], where α denotes the machine

G. Lin (Ed.): COCOA 2012, LNCS 7402, pp. 212–222, 2012.

environment, β denotes the additional constraints on the jobs, and γ denotes the objective function. Weighted completion time minimization is one of the well studied problems in machine scheduling. The classical scheduling problem $1||\sum w_i c_i$ can be solved by a greedy algorithm. In the multiple machines scheduling, two fundamental models were previously studied. In the single machine scheduling model, problem $Pm||\sum w_i c_i$ where the jobs are not allowed to be executed on two different machines, are well studied. While in the traditional parallel scheduling model $Pm|size_i|\sum w_i c_i$, the parallel jobs must be executed on a number $size_i$ of machines at the same time. In other words, each parallel job is assigned to a time $\times$ machine rectangle. (see Figure 1). It formulates the constraint that machines once occupied by a job cannot be released separately before the job finishes. In contrast, we study the unconstrained model where the jobs are allowed to be scheduled arbitrarily at any number of machines at different time. A job is said satisfied once it has been processed for a number of time that it requires regardless it is continuously or cumulatively scheduled. The aim of our fully parallel scheduling (FPS problem for short) problem is to minimize the weighted completion time in the unconstrained model. In FPS problem, we consider time unit to be the processing unit, which applies to all the jobs. Therefore, each job requires an integer number of time units which makes FPS problem nontrivial. Figure 1 illustrates the fundamental differences among non-migration multiple machines scheduling, migration job scheduling and our proposed FPS problem.

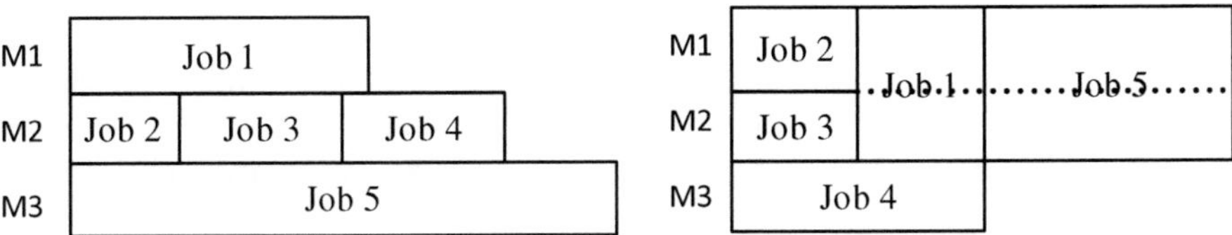

1. Single machine scheduling

2. Traditional parallel scheduling

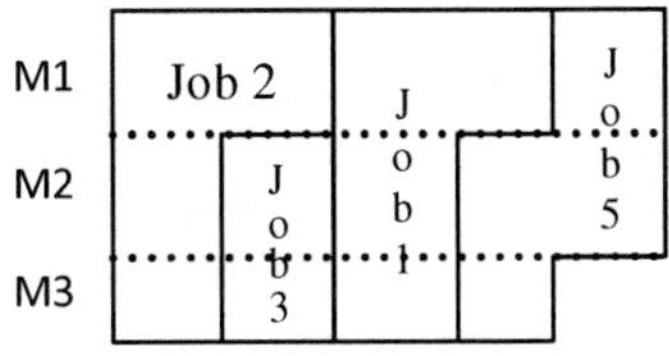

3. Our FPS problem

Fig. 1. Fundamental differences among three scheduling problems

In this paper we present the FPS problem as a machine scheduling problem. We believe it can be easily transformed to other interesting problems with the same characterization.

1.1 Our Results

It is the first time that the machine scheduling problem, without any parallel constraint, is studied. Similar to the other machine scheduling problems, we use the total weighted completion time as the objective in this paper. The objective of FPS problem is to give a schedule that minimizes the total weighted completion time. The main contributions of this paper are the analysis of complexity of this machine scheduling problem and the design of exact algorithm and approximation algorithm to solve it. We first prove that it is NP-hard to obtain the optimal schedule even when the number of machines available at each time unit is a constant. The proof is based on the reduction from 3-partition problem. We then show a greedy algorithm that computes a 2-approximation schedule. At last, we design a $O(n^{d+1})$ polynomial time algorithm to solve this problem when the number of different jobs' sizes that are not multiples of the number of machines is at most d.

1.2 Related Work

In the literature of operations research, resource scheduling has been studied in different settings. Kolisch [14] studied the parallel and serial scheduling method for the classical resource-constrained project problem (RCPSP). In RCPSP where only a limited amount of resources are available, there are two types of constraints: precedence constrains - an activity cannot start until all its predecessors have been finished, and resource constraints - each activity needs a certain type of resources during its execution period. The objective of RCPSP is to minimize the makespan subject to the above two constraints. Kolisch and Hartmann [15] gave an experimental analysis on this problem. Gonçalves *et al.* [9] presented a heuristic algorithm for the resource constrained multi-project scheduling problem (RCMPSP) that is a generalization of RCPSP. More works on resource scheduling in operations research can be found in [2] [13] [11] [12].

Resource allocations are also studied by auctions in economics. One of the most important problems in auction is social welfare maximization. Our FPS problem can be considered as a variant of social welfare maximization problem in the way that each job has a value for a bundle of processing units. Social welfare maximization problem is an NP-hard problem in general combinatorial auctions since it is well-known that computing an optimal solution for social welfare maximization requires an exponential number of queries even in the general queries model [18]. Therefore, most works in this field focus on the approximability of this problem. It is known that there is a $(1-1/e)$-approximation for the general submodular welfare maximization problem in a stronger demand oracle model [4], which was improved to $(1-1/e+\epsilon)$ in [5]. Mirrokni *et al.* [17] gave the tight lower bounds for the welfare maximization via value query model in combinatorial auctions when the valuation function is submodular, subadditive and superadditive.

In the literature of machine scheduling in computer science. There are a lot of studies on algorithms for classical scheduling problems (see [16] for more details).

The most relevant works to ours are as follows. Problem $1||\sum w_i c_i$ can be solved by the Largest Ratio First rule in [20]. When multiple machines are considered, Garey and Johnson [8] showed that problem $Pm||\sum w_i c_i$ becomes NP-hard but solvable in pseudo polynomial time. Skutella and Woeginger [19] gave a PTAS for this problem and Conway, Maxwell and Miller [3] solved the special case of this problem $Pm||\sum c_i$ when all jobs have the same weight/importance. Fishkin *et al.*[6] gave a PTAS for problem $Pm|size_j|\sum w_i c_i$. Further with the introduction of release dates, most of weighted completion time minimization problems are NP-hard. Afratie et al. [1] gave PTASs for some classes of total weighted completion time minimization problems with release dates.

2 Basic Definitions and Notation

Let m denote the number of machines available at every single time $t > 0$. There are n jobs which are indexed as $J_1, J_2, \ldots, J_n$. Job J_i requires $s_i \in \mathbb{Z}^+$ processing units and has weight/importance $w_i \in \mathbb{R}^+$. We also say s_i is the size of job J_i. Without loss of generality, we assume that $\frac{w_1}{s_1} \geq \frac{w_2}{s_2} \geq \ldots \geq \frac{w_n}{s_n}$. Otherwise, it takes $O(n \log n)$ to sort the jobs in non-increasing order of $\frac{w}{s}$. Let $J(k)$ denote the set of jobs whose sizes are k where k is not the multiples of m, that is, $J(k) = \{J_i | s_i = k, k \text{ is not a multiple of } m\}$, and $J(m)$ denotes the set of jobs whose sizes are the multiples of m, that is, $J(m) = \{J_i | s_i = zm, \text{ for some integer } z\}$. Let $d = |\{k | J(k) \neq \emptyset \text{ and } k \text{ is not a multiple of } m\}|$ be the number of different sizes that are not multiples of m.

A schedule is a tuple $(t_{11}, t_{12}, \ldots, t_{1m}, \ldots)$ where job $t_{xy} \in \{1, 2, \ldots n\}$ is scheduled in machine y at time x. A schedule is feasible if $|\{(x, y) | t_{xy} = i\}| = s_i, \forall i \in \{1, 2, \ldots n\}$. We say a job is satisfied if it receives its required processing units. For a certain schedule A, let r_i^A be the time that J_i receives its first processing unit and c_i^A be the time that J_i gets all required number of processing units. We say job J_j starts at time r_i^A and finishes at time c_i^A. We also say that job J_i is scheduled before J_j if $r_i^A < r_j^A$.

Schedule A incurs cost $w_i c_i^A$ for J_i. The objective of the fully parallel scheduling (FPS) problem is to find a feasible schedule that minimizes total weighted completion time $\sum w_i c_i$. Let OPT be the optimal solution to FPS problem. Hence, the minimum total weighted completion time is $\sum w_i c_i^{OPT}$. We say a schedule A is β-approximation if $\sum w_i c_i^A \leq \beta \sum w_i c_i^{OPT}$.

If it is clear in context, the symbol A would be omitted on the above notations.

3 Analysis of Complexity

In this section, we show that FPS problem is NP-hard by a reduction from the well-known 3-partition problem.

Definition 1 (3-partition problem). *Given a positive integer b and a set $S = \{a_1, a_2, \ldots, a_n\}$ of $n = 3k$ positive integers such that $\sum_{j=1}^{n} a_j = kb$. The problem is to determine whether S can be partitioned into k subsets $S_1, S_2, \ldots, S_k$*

such that each subset contains 3 elements and the sum of the numbers in each subset is equal.

Gary and Johnson [7] proved 3-partition problem to be NP-complete.

Definition 2 (Fully scheduled). *For a certain schedule A, we say machines at time t are fully scheduled if the total sizes of jobs that start and finish at time t equals the number of machine, i.e.* $\sum_{J_j | r_j^A = t \text{ and } c_j^A = t} s_j = m$.

Lemma 1. *Assume that there are m machines available at each time, n jobs with $w_i = s_i$ and $\sum s_i = mt$ where $t > 0$, then the total weighted completion time is at least $\frac{1+t}{2}mt$ in any feasible schedule.*

Proof. Let v_i denote the total size of jobs finished at time i in any schedule. Since there are only m machines available at each time, we have $\sum_{i=1}^{j} v_i \leq jm, 0 \leq j \leq t$. The total weighted completion time is $\sum_{i=1}^{t} iv_i$. We also know $\sum_{i=1}^{t} v_i = \sum s_i = mt$ because all the jobs could finish at time t. The minimum total weighted completion time occurs when all machines are *fully scheduled* at every time unit, i.e. $v_i = m$ for all i. Therefore, the total weighted completion time is at least $\sum_{i=1}^{t} im = \frac{1+t}{2}mt$.

Theorem 1. *FPS problem is NP-hard when the number of machines is large, i.e. m is also the input of FPS.*

Proof. We prove this theorem by a reduction from the 3-partition problem to FPS problem. For an arbitrary instance $\mathcal{I}_1$ in 3-partition problem, we create an instance $\mathcal{I}_2$ of FPS problem in the following way. For any positive number $a_i \in S$ in $\mathcal{I}_1$, we create a corresponding job J_i such that $w_i = s_i = a_i$ in $\mathcal{I}_2$. There are b machines available at every time unit in $\mathcal{I}_2$. It is clear that this construction takes polynomial time.

First, we show that any partition solution in $\mathcal{I}_1$ can be mapped to an optimal schedule in $\mathcal{I}_2$. Suppose that a partition $S_1, S_2, \ldots, S_k$ is the solution of instance $\mathcal{I}_1$, then the corresponding schedule A in $\mathcal{I}_2$ assigns the jobs in set $\{J_i | s_i \in S_t\}$ to all the b machines at time t. Since all the jobs finishing on time k has total weight b, the total weighted completion time $\sum w_i c_i^A$ is $bk\frac{k+1}{2}$. Since $w_i = s_i$ in $\mathcal{I}_2$, by Lemma 1, we know that any feasible schedule A' in $\mathcal{I}_2$ would have total weighted completion time at least $bk\frac{k+1}{2}$. Therefore, schedule A is the optimal solution in $\mathcal{I}_2$.

Second, we show that the optimal schedule in $\mathcal{I}_2$ suggests the answer to $\mathcal{I}_1$. Suppose that A is the optimal schedule in $\mathcal{I}_2$, since $w_i = s_i$ in $\mathcal{I}_2$ and $\sum s_i = kb$, any feasible schedule A' in $\mathcal{I}_2$ would have total weighted completion time at least $bk\frac{k+1}{2}$. The case $\sum w_i c_i^{A'} = bk\frac{k+1}{2}$ occurs if and only if the machines are *fully scheduled* at every time unit. Therefore, the answer to 3-partition problem $\mathcal{I}_1$ is to check whether the total weighted completion time $\sum w_i c_i^A$ equals $bk\frac{k+1}{2}$. If it is the case, the partition solution for $\mathcal{I}_1$ can be constructed by assigning the numbers in set $\{s_i | c_i^A = t\}$ to S_t in $\mathcal{I}_1$.

4 Characterizations of the Optimal Schedule

In this section, we make some fundamental observations on the structure of this problem, which will guide us to design and analyze the algorithms in the following sections.

Lemma 2. *There exists an optimal schedule where* $c_i^A \leq r_j^A$ *if* $r_i^A \leq r_j^A$.

Proof. We prove this lemma by construction. Suppose A is an optimal schedule and there are two jobs J_i and J_j in A where J_i is scheduled before J_j and J_j receives processing unit when J_i has not finished, that is, $r_i^A \leq r_j^A$ and $c_i^A > r_j^A$. Then another schedule $A^{'}$ can be constructed as follows: keep processing unit assignments unchanged except J_i and J_j in A, and assign the first s_i processing units assigned to job J_i and J_j to job J_i and remaining s_j processing units to J_j. Note that in $A^{'}$ job J_i will not finish later than in A, that is, $c_i^{A^{'}} \leq c_i^A$, and J_j finishes at the same time as in A. Therefore, $\sum w_i c_i^{A^{'}} \leq \sum w_i c_i^A$. Then the lemma directly follows.

Lemma 2 allows us to only concentrate on the schedules in which all jobs receive the processing units in a non-preemptive way when we design and analyze algorithms for FPS problem since the optimal cost can be computed once the processing order of the jobs are fixed. From now on, we can simply describe a schedule in terms of the processing order of the jobs, i.e. schedule A is an n-tuple $(t_1, \ldots, t_n)$ that specifies the processing order of the jobs where job t_i is the i-th job scheduled in schedule A.

Lemma 3. *In the optimal solution OPT, for any two jobs* $J_i, J_j \in J(m)$, *if* $\frac{w_i}{s_i} > \frac{w_j}{s_j}$, *then* $c_i^{OPT} \leq c_j^{OPT}$.

Proof. We prove this lemma by contradiction. Suppose that OPT is the optimal schedule which minimizes $\sum w_i c_i$ and $c_i^{OPT} > c_j^{OPT}$. By Lemma 2, it is sufficient to only consider the case where $c_j^{OPT} \leq r_i^{OPT}$. Let B be the set of jobs scheduled between c_j^{OPT} and r_i^{OPT}. Since OPT is the optimal solution, we know that advancing B before J_j will not reduce the total weighted completion time. Since s_j is multiples of m, it implies that $\frac{w_j}{s_j/m} \geq \frac{\sum_{b \in B} w_b}{t_1}$ where s_j/m is the time that jobs in the set B advance and t_1 is the time that J_j delays if advancing B before J_j. By the similar argument, advancing J_i before B will not reduce the total weighted completion time, we get $\frac{w_i}{s_i/m} \leq \frac{\sum_{b \in B} w_b}{t_2}$ where s_i/m is the time that jobs in the set B delay and t_2 is the time that J_i advances if advancing J_i before B. Since both s_i and s_j are multiples of m, it is easy to verify that $t_1 = t_2$. Therefore, we have $\frac{w_j}{s_j/m} \geq \frac{\sum_{b \in B} w_b}{t_2} \geq \frac{w_i}{s_i/m}$ which contradicts the assumption $\frac{w_i}{s_i} > \frac{w_j}{s_j}$.

Lemma 4. *In the optimal solution OPT, for any two jobs* J_i, J_j *and* k, *if* $J_i, J_j \in J(k)$ *and* $w_i > w_j$ ($\frac{w_i}{s_i} > \frac{w_j}{s_j}$ *equivalently since* $s_i = s_j$), *then* $c_i^{OPT} \leq c_j^{OPT}$.

Proof. We prove this lemma by contradiction. Suppose OPT is the optimal schedule to minimize $\sum_i w_i c_i$ and job J_i finishes later than J_j in OPT, i.e. $c_i^{OPT} > c_j^{OPT}$. The total weighted completion time can be reduced $(w_i - w_j)(c_i^{OPT} - c_j^{OPT})$ by swapping the processing unit assignments for J_i and J_j in schedule OPT. It contradicts the fact that OPT is the optimal schedule.

We will use Lemma 3 and 4 in Section 6 to design an efficient algorithm to solve FPS problem when d is a constant.

5 2-Approximation Algorithm

In this section we show that the classical Largest Ratio First (LRF) schedule is 2-approximation for total weighted completion time $\sum w_i c_i$ in FPS problem when each job has an arbitrary size and weight/importance.

Definition 3 (Largest Ratio First). *A Largest Ratio First(LRF) schedule is a schedule that assigns time units to the jobs in non-increasing order of* $\frac{w}{s}$.

Theorem 2. *LRF schedule gives 2-approximation for total weighted completion time when each job has an arbitrary size and weight, i.e.* $\sum w_i c_i^{LRF} \leq 2 \sum w_i c_i^{OPT}$.

We prove Theorem 2 by examining the lower bound of the optimal schedule and the upper bound on total weighted completion time $\sum_i w_i c_i$ of LRF schedule. Theorem 2 directly follows by combining the following two lemmas.

Lemma 5. $\sum w_i c_i^{OPT}$ *is at least* $\sum w_i \max(\frac{\sum_{j \leq i} s_j}{m}, 1)$.

Proof. Let A^* denote the optimal schedule in FPS problem. We first prove a lower bound that the minimum total weighted completion time $\sum_i w_i c_i^{A^*}$ is at least the optimal cost in the following minimization problem $\mathcal{P}$. The input consists of the same n jobs as those in our problem and m machines are available at every time unit. Each machine can only be assigned to one job at a time. Instead of minimizing $\sum w_i c_i$, the objective is to find a schedule of jobs S that minimizes $\sum w_i \frac{\sum_{J_j \in F_S(i)} s_j}{m}$, where $F_S(i)$ is the set of jobs scheduled before J_i including J_i itself in sequence S.

It is easy to verify that minimization problem $\mathcal{P}$ is equivalent to problem $1||\sum w_i c_i$ which is the weighted completion time minimization problem in single machine scheduling. Therefore, LRF schedule gives the optimal objective value $\sum w_i \frac{\sum_{j \leq i} s_j}{m}$ for problem $\mathcal{P}$. Since problem $\mathcal{P}$ and the FPS problem share the same set of feasible sequences and c_i^S in FPS problem is greater than or equal to the $\frac{\sum_{J_j \in F_S(i)} s_j}{m}$ in $\mathcal{P}$ for every feasible schedule S, the minimum total weighted completion time $\sum w_i c_i^{A^*}$ is lower bounded by $\sum w_j \frac{\sum_{i \leq j} s_i}{m}$.

Another lower bound $\sum_i w_i$ directly follows since the minimum c_j is at least 1 for any feasible schedule. Combining the two lower bounds completes the proof.

Lemma 6. *$\sum w_i c_i^{LRF}$ is at most $\sum w_i \left(\frac{\sum_{j \leq i} s_j}{m} + \frac{m-1}{m} \right)$.*

Proof. Since the jobs are scheduled by their orders in *LRF* schedule and there are m machines available at every time unit, we have the following

$$\sum w_i c_i^{LRF} = \sum w_i \lceil \frac{\sum_{j \leq i} s_j}{m} \rceil$$

The lemma follows by bounding $\lceil \frac{\sum_{j \leq i} s_i}{m} \rceil$ by $\frac{\sum_{j \leq i} s_j}{m} + \frac{m-1}{m}$.

The following example shows that the analysis for *LRF* schedule is tight. Suppose that there are two jobs and m machines available at each time. Let $w_1 = 1 + \epsilon, s_1 = 1$ and $w_2 = m, s_2 = m$. The total weighted completion time from *LRF* schedule is $(1 + \epsilon) + 2m$ and the minimum total weighted completion time is $m + 2(1 + \epsilon)$. Therefore, *LRF* schedule is $(2 - \epsilon^{'})$ approximation.

6 Jobs with Limited Sizes

We showed that FPS problem is NP-hard in general. In this section we present a polynomial time algorithm with complexity $O(n^{d+1})$ to find the optimal solution in our scheduling problem when d is a constant. Recall that d is the number of different sizes that are not multiples of m. We say that the jobs in the same set $J(k)$ form a group. Therefore, there are $d + 1$ groups. We assume that the jobs in the same group are listed in non-increasing order of $\frac{w}{s}$. Let g_i^j be the j-th job in group i and $|g_i|$ be the size of group i. By Lemma 3 and 4, we know that, for any i, the jobs $\{J_j \in g_i\}$ must be placed in non-increasing order of $\frac{w}{s}$. An optimal schedule can be computed once the processing order of jobs are fixed. Moreover, once we can identify the group where the k-th scheduled job comes from, the optimal schedule can also be fixed. Therefore, by associating the k-th scheduled job to one of the d groups, there is an exponential algorithm with complexity $O((d + 1)^n)$ to compute the optimal solution. For constant value d, the merit of this section is to show that the optimal solution can be computed with a significantly lower complexity $O(n^{d+1})$.

The intuition behind our algorithm for this restricted problem is to construct the optimal schedule step by step by applying Lemma 3, 4 and the non-preemptive property of the optimal solution. Our algorithm assigns the processing order one by one and always favors the schedule with smaller total cost. We present and analyze the computation complexity of our algorithm in terms of a dynamic programming in Algorithm 1. Let $c(a_1, \ldots, a_{d+1})$ denote the minimum weighted completion time where a_i jobs have been scheduled in group i and the number of scheduled jobs in each group cannot exceed the number of jobs in this group, i.e. for all i, $a_i \leq |g_i|$. The dynamic programming is initialized at $c(0, \ldots, 0) = 0$. Recall that a schedule is determined once the execution order of groups is determined by Lemma 4, the corresponding schedule to the minimum weighted completion time can be easily constructed. Moreover, by Lemma 4, it is easy to see that $c(a_1, \ldots, a_{d+1}) = \min_i c(a_1, \ldots, a_i - 1, \ldots, a_{d+1}) + w_{g_i^{a_i}} c_{g_i^{a_i}}$

Algorithm 1:

Input: A set of jobs $J_1, J_2, \ldots, J_n$ are divided into $d+1$ groups based on their sizes

Output: The minimum total weighted completion time $\sum w_i c_i$

Initialization:

// initialize the base case

$c(0, \ldots, 0) = 0$;

$k_i = |g_i|, \forall i$;

Main:

// compute the minimum total weighted completion time by procedure c

$\min \sum w_i c_i = c(k_1, \ldots, k_{d+1})$;

Procedure $c(a_1, \ldots, a_{d+1})$**:**

// H are the indexes of the groups in $c(\cdot)$ which are not zero

$H = \{i | a_i \geq 1\}$;

// append job $g_i^{a_i}$ into the previous schedule $c(a_1, \ldots, a_i - 1, \ldots, a_{d+1})$ and compute the resulting weighted completion time.

// return the one with minimum total weighted completion time

return $\min_{i \in H} c(a_1, \ldots, a_i - 1, \ldots, a_{d+1}) + w_{g_i^{a_i}} c_{g_i^{a_i}}$;

where $c_{g_i^{a_i}}$ is the completion time of the a_i^{th} job in group i by appending it after the previous schedules. By recursively computing $c(\cdot)$, the minimum weighted completion time directly follows.

Lemma 7. *Algorithm* 1 *computes a schedule that minimizes the cost* $\sum w_i c_i$.

Proof. Recall that Lemma 3 and 4 are the necessary conditions for any optimal schedule that minimizes total weighted completion time $\sum w_i c_i$. Algorithm 1 examines all the schedules satisfying Lemma 3 and 4. Therefore, it gives an optimal schedule.

Lemma 8. *Algorithm* 1 *terminates in* $O(n^{d+1})$ *time.*

Proof. We analyze the running time of our algorithm by counting the size of dynamic programming in Algorithm 1. For a particular number $l \leq n$, there are at most $\binom{d}{l+d}$ schedules such at $\sum a_i = l$. Each schedule needs $d+1$ computations to get the minimum. Therefore, there are at most $(d+1)\binom{d}{l+d}$ computations. Since l can only take n different values, the total required computations are bounded by $O(n^{d+1})$. This completes the proof.

7 Conclusions and Discussions

In this paper we consider the weighted completion time as the objective in the fully parallel scheduling problem. We prove that total weighted completion time minimization is NP-hard in general and show that a greedy algorithm is 2-approximation. We also give a polynomial time algorithm to compute the optimal solution when the sizes of the jobs are restricted. It would be interesting to explore other classes of this problem where algorithms can be designed to compute the optimal solution. Besides the weighted completion time, other objectives could be considered such as minimization of maximum lateness when the deadlines are introduced. Furthermore, with the introduction of release time of each job in the online environment, the problem could be more challenging.

References

1. Afrati, F., Bampis, E., Chekuri, C., Karger, D., Kenyon, C., Khanna, S., Milis, I., Queyranne, M., Skutella, M., Stein, C., Sviridenkom, M.: Approximation schemes for minimizing average weighted completion time with release dates. In: Proceedings of the 40th Annual IEEE Symposium on Foundations of Computer Science, pp. 32–43 (1999)
2. Brucker, P., Drexl, A., Mohring, R., Neumann, K., Pesch, E.: Resource-constrained project scheduling: Notation, classification, models, and methods. European Journal of Operational Research 112(1), 3–41 (1999)
3. Conway, R., Maxwell, W., Miller, L.: Theory of scheduling. Dover Publications (2003)
4. Dobzinski, S., Schapira, M.: An improved approximation algorithm for combinatorial auctions with submodular bidders. In: Proceedings of the 17th Annual ACM-SIAM Symposium on Discrete Algorithm, pp. 1064–1073 (2006)
5. Feige, U., Vondrak, J.: Approximation algorithms for allocation problems: Improving the factor of 1-1/e. In: Proceedings of the 47th Annual IEEE Symposium on Foundations of Computer Science, pp. 667–676 (2006)
6. Fishkin, A., Jansen, K., Porkolab, L.: On minimizing average weighted completion time: A ptas for scheduling general multiprocessor tasks. In: Proceedings of the 13th International Symposium on Fundamentals of Computation Theory, pp. 495–507 (2001)
7. Garey, M.R., Johnson, D.S.: Complexity results for multiprocessor scheduling under resource constraints. SIAM Journal on Computing 4, 397 (1975)
8. Garey, M.R., Johnson, D.S.: Computers and intractability: A Guide to the Theory of NP-Completeness. Freeman, New York (1979)
9. Gonçalves, J., Mendes, J., Resende, M.: A genetic algorithm for the resource constrained multi-project scheduling problem. European Journal of Operational Research 189(3), 1171–1190 (2008)
10. Graham, R., Lawler, E., Lenstra, J., Kan, A.: Optimization and approximation in deterministic sequencing and scheduling: a survey. Annals of Discrete Mathematics 5(2), 287–326 (1979)
11. Graves, S.C.: A review of production scheduling. Operations Research, 646–675 (1981)
12. Hartmann, S., Briskorn, D.: A survey of variants and extensions of the resource-constrained project scheduling problem. European Journal of Operational Research 207(1), 1–14 (2010)
13. Herroelen, W., Leus, R.: Project scheduling under uncertainty: Survey and research potentials. European Journal of Operational Research 165(2), 289–306 (2005)

14. Kolisch, R.: Serial and parallel resource-constrained project scheduling methods revisited: Theory and computation. European Journal of Operational Research 90(2), 320–333 (1996)
15. Kolisch, R., Hartmann, S.: Experimental investigation of heuristics for resource-constrained project scheduling: An update. European Journal of Operational Research 174(1), 23–37 (2006)
16. Leung, J.Y.T.: Handbook of Scheduling: Algorithms, Models, and Performance Analysis. CRC Press (2004)
17. Mirrokni, V., Schapira, M., Vondrák, J.: Tight information-theoretic lower bounds for welfare maximization in combinatorial auctions. In: Proceedings of the 9th ACM Conference on Electronic Commerce, pp. 70–77 (2008)
18. Nisan, N., Segal, I.: The communication requirements of efficient allocations and supporting prices. Journal of Economic Theory 129(1), 192–224 (2006)
19. Skutella, M., Woeginger, G.: A ptas for minimizing the weighted sum of job completion times on parallel machines. In: Proceedings of the 31th Annual ACM Symposium on Theory of Computing, pp. 400–407. ACM (1999)
20. Smith, W.E.: Various optimizers for single-stage production. Naval Research Logistics Quarterly 3(1-2), 59–66 (1956)

On Certain Geometric Properties of the Yao-Yao Graphs

Iyad A. Kanj[1] and Ge Xia[2]

[1] School of Computing, DePaul University, 243 S. Wabash Avenue, Chicago, IL 60604-2301
ikanj@cs.depaul.edu

[2] Department of Computer Science, Acopian Engineering Center, Lafayette College, Easton PA 18042
gexia@cs.lafayette.edu

Abstract. We show that, for any constant $\rho > 1$, there exists an integer constant k such that the Yao-Yao graph with parameter k defined on a civilized unit disk graph is a geometric spanner of stretch factor ρ. This improves the results of Wang and Li in several aspects, as described in the paper. We also show that the Yao-Yao graph with parameter $k = 4$ defined on the complete Euclidean graph is not a spanner and is not plane. This partially answers an open problem posed by Demaine, Mitchell and O'Rourke about the spanner properties of Yao-Yao graphs.

Keywords: Yao graphs, Yao-Yao graphs, unit disk graphs, spanners.

1 Introduction

Let $\mathcal{E}$ be the complete Euclidean graph on a set of points S in the plane, and let G be a spanning subgraph of $\mathcal{E}$. Fix an ordering $\prec$ on the edges in G such that shorter edges appear before longer edges and ties are broken arbitrarily. Given an integer parameter $k > 0$, the *Yao graph* [21] with parameter k defined on G, denoted $\overrightarrow{Y_k}(G)$, is constructed as follows. For each point p in G, partition the space into k cones of equal measure whose apex is p (the orientation of the cones is fixed for all points), thus creating k closed cones of angle $2\pi/k$ each. In each cone, choose the smallest edge according to the ordering $\prec$ in G incident to p (if any) and add it to $\overrightarrow{Y_k}(G)$ as a directed edge outgoing from p. The undirected Yao graph with parameter k defined on G, denoted $Y_k(G)$, is the underlying undirected graph of $\overrightarrow{Y_k}(G)$.

The *Yao-Yao graph* with parameter $k > 0$, denoted $\overrightarrow{YY_k}(G)$, is constructed in two stages. The first stage proceeds as in the construction of $\overrightarrow{Y_k}(G)$: each point p in G partitions the space into k cones of equal measure whose apex is p and chooses the smallest edge according to the ordering $\prec$ in G incident to p (if any) in each cone as an outgoing edge from p. In the second stage, for each point $p \in G$, and for each cone defined by p in the construction of $\overrightarrow{Y_k}(G)$, point p keeps *only* the smallest incoming edge in the cone according to the ordering $\prec$.

G. Lin (Ed.): COCOA 2012, LNCS 7402, pp. 223–233, 2012.

The directed edges kept by the points in G in the second stage constitute $\overrightarrow{YY_k}(G)$. $YY_k(G)$ denotes the underlying directed graph of $\overrightarrow{YY_k}(G)$. Clearly, $\overrightarrow{YY_k}(G)$ is a subgraph of $\overrightarrow{Y_k}(G)$, and $YY_k(G)$ is a subgraph of $Y_k(G)$. For simplicity, we write Y_k for $YY_k(\mathcal{E})$ and YY_k for $YY_k(\mathcal{E})$.

The Yao graphs have been extensively studied, and many of their geometric properties have been discovered. In particular, it is known that Y_2 and Y_3 are not geometric spanners [17] (of $\mathcal{E}$), Y_4 is a spanner with stretch factor $8\sqrt{2}(29+23\sqrt{2})$ [2], Y_6 is a spanner with stretch factor 17.7 [6], and that for $k \geq 7$, Y_k is a spanner with stretch factor $(1+\sqrt{2-2\cos(2\pi/k)})/(2\cos(2\pi/k)-1)$ [3]. The question of whether or not Y_5 is a spanner remains open.

Whereas the Yao graphs were extensively studied, little work has been done on the Yao-Yao graphs. One advantage of the Yao-Yao graphs over the Yao graphs is that their maximum degree is bounded: Whereas $Y_k(G)$ can have unbounded degree, the maximum degree of $YY_k(G)$ is at most $2k$. Demaine, Mitchell and O'Rourke [10] asked whether the Yao-Yao graphs are geometric spanners of $\mathcal{E}$.

Wang and Li [19] studied the Yao-Yao graphs. Their study was motivated by the problem of computing bounded-degree spanners. The problem of constructing bounded-degree spanners has been extensively studied within computational geometry (for example, see [1, 4, 7–9, 11, 13, 15, 21], and the following book on spanners [18]). More recently, wireless network researchers have approached the problem as well. Emerging wireless distributed system technologies, such as wireless ad hoc and sensor networks, are often modeled as a *unit disk graph* (UDG) in the Euclidean plane: the points of the UDG correspond to the mobile wireless devices, and its edges connect pairs of points whose corresponding devices are in each other's transmission range equal to one unit. Spanners of UDGs are fundamental to wireless systems because they represent topologies that can be used for efficient unicasting, multicasting, *and/or* broadcasting (see [4, 5, 11, 12, 14, 16, 20], to name a few). For these applications, spanners are typically required to have bounded degree; this requirement is motivated by interference issues and the physical limitations of wireless devices (e.g., [4, 5, 11, 12, 14, 20]).

Wang and Li [19] investigated some geometric properties of the Yao-Yao graphs on unit disk graphs. They proved that if the minimum distance between any two points in a UDG $\mathcal{U}$ is lower-bounded by a positive constant (referred to as a *civilized* UDG), then there exists a constant k such that the Yao-Yao graph of $\mathcal{U}$ with parameter k is a *power spanner* with stretch factor 2. Although Wang and Li [19] claimed a stretch factor of 2, their proof implies a stretch factor of $1+\epsilon$ for any $\epsilon > 0$. A power spanner is a spanner in which the weight of any edge ab is defined to be $||ab||^{\beta}$, where $||ab||$ is the Euclidean distance between a and b and $\beta \in [2,5]$, whereas in a *geometric spanner* the edge-weight is defined to be $||ab||$. Observe that any geometric spanner is also a power spanner.

In this paper we improve Wang and Li's results [19]. We prove that, for any $\rho > 1$, there exists a constant k such that the Yao-Yao graph with parameter k defined on a *civilized* unit disk graph is a geometric spanner with stretch factor ρ. Whereas the general outline of our proof resembles that of Wang and Li [19],

Wang and Li's proof works only for power weights because their proof depends on some properties of power spanners that are not shared by geometric spanners. For example, at several places in their proof they used the fact that in a triangle $\triangle abc$, if the angle $\angle abc$ is not acute, then the power weights satisfy

$$||ab||^\beta + ||bc||^\beta \leq ||ac||^\beta,$$

for $\beta \in [2, 5]$, which is not true for geometric weights (i.e., $\beta = 1$). We need to overcome nontrivial technical issues to prove the results for geometric spanners. Since any geometric spanner is also a power spanner, this improves Wang and Li's results in terms of the spanner type (geometric spanner vs. power spanner). We note that both Wang and Li's results and ours apply to the directed Yao-Yao graphs, and hence obviously apply as well to their undirected counterparts. We also show in the current paper that YY_4 is not a spanner and is not plane. This partially answers (for $k = 4$) an open problem posed by Demaine, Mitchell and O'Rourke [10].

The paper is organized as follows. In Section 2, we introduce the notations and terminology used throughout the paper. In Section 3, we prove that the Yao-Yao graph is a spanner for civilized unit disk graphs. In Section 4 we show that YY_4 is not a spanner, and in Section 5 we show that YY_4 is not plane. We conclude the paper in Section 6.

2 Preliminaries

Given a set of points S in the two-dimensional Euclidean plane, the complete Euclidean graph $\mathcal{E}$ on S is defined to be the complete graph whose point-set is S. Each edge ab connecting points a and b is assumed to be embedded in the plane as the straight line segment ab; we define its *weight* to be the Euclidean distance $||ab||$. We define the *unit disk graph* (UDG) $\mathcal{U}$ whose point set is S to be the subgraph of $\mathcal{E}$ consisting of all edges ab with $||ab|| \leq 1$. We assume in this paper that the UDG U is connected. The unit disk graph $\mathcal{U}$ is said to be *civilized* if the minimum distance between any two points in $\mathcal{S}$ is at least λ, for some pre-specified constant $\lambda > 0$.

Let G be a subgraph of $\mathcal{E}$. The weight of a simple path $a = m_0, m_1, \ldots, m_r = b$ between two points a, b in G is $\sum_{j=0}^{r-1} ||m_j m_{j+1}||$. For two points a, b in G, we denote by $d_G(a, b)$ the weight of a shortest path from a to b in G. A spanning subgraph H of G is said to be a *geometric spanner* of G if there is a constant ρ such that, for every two points $a, b \in G$, $d_H(a, b) \leq \rho \cdot d_G(a, b)$. The constant ρ is called the *stretch factor* of H (with respect to G). The following is a well-known and obvious fact:

Fact 1. A subgraph H of graph G has stretch factor ρ with respect to G if and only if for every edge $xy \in G$, $d_H(x, y) \leq \rho ||xy||$.

A graph embedded in the Euclidean plane is said to be a *plane graph* if its edges do not cross.

3 YY_k Is a Spanner for Civilized UDG

Let $\mathcal{U}$ be a civilized UDG with parameter $\lambda > 0$ defined on a point-set S. Let $\overrightarrow{Y}_k(\mathcal{U})$ be the directed Yao graph with parameter k defined on $\mathcal{U}$, and let $\overrightarrow{YY}_k(\mathcal{U})$ be the directed Yao-Yao graph with parameter k defined on $\mathcal{U}$.

Theorem 1. *For any constant $\rho > 1$, there is a positive integer constant k_0 such that, for all $k \geq k_0$, the stretch factor of $\overrightarrow{YY}_k(\mathcal{U})$ is at most ρ, where k_0 is dependent only on ρ and λ.*

Proof. By Fact 1, it will suffice to prove that there is a constant $c \in (\frac{1}{\rho}, 1)$ (to be determined later) such that the following two statements are true for all points u, v in $\mathcal{U}$.

1. If $\overrightarrow{uv}$ is not in the graph $\overrightarrow{Y}_k(\mathcal{U})$, then there is a directed path from u to v in $\overrightarrow{YY}_k(\mathcal{U})$ whose length is at most $\rho||uv||$.
2. If $\overrightarrow{uv}$ is in the graph $\overrightarrow{Y}_k(\mathcal{U})$, then there is a directed path from u to v in $\overrightarrow{YY}_k(\mathcal{U})$ whose length is at most $c\rho||uv||$.

We proceed by induction on the ordering $\prec$ of the edges uv in U.

For the base case when $||uv||$ is smallest according to $\prec$, it is easy to verify that $\overrightarrow{uv}$ is an edge in $\overrightarrow{YY}_k(\mathcal{U})$ and the statement is true. Assume that the statement is true for all edges $ts \in U$ where $ts \prec uv$. We distinguish the following cases.

1. $\overrightarrow{uv}$ is not in the graph $\overrightarrow{Y}_k(\mathcal{U})$. In this case there is an edge uw in $\mathcal{U}$ such that uv and uw belong to the same cone for point u, $uw \prec uv$, and $\overrightarrow{uw}$ is in the graph $\overrightarrow{Y}_k(\mathcal{U})$. By the inductive hypothesis, there is a directed path from u to w in $\overrightarrow{YY}_k(\mathcal{U})$ whose length is at most $c\rho||uw||$. For k large enough, we have $||wv|| < ||uv||$ because $\angle vuw$ is small and $||uw|| \leq ||uv||$. This means that wv is an edge in U and $wv \prec uv$. By the inductive hypothesis, there is a directed path from w to v in $\overrightarrow{YY}_k(\mathcal{U})$ whose length is at most $\rho||wv||$. Therefore, there is a directed path from u to v in $\overrightarrow{YY}_k(\mathcal{U})$ whose length is at most $c\rho||uw|| + \rho||wv||$. Let $x = \frac{||uw||}{||uv||}$ and $\theta = \angle vuw$. Note that $0 < x < 1$ and $\theta \leq \frac{2\pi}{k}$. Applying trigonometric identities, we have $||wv|| = \sqrt{1 + x^2 - 2x\cos\theta} \cdot ||uv||$. In order to prove the inductive statement, it suffices to show that:

$$\frac{c\rho||uw|| + \rho||wv||}{\rho||uv||} \leq 1$$
$$\Leftrightarrow cx + \sqrt{1 + x^2 - 2x\cos\theta} \leq 1$$
$$\Leftrightarrow \sqrt{1 + x^2 - 2x\cos\theta} \leq 1 - cx$$
$$\Leftrightarrow 1 + x^2 - 2x\cos\theta \leq (1 - cx)^2$$
$$\Leftrightarrow x^2 - 2x\cos\theta \leq (cx)^2 - 2cx$$
$$\Leftrightarrow x - 2\cos\theta \leq c^2x - 2c$$
$$\Leftrightarrow (1 - c^2)x + 2c \leq 2\cos\theta. \quad (1)$$

Since $\frac{1}{\rho} < c < 1$ and $0 < x < 1$, we have

$$(1 - c^2)x < 1 - c^2. \tag{2}$$

From (1) and (2), in order to prove that the inductive statement holds for this case, it suffices to show that:

$$1 - c^2 + 2c = 2 - (1 - c)^2 \leq 2\cos\theta. \tag{3}$$

For any fixed value of $\frac{1}{\rho} < c < 1$, there is a k_0 such that

$$2\cos\frac{2\pi}{k_0} \geq 2 - (1 - c)^2.$$

For all $k \geq k_0$, we have

$$2\cos\theta \geq 2\cos\frac{2\pi}{k} \geq 2\cos\frac{2\pi}{k_0} \geq 2 - (1 - c)^2,$$

i.e., (3) is true and the claim holds.

2. $\overrightarrow{uv}$ is in the graph $\overrightarrow{Y}_k(\mathcal{U})$. If $\overrightarrow{uv}$ is also in $\overrightarrow{YY}_k(\mathcal{U})$ then we are done. If $\overrightarrow{uv}$ is not in $\overrightarrow{YY}_k(\mathcal{U})$, then there is an edge wv in $\mathcal{U}$ such that uv and wv are in the same cone for point v, $wv \prec uv$, and $\overrightarrow{wv}$ is in the graph $\overrightarrow{YY}_k(\mathcal{U})$. For k large enough, we have $||uw|| < ||uv||$ because $\angle wvu$ is small and $||wv|| \leq ||uv||$. By the inductive hypothesis, there is a directed path from u to w in $\overrightarrow{YY}_k(\mathcal{U})$ whose length is at most $\rho||uw||$. Therefore, there is a directed path from u to v in $\overrightarrow{YY}_k(\mathcal{U})$ whose length is at most $\rho||uw|| + ||wv||$. Let $x = \frac{||wv||}{||uv||}$ and $\theta = \angle uvw$. Note that $0 < x < 1$ and $\theta \leq \frac{2\pi}{k}$. Applying trigonometric identities, we have $||uw|| = \sqrt{1 + x^2 - 2x\cos\theta} \cdot ||uv||$. In order to prove the inductive statement, it suffices to show that

$$\begin{aligned}
&\frac{\rho||uw|| + ||wv||}{c\rho||uv||} \leq 1 \\
\Leftrightarrow\ &\frac{1}{c}\sqrt{1 + x^2 - 2x\cos\theta} + \frac{x}{c\rho} \leq 1 \\
\Leftrightarrow\ &\sqrt{1 + x^2 - 2x\cos\theta} \leq c - \frac{x}{\rho} \\
\Leftrightarrow\ &1 + x^2 - 2x\cos\theta \leq (c - \frac{x}{\rho})^2 \\
\Leftrightarrow\ &1 + x^2 - (c - \frac{x}{\rho})^2 \leq 2x\cos\theta \\
\Leftrightarrow\ &\frac{1 + x^2 - (c - \frac{x}{\rho})^2}{x} \leq 2\cos\theta \\
\Leftrightarrow\ &\frac{1 + x^2 - c^2 - \frac{x^2}{\rho^2} + \frac{2cx}{\rho}}{x} \leq 2\cos\theta \\
\Leftrightarrow\ &\frac{1 - c^2}{x} + (1 - \frac{1}{\rho^2})x + \frac{2c}{\rho} \leq 2\cos\theta. \qquad (4)
\end{aligned}$$

Fixing the values of c and ρ, it can be easily verified that the maximum value of $\frac{1-c^2}{x} + (1 - \frac{1}{\rho^2})x$ in the interval $[\lambda, 1]$ is achieved when x is minimized ($x = \lambda$) or maximized ($x = 1$).

(a) If $x = 1$, then plugging it into the LHS of (4) we get

$$\frac{1-c^2}{x} + (1 - \frac{1}{\rho^2})x + \frac{2c}{\rho} = 1 - c^2 + 1 - \frac{1}{\rho^2} + \frac{2c}{\rho} = 2 - (c - \frac{1}{\rho})^2.$$

For any fixed values of $\rho > 1$ and $\frac{1}{\rho} < c < 1$, there is a k_0 such that

$$2\cos\frac{2\pi}{k_0} \geq 2 - (c - \frac{1}{\rho})^2.$$

For all $k \geq k_0$, we have

$$2\cos\theta \geq 2\cos\frac{2\pi}{k} \geq 2\cos\frac{2\pi}{k_0} \geq 2 - (c - \frac{1}{\rho})^2,$$

i.e., (4) is true and the claim holds.

(b) If $x = \lambda$, then plugging it into the LHS of (4) we get

$$\begin{aligned}
\frac{1-c^2}{x} + (1 - \frac{1}{\rho^2})x + \frac{2c}{\rho} &= \frac{(1-c)(1+c)}{\lambda} + (1 - \frac{1}{\rho^2})\lambda + \frac{2c}{\rho} \\
&< \frac{2(1-c)}{\lambda} + (1 - \frac{1}{\rho^2})\lambda + \frac{2c}{\rho} \quad \text{(since } 1 + c < 2\text{)} \\
&= \frac{2(1-c)}{\lambda} + (1 - \frac{1}{\rho^2})\lambda + \frac{2}{\rho} - \frac{2(1-c)}{\rho} \\
&= 2(1-c)(\frac{1}{\lambda} - \frac{1}{\rho}) + (1 - \frac{1}{\rho^2})\lambda + \frac{2}{\rho} \\
&< 2(1-c)(\frac{1}{\lambda} - \frac{1}{\rho}) + 1 - \frac{1}{\rho^2} + \frac{2}{\rho} \\
&= 2(1-c)(\frac{1}{\lambda} - \frac{1}{\rho}) + 2 - (1 - \frac{1}{\rho})^2. \qquad (5)
\end{aligned}$$

Set

$$c = 1 - \frac{(1 - \frac{1}{\rho})^2}{4(\frac{1}{\lambda} - \frac{1}{\rho})}.$$

Because $1 > \lambda > 0$ and $\rho > 1$, we have $0 < \frac{1-\frac{1}{\rho}}{4(\frac{1}{\lambda} - \frac{1}{\rho})} < 1$. Hence $c < 1$ and

$$\begin{aligned}
c &= 1 - \frac{(1 - \frac{1}{\rho})^2}{4(\frac{1}{\lambda} - \frac{1}{\rho})} \\
&= 1 - (1 - \frac{1}{\rho}) \cdot \frac{1 - \frac{1}{\rho}}{4(\frac{1}{\lambda} - \frac{1}{\rho})} \\
&> 1 - (1 - \frac{1}{\rho}) \\
&= \frac{1}{\rho}.
\end{aligned}$$

Therefore $\frac{1}{\rho} < c < 1$ as required.
Substituting the value of c into the RHS of (5) we get

$$2(1-c)(\frac{1}{\lambda}-\frac{1}{\rho})+2-(1-\frac{1}{\rho})^2 = \frac{1}{2}(1-\frac{1}{\rho})^2+2-(1-\frac{1}{\rho})^2$$
$$= 2-\frac{1}{2}(1-\frac{1}{\rho})^2. \quad (6)$$

For any fixed value of $\rho > 1$, there is a k_0 such that

$$2\cos\frac{2\pi}{k_0} \geq 2-\frac{1}{2}(1-\frac{1}{\rho})^2.$$

For all $k \geq k_0$, we have

$$2\cos\theta \geq 2\cos\frac{2\pi}{k} \geq 2\cos\frac{2\pi}{k_0} \geq 2-\frac{1}{2}(1-\frac{1}{\rho})^2. \quad (7)$$

Combining (5), (6) and (7), we have proven (4) and hence the claim holds.

This completes the proof.

4 YY_4 Is Not a Spanner

In this section we show that the YY_4 graph defined on the complete Euclidean graph is not a spanner. We do so by exhibiting, for every stretch factor $\rho > 1$, a set of points in the Euclidean plane whose stretch factor is more than ρ. The set of points consists of two sequences $(p_i)_{i=0}^n$, $(q_i)_{i=0}^n$, and hence, is of cardinality $2n+2$. The two sequences of points are placed in the Euclidean plane such that $d_{YY_4}(p_0, q_0) > \rho$. Since there are four rectilinear cones around each point in YY_4, we will refer to the cones as *quadrants*, and number them in the same way the quadrants around the origin in the Cartesian system are numbered.

Let $\rho > 1$ be given. Choose n large enough so that $\epsilon = 1 - \rho/2n$ satisfies $0 < \epsilon < 1$, and choose $\delta > 0$ such that $\delta < (2\epsilon - \epsilon^2)/(4n)$. The sequence of points $(p_i)_{i=0}^n$ belongs to a ray starting at p_0 whose slope is $-\delta/(1-\epsilon)$, and the sequence of points $(q_i)_{i=0}^n$ belongs to a ray starting at q_0 whose slope is $\delta/(1-\epsilon)$. More precisely, the coordinates of the points in the sequences $(p_i)_{i=0}^n$ and $(q_i)_{i=0}^n$ are defined as follows: p_i has coordinates $((1-\epsilon)i, 1-\delta i)$ and q_i has coordinates $((1-\epsilon)i, \delta i)$, for $i = 0, \ldots, n$. Since points p_i and q_i lie on the same vertical line, we will assume, without loss of generality, that point q_i is in the third quadrant of point p_i (equivalently, point p_i is in the first quadrant of point q_i), for $i = 0, \ldots, n$. This assumption can be justified by perturbing the points in the sequence (q_i) so that each of them is slightly shifted horizontally by a vector $\overrightarrow{v} = (-c, 0)$, where c is a very small positive constant chosen in such a way that it does not affect the desired properties.

Based on the definition of the two sequences $(p_i)_{i=0}^n$, $(q_i)_{i=0}^n$, we have $||p_0q_0|| = 1$ and $||p_iq_i|| = 1 - 2\delta i$; $||p_ip_{i+1}|| = ||q_iq_{i+1}|| = \sqrt{1+\epsilon^2+\delta^2-2\epsilon}$. Moreover, it is not difficult to verify that by the choice of ϵ and δ above we have $||q_iq_{i+1}|| = \sqrt{1+\epsilon^2+\delta^2-2\epsilon} < 1 - 2\delta n \leq 1 - 2\delta i = ||p_iq_i||$. For completeness, we present in the following the details of the edge selection process. First, it is easy to verify that the following holds after the first stage of the edge selection in the construction of YY_4 (refer to Figure 1 for illustration):

- For a point p_i, $i = 0, \ldots, n$, its first quadrant/cone is empty. Its second quadrant is empty if $i = 0$, and contains points $p_0, \ldots, p_{i-1}$ if $i \in \{1, \ldots, n\}$; hence, point p_0 does not select any point in its second quadrant, and for $i \neq 0$ point p_i selects point p_{i-1} in its second quadrant. Its third quadrant contains points $q_0, \ldots, q_i$, and point p_i selects point q_i in this quadrant. Its fourth quadrant is empty if $i = n$, and contains points $q_{i+1}, \ldots, q_n$ plus points $p_{i+1}, \ldots, p_n$ if $i \in \{0, \ldots, n-1\}$; hence, point p_i selects point p_{i+1} in this quadrant if $i \neq n$, and point p_n does not select any point in this quadrant.
- For a point q_i, $i = 0, \ldots, n-1$, its first quadrant contains points $q_{i+1}, \ldots, q_n$ plus points $p_i, \ldots, p_n$, and q_i selects point q_{i+1} in this quadrant; if $i = n$ then its first quadrant contains only p_n, and q_n selects p_n in this quadrant. The second quadrant of q_i is empty if $i = 0$, and contains points $p_0, \ldots, p_{i-1}$ if $i \in \{1, \ldots, n\}$; hence, point q_i selects point p_{i-1} in this quadrant if $i \in \{1, \ldots, n\}$, and if $i = 0$ point q_0 does not select any point in its second quadrant. The third quadrant of q_i is empty if $i = 0$, and otherwise contains points $q_0, \ldots, q_{i-1}$; hence, point q_0 does not select any point in its third quadrant, and for $i \neq 0$ point q_i selects point q_{i-1} in its third quadrant. Its fourth quadrant is empty and q_i does not select any point in its fourth quadrant.

Based on the above, after the second stage of the edge selection in the YY_4 construction, the final edge-set of the YY_4 graph G defined by the two sequences of points $(p_i)_{i=0}^n$ and $(q_i)_{i=0}^n$ is: $\{p_ip_{i+1} : i = 0, \ldots n-1\} \cup \{q_iq_{i+1} : i = 0, \ldots, n-1\} \cup \{p_nq_n\}$. Therefore, the shortest path between points p_0 and q_0 in G has length at least $\sum_{i=0}^{n-1} ||p_ip_{i+1}|| + \sum_{i=0}^{n-1} ||q_iq_{i+1}|| + ||p_nq_n|| > \sum_{i=0}^{n-1} ||p_ip_{i+1}|| + \sum_{i=0}^{n-1} ||q_iq_{i+1}|| > 2n(1-\epsilon) = \rho$. This shows that the stretch factor of G is more than ρ and completes the proof.

5 YY_4 Is Not Plane

In this section we show that the YY_4 graph defined on the complete Euclidean graph is not plane. Consider the set of points $\{u, v, w, r, s, t\}$ in the Euclidean plane, whose coordinates are $u(4, 4)$, $v(1, 7)$, $w(-3/2, 9)$, $r(0, 0)$, $s(-2, 1/2)$, and $t(-5, 1)$. For completeness, we give a detailed description of the edge selection process. First, it is easy to verify that the following holds after the first stage of the edge selection in the construction of YY_4 (refer to Figure 2 for illustration).

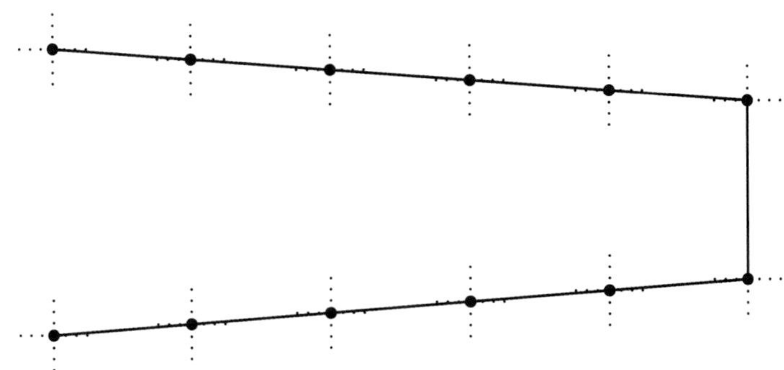

Fig. 1. An illustration of the construction showing that the YY_4 graph is not a spanner. The illustration shows a YY_4 graph whose stretch factor $\rho > 5$. In this illustration n is chosen to be 5, and $\epsilon = 1/2$.

- For point u, its first quadrant/cone is empty. Its second quadrant contains points v, w, and u selects v in its second quadrant. Its third quadrant contains r, s, t, and u selects r in its third quadrant. Its fourth quadrant is empty.
- For point v, its first quadrant is empty. Its second quadrant contains w, and hence v selects w in its second quadrant. Its third quadrant contains r, s, t, and v selects r in its third quadrant. Its fourth quadrant contains u, and v selects u in its fourth quadrant.
- For point w, its first and second quadrants are empty. Its third quadrant contains s, t, and w selects s in its third quadrant. Its fourth quadrant contains u, v, r, and w selects v in its fourth quadrant.
- For point r, its first quadrant contains u, v, and r selects u in its first quadrant. Its second quadrant contains w, s, t, and r selects s in its second quadrant. Its third and fourth quadrants are empty.
- For point s, its first quadrant contains u, v, w, and s selects u in its first quadrant. Its second quadrant contains t, and s selects t in its second quadrant. Its third quadrant is empty. Its fourth quadrant contains r, and s selects r in its fourth quadrant.
- For point t, its first quadrant contains u, v, w, and t selects v in its first quadrant. Its second and third quadrants are empty. Its fourth quadrant contains r, s, and t selects s in its fourth quadrant.

Based on the above, after the second stage of the edge selection in the YY_4 construction, the final edge-set of the YY_4 graph G defined by the set of points $\{u, v, w, r, s, t\}$ is: $\{uv, , vw, ur, rs, st, sw, tv\}$. The two edges sw and tv intersect, and hence G is not a plane graph. It follows that the YY_4 graph defined on a set of points is not necessarily a plane graph.

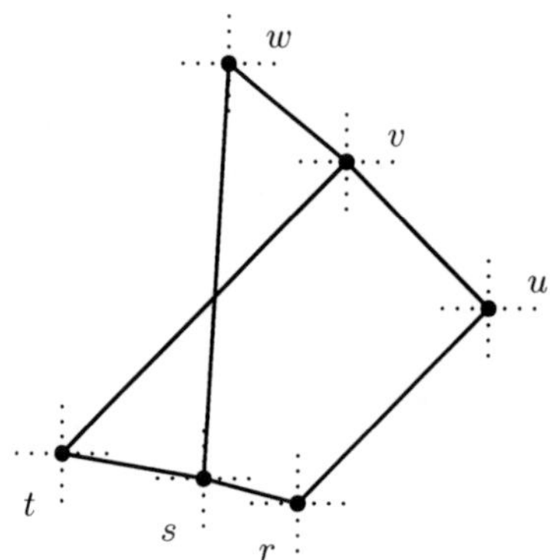

Fig. 2. A set of points for which the YY_4 graph is not plane

6 Concluding Remarks

In this paper we studied some geometric properties of the Yao-Yao graphs. We showed that the Yao-Yao graphs defined on civilized unit disk graphs are spanners with a stretch factor that is arbitrarily close to 1. We also showed that YY_4 is not a spanner and is not plane.

Clearly, the Yao-Yao graphs are less well understood than the Yao graphs. Several questions about the properties of Yao-Yao graphs remain unresolved. For example, are the Yao-Yao graphs geometric spanner of $\mathcal{E}$? More precisely, for what values of k (if any) is YY_k a geometric spanner? We leave those as open problems.

References

1. Althöfer, I., Das, G., Dobkin, D., Joseph, D., Soares, J.: On sparse spanners of weighted graphs. Discrete & Computational Geometry 9, 81–100 (1993)
2. Bose, P., Damian, M., Douïeb, K., O'Rourke, J., Seamone, B., Smid, M., Wuhrer, S.: $\pi/2$-Angle Yao Graphs Are Spanners. In: Cheong, O., Chwa, K.-Y., Park, K. (eds.) ISAAC 2010, Part II. LNCS, vol. 6507, pp. 446–457. Springer, Heidelberg (2010)
3. Bose, P., Damian, M., Douïeb, K., O'Rourke, J., Seamone, B., Smid, M., Wuhrer, S.: $\pi/2$-angle Yao graphs are spanners. CoRR, abs/1001.2913 (2010)
4. Bose, P., Gudmundsson, J., Smid, M.: Constructing plane spanners of bounded degree and low weight. Algorithmica 42(3-4), 249–264 (2005)
5. Bose, P., Morin, P., Stojmenovic, I., Urrutia, J.: Routing with guaranteed delivery in ad hoc wireless networks. Wireless Networks 7(6), 609–616 (2001)
6. Damian, M., Raudonis, K.: Yao Graphs Span Theta Graphs. In: Wu, W., Daescu, O. (eds.) COCOA 2010, Part II. LNCS, vol. 6509, pp. 181–194. Springer, Heidelberg (2010)
7. Das, G., Heffernan, P., Narasimhan, G.: Optimally sparse spanners in 3-dimensional Euclidean space. In: Proceedings of the 19th ACM Symposium on Computational Geometry, pp. 53–62 (1993)

8. Das, G., Narasimhan, G.: A fast algorithm for constructing sparse Euclidean spanners. In: Proceedings of the 20th ACM Symposium on Computational Geometry, pp. 132–139 (1994)
9. Das, G., Narasimhan, G., Salowe, J.: A new way to weigh malnourished Euclidean graphs. In: Proceedings of the Sixth Annual ACM-SIAM Symposium on Discrete Algorithms, pp. 215–222 (1995)
10. Demaine, E., Mitchell, J., O'Rourke, J. (eds.): The open problems project: Problem 70, `http://maven.smith.edu/~orourke/TOPP/P70.html`
11. Gudmundsson, J., Levcopoulos, C., Narasimhan, G.: Fast greedy algorithms for constructing sparse geometric spanners. SIAM Journal on Computing 31(5), 1479–1500 (2002)
12. Kanj, I., Perković, L., Xia, G.: On spanners and lightweight spanners of geometric graphs. SIAM Journal on Computing 39(6), 2132–2161 (2010)
13. Keil, J., Gutwin, C.: Classes of graphs which approximate the complete Euclidean graph. Discrete & Computational Geometry 7, 13–28 (1992)
14. Kranakis, E., Singh, H., Urrutia, J.: Compass routing on geometric networks. In: Proceedings of the 11th Canadian Conference on Computational Geometry, pp. 51–54 (1999)
15. Levcopoulos, C., Lingas, A.: There are planar graphs almost as good as the complete graphs and almost as cheap as minimum spanning trees. Algorithmica 8(3), 251–256 (1992)
16. Li, X.-Y., Calinescu, G., Wan, P.-J., Wang, Y.: Localized delaunay triangulation with application in Ad Hoc wireless networks. IEEE Transactions on Parallel and Distributed Systems 14(10), 1035–1047 (2003)
17. Molla, N.: Yao spanners for wireless ad hoc networks. M.S. Thesis, Department of Computer Science, Villanova University (December 2009)
18. Narasimhan, G., Smid, M.: Geometric Spanner Networks. Cambridge University Press (2007)
19. Wang, Y., Li, X.-Y.: Distributed spanner with bounded degree for wireless ad hoc networks. In: Proceedings of the 16th International Parallel and Distributed Processing Symposium (2002)
20. Wang, Y., Li, X.-Y.: Localized construction of bounded degree and planar spanner for wireless ad hoc networks. Mobile Networks and Applications 11(2), 161–175 (2006)
21. Yao, A.C.-C.: On constructing minimum spanning trees in k-dimensional spaces and related problems. SIAM Journal on Computing 11(4), 721–736 (1982)

Distance-d Independent Set Problems for Bipartite and Chordal Graphs

Hiroshi Eto, Fengrui Guo, and Eiji Miyano

Department of Systems Design and Informatics, Kyushu Institute of Technology, Fukuoka 820-8502, Japan
{eto,guo}@theory.ces.kyutech.ac.jp, miyano@ces.kyutech.ac.jp

Abstract. The paper studies a generalization of the INDEPENDENT SET (IS) problem. A distance-d independent set for a positive integer $d \geq 2$ in an unweighted graph $G = (V, E)$ is a set $S \subseteq V$ of vertices such that for any pair of vertices $u, v \in S$, the distance between u and v is at least d in G. Given an unweighted graph G and a positive integer k, the DISTANCE-d INDEPENDENT SET (DdIS) problem is to decide whether G contains a distance-d independent set S such that $|S| \geq k$. D2IS is identical to the original IS and thus D2IS is in $\mathcal{P}$ for bipartite graphs and chordal graphs. In this paper, we show that for every fixed integer $d \geq 3$, DdIS is $\mathcal{NP}$-complete even for planar bipartite graphs of maximum degree three, and also $\mathcal{NP}$-complete even for chordal bipartite graphs. Furthermore, we show that if the input graph is restricted to chordal graphs, then DdIS can be solved in polynomial time for any even $d \geq 2$, whereas DdIS is $\mathcal{NP}$-complete for any odd $d \geq 3$.

1 Introduction

One of the most important and most investigated computational problems in theoretical computer science is the INDEPENDENT SET problem (IS for short) because of its many applications in scheduling, computer vision, pattern recognition, coding theory, map labeling, computational biology, and some other fields. The input of IS is an unweighted graph $G = (V, E)$ and a positive integer $k \leq |V|$. An *independent set* of G is a subset $S \subseteq V$ of vertices such that, for all $u, v \in S$, the edge (u, v) is not in E. IS asks whether G contains an independent set S having $|S| \geq k$. IS is among the first problems ever to be shown to be $\mathcal{NP}$-complete, and has been used as a starting point for proving the $\mathcal{NP}$-completeness of other problems [10]. Moreover, it is well known that IS remains $\mathcal{NP}$-complete even for substantial restricted graph classes such as cubic planar graphs [9], triangle-free graphs [18], and graphs with large girth [17].

In this paper, we consider a generalization of IS, named the DISTANCE-d INDEPENDENT SET problem (DdIS for short). A distance-d independent set for a positive integer $d \geq 2$ in an unweighted graph $G = (V, E)$ is a set $S \subseteq V$ of vertices such that for any pair of vertices $u, v \in S$, the distance between u and v is at least d in G. For a fixed constant $d \geq 2$, DdIS considered in this paper is formulated as the following class of problems [1]:

G. Lin (Ed.): COCOA 2012, LNCS 7402, pp. 234–244, 2012.

Distance-d Independent Set (DdIS)

Instance: An unweighted graph $G = (V, E)$ and a positive integer $k \leq |V|$.

Question: Does G contains a distance-d independent set of size k or more?

One can see that D2IS is identical to the original IS, and DdIS is equivalent to IS on the $(d-1)$th power graph G^{d-1} of the input graph G.

IS, i.e., D2IS is even $\mathcal{NP}$-complete, and thus it would be easy to show that DdIS is $\mathcal{NP}$-complete in general. Fortunately, however, it is known that if the input graph is restricted to, for example, bipartite graphs [14], chordal graphs [11], circular arc graphs [12], comparability graphs [13], and many other classes [16,15,4], then D2IS admits polynomial-time algorithms. Furthermore, Agnarsson, Damaschke, Halldórsson [1] show the following tractability of DdIS by using the closure property under taking power [7,8,19]:

Fact 1 ([1]). *Let n denote the number of vertices in the input graph G. Then, for every integer $d \geq 2$, DdIS is solvable in $O(n)$ time for interval graphs, in $O(n(\log\log n + \log d))$ time for trapezoid graphs, and in $O(n)$ time for circular-arc graphs.*

This tractability suggests that if we restrict the set of instances to, for example, subclasses of bipartite graphs and chordal graphs, then DdIS for a fixed $d \geq 3$ might be also solvable efficiently. On the other hand, however, we have a "negative" fact that if G is planar/bipartite, then the $(d-1)$th power graph G^{d-1} is not necessarily planar/bipartite. From those points of view, this paper investigates DdIS, namely, our work focuses on the complexity of DdIS on (subclasses of) bipartite graphs and chordal graphs.

Our main results are summarized in the following list:

(i) For every fixed integer $d \geq 3$, DdIS is $\mathcal{NP}$-complete even for planar bipartite graphs of maximum degree three.
(ii) For every fixed integer $d \geq 3$, DdIS is $\mathcal{NP}$-complete even for chordal bipartite graphs.
(iii) For any $\varepsilon > 0$ and fixed integer $d \geq 3$, it is $\mathcal{NP}$-hard to approximate the maximization version of DdIS to within a factor of $n^{1/2-\varepsilon}$ for chordal bipartite graphs of n vertices.
(iv) For every fixed integer $d \geq 3$, DdIS is $\mathcal{W}[1]$-hard with respect to the size k of the distance-d independent set as a parameter for chordal bipartite graphs.
(v) For every fixed even integer $d \geq 2$, DdIS is in $\mathcal{P}$ for chordal graphs.
(vi) For every fixed odd integer $d \geq 3$, DdIS is $\mathcal{NP}$-complete for chordal graphs.
(vii) For any $\varepsilon > 0$ and fixed odd integer $d \geq 3$, it is $\mathcal{NP}$-hard to approximate the maximization version of DdIS to within a factor of $n^{1/2-\varepsilon}$ for chordal graphs of n vertices.

(viii) For every fixed odd integer $d \geq 3$, DdIS is $\mathcal{W}[1]$-hard with respect to the size k of the distance-d independent set as a parameter for chordal graphs.

One can see that the complexity of DdIS depends on the parity of d if the set of input graphs is restricted to chordal graphs.

The organization of the paper is as follows: Section 2 is devoted to our notation and terminology. In Section 3 we prove the $\mathcal{NP}$-completeness of the problem for planar bipartite graphs and for chordal bipartite graphs. In Section 4, we provide tractable and intractable cases for chordal graphs.

2 Preliminaries

Let $G = (V, E)$ be an unweighted graph, where V and E denote the set of vertices and the set of edges, respectively. $V(G)$ and $E(G)$ also denote the vertex set and the edge set of G, respectively. We denote an edge with endpoints u and v by (u, v). For a pair of vertices u and v, the length of a shortest path from u to v, i.e., the distance between u and v is denoted by $dist_G(u, v)$, and the diameter G is defined as $diam(G) = \max_{u,v \in V} dist_G(u, v)$.

A graph G_S is a subgraph of a graph G if $V(G_S) \subseteq V(G)$ and $E(G_S) \subseteq E(G)$. For a subset of vertices $U \subseteq V$, let $G[U]$ be the subgraph induced by U. For a subgraph $G_S = (V_S, E_S)$ of G, if $E_S = V_S \times V_S$, then G_S (or $G[V_S]$) and V_S are called a *clique* and a *clique set*, respectively.

For a positive integer $d \geq 1$ and a graph G, the dth power of G, denoted by $G^d = (V(G), E^d)$, is the graph formed from $V(G)$, where all pairs of vertices $u, v \in G$ such that $dist_G(u, v) \leq d$ are connected by an edge (u, v). Note that $E(G) \subseteq E^k$, i.e., the original edges in $E(G)$ are retained.

A path of length ℓ, denoted by P_ℓ, from a vertex v_0 to a vertex v_ℓ is represented as a sequence of vertices such that $P_\ell = \langle v_0, v_1, \cdots, v_\ell \rangle$. A cycle of length ℓ, denoted by C_ℓ, is similarly written as $C_\ell = \langle v_0, v_1, \cdots, v_{\ell-1}, v_0 \rangle$. An even (odd) cycle (path) is a cycle (path) of even (odd) length. A *chord* of a path (cycle) is an edge between two vertices of the path (cycle) that is not an edge of the path (cycle).

A graph $G = (V, E)$ is *bipartite* if there is a partition of V into two disjoint independent sets V_1 and V_2. A graph G is *chordal* if each cycle in G of length at least four has at least one chord. A graph G is *weakly chordal* if G and its complement $\overline{G}$ contains no induced cycle C_ℓ for $\ell \geq 5$. A graph is *chordal bipartite* if it is simultaneously weakly chordal and bipartite, equivalently a bipartite graph is chordal bipartite if it has no induced cycle C_ℓ for $\ell \geq 6$. A graph $G = (V, E)$ is *split* if there is a partition of V into a clique set V_1 and an independent set V_2 such that $V_1 \cap V_2 = \emptyset$ and $V_1 \cup V_2 = V$. Note that the split graphs are a subclass of the chordal graphs (see, e.g., [5]). A graph is *star* if it is a rooted tree of height one.

The objective of the maximization version of DdIS is to find a distance-d independent set of maximum size in an input graph G, which is denoted by MaxDdIS for short. For the maximization problems, an algorithm ALG is called a σ-approximation algorithm and the approximation ratio of ALG is σ if $OPT(G)/ALG(G) \leq \sigma$ holds for every input G, where $ALG(G)$ and $OPT(G)$ are the numbers of vertices of obtained subsets by ALG and an optimal algorithm, respectively.

3 Bipartite Graphs

First, we consider the class of bipartite graphs. As mentioned in Section 1, D2IS is solvable in polynomial time by using a polynomial time algorithm which finds the maximum complete matching in a given bipartite graph. On the other hand, in this section, we show the $\mathcal{NP}$-completeness of DdIS on planar bipartite graphs and on chordal bipartite graphs when $d \geq 3$.

3.1 Planar Bipartite Graphs

For planar bipartite graphs, we have the following theorem:

Theorem 1. *For every fixed integer $d \geq 3$, DdIS is $\mathcal{NP}$-complete even for planar bipartite graphs of maximum degree three.*

Proof. We first show the $\mathcal{NP}$-completeness of D3IS and then one of the general DdIS for $d \geq 4$. Obviously, DdIS is in $\mathcal{NP}$ for every $d \geq 3$. To show that D3IS is $\mathcal{NP}$-complete, we reduce the $\mathcal{NP}$-complete problem D2IS on any cubic planar graph $G_0 = (V_0, E_0)$ to D3IS on a new planar bipartite graph $G = (V, E)$ of maximum degree three.

Let $V_0 = \{v_1, v_2, \cdots, v_n\}$ and $E_0 = \{e_1, e_2, \cdots, e_m\}$ be vertex and edge sets of G_0. We construct the planar bipartite graph G which consists of (i) n vertices, u_1 through u_n, which are associated with n vertices in V_0, v_1 through v_n, respectively, and (ii) m subgraphs, G_1 through G_m, which are associated with m edges in E_0, e_1 through e_m, respectively. For every i $(1 \leq i \leq m)$, the ith subgraph G_i contains three vertices, $w_{i,0}$, $w_{i,1}$, and $w_{i,2}$ and two edges, $(w_{i,0}, w_{i,1})$ and $(w_{i,1}, w_{i,2})$ such that G_i forms a path P_2 of length 2. (iii) If $e_i = (v_j, v_k) \in E_0$, then we introduce two edges $(w_{i,0}, u_j)$ and $(w_{i,0}, u_k)$.

Just to make the above construction clear, see Figure 1. For example, if the instance G_0 is the left graph, then the reduced graph G is illustrated in the right graph. Clearly, the constructed graph G is planar bipartite and the construction can be accomplished in polynomial time.

For the above construction of G, we show that G has a distance-3 independent set S such that $|S| \geq k + m$ if and only if G_0 has a distance-2 independent set S_0 such that $|S_0| \geq k$.

Suppose that the graph G_0 of D2IS has the distance-2 independent set $S_0 = \{v_{1^*}, v_{2^*}, \cdots v_{k^*}\}$ in G_0, where $\{1^*, 2^*, \cdots, k^*\} \subseteq \{1, 2, \cdots, n\}$. Then, we select subsets of vertices $S' = \{u_{1^*}, u_{2^*}, \cdots, u_{k^*}\}$ and $S'' = \{w_{1,2}, w_{2,2}, \cdots, w_{m,2}\}$ such that $|S'| = k$ and $|S''| = m$. One can see that $S = S' \cup S''$ is a distance-3 independent set in G since the pairwise distance in S' is at least four and the distance between $w_{i,2}$ in S'' and every vertex in S' is at least three for each i.

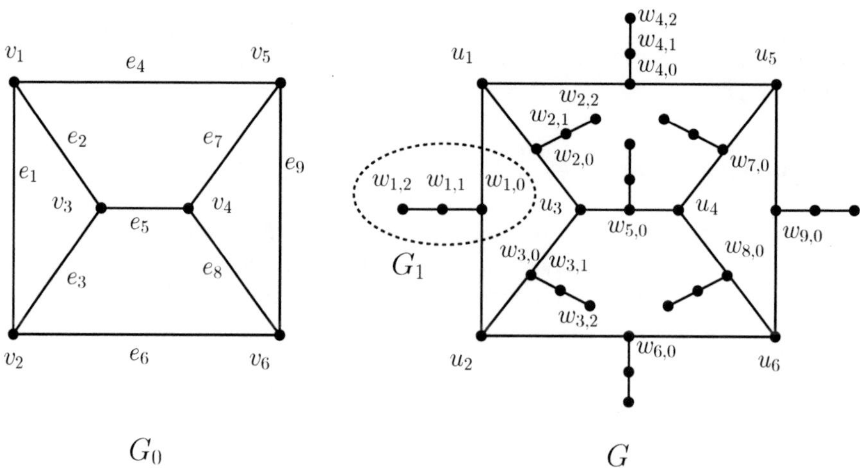

Fig. 1. (Left) graph G_0 of D2IS and (Right) reduced graph G from G_0

Conversely, suppose that the graph G has the distance-3 independent set S such that $|S| \geq k + m$. First, from each subgraph G_i which is the path of length 2, we can select at most one vertex as the distance-3 independent set. Thus, the maximum size of the distance-3 independent set in $V(G_1) \cup V(G_2) \cup \cdots \cup V(G_m)$ is at most m, which means that $|S \cap \{u_1, u_2, \cdots, u_n\}| \geq k$. Let $\{u_{1^*}, u_{2^*}, \cdots, u_{k^*}\}$ be a subset of k vertices in $S \cap \{u_1, u_2, \cdots, u_n\}$. Then, the pairwise distance in the corresponding subset of vertices $\{v_{1^*}, v_{2^*}, \cdots, v_{k^*}\}$ of G_0 is surely at least 2, i.e., G_0 has a distance-2 independent set S_0 such that $|S_0| \geq k$. This completes the proof of the $\mathcal{NP}$-hardness of D3IS.

In the following, we give a brief sketch of the ideas to prove the $\mathcal{NP}$-hardness of DdIS for $d \geq 4$. In the case of D4IS, all we have to do is replace the 2-length path G_i corresponding to the edge e_i with the 3-length path $G_i^{IV} = (\{w_{i,0}, w_{i,1}, w_{i,2}, w_{i,3}\}, \{(w_{i,0}, w_{i,1}), (w_{i,1}, w_{i,2}), (w_{i,2}, w_{i,3})\})$ for each i. See the left graph in Figure 2. In the case of D5IS, G_i is replaced with $G_i^V = (V_i^V, E_i^V)$:

$$V_i^V = \{w_{i,0}^0, w_{i,0}^1, w_{i,0}^2, w_{i,1}, w_{i,2}, w_{i,3}\}$$
$$E_i^V = \{(w_{i,0}^0, w_{i,0}^1), (w_{i,0}^1, w_{i,0}^2), (w_{i,0}^1, w_{i,1}), (w_{i,1}, w_{i,2}), (w_{i,2}, w_{i,3})\}.$$

Then, u_j (u_k) corresponding to the vertex v_j (v_k) is connected to $w_{i,0}^0$ ($w_{i,0}^2$) if $e_i = (v_j, v_k) \in E_0$ (see the center graph in Figure 2). For $d = 6$, we connect one vertex $w_{i,4}$ to the top vertex $w_{i,3}$ of G_i^V (see the right graph in Figure 2). For $d \geq 7$, we only add a similar modification to the subgraph corresponding to the edge e_i. Further details are omitted here. □

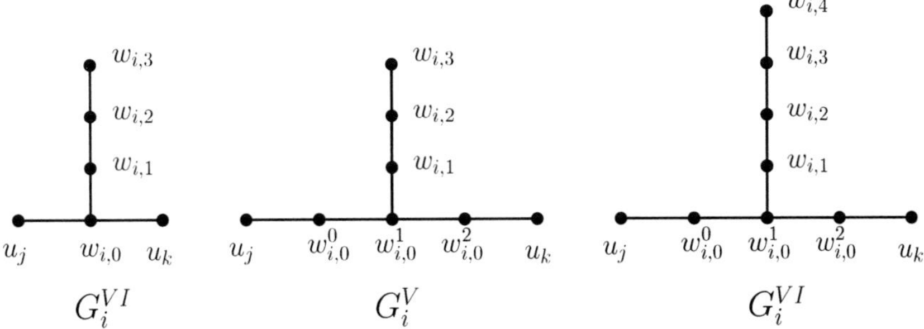

Fig. 2. (Left) subgraphs G_i^{IV} for $d = 4$, (Center) G_i^V for $d = 5$, and (Right) G_i^{VI} for $d = 6$

3.2 Chordal Bipartite Graphs

This subsection investigates chordal bipartite graphs. Note that chordal bipartite graphs are *not* chordal in general (e.g., C_4 is chordal bipartite), but of course bipartite. Thus, D2IS on chordal bipartite graphs is solvable in polynomial time. Unfortunately, however, for $d \geq 3$, DdIS on chordal bipartite graphs is intractable:

Theorem 2. *For every fixed integer $d \geq 3$, DdIS is $\mathcal{NP}$-complete even for chordal bipartite graphs.*

Proof. We only give a reduction from D2IS on general graphs to D3IS on chordal bipartite graphs since the proof for the case $d \geq 4$ is very similar to the case $d = 3$. Given a graph $G_0 = (V_0, E_0)$ of D2IS with n vertices, $V_0 = \{v_1, v_2, \cdots, v_n\}$, and m edges, $E_0 = \{e_1, e_2, \cdots, e_m\}$, we construct the chordal bipartite graph G in the following way. The constructed graph G consists of (i) n vertices, u_1 through u_n, each u_i of which is corresponding to $v_i \in V_0$, (ii) m vertices, w_1 through w_m, each w_i of which is corresponding to $e_i \in E_0$, and (iii) two special vertices α and β. (iv) The vertex α is connected to each $\{\beta\} \cup \{w_1, \cdots, w_m\}$, i.e., the induced graph $G[\{\alpha, \beta\} \cup \{w_1, \cdots, w_m\}]$ is star. (v) If $e_i = (v_j, v_k) \in E_0$, then we add two edges (w_i, u_j) and (w_i, u_k). Figure 3 illustrates the reduced graph G from G_0 in Figure 1. It is clear that this reduction can be done in polynomial time. Note that since all the vertices w_1 through w_m are connected to one vertex α, every cycle C_ℓ for $\ell \geq 6$ has a chord and thus G is chordal bipartite.

For the above construction of G, we show that G has a distance-3 independent set S such that $|S| \geq k+1$ if and only if G_0 has a distance-2 independent set S_0 such that $|S_0| \geq k$.

Suppose that the graph G_0 of D2IS has the distance-2 independent set $S_0 = \{v_{1^*}, v_{2^*}, \cdots v_{k^*}\}$ in G_0, where $\{1^*, 2^*, \cdots, k^*\} \subseteq \{1, 2, \cdots, n\}$. Then, we select subsets of vertices $S = \{u_{1^*}, u_{2^*}, \cdots, u_{k^*}\} \cup \{\beta\}$ of size $k+1$. Note that the distance $dist_G(\beta, u_i)$ for every i is three. Since the distance $dist_{G_0}(v_{i^*}, v_{j^*})$ for

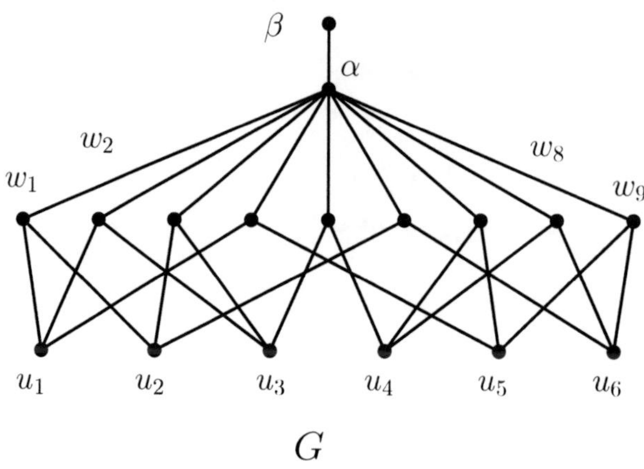

Fig. 3. An illustration of the construction

any pair of vertices $v_{i^*}, v_{j^*} \in S_0$ ($i^* \neq j^*$) is at least two, the shortest path from u_{i^*} to u_{j^*} contains two vertices in $\{w_1, w_2, \cdots, w_m\}$. This means that the distance $dist_G(u_{i^*}, u_{j^*})$ for any $i^* \neq j^*$ is four. Thus, the selected vertex set S of size $k+1$ is a distance-3 independent set.

Conversely, suppose that the graph G has the distance-3 independent set S such that $|S| \geq k+1$. First, take a look at the induced graph $G[\{\alpha, \beta\} \cup \{w_1, \cdots, w_m\}]$. Since its diameter $diam(G[\{\alpha, \beta\} \cup \{w_1, \cdots, w_m\}])$ is two, $S \cap V(G[\{\alpha, \beta\} \cup \{w_1, \cdots, w_m\}]) \leq 1$ holds, i.e., $|S \cap \{u_1, u_2, \cdots, u_n\}| \geq k$ holds. Let $\{u_{1^*}, u_{2^*}, \cdots, u_{k^*}\}$ be a subset of k vertices in $S \cap \{u_1, u_2, \cdots, u_n\}$. Then, the pairwise distance of vertices in $\{v_{1^*}, v_{2^*}, \cdots, v_{k^*}\}$ of G_0 corresponding to $\{u_{1^*}, u_{2^*}, \cdots, u_{k^*}\}$ in G is surely at least 2, i.e., G_0 has a distance-2 independent set S_0 such that $|S_0| \geq k$. This completes the proof of the $\mathcal{NP}$-hardness of D3IS. The proof for the case $d \geq 4$ is omitted. □

Now consider the maximization version Max*Dd*IS of the decision one D*d*IS, which asks for a distance-d independent set of maximum size in an input graph G. The above reduction can preserve the approximation-gap and thus gives us the following inapproximability of Max*Dd*IS.

Corollary 1. *For any $\varepsilon > 0$ and a fixed integer $d \geq 3$, it is $\mathcal{NP}$-hard to approximate* MaxD*d*IS *to within a factor of $n^{1/2-\varepsilon}$ for chordal bipartite graphs.*

Proof. Let $OPT_1(G_0)$ denote the number of vertices of an optimal solution for an input graph G_0 of MaxD2IS. Let $OPT_2(G)$ denote the number of vertices of an optimal solution for an input chordal bipartite graph G of MaxD*d*IS for a fixed $d \geq 3$. Let $g(n)$ be a parameter function of the instance G of D2IS. Note that the reduction described in the proof of Theorem 2 is the following gap-preserving reduction: (1) If $OPT_1(G_0) \geq g(n)$, then $OPT_2(G) \geq g(n)+1$, and (2) if $OPT_1(G_0) < \frac{g(n)}{n^{1-\varepsilon_1}}$ for a positive constant ε_1, then $OPT_2(G) < \frac{g(n)}{n^{1-\varepsilon}} + 1$.

Since $|V(G)| \leq 2n^2$ and so the approximation gap is $\Theta(|V(G)|^{1/2-\varepsilon})$ for any $\varepsilon > 0$, the corollary holds. □

Also, the reduction in the proof of Theorem 2 shows the following *fixed-parameter intractability* of DdIS:

Corollary 2. *For every fixed integer $d \geq 3$, DdIS is $\mathcal{W}[1]$-hard with respect to the size k of the distance-d independent set as a parameter for chordal bipartite graphs. (The proof is omitted.)*

4 Chordal Graphs

In this section we restrict the instances to chordal graphs. In [11], Gavril shows that D2IS admits an efficient algorithm for chordal graphs:

Lemma 1 ([11]). D2IS *is in $\mathcal{P}$ for chordal graphs.*

Recall that if the dth power graph G^d is interval (trapezoid, or circular arc, resp.), then the $(d+1)$th power G^{d+1} is also interval [19] (trapezoid [7], or circular arc [8], resp.) for any integer $d \geq 1$. The class of chordal graphs does *not* satisfy the closure property under the graph power operation, i.e., the square G^2 of a chordal graph G is not necessarily chordal, but it does satisfy the closure property under the graph *odd power* operation:

Lemma 2 ([2,3]). *Let $d \geq 1$ be an odd integer. If G is a chordal graph, then G^d is also chordal.*

Together with Lemma 1, this yields:

Theorem 3. *For every fixed even integer $d \geq 2$,* DdIS *is in $\mathcal{P}$ for chordal graphs.*

Proof. Given a graph G, we first construct the odd power graph G^{d-1} in polynomial time, which must be chordal by Lemma 2. Then, by using a polynomial-time algorithm for D2IS in Lemma 1, we can obtain a solution of DdIS in polynomial time. □

For an odd d, DdIS is hard:

Theorem 4. *For every fixed odd $d \geq 3$,* DdIS *is $\mathcal{NP}$-complete for chordal graphs.*

Proof. Obviously, DdIS on chordal graphs is in $\mathcal{NP}$ for every odd $d \geq 3$. To show that DdIS on chordal graphs is $\mathcal{NP}$-complete, we reduce D2IS on any graph $G_0 = (V_0, E_0)$ to DdIS on a new chordal graph $G = (V, E)$.

Given a graph $G_0 = (V_0, E_0)$ of D2IS with n vertices, $V_0 = \{v_1, v_2, \cdots, v_n\}$, and m edges, $E_0 = \{e_1, e_2, \cdots, e_m\}$, we construct the following chordal graph G: (i) We prepare n paths of length $(d-3)/2$, $G_1 = \langle u_{1,1}, u_{1,2}, \cdots, u_{1,(d-1)/2}\rangle$ through $G_n = \langle u_{n,1}, u_{n,2}, \cdots, u_{n,(d-1)/2}\rangle$, each G_i of which is corresponding to

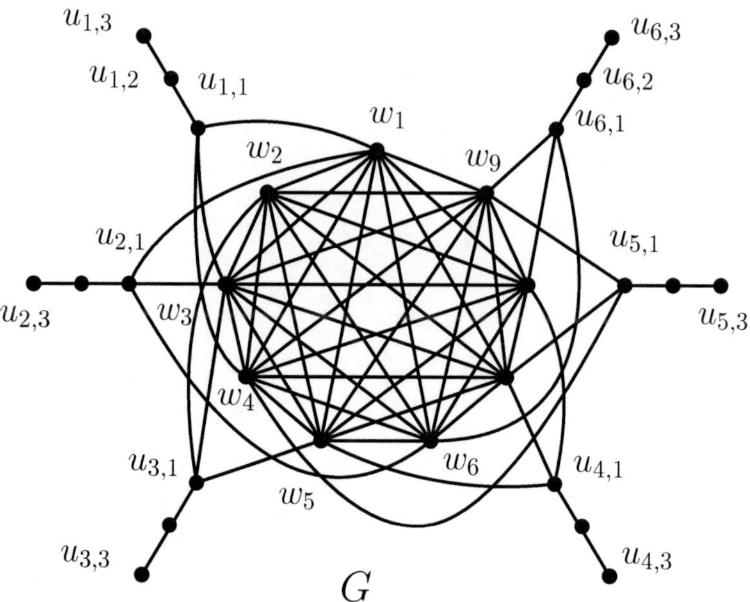

Fig. 4. An illustration of the construction when $d = 7$

$v_i \in V_0$, and (ii) m vertices, w_1 through w_m, each w_i of which is corresponding to $e_i \in E_0$. (iii) All the vertices w_1 through w_m are connected such that $G[\{w_1, \cdots, w_m\}]$ forms a clique of m vertices. (iv) If $e_i = (v_j, v_k) \in E_0$, then we connect w_i to two vertices $u_{j,1}$ and $u_{k,1}$.

Figure 4 illustrates the reduced graph G from G_0 in Figure 1 when $d = 7$. The constructed graph G is chordal since all C_4's in the clique graph $G[\{w_1, \cdots, w_m\}]$ have chords and also $G[\{w_1, \cdots, w_m\} \cup \{v_{i,0}\}]$ contains only cycles C_3's for every i. Also, G can be constructed in polynomial time from G_0.

We show that the reduction satisfies that if G has a distance-d independent set S such that $|S| \geq k$ if and only in G_0 has a distance-2 independent set S_0 such that $|S_0| \geq k$. In the remaining of this proof, the crucial observations are: (1) The distance between any vertex v in $V \setminus \{u_{1,(d-1)/2}, u_{2,(d-1)/2}, \cdots, u_{n,(d-1)/2}\}$ and another vertex u in $V \setminus \{v\}$ is at most $d-1$. (2) The pairwise distance of $\{u_{1,(d-1)/2}, u_{2,(d-1)/2}, \cdots, u_{n,(d-1)/2}\}$ is at most d. The two observations (1) and (2) imply that the distance-d independent set S must be a subset of outside vertices $\{u_{1,(d-1)/2}, u_{2,(d-1)/2}, \cdots, u_{n,(d-1)/2}\}$. (3) If v_j and v_k are two endpoints of single edge e_i in G_0, then there must be a path

$$\langle u_{j,(d-1)/2}, u_{j,(d-3)/2}, \cdots, u_{j,1}, w_i, u_{k,1}, u_{k,2}, \cdots, u_{k,(d-1)/2} \rangle$$

by the above reduction rules. Thus, the distance between $u_{j,d}$ and $u_{k,d}$ in G is $(d-1)/2 \times 2 = d-1$.

Now suppose that the graph G_0 of D2IS has the distance-2 independent set $S_0 = \{v_{1^*}, v_{2^*}, \cdots v_{k^*}\}$ in G_0, where $\{1^*, 2^*, \cdots, k^*\} \subseteq \{1, 2, \cdots, n\}$. Then, we select a subset $S = \{u_{1^*,(d-1)/2}, u_{2^*,(d-1)/2}, \cdots, u_{k^*,(d-1)/2}\}$ of size k. Note that the pairwise distance in S is exactly d.

Conversely, suppose that the graph G has the distance-d independent set $S = \{u_{1^*,(d-1)/2}, u_{2^*,(d-1)/2}, \cdots, u_{k^*,(d-1)/2}\}$ of size k. Then, the pairwise distance in the corresponding subset of vertices $\{v_{1^*}, v_{2^*}, \cdots, v_{k^*}\}$ of G_0 is surely at least 2, i.e., G_0 has a distance-2 independent set S_0 such that $|S_0| \geq k$. □

Corollary 3. D3IS *is $\mathcal{NP}$-complete for split graphs.*

Proof. When $d = 3$ in the proof of Theorem 4, the constructed graph G is a split graph since there is a partition of $V(G)$ into a clique set $\{w_1, w_2, \cdots, w_m\}$ and an independent set $\{u_{1,1}, u_{2,1}, \cdots, u_{n,1}\}$. □

Similarly to the previous section, it can be shown that the reduction in the proof of Theorem 4 can preserve the approximation gap, and also it is an ftp-reduction (details are omitted):

Corollary 4. *For any $\varepsilon > 0$ and fixed odd integer $d \geq 3$, it is $\mathcal{NP}$-hard to approximate* MaxDdIS *to within a factor of $n^{1/2-\varepsilon}$ for chordal graphs.*

Corollary 5. *For every fixed odd integer $d \geq 3$,* DdIS *is $\mathcal{W}[1]$-hard with respect to the size k of the distance-d independent set as a parameter for chordal graphs.*

Acknowledgments. This work is partially supported by KAKENHI 23500020.

References

1. Agnarsson, G., Damaschke, P., Halldórsson, M.H.: Powers of geometric intersection graphs and dispersion algorithms. Discrete Applied Mathematics 132, 3–16 (2004)
2. Agnarsson, G., Greenlaw, R., Halldórsson, M.M.: On powers of chordal graphs and their colorings. Congr. Numer. 144, 41–65 (2000)
3. Balakrishnan, R., Paulraja, P.: Powers of chordal graphs. Australian Journal of Mathematics, Series A 35, 211–217 (1983)
4. Brandstädt, A., Giakoumakis, V.: Maximum weight independent sets in hole- and co-chair-free graphs. Information Processing Letters 112, 67–71 (2012)
5. Brandstädt, A., Le, V.B., Spinrad, J.P.: Graph Classes: A Survey. SIAM (1999)
6. Downey, R.G., Fellows, M.R.: Fixed-parameter tractability and completeness II: On completeness for W[1]. Theoretical Computer Science A 141(1-2), 109–131 (1995)
7. Flotow, C.: On powers of m-trapezoid graphs. Discrete Applied Mathematics 63, 187–192 (1995)
8. Flotow, C.: On powers of circular arc graphs and proper circular arc graphs. Discrete Applied Mathematics 74, 199–207 (1996)
9. Garey, M.R., Johnson, D.S., Stockmeyer, L.: Some simplified $\mathcal{NP}$-complete graph problems. Theoretical Computer Science 1, 237–267 (1976)
10. Garey, M.R., Johnson, D.A.: Computers and intractability - A guide to the theory of $\mathcal{NP}$-completeness (1979)

11. Gavril, F.: Algorithms for minimum coloring, maximum clique, minimum covering by cliques, and maximum independent set of chordal graph. SIAM J. Comput. 1, 180–187 (1972)
12. Gavril, F.: Algorithms on circular-arc graphs. Networks 4, 357–369 (1974)
13. Golumbic, M.C.: The complexity of comparability graph recognition and coloring. Computing 18, 199–208 (1977)
14. Harary, F.: Graph Theory. Addison-Wesley (1969)
15. Lozin, V.V., Milanič, M.: A polynomial algorithm to find an independent set of maximum weight in a fork-free graph. J. Discrete Algorithms 6, 595–604 (2008)
16. Minty, G.J.: On maximal independent sets of vertices in claw-free graphs. J. Combin. Theory Ser. B 28, 284–304 (1980)
17. Murphy, O.J.: Computing independent sets in graphs with large girth. Discrete Applied Mathematics 35, 167–170 (1992)
18. Poljak, S.: A note on stable sets and coloring of graphs. Comment. Math. Univ. Carolin. 15, 307–309 (1974)
19. Raychaudhuri, A.: On powers of interval and unit interval graphs. Congr. Numer. 459, 235–242 (1987)

Domatic Partition on Several Classes of Graphs[⋆,⋆⋆]

Sheung-Hung Poon[1], William Chung-Kung Yen[2], and Chin-Ting Ung[1]

[1] Department of Computer Science,
National Tsing Hua University, Hsinchu, Taiwan
spoon@cs.nthu.edu.tw, s9862657@m98.nthu.edu.tw
[2] Department of Information Management,
Shih Hsin University, Taipei, Taiwan
ckyen001@ms7.hinet.net

Abstract. The *domatic number* of a graph G, denoted by $DN(G)$, is the maximum number k such that V can be partitioned into k disjoint *dominating sets*. The *domatic partition problem* is to find a partition of the vertices of G into $DN(G)$ dominating sets. The k*-domatic partition problem* with fixed k is to find a partition of the vertices of G into k dominating sets. In this paper, we show that 3-domatic partition problem is NP-complete on planar bipartite graphs, and the domatic partition problem is NP-complete on co-bipartite graphs. We further show that the unique 3-domatic partition problem is NP-hard on general graphs. Moreover, we propose an $O(n)$-time algorithm on the 3-domatic partition problem for maximal planar graphs, and $O(n^3)$-time algorithms on the domatic partition problem for P_4-sparse graphs and tree-cographs, respectively.

1 Introduction

A *dominating set* in a graph $G = (V, E)$ is a subset D of the vertex set V such that every vertex $v \in V - D$ is adjacent to a vertex in D. The *domatic number* of a graph G, denoted $DN(G)$, is the maximum number k such that V can be partitioned into k disjoint *dominating sets*. The *domatic partition problem* is to find a partition of the vertices of G into $DN(G)$ dominating sets. The k*-domatic partition problem* with fixed k is to find a partition of the vertices of G into k dominating sets. The domatic partition problems are important and hard problems in the field of graph algorithms. Nearly three decades, there have been many applications and results on these problems. For instance, one application area lies in communication networks [4]. The network is modelled by an undirected graph in which edges represent communication links and vertices represent cities. A *transmitting group* is a set of cities which, acting as transmitting stations, can transmit messages to all cities in the network. Hence, a *transmitting group* in

⋆ This research was supported by National Science Council, Taiwan, under the grant numbers NSC99-2221-E-128-003 and NSC100-2628-E-007-020-MY3.

⋆⋆ Every author contributes to this research equally.

G. Lin (Ed.): COCOA 2012, LNCS 7402, pp. 245–256, 2012.

the network is a *dominating set* in the graph. However, when some party in the *transmitting group* fails to work, the whole network may not be fully served. Consequently, we tend to search for a maximum number of disjoint transmitting groups as backups. We can see that finding a maximum number of disjoint transmitting groups in a communication network is same as finding a domatic partition in the corresponding graph.

It is easy to see that $DN(G) \leq \delta(G) + 1$, where $\delta(G)$ is the minimum degree of G [5]. Graphs for which $DN(G) = \delta(G) + 1$ are called *domatically full*. The domatic partition problem is NP-complete for general graphs [9]. Cockayne and Hedetniemi [5] showed that trees, complete graphs, and maximal outerplanar graphs are domatically full. Bonuccelli [2] showed that the domatic partition problem is NP-complete even for circular-arc graphs and proposed an $O(n^2 \log n)$-time algorithm for proper circular-arc graphs. Bertossi [1] proposed an $O(n^{2.5})$-time algorithm for interval graphs and an $O(n \log n)$-time algorithm for proper interval graphs. Later, Lu et al. [12] and Rao and Rangan [16] independently showed that interval graphs are domatically full and gave an $O(m + n)$-time algorithm for interval graphs. Few years later, Manacher and Mankus [13] proposed an $O(n)$-time algorithm for interval graphs with sorted intervals. Moreover, Peng and Chang [15] showed that strongly chordal graphs are domatically full and proposed an $O(n + m)$-time algorithm for such kind of graphs. Tsui [19] further presented an $O(m+n)$-time algorithm for bipartite permutation graphs. In addition, Rautenbach and Volkmann [17] showed that block-cactus graphs are domatically full in some special cases. On the complexity side, Kaplan and Shamir [11] showed that the domatic partition problem for chordal graphs, co-chordal graphs, split graphs, bipartite graphs, comparability graphs, and uniquely partially orderable graphs are all NP-complete.

The rest of the paper is organized as follows. In Section 2, we show that 3-domatic partition problem is NP-complete on planar-bipartite graphs, and the domatic partition problem is NP-complete on co-bipartite graphs. We also show that the unique 3-domatic partition problem is NP-hard on general graphs. In Section 3, we propose an $O(n)$-time algorithm for the 3-domatic partition problem on maximal planar graphs, and $O(n^3)$-time algorithms for finding domatic partitions on P_4-sparse graphs and tree-cographs, respectively.

2 NP-Hardness

In this section, we show that the domatic partition problem is NP-complete for planar bipartite graphs and co-bipartite graphs, respectively. Moreover, we also show that the unique 3-domatic partition problem is NP-hard.

Planar bipartite graphs. A planar-bipartite graph is a graph that is both planar and bipartite. We show that the 3-domatic partition problem for such graphs is NP-complete.

Theorem 1. *The 3-domatic partition problem for planar-bipartite graphs with vertex degree at most 8 is NP-complete.*

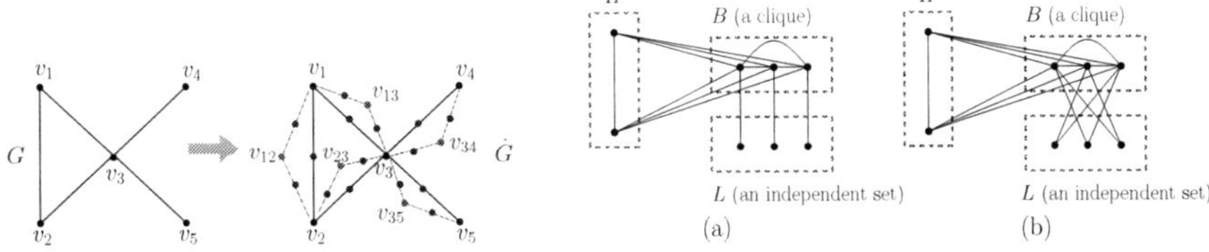

Fig. 1. An example construction for Thereom 1

Fig. 2. (a) A thin spider. (b) A thick spider.

Proof: Clearly, this problem is in NP. We give a reduction from the known NP-complete problem, the planar 3-colorability problem for a planar graph $G = (V, E)$ with vertex degree at most 4 [8]. The planar 3-colorability problem is to decide whether the vertices of G can be colored with three distinct colors such that each pair of adjacent vertices possesses different colors. We then construct a new graph $\hat{G} = (\hat{V}, \hat{E})$ by augmenting some new vertices and edges as follows. First, we add a new vertex v_{ij} for each original edge $(v_i, v_j) \in E$. Then, we connect v_{ij} to v_i and v_j, respectively. Precisely, $\hat{V} = V \cup \{v_{ij} | \ (v_i, v_j) \in E\}$ and $E' = E \cup \{(v_i, v_{ij}),\ (v_j, v_{ij}) | \ (v_i, v_j) \in E\}$. Now for each edge $e \in E'$, we add one new vertex x_e to divide edge e into two new edges, e_1 and e_2. Thus we obtain the final edge set $\hat{E} = \{e_1, e_2 | \ e \in E'\}$. See Figure 1 for an example construction.

Clearly, $\hat{G}$ is a planar-bipartite graph with vertex degree at most 8 and the construction of $\hat{G}$ takes polynomial time in terms of the size of graph G. We then need to claim that G is 3-colorable if and only if $\hat{G}$ has a 3-domatic partition. We first consider the forward direction. Suppose that G is 3-colorable. A 3-coloring of G induces a partition $\Pi = \{V_1, V_2, V_3\}$ where V_i is the set of vertices colored by color i. We then modify partition Π to another partition $\hat{\Pi} = \{\hat{V_1}, \hat{V_2}, \hat{V_3}\}$ as follows. First, we place all vertices of V_i into $\hat{V_i}$. Then, for each v_{ij}, if $v_i \in \hat{V_p}$ and $v_j \in \hat{V_q}$, we then assign v_{ij} to the third class $\hat{V_r}$ where $r \neq p, q$. Since v_{ij} is of degree two, we also have that $p \neq q$. Similarly, for each x_i, we assign x_i to the third class that is distinct from the two classes to which its two neighbors belong. Now, it is clear to see that each $\hat{V_i}$ is a dominating set of $\hat{G}$.

We then consider the backward direction. Suppose that $\hat{G}$ has a 3-domatic partition, $\hat{V_1}, \hat{V_2}$, and $\hat{V_3}$. For a new vertex x_e in $\hat{V}$, since the degree of x_e is 2, x_e and its two neighbors must belong to distinct dominating sets. Now, we color each vertex v of G by index number of the partition subset to which v belongs. Such a coloring for V is thus a 3-coloring for G. □

Co-bipartite graphs. A *co-bipartite graph* is the complement of a bipartite graph. We show, in this section, that the domatic partition problem for co-bipartite graphs is NP-complete.

Theorem 2. *The domatic partition problem for co-bipartite graphs is NP-complete.*

Proof: Clearly, this problem is in NP. We give a reduction from the set cover problem. The set cover problem is defined as follows: Let $A = \{a_1, a_2, a_3, ..., a_n\}$ be a finite set of size n and let $S = \{s_1, s_2, s_3, ..., s_p\}$ be a subset collection of A, that is $s_i \subseteq A$ and $|S| = p$. Then a subset S' of S is a set cover of A if $\bigcup_{s_i \in S'} s_i = A$. Given an integer $k \leq |S|$, the set cover problem is to check if there exists a set cover S', $|S'| \leq k$. We give one more restriction $n \leq p$ on the set cover problem instance. The reason is that if $n > p$, we can add some more occurrences of s_p into set S until $n = p$. The newly created set cover problem instance is equivalent to the original set cover problem. Thus the set cover problem is still NP-complete even under the restriction that $n \leq p$.

Given a set cover problem instance as described above, we will construct a co-bipartite graph $G = (V_1, V_2, E)$ as follows. First, we place all elements of A into V_1, and place all elements of S to V_2. If s_j contains a_i, we construct a new edge connecting a_i and s_j. See Figure 3(a). Second, we add $|p - n|$ new vertices $X = \{x_i | 1 \leq i \leq |p - n|\}$ into V_1, and we then connect each vertex $x_i \in X$ to all elements in S. See Figure 3(b). Furthermore, we add k isolated vertices $Y = \{y_j | 1 \leq j \leq k\}$ into V_2. See Figure 3(c). Finally, we add edges to form cliques on vertex sets V_1 and V_2, respectively. See Figure 3(d) for a complete constructed example for G.

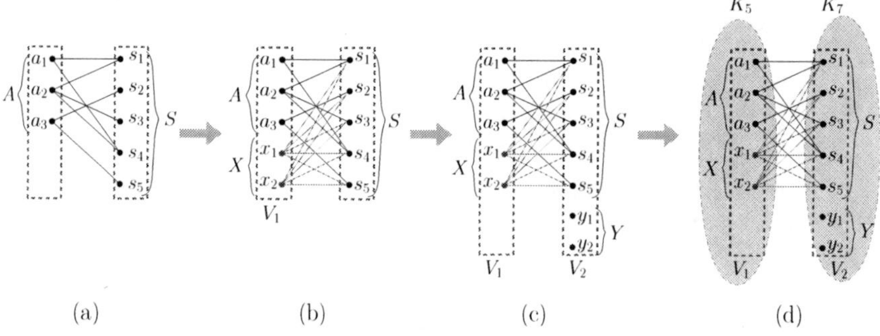

Fig. 3. We assume that $n = 3$, $p = 5$ and $k = 2$. $A = \{a_1, a_2, a_3\}$ and $S = \{s_1, s_2, s_3, s_4, s_5\}$, where $s_1 = \{a_1, a_2\}$, $s_2 = \{a_3\}$, $s_3 = \{a_2\}$, $s_4 = \{a_1, a_2\}$, and $s_5 = \{a_3\}$. $X = \{x_1, x_2\}$ and $Y = \{y_1, y_2\}$. Also $V_1 = A + X$, and $V_2 = S + Y$. In figure (d), $G[V_1] = K_5$ and $G[V_2] = K_7$, where $G[V_i]$ is the induced subgraph of G with vertex set V_i.

Clearly, G is a co-bipartite graph and the construction of G takes polynomial time in terms of n and p. Now, we need to claim that there is a set cover S' of A such that $|S'| \leq k$ if and only if there exist a domatic partition which contains at least $\beta = p + 1$ dominating sets. We first consider the proof for the forward direction. Suppose that there is a set cover S' of A such that $|S'| \leq k$. Since S' is a set cover and it is also a subset of V_2, S' is clearly a dominating set of G. Moreover, since V_1 and V_2 are two cliques, we can find p more dominating sets, each of which is formed by picking one vertex from V_1 and another vertex from $V_2 - S'$. Thus we obtain $p + 1$ dominating sets.

Next we consider the proof for backward direction. Assume that we obtain $\beta = p + 1$ dominating sets in G. We need to show that there is a set cover S' of A such that $|S'| \le k$. First, we observe that we have k vertices y_i in Y that do not connect to any vertex in V_1. Thus in order to dominate the vertices in Y, every dominating set must contain at least one vertex in V_2. Among these β dominating sets, suppose that there does not exist any dominating set that is completely contained in V_2, i.e., each of these β dominating sets contains at least one vertex in V_1. Then we could only obtain at most p dominating sets. This contradicts to our assumption. Therefore, there exists one dominating set D that is completely contained in V_2. We can simply assume that D does not contain any vertex in Y. If D contain some vertices in Y, we can just delete them and $D - Y$ is still a dominating set. It is clear that D is a set cover of A. If $|D| > k$, then $|V_2 - D| < p$ and we can only obtain at most p dominating sets in total. This is again a contradiction. Hence, we have shown that the set cover D has size $|D| \le k$. □

Unique 3-domatic partition problem. The *unique 3-domatic partition problem* on a graph G is to decide whether there is a unique 3-domatic partition in G. We show that the unique 3-domatic partition problem on general graphs is NP-hard. We reduce the unique 3-colorability problem [14] to this problem. Its proof is omitted here.

Theorem 3. *The unique 3-domatic partition problem is NP-hard.*

3 Domatic Partition Algorithms

In this section, we present three polynomial time algorithms for domatic partition problems on P_4-sparse graphs, tree-cographs, and maximal planar graphs, respectively.

3.1 Domatic Partition on P_4-Sparse Graphs

A *P_4-sparse graph* is a graph in which every set of five vertices induces at most one P_4. Such graphs are super class of cographs. Hoáng [10] showed that a graph G is a P_4-sparse graph if and only if G has a corresponding decomposition tree $\mathcal{T}$ which can be constructed from vertices by performing a finite sequence of union, join, thin spider and thick spider operations, which are denoted by operators ⓪, ①, ②, and ③, respectively. Suppose that we are given a P_4-sparse graph G. Our algorithm uses dynamic programming strategy by performing one of these four operations on each internal node of the decomposition tree of G in a bottom-up fashion. In the following, we introduce these four operations, respectively. Given two disjoint graphs $G_1 = (V_1, E_1)$ and $G_2 = (V_2, E_2)$ with $|V_1| = n_1$ and $|V_2| = n_2$, we define *union*, and *join* operations as the following formulas:

- $G_1$⓪$G_2 = (V_1 \cup V_2, E_1 \cup E_2)$;
- $G_1$①$G_2 = (V_1 \cup V_2, E_1 \cup E_2 \cup \{xy|\ x \in V_1, y \in V_2\})$.

A *spider* is a graph whose vertices can be partitioned into three vertex sets, H, B, and L such that B is a clique with k vertices, L is an independent set with k vertices, H is a general graph with m vertices, and H and B join together. Let f be a bijection from L to B. when each vertex v in L is only connected to one vertex $f(v)$ in B, we call such a graph a *thin spider*. See Figure 2(a) for an example. On the other hand, when each vertex v in L is connected to all vertices in $B - f(v)$, we call such a graph a *thick spider*. See Figure 2(b) for an example. We now have a close look on how spider operations appear in the decomposition tree T in Hoáng's construction [10]. Consider a spider S with corresponding vertex sets, H, B, and L, and edge set E. In T, the spider operation for S involves two subgraphs G_1 and G_2, which are given as follows: $G_1 = (V_1, \phi)$ and $G_2 = (V_2, E_2)$, where $V_1 = L - \{v\}$ and $V_2 = \{v\} \cup H \cup B$ for some $v \in L$, and E_2 contains the edges in the induced subgraph $S[V_2]$ of spider S. We also note that $|L| = |V_1| + 1 = |B|$, and every vertex in H is adjacent to every vertex in B. Now we can examine the neighborhood $N(v)$ of vertex v to determine whether the corresponding spider S is thin or thick. If $N(v) = \{v'\}$ for some vertex $v' \in B$, then S is a thin spider, and we need to perform thin spider operation ②; otherwise if $N(v) = B - \{v'\}$ for some vertex $v' \in B$, then S is a thick spider, and we need to perform thick spider operation ③; Let the bijection $f : L \to B$ be the corresponding bijection for spider S. Note that $f(v) = v'$. We then describe the thin and thick spider operations, ② and ③, as follows.

- $G_1 ② G_2 = (V_1 \cup V_2, E_2 \cup E')$ with $E' = \{xf(x)|\ x \in V_1\}$;
- $G_1 ③ G_2 = (V_1 \cup V_2, E_2 \cup E')$ with $E' = \{xz|\ x \in V_1, z \in B - \{f(x)\}\}$.

In our dynamic programming algorithm, we need to compute the result for one of the above four operations corresponding to each internal node in the decomposition tree T of G. In order to compute the results for the operations, we need to introduce the following new concept. For any integer m, an *m-vertex domatic partition* of a graph G is a collection $D_1, D_2, ...D_k$ of k pairwise disjoint dominating sets of G such that $|D_1 \cup D_2 \cup ... \cup D_k| \leq m$. The *$m$-vertex domatic number* $DN(G|m)$ of G is the maximum k such that there are k pairwise disjoint dominating sets for m-vertex domatic partition problem. Note that $DN(G) = DN(G|n)$ for graph G of n vertices. Now we consider an internal node v in the decomposition tree T of G. Let G_v be subgraph corresponding to the subtree v in the decomposition tree. We denote the vertex set of graph G_v by $V(G_v)$. We plan to store at v all values of $DN(G_v|m)$, where $m = 1, 2, \ldots, |V(G_v)|$. Let G_1 and G_2 be the subgraphs corresponding to two child subtrees of v. Then in our dynamic programming, we need to compute $DN(G_v|m)$ from $DN(G_1|i)$ and $DN(G_2|j)$ for varying values of i, j and m when G_v is obtained by performing the corresponding operation τ_v at v on G_1 and G_2. We need to divide into four cases, each of which corresponds to one of the four aforementioned operations. We first consider the case that τ_v is the join operation. This case has been done by Chang [3], who obtained the formula to compute the result m-vertex domatic number $DN(G_v|m)$ for the join operation. See the following lemma. In the following, we let n_1 (resp. n_2) be the number of vertices of G_1(resp. G_2).

Lemma 1. *[3] Suppose that G_1, G_2 does not contain any dominating vertex. Then*

$$DN(G_1 \textcircled{1} G_2|m) = \begin{cases} \lfloor \frac{m}{2} \rfloor, & \text{if } 0 \le m \le 2n_1 \\ n_1 + DN(G_2|m - 2n_1), & \text{if } 2n_1 < m \le n_1 + n_2. \end{cases}$$

Next, we consider the case that τ_v is the union operation. It is clear that the domatic number $DN(G_1 \textcircled{0} G_2)$ of the union of G_1, G_2 is the minimum number of $DN(G_1)$ and $DN(G_2)$. By this observation, we obtain the following lemma.

Lemma 2. *For $1 \le m \le n_1 + n_2$,*

$$DN(G_1 \textcircled{0} G_2|m) = \max\{\min(DN(G_1|i),\ DN(G_2|m - i)) :\ \forall i = 1, 2, \ldots, m\}.$$

Since cographs can be constructed using these two operations only, thanking to these two formulas, we thus obtain a dynamic programming algorithm for the domatic partition problem on cographs. Now, we proceed to derive the formulas for the two remaining operations, the thin and thick spider operations. The proof of Lemma 4 about thick spider is omitted here.

Lemma 3. *Let thin spider G_v be the corresponding subgraph at node v of T, and $G_1 = (V_1, \phi)$ and $G_2 = (V_2, E_2)$ be its two disjoint child subgraphs such that $G_v = G_1 \textcircled{2} G_2$, where H, B, and L are, respectively, the head, body, and leg of G_v, and $V_1 = L - \{v\}$ and $V_2 = \{v\} \cup H \cup B$ for some $v \in L$. Also let $n_1 = |V_1|$, $n_2 = |V_2|$, and $k = |L| = |B| = n_1 + 1 \ge 2$. Then*

$$DN(G_v|m) = \begin{cases} 0, & \text{if } 0 \le m < k \\ 1, & \text{if } k \le m < 2k \\ 2, & \text{if } 2k \le m \le n_1 + n_2. \end{cases}$$

Proof: First we note that $n_2 = 1 + |H| + |B| > k + 1$, L is an independent set and B is a clique. Let the bijection $f : L \to B$ be the corresponding bijection for spider G_v. Since the edges $(v, f(v))$ for all vertices $v \in L$ are the only edges between L and B, if we want to dominate all vertices in L, we have to select either v or $f(v)$ for each edge $(v, f(v))$ between L and B. As $|L| = k$, any dominating set of spider G_v contains at least k vertices. Thus if $0 \le m < k$, there cannot be any dominating set of size m for spider G_v.

Next we consider the case that $k \le m < 2k$. We can form a dominating set D of size k as follows: $D = \{f(v) : v \in L\}$. Since $m - k < k$, we cannot form another dominating set of size $m - k$. Thus we can only obtain one dominating set. Lastly, we consider the case that $2k \le m \le n_1 + n_2$. First of all, as the minimum vertex degree of G_v is one, there exist at most two disjoint dominating sets in G_v. We can form two disjoint dominating sets D_1, D_2 by taking vertices in turn from B and L. $|D_1| = |D_2| = k$. Hence, for this case, the domatic number is just two. □

Lemma 4. *Let thick spider G_v be the corresponding subgraph at node v of T, and $G_1 = (V_1, \phi)$ and $G_2 = (V_2, E_2)$ be its two disjoint child subgraphs such that $G_v = G_1 \textcircled{3} G_2$, where H, B, and L are, respectively, the head, body, and leg of*

G_v, and $V_1 = L - \{v\}$ and $V_2 = \{v\} \cup H \cup B$ for some $v \in L$. Also let $n_1 = |V_1|$, $n_2 = |V_2|$, and $k = |L| = |B| = n_1 + 1 \geq 2$. Then

$$DN(G_1 ③ G_2|m) = \begin{cases} \lfloor \frac{m}{2} \rfloor, & \text{if } 0 \leq m \leq 2k \\ k, & \text{if } 2k < m \leq n_1 + n_2. \end{cases}$$

The dynamic programming runs from the leaves of decomposition tree $\mathcal{T}$ in a bottom-up fashion. Whenever we encounter an internal node v labeled ⓪, ①, ②, or ③, we perform the specific operation using the corresponding lemma among Lemmas 1—4. In fact, we compute and store the domatic numbers $DN(G_v|m)$ for all $m = 1, 2, \ldots, |V(G_v)|$. In order to compute one value $DN(G_v|m)$ at node v from its two child subgraphs, G_1 and G_2, it takes $O(1)$ time if the operation is join, thin spider or thick spider whereas it takes $O(|V(G_v)|) = O(n)$ time if the operation is union. Since there are $O(n)$ nodes in $\mathcal{T}$ and we store $O(n)$ domatic numbers at each node of $\mathcal{T}$, the dynamic programming takes $O(n^3)$ time. In the above description, we only compute and store all domatic numbers $DN(G_v|m)$ at any node v in T. In fact, we can even compute and store the corresponding m-vertex domatic partitions at v as well. The time complexity still stays the same. We thus have an $O(n^3)$-time algorithm for finding the domatic partitions on P_4-sparse graphs, which is stated as the following theorem. Its detailed proof is omitted here.

Theorem 4. *There is an $O(n^3)$-time algorithm for the domatic partition problem on P_4-sparse graphs.*

3.2 Domatic Partition on Tree-Cographs

In this section, we will present an $O(n^3)$-time algorithm to solve domatic partition problem for tree-cographs. *Tree-cographs* are defined recursively as follows: (*i*) trees are tree-cographs; (*ii*) the union of tree-cographs is still a tree-cograph; (*iii*) the complement of a tree-cograph is also a tree-cograph [18]. We observe that the complement of an union operation $G_1 ⓪ G_2$ of two subgraphs G_1 and G_2 is the join of their complements, i.e., $\overline{G_1 ⓪ G_2} = \overline{G_1} ① \overline{G_2}$. Symmetrically, $\overline{G_1 ① G_2} = \overline{G_1} ⓪ \overline{G_2}$. Thus any tree-cograph G can be represented by a corresponding decomposition tree $\mathcal{T}$ such that any internal node of $\mathcal{T}$ is labeled as an union node ⓪ or a join node ①, and the leaves of $\mathcal{T}$ stores the trees and the complements of trees. Our algorithm uses dynamic programming strategy by performing the union or join operation on each internal node of the decomposition tree $\mathcal{T}$ of G in a bottom-up fashion. Since we have known how to perform an union or join operation on two subgraphs in previous section. Therefore, in this section, we only need to compute the *m-vertex domatic number* $DN(H|m)$, $m = 1, 2, \ldots, |V(H)|$, where H is a tree or the complement of a tree. In the following two subsections, we present the algorithms for computing the m-vertex domatic numbers for a tree and the complement of a tree, respectively.

Domatic partition on trees. Since $\delta(T) = 1$ for any tree T, by either selecting all vertices on odd levels or on even levels, we obtain two disjoint dominating sets for T. Thus the domatic number of tree T is always equal to two, and consequently $DN(T|m)$ only have three possible choices, 0, 1, or 2. Let $\sigma_1(T)$ be the size of minimum dominating set of T, and let $\sigma_2(T)$ be the minimum size of the union of any two disjoint dominating sets of T. Goodman [6] presented a linear time algorithm to compute $\sigma_1(T)$ and the corresponding minimum dominating set for a tree. $DN(T)$ is at most two. In fact, we can always form two disjoint dominating sets in T. On the other hand, in the following, we show how to compute $\sigma_2(T)$ on tree T in linear time.

Let D_1, D_2 be the two disjoint dominating sets in T realizing the optimal value $\sigma_2(T)$. In order to compute $\sigma_2(T)$, our algorithm uses dynamic programming, which runs in bottom-up fashion by starting from the leaves of tree T. At any internal node v, we assume that v has k children $c_1, c_2, \ldots, c_k$ and one parent p. Let T_v be the subtree rooted at v. See Figure 4 for illustration. At node v, three possible domination situations can happen: v belongs to either D_1 or D_2, or none of them. At its parent node p, the same three domination situations can happen. Thus by considering the domination situations of v and p by D_1 and D_2 altogether, we have nine different cases in total. When the dynamic programming runs up to this node v, we need to compute and store the optimal domination values $\sigma_2(T_v)$ on subtree T_v for all these nine possible restricted domination situations on nodes v and p. Its full proof omitted here. As an example, we will consider how to deal with the case that v and p both belong to D_1. See Figure 4. Since v is already dominated by D_1, at least one child of v has to lie in D_2. First, for each child c_i of v, we compute the minimum m_i among the three domination values $\sigma_2(T_{c_i})$ of T_{c_i} under the restrictions that the parent v of c_i belongs to D_1 and c_i may belong to either D_1 or D_2, or none of them. We examine the domination situations corresponding to m_i for all children c_i of v, and if there is some child c_i which belongs to D_2, then we are done for this case. Otherwise, we need to change the domination situation of some child c_j to a new one such that c_j belongs to D_2 in this new domination situation. We traverse all children of v to search for a child which has minimum absolute difference Δ among the difference values between m_i and the minimum domination value $\sigma_2(T_{c_i})$ under the restriction that c_i belongs to D_2 and the parent v of c_i belongs to D_1. We denote such a child corresponding to minimum difference Δ by c_j. Then we can simply change the domination situation for subtree T_{c_j} by the domination situation corresponding to the minimum difference Δ. The proof for the rest of the cases is omitted. We thus obtain the following lemma.

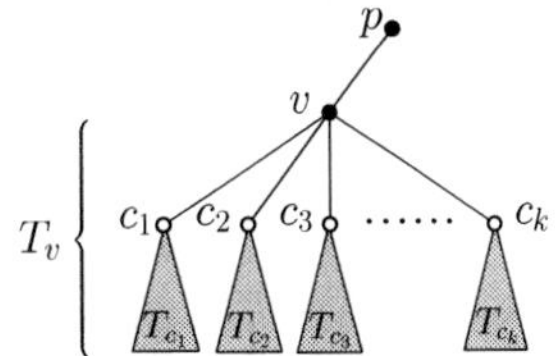

Fig. 4. The subtree T_v at node v

Lemma 5. *There is a linear time algorithm to compute $\sigma_2(T)$ for a tree T.*

With these two values $\sigma_1(T)$ [6] and $\sigma_2(T)$ (from Lemma 5), we can compute the m-vertex domatic numbers for tree T, as stated in the following lemma.

Lemma 6. *Let T be a tree with $|V(T)| \geq 2$. Then*

$$DN(T|m) = \begin{cases} 0, & \text{if } 0 \leq m < \sigma_1(T) \\ 1, & \text{if } \sigma_1(T) \leq m < \sigma_2(T) \\ 2, & \text{if } \sigma_2(T) \leq m \leq |V(T)| \end{cases}$$

Domatic partition on complements of trees. In this part, we first show that the domatic number $DN(\overline{T})$ of the complement $\overline{T}$ of a tree T can be computed in $O(n^{2.38}$ time. Its proof is omitted here. With this domatic number, we immediately obtain the following formula for computing the m-vertex domatic number on the complement $\overline{T}$, where $m = 1, 2, \ldots, |V(T)|$.

Lemma 7. *Let $\overline{T}$ be the complement of a tree T. Then*

$$DN(\overline{T}|m) = \begin{cases} \lfloor \frac{m}{2} \rfloor, & \text{if } 0 \leq m < 2 \cdot DN(\overline{T}) \\ DN(\overline{T}), & \text{if } 2 \cdot DN(\overline{T}) \leq m \leq |V(T)|. \end{cases}$$

Our dynamic programming algorithm runs in a bottom-up fashion on the decomposition tree $\mathcal{T}$ of the given tree-cograph G. Thus we obtain the following theorem, whose proof is omitted here.

Theorem 5. *There is an $O(n^3)$-time algorithm for the domatic partition problem on tree-cographs.*

3.3 3-Domatic Partition on Maximal Planar Graphs

In Section 2, we have shown that the 3-domatic partition problem on general planar bipartite graphs is NP-complete. To the other end, we find a surprising result that the 3-domatic partition problem on maximal planar graphs can in fact be solved in $O(n)$ time. A *maximal planar graph* G is a planar graph to which no new edge can be added without violating the planarity of G. In other words, a graph G is a maximal planar graph if every face is bounded by three edges. Fraysseix, Pach, and Pollack [7] showed that in any maximal planar graph G, there exists a labeling of vertices, called *canonical labeling*, which is defined in Definition 1. Moreover, they also showed that such a labeling can be computed in $O(n)$ time.

Definition 1. *Let G be a maximal planar graph of n vertices with exterior face uvw. Then the canonical labeling of vertices of G is a labeling of vertices $v_1 = u, v_2 = v, v_3, ..., v_n = w$ satisfying following conditions for each k $(4 \leq k \leq n)$.*

(i) Boundary of exterior face of subgraph G_{k-1} of G induced by $\{v_1, v_2, ..., v_{k-1}\}$ is a cycle C_{k-1} containing edge uv.
(ii) v_k is in the exterior face of G_{k-1}, and its neighbors in G_{k-1} are some (at least two) consecutive vertices along the path $C_{k-1} - uv$.

See Figure 5(a) for an example of a canonical labeling. Such a labeling is the main technique we use to derive an $O(n)$-time algorithm for the 3-domatic partition problem on maximal planar graphs. We state our result in the following theorem.

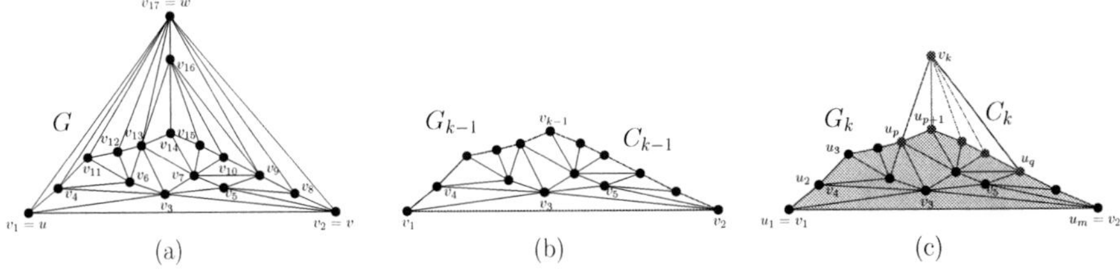

Fig. 5. (a) A canonical labeling $v_1 = u, v_2 = v, v_3, ..., v_n = w$; (b) G_{k-1} and C_{k-1}; (c) G_k and C_k

Theorem 6. *There is an $O(n)$-time algorithm for the 3-domatic partition problem on maximal planar graphs.*

Proof: Suppose that G is a maximal planar graph with a *canonical labeling* $v_1, v_2, \ldots, v_n$. Let G_k denote the subgraph induced by $\{v_1, v_2, ..., v_k\}$, and C_k denote the outer boundary of G_k. We conduct our proof by applying induction on $n \geq 3$. In the basis step, the graph G_3 is a triangle $\triangle v_1v_2v_3$, and by setting the three dominating sets $D_1 = \{v_1\}, D_2 = \{v_2\}$, and $D_3 = \{v_3\}$, we obtain the 3-domatic partition $\{D_1, D_2, D_3\}$ of G_3. In the hypothesis step, we assume that there exists a 3-domatic partition $\Pi = \{D_1, D_2, D_3\}$ for graph G_{k-1}, $k \geq 4$, such that each pair of consecutive vertices on cycle C_{k-1} belongs to distinct sets in partition Π.

In the inductive step, we need to consider how to obtain a 3-domatic partition for the graph G_k. See Figure 5(c) for an example of G_k. We know that G_{k-1} is obtained by removing vertex v_k from G_k. Thus we plan to extend the domination on G_{k-1} to the graph G_k by re-inserting vertex v_k back into graph G_{k-1}. By induction hypothesis, we know that there exists a 3-domatic partition $\Pi = \{D_1, D_2, D_3\}$ for graph G_{k-1}, $k \geq 4$, such that each pair of consecutive vertices on cycle C_{k-1} belongs to distinct sets in partition Π. Now we only need to decide to which set among D_1, D_2, and D_3 vertex v_k belongs. Since v_k is only adjacent to some consecutive vertices on C_{k-1}, the answer to this question depends only on the domination situation of the vertices on C_{k-1} adjacent to v_k. Thus we first let $C_{k-1} = u_1u_2 \ldots u_m$ such that $u_1 = v_1, u_m = v_2$, as shown in Figure 5(c). We also let $u_p, u_{p+1}, \ldots, u_q$, $1 \leq p < q \leq m$, be the consecutive sequence of vertices on C_{k-1} adjacent to v_k. Since each pair of consecutive vertices on cycle C_{k-1} belongs to distinct sets in partition Π, the neighbors of v_k cannot all belong to some same set in Π. Now, we need to consider two cases for our proof. First, if the neighbors of v_k belong to either one of two sets in partition Π, say D_i, D_j $(i \neq j)$, then we only need to place v_k in the third set D_k which is distinct from D_i, D_j, i.e., $k \neq i, j$. In another situation, the neighbors of v_k belong to either one of all three sets in Π. For this case, we suppose that $u_p \in D_i$ and $u_q \in D_j$, where $D_i, D_j \in \Pi$. Then we can we can simply place v_k in another set D_k $(k \neq i, j)$ in partition Π. Thus the updated partition $\Pi = \{D_1, D_2, D_3\}$ is a 3-domatic partition for graph G_k satisfying the condition that each pair of consecutive vertices on cycle C_{k-1} belongs to distinct sets in partition Π.

In our algorithm, we only need to traverse all edges in G according to the order of the canonical labeling on vertices of G. As any planar graph has $O(n)$ edges, the time complexity of our algorithm is thus $O(n)$. □

References

1. Bertossi, A.A.: On the domatic number of internal graphs. Information Processing Letters 28, 275–280 (1988)
2. Bonuccelli, M.A.: Dominating sets and domatic number of circular arc graphs. Discrete Applied Mathematics 12, 203–213 (1985)
3. Chang, G.J.: The domatic number problem. Discrete Mathematics 125(1-3), 115–122 (1994)
4. Cockayne, E.J., Hedetniemi, S.T.: Optimal domination in graphs. IEEE Transactions on Circuits and Systems 22, 855–857 (1975)
5. Cockayne, E.J., Hedetniemi, S.T.: Towards a theory of domination in graphs. Networks 7, 247–261 (1977)
6. Cockayne, E.J., Hedetniemi, S.T.: A linear algorithm for the domination number of a tree. Information Processing Letters 4, 41–44 (1975)
7. Fraysseix, H., Pach, J., Pollack, R.: How to draw a planar graph on a grid. Combinatorica 10, 41–51 (1990)
8. Garey, M.R., Johnson, D.S., Stockmeyer, L.: Some simplified np-complete graph problems. Theoretical Computer Science 1, 237–267 (1976)
9. Garey, M.R., Johnson, D.S.: Computers and intractability: A guide to the theory of np-completeness (1979)
10. Hoang, C.: Ph.d. thesis. McGill University, Montreal, Canada (1985)
11. Kaplan, H., Shamir, R.: The domatic number problem on some perfect graph families. Information Processing Letters 49, 51–56 (1994)
12. Lu, T.L., Ho, P.H., Chang, G.J.: The domatic number problem in interval graphs. SIAM Journal on Discrete Mathematics 3, 531–536 (1990)
13. Manacher, G.K., Mankus, T.A.: Finding a domatic partition of an interval graph in time o(n). SIAM Journal on Discrete Mathematics 9, 167–172 (1996)
14. Dailey, D.P.: Uniqueness of colorability and colorability of planar 4-regular graphs are np-complete. Discrete Mathematics 30, 289–293 (1980)
15. Peng, S.L., Chang, M.S.: A simple linear time algorithm for the domatic partition problem on strongly chordal graphs. Information Processing Letters 43, 297–300 (1992)
16. Rao, A.S., Rangan, C.P.: Linear algorithm for domatic number problem on interval graphs. Information Processing Letters 33, 29–33 (1989)
17. Rautenbach, D., Volkmann, L.: The domatic number of block-cactus graphs. Discrete Mathematics 187, 185–193 (1998)
18. Tinhofer, G.: Strong tree-cographs are birkhoff graphs. Discrete Applied Mathematics 22, 275–288 (1988)
19. Tsui, K.W.: On the domatic number of bipartite permutation graphs. Master thesis, Dept. of Computer Science, National Tsing Hua University, Hsinchu, Taiwan, R.O.C (2010)

Online Bottleneck Matching

Barbara M. Anthony[1] and Christine Chung[2]

[1] Math and Computer Science Department, Southwestern University, Georgetown, TX, USA
anthonyb@southwestern.edu

[2] Department of Computer Science, Connecticut College, New London, CT, USA
cchung@conncoll.edu

Abstract. We consider the online bottleneck matching problem, where k server-vertices lie in a metric space and k request-vertices that arrive over time each must immediately be permanently assigned to a server-vertex. The goal is to minimize the maximum distance between any request and its server. Because no algorithm can have a competitive ratio better than $O(k)$ for this problem, we use resource augmentation analysis to examine the performance of three algorithms: the naive GREEDY algorithm, PERMUTATION, and BALANCE. We show that while the competitive ratio of GREEDY improves from exponential (when each server-vertex has one server) to linear (when each server-vertex has two servers), the competitive ratio of PERMUTATION remains linear. The competitive ratio of BALANCE is also linear with an extra server at each server-vertex, even though it has been shown that an extra server makes it constant-competitive for the min-weight matching problem.

1 Introduction

We consider the online bottleneck matching problem, where we are given k server-vertices located in a metric space, and k request-vertices that arrive over time. As each request-vertex arrives, it must be immediately and permanently matched to a server-vertex. Our goal is to minimize the maximum distance between any request-vertex and its assigned server-vertex.

The standard technique for studying algorithms for online problems is competitive analysis. The *competitive ratio* of an algorithm is the worst-case ratio of the cost of the algorithm's solution to the cost of the optimal offline solution (which knows all request locations in advance). Kalyansundaram and Pruhs [4] proposed an algorithm, PERMUTATION, in the context of the corresponding online *min-weight* matching problem, where the goal is to minimize the total (or average) distance between request-vertices and server-vertices. Without proof, [4] mentioned that PERMUTATION achieves a competitive ratio of $2k - 1$ for the online bottleneck matching problem. Idury and Schäffer [3] then proved that no algorithm can achieve a competitive ratio better than approximately $1.5k$. The basic GREEDY algorithm, which assigns each arriving request to the nearest available server-vertex, has a competitive ratio that is $\Omega(2^k)$ (see Section 2).

The prohibitive general lower bound on the problem and the exceedingly poor performance of a simple and natural algorithm like GREEDY motivate us to consider a benchmark that is less formidable than the optimal solution, in order to attain a more

G. Lin (Ed.): COCOA 2012, LNCS 7402, pp. 257–268, 2012.

informative analysis of these algorithms. Specifically, we employ a *weak adversary model* of analysis in pursuit of further insight on the performance of these (and related) algorithms for the bottleneck matching problem. The weak adversary, or *resource augmentation*, model of analysis has long been used effectively in the study of matching and scheduling problems (e.g., [5,6,8,1]). Results obtained under this model can be viewed as "bicriteria" results, which have also become an informative and successful approach in other sub-fields of algorithms (e.g., [9,2]).

In our setting with resource augmentation, we ask how well the online algorithm performs when it has multiple servers (namely two) per server-vertex, while the optimal offline solution only has one; thus the online algorithm can service twice as many request-vertices with each server-vertex. Following [5], we will use the term *halfOPT-competitive ratio* to refer to the competitive ratio of an online algorithm with server-vertices that have two servers when compared with an optimal offline solution with each server-vertex having a single server.

Resource augmentation was used to study the corresponding online min-weight matching problem in [5]. They showed that by having two servers per server-vertex, the competitive ratio of GREEDY improves from $\Theta(2^k)$ to a halfOPT-competitive ratio of $\Theta(\log k)$. They then proposed an algorithm BALANCE, which is a modified form of GREEDY that is more judicious in its use of the additional server at each server-vertex. They show that BALANCE has a halfOPT-competitive ratio of $O(1)$.

Our results for the online bottleneck matching problem for $k \geq 2$ are as follows. (Naturally, when there is a single request-vertex and server-vertex ($k = 1$) the algorithms all perform optimally.)

1. GREEDY has a competitive ratio of at least 2^{k-1}, and at most $k2^{k-1}$.
2. PERMUTATION (proposed in [4] and [7]) is $(2k - 1)$-competitive, and this is tight. This is comparable to its performance for the min-weight objective, for which it is also $(2k - 1)$-competitive. This $O(k)$ upper bound on the ratio is asymptotically tight with the $\Omega(k)$ general lower bound for the problem of [3].
3. GREEDY has a halfOPT-competitive ratio of no more than $(k - 1)$. Note that this is an exponential improvement in competitive ratio from simply having two servers available per server-vertex.
4. GREEDY has a halfOPT-competitive ratio of at least $(k + 1)/2$. Interestingly, this is still exponentially worse than its performance for the corresponding min-weight problem, where it has a halfOPT-competitive ratio of $2 \log k$ [5].
5. BALANCE (proposed in [5]), a modified form of GREEDY designed for the setting of multiple servers per server-vertex, has a halfOPT-competitive ratio of $k - 1$.
6. BALANCE has a halfOPT-competitive ratio of at least $(\frac{1}{c} + 1)^{\log(k+1)-1} = \Omega(k)$. This is in contrast with the fact that BALANCE has a halfOPT-competitive ratio of $O(1)$ for the corresponding min-weight problem [5].
7. PERMUTATION has a halfOPT-competitive ratio of k and this is tight. (Note that having two servers per server-vertex does not improve PERMUTATION's asymptotic performance guarantee, as it did so dramatically with GREEDY.)

A table summarizing these and related results is shown below.

Table 1. Lower bounds and upper bounds for the various algorithms. All bottleneck objective results are from the present work, though the PERMUTATION bounds without resource augmentation were hinted at in [4]. The result marked by † is immediate from the corresponding bound without resource augmentation. BALANCE is only defined in the resource augmentation setting.

Objective	Adversary	GREEDY LB	GREEDY UB	PERMUTATION LB	PERMUTATION UB	BALANCE LB	BALANCE UB
min-bottleneck	OPT	2^{k-1}	$k2^{k-1}$	$2k-1$	$2k-1$	N/A	N/A
	halfOPT	$(k+1)/2$	$k-1$	k	k	$\Omega(k)$	$k-1$
min-weight	OPT [4]	2^k-1	2^k-1	$2k-1$	$2k-1$	N/A	N/A
	halfOPT	$\Theta(\log k)$ [5]		$O(1)$	$2k-1^{\dagger}$	$\Theta(1)$ [5]	

While resource augmentation has the potential to improve the competitive ratio, these results suggest that in some sense the bottleneck objective is more difficult than the total distance objective. Resource augmentation greatly helps GREEDY for the minimum weight objective, but none of the three algorithms break the $\Omega(k)$ barrier for the bottleneck objective. Perhaps this can be explained by noting that for the minimum weight objective, any sub-optimal assignment is mitigated by the total cost, whereas with the bottleneck objective, a poor assignment can dominate, even with resource augmentation. Our results suggest that GREEDY can be a reasonable choice of algorithm for the bottleneck objective with resource augmentation, due to its relative simplicity, and comparable performance to BALANCE and PERMUTATION, despite its decay in performance as its adversary gets stronger.

Section 2 provides some results for the algorithms without resource augmentation. We then consider three algorithms with resource augmentation: GREEDY (Section 3), BALANCE (Section 4), and PERMUTATION (Section 5).

2 Preliminaries

Formally, the online bottleneck matching problem is as follows: Given a collection $S = \{s_1, s_2, \ldots, s_k\}$ of server-vertices in a metric space M, the online algorithm A sees over time a sequence of request-vertices $R = \{r_1, r_2, \ldots, r_k\}$ also in M. When request-vertex r_i arrives, algorithm A must assign a server-vertex $s_{\sigma(i)}$ to service that request, with cost equaling the distance $d(r_i, s_{\sigma(i)})$ (we use the terms cost and distance interchangeably). Once an assignment is made, it cannot be changed. While A does not know the sequence of requests in advance, its goal is to minimize the bottleneck distance of the overall assignment, that is minimize $\max_i \ d(r_i, s_{\sigma(i)})$. We refer to the assignment (or "matching") that optimizes this objective as OPT. As is typical of online problems, we use competitive analysis, and seek to minimize the worst-case ratio of the online bottleneck cost to the optimal (offline) bottleneck cost. An online algorithm is α-competitive if this ratio is at most α for all possible instances. We use $cost(\cdot)$ to represent the bottleneck weight of a particular assignment, e.g. $cost$(OPT). Throughout the paper, $\epsilon > 0$ represents an arbitrarily small constant, typically used to break ties when assigning requests to servers.

We now prove a few basic results about the online bottleneck matching problem without resource augmentation that have been hinted at in the existing literature (e.g., see the Conclusion of [4]). We consider both the standard GREEDY algorithm, as well as PERMUTATION, introduced by Kalyanasundaram and Pruhs (a similar algorithm was also studied by [7]). Note that the algorithm BALANCE is only defined when there are multiple servers per server-vertex.

2.1 Analysis of GREEDY

As its name suggests, GREEDY assigns the nearest available server at a server-vertex to each request-vertex as it arrives. While this algorithm can perform well on some instances, GREEDY is exponentially bad against OPT. In fact, this can be exhibited by the same instance of [4] that demonstrates GREEDY is exponentially bad against OPT for the corresponding objective of minimizing total weight.

The proofs of the following two theorems can be found in the full version of the paper.

Theorem 1. *The competitive ratio of* GREEDY *is at least* 2^{k-1} *for the bottleneck matching problem.*

Theorem 2. *The competitive ratio of* GREEDY *is at most* $k2^{k-1}$ *for the bottleneck matching problem.*

2.2 Analysis of PERMUTATION

Informally, PERMUTATION assigns requests as follows. Note that the assignment of request-vertices to server-vertices is a matching. To choose a server for request r_i, consider the optimal matching of the first i requests, and the optimal matching of the first $i-1$ requests. There is exactly one server that is matched in the former scenario and not in the latter. PERMUTATION matches that server to the current request r_i. Observe that PERMUTATION guarantees that if a request arrives at an unused server-vertex, it is matched to the server at that server-vertex.

More formally, as defined in [4], let $R_i \subset R$ be the first i request-vertices. A *partial matching* of R_i is a perfect matching of R_i with a subset of the servers of S. Let M_0 and P_0 be empty. Define M_i to be the edges that form a minimal weight partial matching on R_i where the number of edges in $M_i - M_{i-1}$ is minimized, choosing arbitrarily if multiple such matchings exist. Let $S_i \subset S$ be the server-vertices incident to an edge in M_i. Let P_i denote the partial matching constructed by PERMUTATION after the first i requests. PERMUTATION constructs P_{i+1} by computing M_{i+1}, assigning r_{i+1} to the unique server-vertex $s \in S_{i+1} - S_i$, and adding that edge to the matching P_i.

We now show that PERMUTATION is $(2k-1)$-competitive, which was stated without proof in the Conclusion of the preliminary version of [4]. The proof, given in the full version of the paper, is similar to the proof in [4] which shows that PERMUTATION is $(2k-1)$-competitive for the online minimum-weight matching problem.

Theorem 3. PERMUTATION *is* $(2k-1)$*-competitive for the bottleneck matching problem.*

Theorem 4. *The competitive ratio of* PERMUTATION *is at least* $2k-1$ *for the bottleneck matching problem.*

Proof. Let M be a subspace of the real line, with the standard distance metric. Set $s_i = i$ for $1 \leq i \leq k$. Let $r_i = i + .5 + \epsilon$ for $1 \leq i \leq k$. PERMUTATION matches r_i to s_{i+1} when it exists, and matches the final request, r_k, to s_1, for a bottleneck cost of $k - .5 + \epsilon$. OPT assigns r_i to s_i, so all edges have a cost of $.5 + \epsilon$, which is thus $cost$(OPT). Thus, the performance on this instance is $(2k - 1 + 2\epsilon)/(1 + 2\epsilon)$, which approaches $2k - 1$. (The ϵ could be removed if ties can be broken arbitrarily.)

Resource augmentation was used in [5] to show that, for the min-weight objective, GREEDY has a halfOPT-competitive ratio of $O(\log k)$, in contrast with its $\Omega(2^k)$ competitive ratio without resource augmentation. Motivated in part by these results, we turn to a resource augmentation setting for the bottleneck objective.

3 Bicriteria Analysis of GREEDY

Noting that "the poor competitive ratio of an intuitive greedy algorithm may not reflect the fact that it may perform reasonably well on 'normal' inputs", [5] adopts a *weak adversary model*, in which the adversary has fewer resources than the online algorithm. Their work address the online transportation problem, which is a generalization of the min-weight matching problem. We perform a similar analysis for the bottleneck matching problem, and show that the improvement for GREEDY is more limited for our objective.

While each server-vertex in OPT can service exactly one request, the online algorithm can assign requests to two servers at each server-vertex. Thus, as in [5] we say that the *halfOPT-competitive ratio* of an online algorithm A is the supremum over all instances I with at most k requests of $A(I)/OPT(I)$ where A has two servers available at each server-vertex, while OPT only has one.

We now show that the halfOPT-competitive ratio for GREEDY is linear in the number of requests. Since each server-vertex s_i has two servers in the online setting, we denote them by s_i^1 and s_i^2 as needed. Without loss of generality, we assume that s_i^2 is not used unless s_i^1 is already in use. The adversary has only s_i^1 available to it. We first prove a lemma about the *response graph* G, defined in [5] to be $G = (S \cup R, E)$, where E is the set of edges that includes the online edge $(r_i, s_{\sigma(i)})$ and adversary edge (r_i, s_i) for each request r_i. The proof of the following lemma is found in the full version of the paper.

Lemma 1. *Each connected component of G contains exactly one cycle.*

Theorem 5. *The halfOPT-competitive ratio of* GREEDY *for the bottleneck matching problem is at most* $k-1$ *for* $k \geq 2$ *server-vertices.*

Proof. Let (r_i, s_j) be the online bottleneck edge in the response graph, G. (If there are multiple edges with the maximum bottleneck cost, pick one arbitrarily.) Let (r_i, s_i) be the edge in OPT that serves request r_i. If $s_i = s_j$ then we're done. So we only consider the case that $s_i \neq s_j$. Now consider the connected component containing r_i. By Lemma 1 this connected component has exactly one cycle. Note that this cycle may have trees joined to it at the vertices on the cycle. Observe that all such junctions must represent a server-vertex, since each request can have at most two incident edges in the response graph, one for the online edge and one for the optimal edge. Consider separately the cases when r_i lies on the cycle, and when it does not.

If r_i is a vertex on the cycle, then since only server-vertices can be junctions, both the online and offline edges incident on r_i must lie on the cycle. Removing the online edge (r_i, s_j) from the cycle yields a tree which can be rooted at r_i. Since there are k request-vertices and k server-vertices, there are at most $2k$ vertices in the tree. Furthermore, the tree contains alternating levels of server-vertices and request-vertices. Each request-vertex has one child (the server-vertex chosen for it by OPT), and each server-vertex can have up to two children (the online edges).

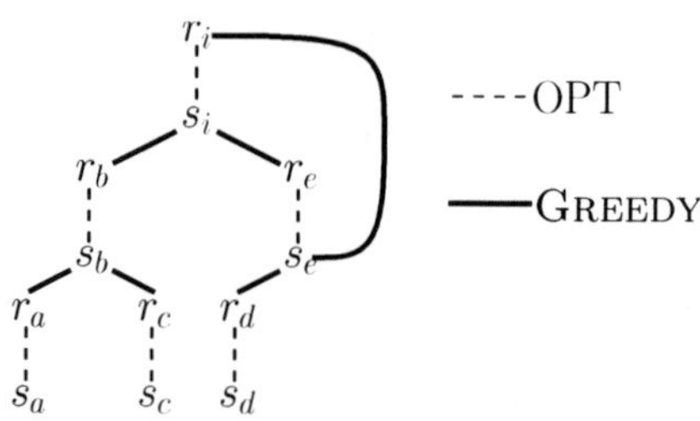

Fig. 1. An example response graph

To upper bound the cost of the edge (r_i, s_j), it suffices to upper bound the distance of the shortest path from r_i to some server-vertex s_a with s_a^2 unused, since GREEDY picked s_j instead of s_a. Since r_i is the root of the tree, it suffices to find the cost of a path requiring the minimum number of edges that must be traversed to arrive at a leaf. Consider a version of the tree where the edges from a request to its child are contracted, thus resulting in a binary tree T with at most k vertices. Let $k_T \leq k$ refer to the number of vertices in the contracted tree. Since a full binary tree would have $\log(k_T)$ levels, a leaf of T, which may or may not be full, is reachable in at most $\log(k_T)$ edges. Uncontracting the edges (at most one per server-vertex) indicates that in the original graph, there are at most $\log(k_T)$ optimal and $\log(k_T) - 1$ online edges between the root and some leaf, call it s_a.

Now consider the cost of the path in the tree from r_i to server-vertex s_a. By definition, any edge used in OPT must have cost at most $cost(OPT)$. Since all leaves of the tree are incident only with one edge, an OPT edge, the edge (r_a, s_a) is an edge in OPT, and thus has cost at most $cost(OPT)$. Proceeding from s_a to the root, the next edge on the path is an online edge, call it (r_a, s_b). GREEDY chose to assign r_a to s_b rather than s_a which had a server available, and thus has cost at most the cost of the edge from (r_a, s_a), which is again at most $cost(OPT)$. The next edge in the path, (r_b, s_b)

is an edge in OPT, and thus has cost at most $cost(OPT)$. The next edge, the online edge (r_b, s_c) again was again chosen by GREEDY over the edge (r_b, s_a) and thus has cost at most the distance in the tree from r_b to s_a, which is bounded by the three edges previously mentioned in the path, for a total cost of at most $3 \cdot cost(OPT)$. This process continues, with successive edges in OPT having cost at most $cost(OPT)$ and successive online edges having cost at most $(2^h - 1) \cdot cost(OPT)$ where h represents the height of the request in the tree with the online edges contracted. As the edge incident to r_i in the subtree is an edge in OPT, the final edge in the path from s_a to r_i has cost at most $cost(OPT)$. Thus, the total cost of the path is at most $cost(OPT)$ for each of the $\log(k)$ edges in OPT and $\sum_{h=1}^{\log(k)-1}(2^h - 1) \cdot cost(OPT)$ for the online edges, giving $\sum_{h=0}^{\log(k-1)} 2^h \cdot cost(OPT) = (2^{\log(k)} - 1) \cdot cost(OPT) = (k-1) \cdot cost(OPT)$. Hence, since GREEDY assigned r_i to s_j instead of s_a, the online bottleneck edge cost is at most $(k - 1) \cdot cost(OPT)$.

Now consider the case where r_i does not lie on the cycle. Removing (r_i, s_j) from the response graph partitions the original connected component into two connected components, with r_i and the original cycle now in separate connected components. As the original connected component contained exactly one cycle, the connected component rooted at r_i is a tree. By the same process, the upper bound on the distance from r_i to some leaf server-vertex s_a is at most $(k - 1) \cdot cost(OPT)$, completing the proof.

The example used in [5] to provide a lower bound for GREEDY for the online transportation problem gives a lower bound of $k/2$ for GREEDY in this setting. We prove a slightly improved lower bound of $(k + 1)/2$ in Corollary 1 in Section 4.

4 Bicriteria Analysis of BALANCE

In this section we consider the BALANCE algorithm detailed in [5]. We first define some convenient notation for our resource augmentation model. As in the previous section, each server-vertex s_i in S is said to have a primary server s_i^1 and a secondary server s_i^2. Thus, while there are k vertices in S, one for each request in R, the online algorithm effectively has $2k$ servers to choose from. For BALANCE, the *pseudo-distance* from a request r_i to a primary server s_j^1 is the actual distance $d(r_i, s_j)$, while the *pseudo-distance* from the same request r_i to the secondary server s_j^2 is $c \cdot d(r_i, s_j)$, for a constant $c > 1$. (In [5], a $c > 11$ was specified.) BALANCE then uses GREEDY to assign arriving requests to servers, based on their pseudo-distances. (Thus BALANCE with $c = 1$ is precisely GREEDY.) Note also that BALANCE only applies in the resource augmentation setting because it uses primary and secondary servers explicitly.

We begin with a lower bound on the halfOPT-competitive ratio of BALANCE.

Theorem 6. *The halfOPT-competitive ratio of* BALANCE *for the bottleneck matching problem is at least* $(\frac{1}{c} + 1)^{\log(k+1)-1} = \Omega(k)$*, where k is the number of requests and c is the constant in the definition of* BALANCE.

Proof. Consider the following example on the line, where at each location the number of requests and server-vertices are powers of two. Let $L_0, L_1, L_2, \ldots, L_m$ be the $m+1$ server-vertex locations, where L_i has 2^{m-i} server-vertices. Similarly, the $m+1$ request locations are $R_0, R_1, R_2, \ldots, R_m$ where R_i has 2^{m-i} requests. Let $L_0 = -c$, $R_0 = 0$, and for $1 \leq i \leq m$, $L_i = R_i$.

We now determine the most extreme placement for the server-vertices so that OPT will assign requests at R_i to servers at L_i but that BALANCE will choose not to send any requests to L_0 until the final request. Thus OPT will have a bottleneck cost of c while BALANCE will pay c plus the location of the final server. Since c is fixed, the ratio will grow with the location L_m.

We break ties at our convenience. (Alternatively, a small $\epsilon > 0$ could be used to perturb the locations slightly to enforce such choices.) L_1 must be at 1 so that the secondary servers at L_1 (with a cost of $c \cdot 1$) are equally desirable as the primary servers at L_0 (cost of c) for the requests at R_0. L_2 must be chosen so that the requests at R_1 consider the secondary servers at L_2 (with cost $c \cdot d(L_1, L_2)$) as desirable as the primary servers at $L_0 = -c$, with cost $c + 1$. Thus, $d(L_1, L_2) = \frac{c+1}{c}$, placing L_2 at $2 + \frac{1}{c}$. Repeating this process, L_i can be placed at $\sum_{j=1}^{i} \binom{i}{j} \frac{1}{c^{j-1}}$ for all $1 \leq i \leq m$.

We now find a closed form for the location of server L_m, as shown in (1).

$$L_m = \sum_{j=1}^{m} \binom{m}{j} \frac{1}{c^{j-1}} = c \sum_{j=1}^{m} \binom{m}{j} \frac{1}{c^j} = c \left(\sum_{j=0}^{m} \binom{m}{j} \frac{1}{c^j} \right) - c \binom{m}{0} \frac{1}{c^0}. \quad (1)$$

Using the binomial theorem on the summation gives the expression $c(\frac{1}{c}+1)^m - c$. Thus, if L_m is the rightmost server, the bottleneck distance from L_0 to L_m is $c(\frac{1}{c} + 1)^m$.

Note that the total number of requests is $k = \sum_{i=0}^{m} 2^i = 2^{m+1} - 1$. Thus $m = \log(k+1) - 1$. Thus the bottleneck cost for BALANCE is $c(\frac{1}{c} + 1)^{\log(k+1)-1}$ where k is the number of servers/requests, and the bottleneck cost for OPT is c. If c is a fixed constant, then the lower bound on the competitive ratio is $(\frac{1}{c} + 1)^{\log(k+1)-1}$.

Corollary 1. *The halfOPT-competitive ratio of* GREEDY *for the bottleneck matching problem is at least* $\frac{k+1}{2}$, *where* k *is the number of servers.*

Proof. Noting that $c = 1$ is precisely GREEDY, observe that if $c = 1$ this gives a competitive ratio of $2^{\log(k+1)-1} = \frac{k+1}{2}$.

We now show that the upper bound on the halfOPT-competitive ratio of BALANCE is a matching $O(k)$.

Theorem 7. BALANCE *has a halfOPT-competitive ratio of* k *for the bottleneck matching problem.*

Proof. The same argument as for the GREEDY upper bound (Theorem 5) applies. Note that it holds because the server-vertex s_a used in the argument is a leaf of the tree, which means the online algorithm has not used either of its servers. Thus the pseudo-distance to that vertex in BALANCE is the same as the original distance in GREEDY.

5 Bicriteria Analysis of PERMUTATION

We next consider PERMUTATION with resource augmentation. As before, each server-vertex s_i has two servers in the online setting, the primary server s_i^1 and the secondary server s_i^2. Without loss of generality, we assume that a secondary server can only be used if the corresponding primary server is used. Again, we compare PERMUTATION to OPT which can serve exactly one request per server-vertex.

We now note how the definition of PERMUTATION from Section 2.2 applies to the resource augmentation setting. Let S^{aug} be the set of $2k$ servers available to the online algorithm. Then a *partial matching* of the first i requests is a perfect matching of these requests with a subset of S^{aug}. Define M_i to be the set of edges in a minimal weight partial matching of the first i requests that is "most similar" to M_{i-1}, in the sense that the number of edges in $M_i - M_{i-1}$ is minimized. Let $S_i \subset S^{aug}$ be the set of servers incident to an edge in M_i. By convention, M_0 is empty.

Suppose that PERMUTATION services request r_i with a server s_j^x at vertex s_j. Then define M' to be the union of M_{i-1} with the edge (r_i, s_j^x). Let P_i denote the partial matching constructed by PERMUTATION for the first i requests.

Intuitively, it may seem that PERMUTATION should benefit substantially from resource augmentation; the availability of a secondary server seemingly allows the algorithm to 'correct' itself if a request arrives and finds that the primary server it would have used in OPT was already in use. Yet, PERMUTATION has a halfOPT-competitive ratio of k and this is tight, as illustrated by the following lower bound instance and a matching upper bound guarantee. This is in comparison with its competitive ratio of $2k-1$ in the absence of resource augmentation.

Theorem 8. PERMUTATION *has a halfOPT-competitive ratio of $\Omega(k)$ for the bottleneck matching problem.*

Proof. Fix a small constant $\epsilon > 0$. Without loss of generality, let k be odd. Consider the following instance, as depicted in Figure 2 for $k = 9$. Server vertices and requests s_i, r_i for $1 \le i \le k$ with i odd are placed along the line, in the order $s_1, r_1, s_3, r_3, \ldots, s_k, r_k$ where the distance between s_i and r_i is $1+\epsilon$, and the distance between r_i and s_{i+2} is 1. For each $i \ge 3$, let request r_{i-1} be 1 away from s_i, and let server-vertex s_{i-1} be at a distance of $1+2\epsilon$ from r_{i-1}. All other distances are additive based on this graph.

Since PERMUTATION assigns requests based on M_i, note that M_1 assigns r_1 to s_3^1. Thus, PERMUTATION does the same. In M_2, this assignment remains, and r_2 is assigned to s_3^2, and again PERMUTATION behaves identically. In general, M_j for $j < k$ behaves as follows: if j is odd, r_j is assigned to s_{j+2}^1 and if j is even, r_j is assigned to s_{j+1}^2. PERMUTATION's assignments are identically M_j for $j < k$. Naturally, this pattern cannot continue for request r_k; observe that M_k that shares only about half of its edges with M_{k-1}. In particular, M_k assigns r_i to s_i^1 for i odd, and assigns r_j to s_{j+1}^2 for j even. Thus, PERMUTATION assigns the final request r_k to the only server used in M_k that was not used in M_{k-1}, that is, s_1^1. Hence, PERMUTATION assigns r_k to s_1, for a bottleneck cost of $k + \frac{k+1}{2}\epsilon$ (its other assignments all have cost 1).

Observe that OPT matches each r_i to its corresponding s_i, for a bottleneck cost of $1+2\epsilon$. Hence, PERMUTATION has a halfOPT-competitive ratio of $\Omega(k)$.

We now develop a sequence of lemmas which show that $cost$(PERMUTATION) is at most $O(k) \cdot cost(OPT)$ for any instance. As in [4], let $H := M_i \oplus M'$. For convenience, we say that a server is *in* H if there is an edge in H incident on the server. Lemma 3,

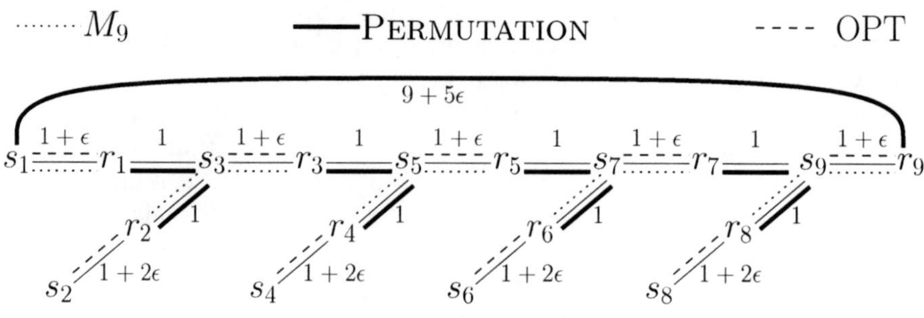

Fig. 2. Even with resource augmentation, PERMUTATION's cost can still be $k \cdot cost(\text{OPT})$

which says that any given server-vertex appears at most once in H, uses a "displacement sequence" in its proof which provides some intuition for the choice of H.

Lemma 2. *The servers used in M' are exactly the servers used in M_i.*

Proof. The name PERMUTATION in [4] comes from maintaining the invariant that "for all i, the vertices in S incident to an edge in M_i are exactly the vertices in S that are incident to an edge in P_i." By Lemma 3.2 of [4], S_i and S_{i-1} differ by exactly one server. Thus, by definition of how PERMUTATION chooses s_j^x, at each step i, M_i and $M_{i-1} \cup (r_i, s_j^x)$ have used the same servers.

Corollary 2. *H is a single alternating cycle.*

Proof. As in [4], this follows immediately from server vertices in M_i and M' being identical (Lemma 2).

Lemma 3. *If s_ℓ^1 is in H, then s_ℓ^2 is not in H, and if s_ℓ^2 is in H, then s_ℓ^1 is not in H.*

Proof. Suppose for the sake of a contradiction that H contains both the primary server and corresponding secondary server for some s_ℓ. By Lemma 2, s_ℓ^1 and s_ℓ^2 must each be used in both M_i and in M'. Let requests r_a and r_b be assigned to s_ℓ^1 and s_ℓ^2, respectively, by matching M_i. Let requests r'_a and r'_b be assigned to s_ℓ^1 and s_ℓ^2, respectively, in M'. To prove the lemma, it suffices to prove the following claim.

Claim: if $r'_a \neq r_a$ and $r'_a \neq r_b$, then $r'_b = r_a$ or $r'_b = r_b$. In other words, at least one of the two requests matched to a server of s_ℓ in M_i must also be matched to a server of s_ℓ in M'. Assume not. So $r'_a \neq r_a$ and $r'_a \neq r_b$, and $r'_b \neq r_a$ and $r'_b \neq r_b$. Let s_j be the server-vertex assigned to r_i in M'.

Case $s_\ell \neq s_j$. Then, since $M' = M_{i-1} \cup (r_i, s_j)$, in M_{i-1} we must also have $r'_a \rightarrow s_\ell$ and $r'_b \rightarrow s_\ell$, where "$\rightarrow$" means "is assigned to." So upon the arrival of r_i, in the transition from M_{i-1} to M_i, both r'_a and r'_b were displaced by r_a and r_b.

Define the *displacement sequence of* r_i to be a sequence of server vertices and requests affected by the arrival of r_i, written as follows:

$$r_i \longrightarrow s_i \dashleftarrow r_1 \longrightarrow s_1 \dashleftarrow r_2 \longrightarrow s_2 \ldots$$

where forward-edges are from M_i and backward edges are from M_{i-1}. Here, r_1 is a request that was "displaced" from s_i upon the arrival of r_i; it was displaced to server-vertex s_1. Then r_2 is a request that was displaced from s_1 by r_1, and s_2 is the server-vertex it was displaced to, and so forth. Note that each server-vertex in this sequence can only have one incoming backward edge because it only has one incoming forward edge. Further note that if a server-vertex is not in the displacement sequence of r_i, then it must be matched to the same requests as it was in M_{i-1}, since otherwise the optimality of M_{i-1} or M_i or the assumption that M_i is the most similar optimal matching to M_{i-1} would be violated. So s_ℓ must be in the displacement sequence of r_i. Since s_ℓ has two displaced requests, r'_a and r'_b, then s_ℓ must appear twice in the sequence. But if it appears twice in the sequence, then there is a "cycle" in the sequence. Consider the displacements just in this cycle. The total cost of the forward edges in the cycle must be lower than the total cost of the backward edges, otherwise this cycle would not be present in the displacement sequence of r_i, it would just be cut out altogether (by optimality of M_i). But if the total cost of the forward edges is less than the backward edges, then M_{i-1} was not optimal.

Case $s_\ell = s_j$. Without loss of generality, let us assume that $r'_a = r_i$. Thus in M_{i-1}, only one request was assigned to s_ℓ and it was r'_b. So upon arrival of r_i, r_a was assigned to s_ℓ and r'_b was replaced by r_b. This means in the displacement sequence of r_i, s_ℓ again must appear twice, giving the same contradiction as in the previous case.

Now consider the server-vertices M_i uses exactly once (i.e. only their primary servers). The next lemma says at most one of these server-vertices can appear in H. The proof appears in the full version of the paper.

Lemma 4. *Let s_i^1 be in H. If an edge of M_i is not incident on s_i^2, then for all other servers s_j^1 in H, an edge of M_i must be incident on s_j^2.*

Theorem 9. PERMUTATION *has a halfOPT-competitive ratio of $O(k)$ for the bottleneck matching problem.*

Proof. Let αk be the number of primary servers used by PERMUTATION. (This is the same as the number of primary servers used by M_k.) Since a secondary server is only used if its corresponding primary server is used, there are $(1-\alpha)k$ server-vertices with neither their primary nor secondary server used. Since exactly k requests are served, there must be $(1-\alpha)k$ secondary servers used. Together these guarantee $1 \geq \alpha \geq \frac{1}{2}$.

Let the bottleneck edge of the final PERMUTATION assignment be (r_i, s_j). Now consider the graph of H after the arrival of r_i. Recall that by Corollary 2, H is a single alternating cycle. As in [4], by the triangle inequality, the weight of the newest edge (r_i, s_j) is at most the aggregate weight of the edges in H minus its weight $d(r_i, s_j)$. Thus, if we can bound the number of edges in H by n, then the bottleneck edge for PERMUTATION is at most $n-1$ times the bottleneck edge in M_i, as the cost of the bottleneck edge only increases from M_{i-1} to M_i.

If for every primary server that is used in M_i, the corresponding secondary server is also used in M_i, i.e., $\alpha = \frac{1}{2}$, then by Lemmas 3 and 4, H is an alternating cycle with at most $k/2$ server vertices (and the same number of requests), for at most k edges. If instead the number of primary servers used exceeds the number of secondary servers used, then $\alpha k - (1-\alpha)k \geq 1$ which guarantees that $\alpha \geq \frac{k+1}{2k}$. By Lemmas 3 and 4, H contains at most $(1-\alpha)k+1$ servers, and thus the number of edges in H is maximized when α is as small as possible. Plugging in the lower bound on α gives $\frac{k+1}{2}$ servers, guaranteeing at most $k+1$ edges in H. Hence, in either case, PERMUTATION costs at most k more than the bottleneck edge in M_i; the optimality of M_k and the bottleneck edge of M_i monotonically non-decreasing as i increases complete the proof.

6 Conclusion

Resource augmentation results in a substantial improvement in the performance of the GREEDY algorithm for the bottleneck matching problem, from an exponential lower bound to a guarantee linear in the number of requests. While still exponentially worse than its performance for the objective of minimizing total distance, it is a natural algorithm that is easy to implement. Two algorithms that perform notably better than GREEDY for the min-weight objective (PERMUTATION and BALANCE) also have linear competitive ratios for the bottleneck objective with resource augmentation. These results suggest that in some sense the bottleneck objective is more difficult than the total distance objective, as none of the three algorithms break the $\Omega(k)$ barrier for the bottleneck objective. Determining if the lower bound (under resource augmentation) is in fact $\Omega(k)$ remains an open question.

References

1. Chung, C., Pruhs, K., Uthaisombut, P.: The Online Transportation Problem: On the Exponential Boost of One Extra Server. In: Laber, E.S., Bornstein, C., Nogueira, L.T., Faria, L. (eds.) LATIN 2008. LNCS, vol. 4957, pp. 228–239. Springer, Heidelberg (2008)
2. Hartline, J.D., Roughgarden, T.: Simple versus optimal mechanisms. In: ACM Conference on Electronic Commerce, pp. 225–234 (2009)
3. Idury, R., Schaffer, A.: A better lower bound for on-line bottleneck matching (1992) (manuscript)
4. Kalyanasundaram, B., Pruhs, K.: Online weighted matching. J. Algorithms 14(3), 478–488 (1993); Preliminary version appeared in SODA, pp. 231–240 (1991)
5. Kalyanasundaram, B., Pruhs, K.: The online transportation problem. SIAM J. Discrete Math. 13(3), 370–383 (2000); Preliminary version appeared in ESA, pp. 484–493 (1995)
6. Kalyanasundaram, B., Pruhs, K.: Speed is as powerful as clairvoyance. J. ACM 47, 617–643 (2000); Preliminary version appeared in FOCS, pp. 214–221 (1995)
7. Khuller, S., Mitchell, S.G., Vazirani, V.V.: On-line algorithms for weighted bipartite matching and stable marriages. Theor. Comput. Sci. 127, 255–267 (1994)
8. Phillips, C.A., Stein, C., Torng, E., Wein, J.: Optimal time-critical scheduling via resource augmentation. Algorithmica 32(2), 163–200 (2002); Preliminary version appeared in STOC, pp. 140–149 (1997)
9. Roughgarden, T., Tardos, É.: How bad is selfish routing? J. ACM 49(2), 236–259 (2002); Preliminary version appeared in STOC, pp. 140–149 (1997); Preliminary version appeared in FOCS, pp. 93–102 (2000)

Streaming with Minimum Space: An Algorithm for Covering by Two Congruent Balls

Chung Keung Poon[1] and Binhai Zhu[2]

[1] Department of Computer Science, City University of Hong Kong, Kowloon, Hong Kong
csckpoon@cityu.edu.hk

[2] Department of Computer Science, Montana State University, Bozeman, MT 59717, USA
bhz@cs.montana.edu

Abstract. In this paper we design a simple streaming algorithm for maintaining two smallest balls (of equal radius) in d-dimension to cover a set of points in an on-line fashion. Different from most of the traditional streaming models, at any step we use the minimum amount of space by only storing the locations and the (common) radius of the balls. Previously, such a geometric algorithm is only investigated for covering with one ball (one-center) by Zarrabi-Zadeh and Chan. We give an analysis of our algorithm, which is significantly different from the one-center algorithm due to the obvious possibility of grouping points wrongly under this streaming model. We obtain upper bounds of 2 and 5.708 for the case of $d = 1$ and $d > 1$ respectively. We also present some lower bounds for the corresponding problems.

1 Introduction

In the past years we have seen huge progress made on streaming algorithms for geometric problems. Streaming algorithms for these problems (like convex hull), are mostly based on the concept of core-sets ([4,3]) and the slightly later concept of *extent* ([1]) and *blurred ball cover* ([2]); namely, given any ϵ, store $O(poly(\frac{1}{\epsilon}))$ of points such that applying a standard (convex hull) algorithm on these points would give a $(1+\epsilon)$-approximation for the original (convex hull) problem. Most of these algorithms then go for improving the space complexity for the core-sets or extents, the detailed list of references can be found, for example, in [6,15].

1.1 Streaming with Minimum Space

One different but related problem was due to Zarrabi-Zadeh and Chan [16], who tried to maintain the minimum enclosing ball for a set of points (in fixed d-dimension) while only storing the location and radius of the current ball; moreover, the data will be scanned in exactly one pass. They analyzed a simple algorithm, i.e., if p_i is inside the current ball D_{i-1}, return D_{i-1}; otherwise, return the smallest ball enclosing $D_{i-1} \cup \{p_i\}$. Each update obviously takes $O(d)$ time. They

G. Lin (Ed.): COCOA 2012, LNCS 7402, pp. 269–280, 2012.

proved that the algorithm admits an approximation ratio of 1.5, which is tight (for the algorithm). They also gave a 1.207 lower bound on any deterministic algorithm that only stores the center and radius of the enclosing ball.

As mentioned in [16], there is a one-pass, $O((1/\epsilon)^{\lfloor d/2 \rfloor})$ space, $O((1/\epsilon)^{\lfloor d/2 \rfloor} n)$ time streaming algorithm that achieves an approximation factor of $1+\epsilon$ by storing extreme points along $O((1/\epsilon)^{\lfloor d/2 \rfloor})$ directions. Agarwal and Sharathkumar [2] obtained another streaming algorithm that reduces the space complexity to $O((d/\epsilon^3)\log(1/\epsilon))$ but has a larger approximation factor of $\frac{\sqrt{3}+1}{2}+\epsilon \approx 1.366+\epsilon$. Chan and Pathak [7] gave a tighter analysis and showed that the algorithm is 1.22-approximate. At the extreme, the algorithm of Zarrabi-Zadeh and Chan shows that ratio 1.5 is achievable using just the minimum amount of storage (and in a simple and natural way).

In this paper, we follow the footsteps of [16] and study the two-center problem under the same specialized streaming model. In other words, at any step i we only update/store the locations of the two balls and their common radius r_i (using $O(d)$ time), and we scan the data points in one pass. We present a simple streaming algorithm under this model and show that it is 2- and 5.708-approximate in the 1d and general d-dimension cases respectively.

Like the 1-center algorithm in [16], our algorithm does not assume any knowledge on the locations of points within the balls. To ensure the continued coverage of existing points, any covered volume at a step will remain covered in subsequent steps. We show that any deterministic algorithm, storing only the locations and radius of the two balls and guaranteeing coverage of existing balls in subsequent steps, has approximation ratio at least 1.5 and 1.604 for the 1d and d-dimension cases respectively.

1.2 Related Work

Our problem can be viewed as a variant of clustering problems, the general goal of which is to group the input points into clusters so as to minimize certain objective/cost function. There has been a huge body of work on clustering in the literature. Our investigation here is related to the *k-center problem*, which is to find k cluster centers so that the maximum distance of points from their nearest center is minimized. Our problem is the special case where $k = 2$.

When the input points are from a metric space, the Doubling Algorithm of Charikar et al. [8] achieves an approximation ratio of 8 using $O(k)$ space. McCutchen and Khuller [13], and Guha [10] presented $(2+\epsilon)$-approximate algorithms that use $O((k/\epsilon)\log(1/\epsilon))$ space. These streaming algorithms either have worse approximation ratio or use more space than our algorithm. Charikar et al. also proposed the Center-Greedy and Diameter-Greedy Algorithms which are 3-approximate for $k = 2$. However, their analysis assumes that the points within each cluster are stored. For d-dimensional Euclidean space, Charikar et al. proposed the Clique Partition Algorithm that has approximation ratio $4(1+\sqrt{d/(2d+2)})$, which is 6 for $d = 1$, 6.3 for $d = 2$ and 6.83 asymptotically for large d. Their algorithm has worse approximation ratio than ours although it works for general k.

Regarding lower bounds, Charikar et al. gave several results, assuming that the algorithms cannot split existing clusters. Note that this is a stronger assumption than ours because we only require an existing ball to remain covered in subsequent steps but possibly by more than one ball. Their other lower bounds and those given by Guha [10] are for the metric case, which do not apply to our model.

Besides k-center, there are other related clustering problems. In particular, the k-median (resp. k-means) problem is to find k cluster centers so that the sum of distances (resp. sum of squares of distances) from the input points to their nearest cluster center is minimized. For these problems, there are streaming algorithms that achieve $(1+\epsilon)$-approximation using $(1/\epsilon)^{O(d)}$ space, (e.g. [12,11]). Finally, we comment that our problem specialized to the case of $d = 2$ is the planar two-center problem. This problem under the ordinary (non-streaming) model is well-studied, with almost linear time solutions, see [5,9,14].

1.3 Some Intuition

While the algorithm for the 1-center case seems very natural, its extension to the 2-center case is not straightforward. Consider an arbitrary step in which the algorithm maintains two balls (their centers and the common radius) while a new point p arrives. Naturally, if p is not too far away from one of the balls, we can group p into the nearer ball by enlarging the radius. Otherwise, we merge the two existing balls into one and create a new ball to cover p. To materialize this idea, one needs to have a criteria to determine when to merge the existing balls; and some rules to choose the centers. One can easily come up with a few simple heuristics that look promising. For example, Center-Greedy restricts the centers to be chosen from the input points seen so far. It treats the new point p as a ball of radius 0 and merges the two balls whose centers are closest. Diameter-Greedy minimizes the increase in radius in each step. In the full paper, we show that they have approximation ratios at least 5.8 in the worst case.

Our algorithm allows the centers to be arbitrary points in the Euclidean space and minimizes the increase in radius greedily. Note that there is still freedom in the choice of the centers of the new balls. We will try to keep the two balls close and yet we do not always minimize the distance between their centers. Our algorithm is easy to implement but the analysis is non-trivial. One source of difficulty is due to the fact that the centers of balls need not be chosen from the input points. The 1-center case presented in [16] is already highly non-trivial. Added to that is the difficulty in grouping points when we forgot most of the information about the input points seen so far. We overcome these difficulties by exploiting the geometric properties preserved at each step. These properties might be useful for other on-line geometric problems.

2 Some Notations and the Algorithm

Our objective is to minimize the common radius of the two balls that cover all the input points. The performance is measured by the *approximation ratio*, defined as the radius of the balls returned by the algorithm over the optimal two-center radius.

We next present a few notations. For $i = 1, 2, \ldots, n$, at each step i a point p_i is added (scanned). Let k_{i-1} be the distance of p_i from (the surface of) the nearer ball. Let r_i and d_i be respectively the radius of the balls and the distance between the two centers after processing point p_i. Denote by $B_{i,1}$ and $B_{i,2}$ the two balls maintained by our algorithms after p_i is processed; and $c_{i,1}$, $c_{i,2}$ their centers respectively.

Our algorithm, which works for all d-dimensions, is presented as follows. At step $i \geq 3$, initially the radius of the two balls and distances between their centers are r_{i-1} and d_{i-1} respectively. Without loss of generality, we assume that the new point p_i is never inside $B_{i-1,1}$ or $B_{i-1,2}$ as we do not need to update anything in such a case. We compute the distance, k_{i-1}, of the new point p_i from the nearer ball.

*Case (1): $k_{i-1} > d_{i-1}$ (**merge** step).* Merge the existing balls $B_{i-1,1}$ and $B_{i-1,2}$, i.e., enclose them with a new ball $B_{i,1}$ of minimum radius ($= r_{i-1} + d_{i-1}/2$). Create another ball $B_{i,2}$ with the same radius and with p_i on its surface such that its center $c_{i,2}$ is as close to $c_{i,1}$ as possible. Update $r_i = r_{i-1} + d_{i-1}/2$. In 1d, we additionally have $d_i = k_{i-1}$.

*Case (2): $k_{i-1} \leq d_{i-1}$ (**walk** step).* Move the centers of the two balls $B_{i-1,1}$ and $B_{i-1,2}$ towards p_i by a distance of $k_{i-1}/2$. Enlarge the common radius by $k_{i-1}/2$ and update $r_i = r_{i-1} + k_{i-1}/2$. Clearly, the nearer ball will touch p_i. Moreover, $B_{i-1,1}$ (resp. $B_{i-1,2}$) is enclosed by $B_{i,1}$ (resp. $B_{i,2}$). In 1d, we have $d_i = d_{i-1}$ if p_i is outside of the convex hull of $B_{i-1,1}$ and $B_{i-1,2}$ (denoted as $CH(B_{i-1,1}, B_{i-1,2})$); otherwise, we have $d_i = d_{i-1} - k_{i-1}$.

3 Analysis for the 1d Case

Note that a one-dimensional (1d) ball is just an interval. To facilitate our discussion, we assume in this section that $B_{i,1}$ and $B_{i,2}$ lie horizontally with $B_{i,1}$ on the left and $B_{i,2}$ on the right. We analyze the approximation ratio as follows. After step 2, we have $r_2 = 0$ and $d_2 > 0$. For $i \geq 3$, we have $r_i = r_{i-1} + \frac{\min\{k_{i-1}, d_{i-1}\}}{2}$ $= \frac{1}{2}(\min\{k_2, d_2\} + \cdots + \min\{k_{i-1}, d_{i-1}\})$ and $d_i = \max\{k_{i-1}, d_{i-1}\}$ if p_i is outside of $CH(B_{i-1,1}, B_{i-1,2})$; and $d_i = d_{i-1} - k_{i-1}$ otherwise.

Call the interval between two adjacent points a *segment*. Then after step i, there are $i-1$ segments. Due to space limitations, proofs of the following lemmas are omitted.

Lemma 1. *For all i, there exist a point on the left boundary of $B_{i,1}$ and another point on the right boundary of $B_{i,2}$.*

Lemma 2. *After step i, d_i is no shorter than any segment.*

Theorem 1. *In 1d the approximation ratio of the algorithm is 2.*

Proof. Consider step $i+1$ for an arbitrary $i \geq 2$. Denote by S_j, $1 \leq j \leq i-1$, the j-th segment at the beginning of step $i+1$, indexed from left to right; and denote by s_j its corresponding length. We consider three cases according to whether p_{i+1} is inside or outside the convex hull of $B_{i,1}$ and $B_{i,2}$ and whether $k_i > d_i$ or not.

Case 1: p_{i+1} outside of $CH(B_{i,1}, B_{i,2})$ and $k_i > d_i$. Assume without loss of generality that p_{i+1} is on the right of S_1. By construction of the algorithm, $B_{i,1}$ and $B_{i,2}$ are merged into $B_{i+1,1}$. So, the new radius is $r_{i+1} = \frac{s_1+\cdots+s_{i-1}}{2}$. On the other hand, the optimal solution chooses the best segment S_j to leave out such that the radius $\max\{\frac{s_1+\cdots+s_{j-1}}{2}, \frac{s_{j+1}+\cdots+s_{i-1}+k_i}{2}\}$ is minimized. Let r^*_{i+1} be the the optimal radius for the points $p_1, p_2, \ldots, p_{i+1}$. Then $r^*_{i+1} \geq \min_j\{\max\{\frac{s_1+\cdots+s_{j-1}}{2}, \frac{s_{j+1}+\cdots+s_{i-1}+k_i}{2}\}\} \geq \min_j\{\frac{1}{2}\frac{k_i+s_1+\cdots+s_{i-1}-s_j}{2}\} \geq \frac{1}{2}\frac{s_1+\cdots+s_{i-1}}{2}$ where the last inequality is due to $k_i > d_i \geq s_j$ for any j (Lemma 2).

Case 2: p_{i+1} outside of $CH(B_{i,1}, B_{i,2})$ and $k_i \leq d_i$. Assume without loss of generality that p_{i+1} is on the left of S_1. There is always a point on the left boundary of $B_{i,1}$ (by Lemma 1) while the rightmost point of $B_{i,1}$ may be at some distance D_1 from its right boundary. Then the diameter of $B_{i,1}$ is $2r_i = s_1+\cdots+s_j+D_1$ for some j. By design of the algorithm, we enlarge $B_{i,1}$ to touch p_{i+1}. Therefore, $r_{i+1} = \frac{k_i+s_1+\cdots+s_j+D_1}{2}$. On the other hand, the optimal solution leaves out one segment $S_{j'}$ between the leftmost and rightmost point which are at a distance $k_i+s_1+\cdots+s_j+D_1+d_i$ apart. Therefore, the optimal radius for $p_1, p_2, \ldots, p_{i+1}$ is $r^*_{i+1} \geq \min_{j'}\{\frac{1}{2}\frac{k_i+(s_1+\cdots+s_j+D_1+d_i)-s_{j'}}{2}\}$ where $1 \leq j' \leq j+1$. Since $d_i \geq s_{j'}$ for all j', we have $r^*_{i+1} \geq \frac{1}{2}\frac{k_i+s_1+\cdots+s_j+D_1}{2}$.

Case 3: p_{i+1} inside $CH(B_{i,1}, B_{i,2})$ Without loss of generality, assume p_{i+1} is closer to $B_{i,1}$ than to $B_{i,2}$. In the optimal solution, the point p_{i+1} is either in $B_{i+1,1}$ or $B_{i+1,2}$. In any case, $r^*_{i+1} \geq r_{i+1}$. □

Note that the bound in Theorem 1 is tight. Consider for instance the four points, $p_1 = 0$, $p_2 = 1$, $p_3 = 10$, $p_4 = 11$. The optimal radius is $1/2$ and the solution returned by our algorithm has radius 1.

4 Analysis for the d-Dimensional Case

Now, we turn to the case of d-dimension where $d \geq 2$. Let B^*_1 and B^*_2 be the optimal balls. So, points are taken from $B^*_1 \cup B^*_2$. For the purpose of analysis, we can assume, by appropriate scaling, that B^*_1 and B^*_2 are unit balls. Therefore, the optimal radius is 1. We will prove that for any sequence of points from $B^*_1 \cup B^*_2$, our algorithm will cover them with two balls of radius at most 5.708. However, we emphasize that our algorithm does not know the optimal radius *a priori*.

We start with a simple idea. First, note that at any step i, the centers $c_{i,1}$ and $c_{i,2}$ of $B_{i,1}$ and $B_{i,2}$ must be within the convex hull of B^*_1 and B^*_2. Therefore, if d^* denotes the distance between c^*_1 and c^*_2, the center of the optimal balls B^*_1

and B_2^*, the distance of any point in $B_1^* \cup B_2^*$ from $B_{i,1} \cup B_{i,2}$ is at most $2+d^*-r_i$ where r_i is the current radius of $B_{i,1}$ and $B_{i,2}$. The next step can only increase the radius by at most half of this distance, i.e., $r_{i+1} - r_i \leq (2 + d^* - r_i)/2$. After that, the distance of any point in $B_1^* \cup B_2^*$ from $B_{i+1,1} \cup B_{i+1,2}$ is reduced to at most $2 + d^* - r_{i+1}$. Observe that in this process, the upper bound on the maximum distance of a point in $B_1^* \cup B_2^*$ from the balls maintained by our algorithm decreases by the same amount as the increase in the radius of the balls. Moreover, r_i plus this maximum distance is at most $2 + d^*$. Hence, the final radius is at most $2 + d^*$, implying a (weak) approximation ratio of $2 + d^*$.

To establish a stronger approximation ratio, we prove similar but stronger invariants. To this end, we need the following more refined definitions of distances. For $j = 1, 2$, let $z_{i,j,1}$ (resp. $z_{i,j,2}$) be the maximum distance of a point in B_j^* from $B_{i,1}$ (resp. $B_{i,2}$); and let $y_{i,j} = \min\{z_{i,j,1}, z_{i,j,2}\}$. Thus, $y_{i,j}$ is an upper bound on the maximum distance of a point in B_j^* from $B_{i,1} \cup B_{i,2}$.

Also, recall that if step i is a "merge", we assume $B_{i-1,1}$ and $B_{i-1,2}$ are merged into $B_{i,1}$ while a new ball $B_{i,2}$ is created to touch the point p_i causing the merge. We use $(1+x)B_1^*$ (resp. $(1-x)B_1^*$) to represent the ball obtained by enlarging (reducing) B_1^* by a factor of $1 + x$ (resp. $1 - x$) at the same center.

4.1 The Invariants

We first make a simple observation.

Lemma 3. *At any step i, the radius increases by at most 1.*

Proof. This is clearly true if both B_1^* and B_2^* overlap with at least one of $B_{i,1}$ and $B_{i,2}$. If it is not the case, then all the scanned points $p_1, p_2, ..., p_i$ are from one of B_1^* or B_2^*. Without loss of generality, let all of them be from B_1^*. Then the centers of both $B_{i,1}$ and $B_{i,2}$ are within B_1^*, i.e., the centers are at distance at most 2 from each other. Hence the increase in radius is at most $d_i/2 \leq 1$. □

Next, we prove some invariants which form the basis for our final proof. As they apply to both B_1^* and B_2^*, we drop the subscript and simply write B^*. Likewise, we will drop the subscript j in c_j^*, $z_{i,j,1}$ and $y_{i,j}$, etc.

Lemma 4. *Suppose after step i, both centers, $c_{i,1}$ and $c_{i,2}$, of $B_{i,1}$ and $B_{i,2}$ are in B^*. Then $y_i + r_i + d_i/2 \leq 3\sqrt{3}/2 \approx 2.598$.*

Proof. Consider the plane containing $c_{i,1}$, $c_{i,2}$ and c^* (the center of B^*). The intersection of the plane and the ball B^* forms a circle. Without loss of generality, assume $c_{i,1}$ and $c_{i,2}$ lie on the perimeter of the circle. See Figure 1. (Otherwise, we can move $c_{i,1}$ or $c_{i,2}$ to the perimeter along the radial line from c^* to their original position. This will not decrease the distances d_i, $z_{i,1}$ and $z_{i,2}$.)

Consider a point q at equal distance L from $c_{i,1}$ and $c_{i,2}$. Recall that the distance between $c_{i,1}$ and $c_{i,2}$ is d_i. We want to find an upper bound on $L+d_i/2$. By basic geometry, $L^2 = \left(\frac{d_i}{2}\right)^2 + \left(1 + \sqrt{1 - (\frac{d_i}{2})^2}\right)^2 = 2 + 2\sqrt{1 - \frac{d_i^2}{4}}$. Hence $d_i/2 = L\sqrt{1 - L^2/4}$ and so $L + d_i/2 = L(1 + \sqrt{1 - L^2/4})$ which attains a

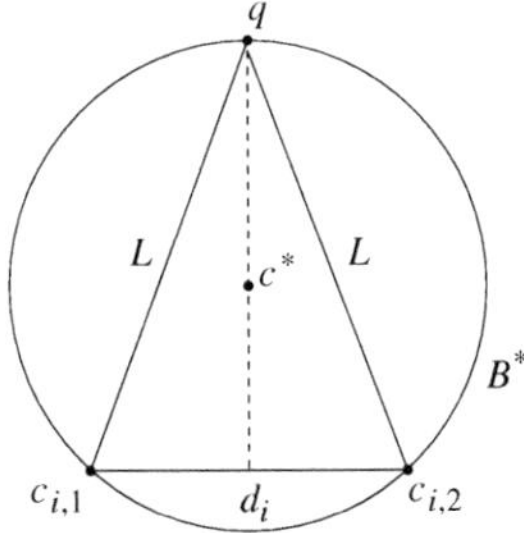

Fig. 1. Both $c_{i,1}$ and $c_{i,2}$ in B^*

maximum of $3\sqrt{3}/2$ when $L = \sqrt{3}$. Now observe that $y_i + r_i$ is at most L. Hence $y_i + r_i + d_i/2 \leq 3\sqrt{3}/2 \approx 2.598$. □

Lemma 5. *Suppose step i is a merge and p_i is in B^*. If $r_i \geq 1.611$ after the move in step i, then $y_i + r_i \leq y_{i-1} + r_{i-1}$.*

Proof. Recall that $z_{i-1,1}$ (resp. $z_{i-1,2}$) is the maximum distance of a point in B^* from $B_{i-1,1}$ (resp. $B_{i-1,2}$); and $y_{i-1} = \min\{z_{i-1,1}, z_{i-1,2}\}$. We will show that $z_{i,2} + r_i \leq \min\{z_{i-1,1}, z_{i-1,2}\} + r_{i-1}$ where $z_{i,2}$ is the maximum distance of a point in B^* from $B_{i,2}$ (i.e., the ball that touches p_i after the merge step). Hence $y_i + r_i \leq z_{i,2} + r_i \leq y_{i-1} + r_{i-1}$.

We first consider $B_{i-1,1}$ and will prove that $z_{i,2}+r_i \leq z_{i-1,1}+r_{i-1}$. (Applying the same argument on $B_{i-1,2}$, we can prove that $z_{i,2} + r_i \leq z_{i-1,2} + r_{i-1}$. Thus we omit the detail.) Draw a line connecting $c_{i-1,1}$ and c^*, the center of $B_{i-1,1}$ and B^* respectively. See Figure 2. Consider the situation before the move in step i. Let x be the distance of $c_{i-1,1}$ from B^*. (Note: $x \leq 0$ if $c_{i-1,1}$ lies within B^*.) Then $x + 2 = r_{i-1} + z_{i-1,1}$.

For convenience, denote by L the distance between $c_{i-1,1}$ and p_i. Since p_i lies within B^*, we have $L \leq x + 2$. Let C be the ball of radius $d_{i-1}/2$ centered at $c_{i-1,1}$. Recall that $c_{i,1}$ denotes the center of $B_{i,1}$ after the move in step i. Therefore, $c_{i,1}$ lies on the surface of C while the newly created ball $B_{i,2}$ will touch p_i and have its center lying on the line $\overline{c_{i,1}p_i}$.

Let q be the intersection of the line $\overline{c_{i-1,1}p_i}$ and the ball C. Then $dist(q,p_i) \geq r_{i-1} + k_{i-1} - d_{i-1}/2 > r_{i-1} + d_{i-1}/2 = r_i$. Hence the center of the newly created ball $B_{i,2}$ must lie within the conical sector of the ball of radius $dist(q,p_i)$, centered at p_i and with the (extended) lateral surface tangential to the ball C. If the whole cone lies within the ball $(1+x)B^*$, then the maximum distance of a point in B^* from $B_{i,2}$ is $z_{i,2} \leq x+2-r_i$. Therefore, $z_{i,2}+r_i \leq x+2 = z_{i-1,1}+r_{i-1}$. Repeating the same argument for $B_{i-1,2}$, we get that $z_{i,2} + r_i \leq z_{i-1,2} + r_{i-1}$. Therefore, $z_{i,2}+r_i \leq \min\{z_{i-1,1}, z_{i-1,2}\}+r_{i-1}$ as required and the lemma follows.

To see that the "cone containment property" holds, refer to Figure 2. Let θ be the angle $\angle p_i c_{i-1,1} c^*$. Then

$$\theta \leq \sin^{-1}\left(\frac{1}{L}\right). \tag{1}$$

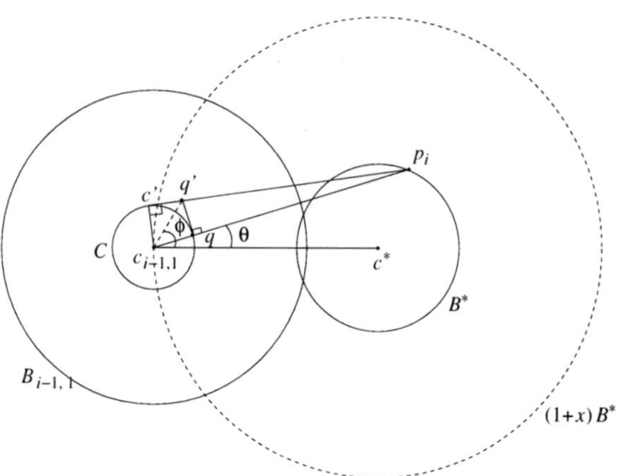

Fig. 2. A merge step

Consider the hyper-plane containing the point q and perpendicular to the line $\overline{p_i c_{i-1,1}}$. Let q' be an arbitrary point on the intersection of this hyper-plane and the cone. Consider the plane containing q', q and p_i. (Note that this plane may not be the same as the plane that contains p_i, $c_{i-1,1}$ and c^*.) The projection of the ball C on this plane forms a circle. Let c' be the intersection of this circle and the line $\overline{q'p_i}$. We let the length of $\overline{qq'}$ be $\eta d_{i-1}/2$ for some $\eta < 1$. By considering the right-angled triangles $\triangle p_i c' c_{i-1,1}$ and $\triangle p_i q q'$, we have $\eta = \sqrt{\frac{L-d_{i-1}/2}{L+d_{i-1}/2}}$. Also, $dist(c_{i-1,1}, q') = (d_{i-1}/2)\sqrt{1+\eta^2}$.

Denote by ϕ the angle $\angle q' c_{i-1,1} c^*$. Then $\phi \leq \theta + (\pi/4)$. Figure 2 shows that q' lies within $(1+x)B^*$. Suppose to the contrary that q' lies outside $(1+x)B^*$. Then $(1+x)B^*$ must have radius less than $(dist(c_{i-1,1}, q')/2)/\cos\phi = \frac{d_{i-1}\sqrt{1+\eta^2}}{4\cos\phi}$. That is, $1 + x < \frac{d_{i-1}\sqrt{1+\eta^2}}{4\cos\phi}$. By the previous observation that $L \leq 2 + x$, we have $L - 1 < d_{i-1}\left(\frac{\sqrt{1+\eta^2}}{4\cos\phi}\right)$, or

$$\theta > \cos^{-1}\left(\frac{d_{i-1}\sqrt{1+\eta^2}}{4(L-1)}\right) - \frac{\pi}{4}. \quad (2)$$

Combining inequality (1) and (2), we have

$$\cos^{-1}\left(\frac{d_{i-1}\sqrt{1+\eta^2}}{4(L-1)}\right) - \frac{\pi}{4} < \sin^{-1}\left(\frac{1}{L}\right). \quad (3)$$

By numerical computation with a computer, when $L \geq 2.586$, inequality (3) cannot be satisfied for any d_{i-1}. (Note that we need only consider $d_{i-1} \leq 2$ because if $B_{i-1,1}$ or $B_{i-1,2}$ touches B^*, then $d_{i-1} < k_{i-1} \leq 2$. Otherwise all points $p_1, \ldots, p_{i-1}$ must be from the other optimal ball and hence $d_{i-1} \leq 2$.)

When $L < 2.586$, there is a lower bound, denoted $\tilde{d}(L)$, on d_{i-1} below which inequality (3) cannot be satisfied. Since $r_i + d_{i-1}/2 < L$, $r_i < L - \tilde{d}(L)/2$. Again, by numerical computation $L - \tilde{d}(L)/2 \leq 1.611$ for any $L < 2.586$. Hence when $r_i \geq 1.611$, inequality (3) cannot be satisfied and so the cone containment holds. □

Lemma 6. *Suppose step i is a walk and p_i is in B^*. If the center $c_{i-1,1}$ of $B_{i-1,1}$ is on or outside the surface of $(1-x)B^*$ for some $0 \leq x \leq 1$ and $r_{i-1} \geq \sqrt{x(2-x)}$ before the move in step i, then $z_{i,1} + r_i \leq z_{i-1,1} + r_{i-1}$. The same conclusion holds for $B_{i-1,2}$.*

Proof. First, consider the case where $c_{i-1,1}$ is on the surface of $(1-x)B^*$. Consider the plane containing p_i, $c_{i-1,1}$ and c^*. Suppose p_i is on the surface of B^*. Let θ be the angle $\angle c^* c_{i-1,1} p_i$ and ϕ be the angle $\angle c^* p_i c_{i-1,1}$. See Figure 3.

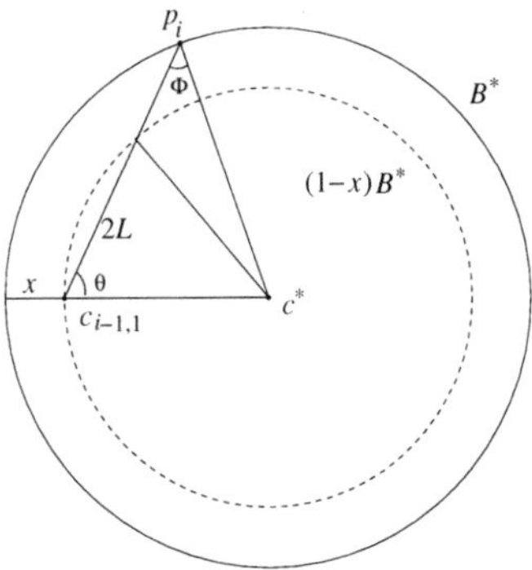

Fig. 3. A walk step

Let $L = (1-x)\cos\theta$. Then $(1-x)\sin\theta = \sqrt{(1-x)^2 - L^2}$. Hence $\cos\phi = \sqrt{1 - ((1-x)^2 - L^2)} = \sqrt{2x - x^2 + L^2}$. Then $dist(c_{i-1,1}, p_i) = L + \cos\phi = L + \sqrt{2x - x^2 + L^2}$. By construction of the algorithm, after the walk in step i, the center of $B_{i,1}$ will be on the line segment $\overline{c_{i-1,1}p_i}$ and at distance $k_{i-1}/2$ from $c_{i-1,1}$. If $2L \geq k_{i-1}/2$, then the center of $B_{i,1}$ is still in $(1-x)B^*$. This is satisfied if $2L \geq (dist(c_{i-1,1}, p_i) - r_{i-1})/2$. But this condition is always satisfied when $r_{i-1} \geq \sqrt{2x - x^2}$ because $3L + r_{i-1} - \sqrt{2x - x^2 + L^2} \geq 3L + \sqrt{2x - x^2} - \sqrt{2x - x^2 + L^2} \geq 3L + \sqrt{2x - x^2} - \sqrt{2x - x^2} - L \geq 0$. Hence $z_{i,1} + r_i \leq z_{i-1,1} + r_{i-1}$.

Now suppose $c_{i-1,1}$ is between B^* and $(1-x)B^*$. Then $c_{i-1,1}$ is on the surface of $(1-x')B^*$ for some $0 \leq x' < x$. Since $\sqrt{x(2-x)}$ is an increasing function for $0 \leq x \leq 1$, we have $r_{i-1} \geq \sqrt{x'(2-x')}$ as well. Therefore, we can apply the previous argument. Finally, if $c_{i-1,1}$ is lying on the surface of $(1+x)B^*$ for some $x \geq 0$, then $c_{i,1}$ is always within $(1+x)B^*$. □

Corollary 1. *Suppose step i is a walk and p_i is in B^*. If $r_{i-1} \geq 1$, then $y_i + r_i \leq y_{i-1} + r_{i-1}$.*

Proof. If $r_{i-1} \geq 1$, then $r_{i-1} \geq \sqrt{x(2-x)}$ for any $0 \leq x \leq 1$. Hence $z_{i,1} + r_i \leq z_{i-1,1} + r_{i-1}$ and $z_{i,2} + r_i \leq z_{i-1,2} + r_{i-1}$. Therefore, $y_i + r_i \leq y_{i-1} + r_{i-1}$. □

4.2 An Upper Bound of 5.708

Theorem 2. *In d-dimension, the approximation ratio of the algorithm is at most 5.708.*

Proof. We consider the earliest step i such that $r_i \geq 1.708$. By Lemma 3, we have $0.708 \leq r_{i-1} < 1.708$. Without loss of generality, assume the first point p_1 comes from B_1^*. Then there are two cases.

Case 1: $p_1, \ldots, p_{i-1}$ are all in B_1^*.
If p_i is in B_2^*, then after step i, $y_{i,2} \leq 2$. Next we bound $y_{i,1}+r_i$. Since $p_1, \ldots, p_{i-1}$ are all in B_1^*, both $c_{i-1,1}$ and $c_{i-1,2}$ are in B_1^*. Then by Lemma 4, we have $y_{i-1,1}+r_{i-1}+d_{i-1}/2 \leq 2.598$. Since $r_i \leq r_{i-1}+d_{i-1}/2$ and $y_{i,1} \leq y_{i-1,1}$, we have $y_{i,1} + r_i \leq 2.598$. As $r_{i'} \geq 1.708$ for all $i' \geq i$, we can repeatedly apply Lemma 5 and Corollary 1 to bound the increase in radius after step i. More precisely, the lemma and corollary guarantee that $y_{i',1} + y_{i',2} + r_{i'} \leq y_{i,1} + y_{i,2} + r_i$. Therefore, the final radius is at most $y_{i,1} + y_{i,2} + r_i \leq 2.598 + 2 = 4.598$.

If p_i is still in B_1^*, then we look for the earliest step i' such that $p_{i'}$ is in B_2^* and apply the above argument.

Case 2: some points among $p_1, \ldots, p_{i-1}$ are in B_2^*.
Then both $y_{i-1,1}$ and $y_{i-1,2}$ are at most 2. Consider the move in step i and let p_i be in B_j^* (where $j = 1$ or 2). If the move is a merge, then by Lemma 5, $y_{i,j} + r_i \leq y_{i-1,j} + r_{i-1} < 2 + 1.708$. Hence $y_{i,1} + y_{i,2} + r_i < 5.708$.

Now we assume that the move is a walk. Suppose the center of either $B_{i-1,1}$ or $B_{i-1,2}$ is within $(B_j^* - x)$ where $x = 1 - \sqrt{1 - 0.708^2} \approx 0.294$. Then $y_{i-1,j} + r_{i-1} \leq 2 - x$. Hence $y_{i,j} \leq y_{i-1,j} \leq 2 - x - r_{i-1} \approx 0.998$. As at each step the radius increases by at most 1 (Lemma 3), we have $r_i < 2.708$. Hence $y_{i,j} + r_i < 0.998 + 2.708 \approx 3.706$ and $y_{i,1} + y_{i,2} + r_i < 3.706 + 2 = 5.706$. Suppose both the centers of $B_{i-1,1}$ and $B_{i-1,2}$ are on the surface of or outside $(B_j^* - x)$, then $r_{i-1} \geq 0.708 \geq \sqrt{x(2-x)}$. Applying Lemma 6 twice and using $y_{i,j} = \min\{z_{i,j,1}, z_{i,j,2}\}$, we have $y_{i,j}+r_i \leq y_{i-1,j}+r_{i-1} < 2+1.708 = 3.708$ and so $y_{i,1}+y_{i,2}+r_i \leq 5.708$.

Thus, whether step i is a merge or walk, we have $y_{i,1} + y_{i,2} + r_i \leq 5.708$. By repeatedly applying Lemma 5 and Corollary 1 from step $i+1$ onwards, the final radius is at most $y_{i,1} + y_{i,2} + r_i \leq 5.708$.

Summarizing case 1 and 2, the approximation ratio of our algorithm is at most 5.708. □

5 Lower Bounds for the Problems

We first state our computation model precisely. We assume that the radius and centers are stored as real numbers of infinite precision. However, such an algorithm can make use of the infinite precision power to encode the locations of the points within the two balls. To prevent such pecularities, we also assume that in each step of any such algorithm, the old balls will be covered by the new balls.

Theorem 3. *Under the special streaming model that at any step only the radius and locations of two balls are remembered and the assumption that the regions of the old balls have to be covered by the new balls, there is a lower bound of 1.5 in 1d and a lower bound of 1.604 in 2d, on the approximation ratio of any deterministic algorithm for two-center covering.*

For the 1d case, consider the following input. The adversary first specifies three points, $p_1 = 0$, $p_2 = 1$ and $p_3 = 10$. If the algorithm puts p_2 and p_3 in the same interval, then the approximation ratio is greater than 2. So, we assume the algorithm puts p_1 and p_2 in one interval and p_3 in another. Let D be the length of the interval. Thus $D \geq 1$. Let p_3 be at distance α from the closer boundary of its enclosing interval. So, $\alpha \leq D/2$. Now, the adversary adds point p_4 so that it is at distance 1 from p_3 and at distance $1 - \alpha$ from the boundary of the interval containing p_3. Then the optimal algorithm can still cover everything with intervals of length 1. However, the algorithm does not remember the location of the points within the intervals. So, it has to increase the length of the interval to $D+1-\alpha \geq D/2+1 \geq 1.5$. Hence the approximation ratio is at least $1.5/1 = 1.5$.

For the d-dimensional case, consider the following input in which all the points lie on a 2d plane. The adversary first specifies three points: $p_1 = (0, -1)$, $p_2 = (0, 1)$ and $p_3 = (100, 0)$ (i.e., far away from p_1, p_2). The algorithm has to enclose p_1 and p_2 by one circle and p_3 by another. Then the adversary presents $p_4 = (-1 - \sqrt{2}, 0)$. Again, the algorithm should enclose p_1, p_2 and p_4 by one circle. Let D be the diameter at this moment. Note that p_1, p_2, p_4 is the collection of three points used in [16] to prove a lower bound of $2 + \sqrt{2}$ on the diameter of the smallest enclosing circle. Thus $D \geq 2 + \sqrt{2}$.

Now consider the other circle containing p_3. Suppose p_3 is at distance α from the nearest point on the perimeter. (So, $\alpha \leq D/2$.) Then, the adversary gives p_5 at distance $2\sqrt{2} - \alpha$ from the circle so that p_5, p_3 and the center of circle are colinear. The algorithm has to enlarge the diameter of the circle to $D + 2\sqrt{2} - \alpha \geq D/2 + 2\sqrt{2} \geq 1 + 1/\sqrt{2} + 2\sqrt{2}$ while optimal diameter is $2\sqrt{2}$. Hence the approximation ratio is at least $1 + \frac{1+1/\sqrt{2}}{2\sqrt{2}} \approx 1.604$.

6 Concluding Remarks and Open Problems

One can construct an input for $d > 1$ on which our algorithm has approximation ratio 2.5. By numerical computation, one can even improve the lower bound to 2.668. However, we believe that our algorithm has approximation ratio much smaller than 5.708. Thus, an open problem is to provide a tighter analysis of our algorithm. It is also interesting to design algorithms with better approximation ratio for the problem (still under the same model). In fact, in this case even for covering with one-center, the lower bound for the problem is 1.207 while the upper bound is only 1.5, see [16]. Another direction is to extend our algorithm to the general k-center covering problem where $k \geq 2$.

References

1. Agarwal, P., Har-Peled, S., Varadarajan, K.: Approximating extent measures of points. J. ACM 51(4), 606–635 (2004)
2. Agarwal, P., Sharathkumar, R.: Streaming algorithms for extent problems in high dimensions. In: Proc. 21st Annual ACM-SIAM Symposium on Discrete Algorithms, SODA 2010, pp. 1481–1489 (2010)
3. Bădoiu, M., Clarkson, K.: Smaller core-sets for balls. In: Proc. 14th Annual ACM-SIAM Symposium on Discrete Algorithms, SODA 2003, pp. 801–802 (2003)
4. Bădoiu, M., Har-Peled, S., Indyk, P.: Approximate clustering via core-sets. In: Proc. 34th Annual ACM Symposium on Theory of Computing, STOC 2002, pp. 250–257 (2002)
5. Chan, T.: More planar two-center algorithms. Comput. Geom. Theory Appls. 13, 189–198 (1999)
6. Chan, T.: Dynamic coresets. In: Proc. 24th Annual ACM Symposium on Computational Geometry, SoCG 2008, pp. 1–9 (2008)
7. Chan, T.M., Pathak, V.: Streaming and Dynamic Algorithms for Minimum Enclosing Balls in High Dimensions. In: Dehne, F., Iacono, J., Sack, J.-R. (eds.) WADS 2011. LNCS, vol. 6844, pp. 195–206. Springer, Heidelberg (2011)
8. Charikar, M., Chekuri, C., Feder, T., Motwani, R.: Incremental clustering and dynamic information retrieval. SIAM Journal on Computing 33(6), 1417–1440 (2004)
9. Eppstein, D.: Faster construction of planar two-centers. In: Proc. 8th Annu. ACM-SIAM Sympos. Discrete Algo., pp. 131–138 (1997)
10. Guha, S.: Tight results for clusering and summarizing data streams. In: Proc. of 12th International Conference on Database Theory, pp. 268–275. ACM (2009)
11. Har-Peled, S., Kushal, A.: Smaller coresets for k-median and k-means clustering. Discrete Computational Geometry 37, 3–19 (2007)
12. Har-Peled, S., Mazumdar, S.: Coresets for k-means and k-median clustering and their applications. In: Proc. 36th Annual ACM Symposium on Theory of Computing, STOC 2004, pp. 291–300 (2004)
13. McCutchen, R.M., Khuller, S.: Streaming Algorithms for k-Center Clustering with Outliers and with Anonymity. In: Goel, A., Jansen, K., Rolim, J.D.P., Rubinfeld, R. (eds.) APPROX and RANDOM 2008. LNCS, vol. 5171, pp. 165–178. Springer, Heidelberg (2008)
14. Sharir, M.: A near-linear algorithm for the planar 2-center problem. Discrete Comput. Geom. 4, 125–134 (1997)
15. Zarrabi-Zadeh, H.: An Almost Space-Optimal Streaming Algorithm for Coresets in Fixed Dimensions. In: Halperin, D., Mehlhorn, K. (eds.) ESA 2008. LNCS, vol. 5193, pp. 817–829. Springer, Heidelberg (2008)
16. Zarrabi-Zadeh, H., Chan, T.: A simple streaming algorithm for minimum enclosing balls. In: Proc. 18th Canadian Conference on Computational Geometry, CCCG 2006, pp. 139–142 (2006)

Online Joint Pricing and Booking Policies in Airline Revenue Management

Guanqun Ni[1,2,3,⋆] and Yinfeng Xu[1,2,3]

[1] School of Management, Xi'an Jiaotong University, Xi'an, 710049, China
[2] State Key Lab for Manufacturing Systems Engineering, Xi'an, 710049, China
[3] Ministry of Education Key Lab for Process Control & Efficiency Engineering, Xi'an, 710049, China
guanqun.ni@stu.xjtu.edu.cn, yfxu@mail.xjtu.edu.cn

Abstract. We introduce a model considering the single-leg revenue management problem from the perspective of online algorithms and competitive analysis. In this model, the price and limitation of bookings are both decision variables. Assuming the process of the customers meets the low-before-high manner and the expected value of customer falls in a closed interval, we analyze the upper bound for deterministic online joint pricing and booking policy and propose one optimal policy for each case.

Keywords: joint pricing and booking, revenue management, online algorithms, competitive analysis.

1 Introduction

In practice, airlines usually make series pricing decisions according to the time before the departure. One airline needs to decide when to adjust price of ticket based on the quantity sold out at prior price and how much to change the price. In the field of revenue management (RM), the first question is quantity-based and the second question is price-based [7]. Most researches on quantity-based and price-based control policy have been carried out under different assumptions independently in airline industry, while limited research has been studied based on integration of pricing and booking [5]. This motivates us to investigate the joint pricing and booking control policy in RM. At the same time, a critical assumption made in most academic studies of RM problems is that the demand function is known to the decision maker. This assumption of "full information" endows the decision maker with knowledge that she/he does not typically possess in practice [2]. Moreover, the airline industry itself is undergoing substantial upheaval, which is leading toward new approaches to defining the airline product [1]. These reasons motivate us to investigate the joint pricing and booking control policy from the perspective of online algorithms and competitive analysis (see [3] for an extensive discussion and [4] for a recent survey) with uncertain demand.

⋆ Corresponding author.

G. Lin (Ed.): COCOA 2012, LNCS 7402, pp. 281–290, 2012.

In [1], the authors first analyzed the online booking policy in RM with uncertain demand from the perspective of competitive analysis and a class of nested protection-level policies is proposed based on fare-class. In their paper, the classes of fares are fixed as exogenous variables and the fare class of request is also known to the airline. This assumption is adopted in most quantity-based RM models. While, in practice the airline usually do not know the customer's expected value whatever the fare she/he pays for the booking is. Moreover, in many price-based models, the possible price values are approximated to a finite set based on customer classes [5]. This assumption is technical and in such a formulation the pricing policy would be handled. However, the assumption about the fare classes may not cover all the possible kinds of customers. A nature way to introduce the price decision variable would be as a real number on some interval. In our model, the airline don't known the customer's expected value exactly facing booking and can choose any price from a given fare interval.

The problem considered in this paper can be described as follows. One flight of the airline has n seats. Before the passenger arriving the airline must post a price f and the passenger will book the ticket if her/his valuation is at least the price. We assume that the expected value of every passenger falls in the closed interval: $[f_0, f_m]$, and the total number of passengers who want to book a seat is no less than n. That is, if the fare of seats is set to be f_0, the airline may sell out all the seats which reflects the situation where *cheap ticket is in short supply*, especially for "cheap flight" in a boom season. In this paper, we call this assumption "CTSS" in short. From a practical point of view, we assume the process of the passengers meets the low-before-high (LBH) manner [6]; i.e., demand in lower fare class arrives earlier. Although demand meets the LBH manner, we assume, the demand or demand function for any kind of fare is unknown and there is no any information about demand arrive rate. Thus, the airline has to decide when to adjust price based on the quantity sold out at prior price and how much to change the price facing the online information of demand. In practice, if the times of price changing are limited to no more than k (more times more adjusting cost), the airline has to answer the two questions: (1) what the ith price should be; and (2) when to change the $(i-1)th$ price to the ith price? Additionally, we consider the continuous problem, where the value of decision variable is not restricted to integer [1] in our model.

Our approach is general in that we do not restrict ourselves to any class of algorithms, but rather seek the best possible policy. Nonetheless, our results show that the optimal online policies come from a class, namely, *dynamic pricing policies with booking limited.* These policies design series of fares and define limited bookings for each fare. In this paper we define $P(F, Q)$ a unique policy, where $F = (f_1, .., f_k)$ and $Q = (q_0, q_1, ..., q_{k-1})$ are the decision vectors of fare and quantity, respectively. For example, a policy $P(F, Q)$ with $F = (600, 750)$ and $Q = (100, 50)$ for a flight with 180 seats, the lowest fare $f_0 = 450$ and the highest fare $f_m = 800$ means: at the beginning, the airline sells tickets at 450 and the bookings for this fare are limited to 100. If the quantity of seats sold out at 450 equals 100, the airline adjusts the fare from 450 to $f_1 = 600$ and the bookings for fare f_1 are limited to 50. Otherwise, do not change the fare. If the quantity of seats sold out at 600 equals 50, the airline adjusts the fare from 600 to $f_2 = 750$ and then sells the remaining (30) seats at fare f_2 till all the seats are sold out or the flight departs. In this paper all policies are described as the form $P(F, Q)$. From the analysis in Section 2, we could compute that the competitive ratio of policy $P(F, Q)$ with $F = (600, 750)$ and $Q = (100, 50)$ for this example is $c = 0.56$, and the optimal online policy is $P^*(F^*, Q^*)$ with $F^* = (545.12, 660.35)$ and $Q^* = (133.33, 23.33)$ whose competitive ratio is $c^* = 0.61$.

2 Online Joint Pricing and Booking Policies

As described above, the airline should decide: (1) what the *ith* price should be; and (2) when to change the $(i-1)th$ price to the *ith* price? Before answering these questions in detail, we first prove two lemmas used in following analysis from the perspective of online algorithms and competitive analysis.

For any instance, the airline always needs choose a fare as the initial price. Without loss of generality, we let f_0' be the initial fare. Then, we have the following Lemma 1.

Lemma 1. *For the continuous online joint pricing and booking problem, the initial fare f_0' is just equal to f_0. Otherwise, there is no deterministic online policy whose competitive ratio is larger than 0.*

Proof. If the initial fare f_0' is not equal to f_0, it will only be the case that $f_0' > f_0$ because the expected value of every passenger falls in the closed interval: $[f_0, f_m]$. For any online policy P with $f_0' > f_0$, we consider the following instance. There will be more than n requests coming, and all the requests' expected value is equal to f_0. The resulting revenue of policy P is 0 when applied to this instance. And the revenue of an optimal offline policy is $f_0 n$. Thus, the competitive ratio is 0. This shows that no deterministic online policy with $f_0' > f_0$ can guarantee for all instances that its competitive ratio is larger than 0.

This gives us the Lemma 1. □

From a practical point of view, all kinds of price adjusting may be an optimal policy, pricing markup or markdown. Nonetheless, the following Lemma 2 shows that we need only research the optimal policy in the class of markup policies as long as there is an optimal online policy indeed.

Lemma 2. *If there is an optimal online policy $P(F, Q)$ with the case that $f_i > f_j$ for some $i < j$, we revise the vector of $F = (f_1, ..., f_k)$ to a non-descending alternative vector $F' = (f_1', ..., f_k')$ with $f_i' \le f_j'$ for $\forall i < j$. Then the alternative policy $P'(F', Q)$ is still optimal.*

Proof. For policy $P(F, Q)$, we assume, without loss of generality, $f_i > f_{i+1}$ for some $i \ge 1$. Let $p_i (\le n)$ be the quantity of bookings accepted with fare f_i. If it is the case that $p_i > 0$, we could adjust $f_{i+1} = f_i$. And the revenue will not decrease because of LBH manner. If it is the case that $p_i = 0$, then there will not be request whose expected value is equal to or bigger than f_i based on LBH manner. Thus, we could adjust $f_i = f_{i+1}$ and the revenue will also not decrease. And following this analysis, we could prove the lemma for general case where $f_i > f_j$ for $\forall i < j$.

This gives us the Lemma 2. □

Based on the Lemma 2, we only research the optimal one in the class of policies with $f_i > f_j$ for $\forall i > j$.

2.1 Optimal Online Policy for Case $k = 1$

In this subsection, we first design an online policy $P_1(F_1, Q_1)$ for $k = 1$ where the times of price changing are no more than one and then prove the policy has the best competitive ratio.

We describe the online joint pricing and booking policy $P_1(F_1, Q_1)$ as follows.

Policy $P_1(F_1, Q_1)$ with $F_1 = (\sqrt{f_0 f_m})$, and $Q_1 = (\frac{n}{2-\sqrt{f_0/f_m}})$: *at the beginning, the airline sells its seats at fare f_0. When the quantity of seats sold out at fare f_0 equals $q_0 = \frac{n}{2-\sqrt{f_0/f_m}}$, the airline adjusts the price to fare $f_1 = \sqrt{f_0 f_m}$ and sells the remaining seats at fare f_1. Otherwise, the airline doesn't change the price till the departure time.*

Theorem 1. *For the continuous problem with $k = 1$, the online joint pricing and booking policy $P_1(F_1, Q_1)$ has the best competitive ratio $c = \frac{1}{2\sqrt{f_m/f_0}-1}$.*

Before proving this theorem, we need the following Lemma 3.

Lemma 3. *Let $u_1 = \frac{ax+y(n-x)}{bn}$ and $u_2 = \frac{ax}{yn}$, where $0 \le x \le n$, and $0 < a < y \le b$. Define the function $v = \min\{u_1, u_2\}$, we have $(x = \frac{n}{2-\sqrt{a/b}}, y = \sqrt{ab}) = \arg\max v$.*

Proof. From the definitions of u_1 and u_2 , we have $u_1 = \frac{u_2 y(1-y/a)+y}{b}$ and $\frac{\partial u_1}{\partial u_2} = \frac{y(1-y/a)}{b} < 0$, for $0 < a < y \le b$. Thus, u_1 and u_2 are inversely related, that is, $u_1 = u_2$ is the necessary condition when v gets maximal value.

Let $u_1 = u_2 = u$, we have $x = \frac{y^2 n}{ab+y^2-ay}$, $u = \frac{ay}{ab+y^2-ay}$ and $\frac{du}{dy} = \frac{a(ab-y^2)}{(ab+y^2-ay)^2}$. Because $0 < a < y \le b$, u reaches maximum value when $y = \sqrt{ab}$, that is, when $x = \frac{y^2 n}{ab+y^2-ay}$ and $y = \sqrt{ab}$, the function $v = \min\{u_1, u_2\}$ gets maximum value. This gives us the Lemma 3. □

Now we prove the Theorem 1 as follows.

Proof. Consider any demand instance I; after applying the online pricing and booking policy, let R' be the total revenue of all requests accepted. Let R^* be the revenue achieved by the application of an optimal offline policy.

For any general online policy $P(F,Q)$ with $F=(f_1)$ and $Q=(q_0)$, let q_1 be the quantity of seats sold out at fare f_1. We know that any $q_0 \le n$ can be met because of CTSS assumption. Thus, the online decision-maker will only face two possible cases: (1) $q_1 = n - q_0$, and (2) $q_1 < n - q_0$.

Case 1. $q_1 = n - q_0$. In this case, $R' = f_0 q_0 + f_1 q_1 = f_0 q_0 + f_1(n - q_0)$. Based on LBH assumption, the "imaginary adversary" [1,3,4] will generate one of worst requests sequences: there are n requests with expected value f_1 followed by n requests with expected value f_m. Thus, the revenue achieved by online policy is still R', while the optimal revenue achieved by an offline policy will be $R^* = f_m n$. Thus, we have that the ratio for Case 1 is

$$c_1 = \frac{R'}{R^*} = \frac{f_0 q_0 + f_1(n - q_0)}{f_m n}. \tag{1}$$

Case 2. $q_1 < n - q_0$. In this case, there will be two sub-cases: (2.1) $q_1 = 0$, and (2.2) $q_1 > 0$.

Sub-case 2.1. $q_1 = 0$. In this case, $R' = f_0 q_0$. And there is no request with expected value equal to or bigger than fare f_1 based on LBH assumption. The "imaginary adversary" will generate one of worst requests sequences: there are only n requests with expected value equal to $(f_1 - \varepsilon)$ where $\varepsilon \to 0^+$. Thus, the revenue achieved by online policy is still R', while the optimal revenue achieved by an offline policy will be $R^* = (f_1 - \varepsilon)n \to f_1 n$. Thus, we have that the ratio for Case 2.1 is

$$c_{21} = \frac{R'}{R^*} = \frac{f_0 q_0}{f_1 n}. \tag{2}$$

Sub-case 2.2. $q_1 > 0$. In this case, $R' = f_0 q_0 + f_1 q_1$. And because of $q_1 < n - q_0$, the "imaginary adversary" will generate one of worst requests sequences: there are n requests with expected value $(f_1 - \varepsilon)$ where $\varepsilon \to 0^+$ followed by q_1 requests with expected value equal to f_m. Thus, the revenue achieved by online policy is still R', while the optimal revenue achieved by an offline policy will be $R^* = f_m q_1 + (f_1 - \varepsilon)(n - q_1) \to f_1 n + (f_m - f_1) q_1$. Thus, we have that the ratio for Case 2.2 is

$$c_{22} = \frac{R'}{R^*} = \frac{f_0 q_0 + f_1 q_1}{f_1 n + (f_m - f_1) q_1}. \tag{3}$$

Now we analyze the relationship between c_{21} and c_{22} in Case 2.

If it is the case that $\frac{f_0 q_0}{f_1 n} \le \frac{f_1 q_1}{(f_m - f_1) q_1}$, we have $c_{21} \le c_{22}$. Otherwise, it is the case that $\frac{f_0 q_0}{f_1 n} > \frac{f_1 q_1}{(f_m - f_1) q_1}$, and we have $c_{21} > c_{22}$. At the same time we find that $c_{22} = \frac{f_0 q_0 + f_1 q_1}{f_1 n + (f_m - f_1) q_1}$ is monotonically decreasing in q_1. Thus, the "imaginary adversary" will design one worst sequence so that q_1 is infinitely close to its upper bound $(n - q_0)$ to make sure that c_{22} reaches its minimum. If the "imaginary adversary" choose such worst sequence, it is the case that

$$c_{22} \to \frac{f_0 q_0 + f_1(n - q_0)}{f_m n} = c_1. \tag{4}$$

Generally speaking, for the continuous problem with $k = 1$, any online policy $P(F, Q)$ with $F = (f_1)$ and $Q = (q_0)$ will only face two kinds of possible worst instances which generate two possible worst performances. Thus, there will be two possible minimal ratios:

$$c_1 = \frac{f_0 q_0 + f_1(n - q_0)}{f_m n} \tag{5}$$

and

$$c_2 = \frac{f_0 q_0}{f_1 n}, \tag{6}$$

and the online decision-maker need only choose appropriate q_0 and f_1 based on the tradeoff between c_1 and c_2.

Then following the Lemma 3, we complete the proof. □

2.2 Optimal Online Policy for Case $k > 1$

In this subsection, we mainly research the continuous problem with $k > 1$. We also design an online pricing and booking policy $P_k(F_k, Q_k)$ and then prove the policy has the best competitive ratio.

Based on the description about policy $P(F, Q)$ in Section 1, we describe the online joint pricing and booking policy $P_k(F_k, Q_k)$ for $k > 1$ concisely as follows.

> **Policy $P_k(F_k, Q_k)$ with $F_k = (f_1, f_2, ..., f_k)$ and $Q_k = (q_0, q_1, ..., q_{k-1})$** *defines* $q_0 = \frac{n}{1+k-k(\frac{f_0}{f_m})^{1/(k+1)}}$ *the bookings limited for fare* f_0, $q_i = \frac{n}{k+[1-(\frac{f_0}{f_m})^{1/(k+1)}]^{-1}}$ *the bookings limited for fare* f_i *where* $1 \le i \le k-1$, *and the fare* $f_i = f_0^{(k+1-i)/(k+1)} f_m^{i/(k+1)}$ *for* $1 \le i \le k$.

Theorem 2. *For the continuous problem with $k > 1$, the online joint pricing and booking policy $P_k(F_k, Q_k)$ has the best competitive ratio* $c = \frac{1}{(1+k)(\frac{f_m}{f_0})^{1/(k+1)} - k}$.

Before proving this theorem, we need the following Lemma 4.

Lemma 4. *Let* $u_1 = \frac{a x_0}{y_1 n}$, $u_2 = \frac{a x_0 + y_1 x_1}{y_2 n}$, ..., $u_i = \frac{a x_0 + \sum_{j=1}^{i-1} y_j x_j}{y_i n}$ *for* $2 \le i \le k$ *and* $u_{k+1} = \frac{a x_0 + \sum_{j=1}^{k-1} y_j x_j + f_k(n - \sum_{j=0}^{k-1} x_j)}{bn}$, *where* $0 \le x_i \le n$, $\sum_{i=0}^{k-1} x_i \le n$ *for* $0 \le i \le k-1$ *and* $0 < a < y_1 < y_2 < ... < y_k \le b$. *Define the function* $v = \min\{u_1, u_2, ..., u_k, u_{k+1}\}$, *we have* $(x_0 = \frac{n}{1+k-k(\frac{a}{b})^{1/(k+1)}}, x_i = \frac{n}{k+[1-(\frac{a}{b})^{1/(k+1)}]^{-1}}, y_i = a^{(k+1-i)/(k+1)} b^{i/(k+1)}) = \arg\max v$.

Proof. For convenience, we let $y_0 = a$ and $y_{k+1} = b$. From the definition of u_i for $1 \le i \le k+1$, we have $x_0 = \frac{n y_1 u_1}{y_0}$, $x_i = \frac{n(y_{i+1} u_{i+1} - y_i u_i)}{y_i}$ for $1 \le i \le k-1$,

$u_{k+1} = \frac{y_k u_k + y_k(1 - \frac{y_1 u_1}{y_0} - \sum_{i=1}^{k-1} \frac{y_{i+1}u_{i+1} - y_i u_i}{y_i})}{b}$, and $\frac{\partial u_{k+1}}{\partial u_i} = (1 - \frac{y_i}{y_{i-1}})\frac{y_k}{b} < 0$ for $1 \leq i \leq k$. Thus, u_{k+1} and any u_i for $1 \leq i \leq k$ are inversely related, that is, $u_{k+1} = u_i$ for $1 \leq i \leq k$ is the necessary condition when v gets maximal value.

Let $u_{k+1} = u_i = u$ for $1 \leq i \leq k$, we have $x_0 = \frac{ny_1u}{y_0}$, $x_i = \frac{n(y_{i+1} - y_i)u}{y_i}$ for $1 \leq i \leq k-1$ and $u = \frac{y_k}{b + y_k[\sum_{i=0}^{k-1}(y_{i+1}/y_i) - k]}$. Then $\frac{\partial u}{\partial y_i} = \frac{y_k^2}{\{b + y_k[\sum_{i=0}^{k-1}(y_{i+1}/y_i) - k]\}^2}(\frac{y_{i+1}}{y_i^2} - \frac{1}{y_{i-1}})$ for $1 \leq i \leq k-1$, and $\frac{\partial u}{\partial y_k} = \frac{1}{\{b + y_k[\sum_{i=0}^{k-1}(y_{i+1}/y_i) - k]\}^2}(b - \frac{y_k^2}{y_{k-1}})$.

Because $0 < a < y_1 < y_2 < ... < y_k \leq b$, u reaches its maximum value when $y_i^2 = y_{i-1}y_{i+1}$ for $1 \leq i \leq k$, that is, when $y_i = a^{(k+1-i)/(k+1)}b^{i/(k+1)}$, the function $v = \min\{u_1, u_2, ..., u_k, u_{k+1}\}$ gets maximum value. At the same time, we could compute that $x_i = [1 - (\frac{a}{b})^{1/(k+1)}]x_0$ for $1 \leq i \leq k-1$ and $x_0 = \frac{ny_1u}{a}$. Further, let $u_1 = u_{k+1}$, we get that $x_0 = \frac{n}{1 + k - k(\frac{a}{b})^{1/(k+1)}}$ and $x_i = \frac{n}{k + [1 - (\frac{a}{b})^{1/(k+1)}]^{-1}}$ for $1 \leq i \leq k-1$.

This gives us the Lemma 4. □

Now we prove the Theorem 2 as follows.

Proof. Consider any demand instance I; after applying the online control policy, let R' be the total revenue of all requests accepted. Let R^* be the revenue achieved by the application of an optimal offline policy.

For the continuous problem with $k > 1$, consider any general online policy $P_k(F_k, Q_k)$ with $F = (f_1, f_2, ..., f_k)$ and $Q = (q_0, q_1, ..., q_{k-1})$, let p_i be the quantity of seats sold out at fare f_i. We know that any $q_0 \leq n$ can be met because of CTSS assumption, that is, it is always the case that $p_0 = q_0$. Thus, the online decision-maker will only face two possible cases after accepting q_0 bookings with fare f_0: (1) the quantity of seats sold out at fare f_1 is less than q_1, that is, $p_1 < q_1$ and (2) $p_1 = q_1$.

Case 1. $p_1 < q_1$. In this case, there will be two possible cases: (1.1) $p_1 = 0$, and (1.2) $0 < p_1 < q_1$.

Case 1.1. $p_1 = 0$. In this case, $R' = f_0q_0$. And there is no request with expected value equal to or bigger than fare f_1 based on LBH assumption. The "imaginary adversary" will generate one of worst requests sequences: there are only n requests with expected value equal to $(f_1 - \varepsilon)$ where $\varepsilon \to 0^+$. Thus, the revenue achieved by online policy is still R', while the optimal revenue achieved by an offline policy will be $R^* = (f_1 - \varepsilon)n \to f_1n$. Thus, we have that the ratio for Case 1.1 is

$$c_{11} = \frac{R'}{R^*} = \frac{f_0q_0}{f_1n}. \tag{7}$$

Case 1.2. $0 < p_1 < q_1$. In this case, $R' = f_0q_0 + f_1p_1$. And because of $p_1 < q_1$, the "imaginary adversary" will generate one of worst requests sequences: there are n requests with expected value $(f_1 - \varepsilon)$ where $\varepsilon \to 0^+$ followed by p_1 requests with expected value equal to f_m. Thus, the revenue achieved by online policy is still R', while the optimal revenue achieved by an offline policy will be $R^* =$

$f_m p_1 + (f_1 - \varepsilon)(n - p_1) \to f_1 n + (f_m - f_1)p_1$. Thus, we have that the ratio for Case 1.2 is

$$c_{12} = \frac{R'}{R^*} = \frac{f_0 q_0 + f_1 p_1}{f_1 n + (f_m - f_1)p_1}. \tag{8}$$

Consider the relationship between the value of c_{12} and p_1. If the value of c_{12} is monotonically increasing in p_1, the "imaginary adversary" will choose some instance so that

$$c_{12} \to \frac{f_0 q_0}{f_1 n} = c_{11}. \tag{9}$$

Otherwise, the "imaginary adversary" will choose some instance so that the value of p_1 reaches its upper bound q_1 to get possible worst performance, and for this instance, it degenerates into the following Case 2.

Case 2. $p_1 = q_1$. In this case, the online decision-maker will also face two possible cases: (2.1) $p_2 < q_2$ and (2.2) $p_2 = q_2$.

Case 2.1. $p_2 < q_2$. In this case, there will be two sub-cases: (2.1.1) $p_2 = 0$, and (2.1.2) $0 < p_2 < q_2$.

Sub-case 2.1.1. $p_2 = 0$. In this sub-case, $R' = f_0 q_0 + f_1 q_1$. And there is no request with expected value equal to or bigger than fare f_2 based on LBH assumption. The "imaginary adversary" will generate one of worst requests sequences: there are only n requests with expected value equal to $(f_2 - \varepsilon)$ where $\varepsilon \to 0^+$. Thus, the revenue achieved by online policy is still R', while the optimal revenue achieved by an offline policy will be $R^* = (f_2 - \varepsilon)n \to f_2 n$. Thus, we have that the ratio for Case 2.1.1 is

$$c_{211} = \frac{R'}{R^*} = \frac{f_0 q_0 + f_1 q_1}{f_2 n}. \tag{10}$$

Sub-case 2.1.2. $0 < p_2 < q_2$. In this case, $R' = f_0 q_0 + f_1 q_1 + f_2 p_2$. And because of $p_2 < q_2$, the "imaginary adversary" will generate one of worst requests sequences: there are n requests with expected value $(f_2 - \varepsilon)$ where $\varepsilon \to 0^+$ followed by p_2 requests with expected value equal to f_m. Thus, the revenue achieved by online policy is still R', while the optimal revenue achieved by an offline policy will be $R^* = f_m p_2 + (f_2 - \varepsilon)(n - p_2) \to f_2 n + (f_m - f_2)p_2$. Thus, we have that the ratio for Case 2.1.2 is

$$c_{212} = \frac{R'}{R^*} = \frac{f_0 q_0 + f_1 q_1 + f_2 p_2}{f_2 n + (f_m - f_2)p_2}. \tag{11}$$

The same with the analysis for Case 1.2, based on the relationship between c_{212} and p_2, the value of c_{212} may be equal to c_{211} or it degenerates into the following Case 2.2.

Case 2.2. $p_2 = q_2$. In this case, the online decision-maker will also face two possible cases: (2.2.1) $p_3 < q_3$ and (2.2.2) $p_3 = q_3$.

Case 2.2.1. $p_3 < q_3$. In this case, there will be two sub-cases: (2.2.1.1) $p_3 = 0$, and (2.2.1.2) $0 < p_3 < q_3$.

Similar to above analysis, we could get that the ratios for Case 2.2.1.1 and Case 2.2.1.2 are

$$c_{2211} = \frac{R'}{R^*} = \frac{f_0q_0 + f_1q_1 + f_2q_2}{f_3n} \tag{12}$$

and

$$c_{2212} = \frac{R'}{R^*} = \frac{f_0q_0 + f_1q_1 + f_2q_2 + f_3p_3}{f_3n + (f_m - f_3)p_3}, \tag{13}$$

respectively, and the value of c_{2212} may be equal to c_{2211} or it degenerates into the Case 2.2.2 where $p_3 = q_3$. Further, we may analyze that there will be only $(k+1)$ possible worst-performances:

$$c_i = \frac{\sum_{j=0}^{i-1} f_jq_j}{f_in} for 1 \leq i \leq k, \tag{14}$$

and

$$c_{k+1} = \frac{\sum\limits_{j=0}^{k-1} f_jq_j + f_k(n - \sum\limits_{j=0}^{k-1} q_j)}{f_mn} \tag{15}$$

for one possible worst case where $p_i = q_i$ for $0 \leq i \leq k-1$ and $p_k > 0$.

Then following Lemma 4, we complete the proof. □

For the value of optimal competitive ratio, we could find that it is increasing in k and decreasing in f_m/f_0. In practice, the airline could adjust price a little frequently if the discount rate f_0/f_m is fixed, or increase the discount rate to gain a better performance from the perspective of online algorithms and competitive analysis.

3 Conclusion

Using the perspective of online algorithms and competitive analysis, we have proposed new revenue management policies considering uncertain demand. These policies integrate the pricing and booking control. In these policies, we could choose any value on some interval as the price decision variable. The present approach and policies may be useful in high-risk situations, possibly as a safeguard for more traditional RM policies.

In this paper, all the analysis are based on CTSS assumption. Otherwise, if there will be less than n passengers who want to book a ticket, it is obvious that there is no deterministic online policy whose competitive ratio is bigger than f_0/f_m. In other words, it is trivial to propose online policy without CTSS assumption in our model.

Acknowledgements. The authors would like to acknowledge the financial support of Grants (No. 71071123, IRT1173, and 60921003) from NSF of China.

References

1. Ball, M.O., Queyranne, M.: Toward Robust Revenue Management: Competitive Analysis of Online Booking. Operations Research 57(4), 950–963 (2009)
2. Besbes, O., Zeevi, A.: Dynamic Pricing Without Knowing the Demand Function: Risk Bounds and Near-Optimal Algorithms. Operations Research 57(6), 1407–1420 (2009)
3. Borodin, A., El-Yaniv, R.: Online Computation and Competitive Analysis. Cambridge University Press, Cambridge (1998)
4. Chrobak, M.: Online Algorithms (column 13). ACM SIGACT News 39, 96–121 (2008)
5. Feng, Y., Xiao, B.: Integration of Pricing and Capacity Allocation for Perishable Products. European Journal of Operational Research 168(1), 17–34 (2006)
6. Littlewood, K.: Forecasting and Control of Passenger Bookings. AGIFORS Annual Sympos. Proc. 12, 95–117 (1972)
7. Talluri, K., van Ryzin, G.: The Theory and Practice of Revenue Management. Springer, New York (2005)

Minimizing Total Weighted Completion Time with Unexpected Machine Unavailability*

Yumei Huo[1], Boris Reznichenko[1], and Hairong Zhao[2]

[1] Department of Computer Science,
College of Staten Island, CUNY, Staten Island, New York 10314, USA
yumei.huo@csi.cuny.edu, boris.reznichenko@cix.csi.cuny.edu
[2] Department of Mathematics, Computer Science & Statistics,
Purdue University Calumet, Hammond, IN 46323, USA
hairong@purduecal.edu

Abstract. In the past two decades, scheduling with machine availability constraints has received more and more attention. Until now most research has focused on the setting where all machine unavailability information is known at the beginning of scheduling horizon. In real world, this is impractical in some cases.

In this article, we consider the situation where the scheduler has to make scheduling decisions without any knowledge of the machine unavailable intervals. In particular, we study the problem of minimizing the total weighted completion time. When there are two or more unavailable intervals on a single machine, Fu et al. (2009) have shown that the problem is exponentially inapproximable even when jobs' weights are equal to their processing times and one has full knowledge of unavailability. So in this paper we consider the scheduling problem on a single machine with a single unavailable period. And we assume that every job has a weight proportional to its processing time. Based on whether the unavailable interval is due to a breakdown or an emergent job, we have breakdown model and emergent job model. We first show that no $\frac{\sqrt{5}+1}{2}$-competitive online algorithm exists for breakdown model, and no $\frac{11-\sqrt{2}}{7}$-competitive online algorithm exists for emergent job model. Then we show that the simple LPT (Largest Processing Time first) rule can give a 2-competitive ratio and 9/5-competitive ratio for breakdown model and emergent job model, respectively. We show the ratios are tight by examples. For offline case, we show that First Fit LPT (FF-LPT) rule can give a tight approximation ratio of 2 and 4/3 for breakdown model and emergent job model, respectively. Finally, our experimental results show that in practice, both LPT and FF- LPT perform very well and the performance improves when the number of jobs n increases. When $n \geq 50$, the worst error ratio of LPT is about 8.7%, and the worst error ratio of FF-LPT is about 0.7%. So in both cases, the error ratio is quite far from the theoretical bound.

* This work is supported by PSC-CUNY Research Award.

G. Lin (Ed.): COCOA 2012, LNCS 7402, pp. 291–300, 2012.

1 Introduction

In the past two decades, scheduling with machine availability constraints has received more and more attention. Until now most research has focused on the setting where all machine unavailabilities are known at the beginning of the scheduling horizon. However, this may not be practical in some cases. Unknown machine unavailability could result from an unexpected breakdown or emergent(high priority) jobs. Unexpected breakdown becomes a major issue with the advent of parallel/distributed computing where the processing power (idle interval of machines) is donated by volunteers. On the operational level of production scheduling, unknown machine unavailability can come from emergent jobs which have to be processed immediately. A similar problem occurs in operating systems.

In this article, we study scheduling problems without such knowledge, including the start time and the length of the unavailable intervals. In other words, the scheduler has to schedule tasks on machines in an online fashion.

We consider the problem of minimizing the total weighted completion time. When there are two or more unavailable intervals on a single machine, Fu et al. [7] have shown that the problem is exponentially inapproximable even when jobs' weights are equal to their processing times and one has full knowledge of unavailability. So in this paper we concentrate on a single machine with a single unavailable period. And we assume that the weight of each job is proportional to its processing time. i.e., for any job J_i with processing time p_i and weight w_i, $w_i = rp_i$, where r is a given constant number. Without loss of generality we can assume $r = 1$. This assumption is also common in reality. Arkin and Roundy [3] noted that typically, jobs that have longer processing times are larger jobs that have higher selling prices and, consequently, a higher priority. In many systems both the processing time and weight of a job are nearly proportional to the job's dollar value, resulting in weights that are proportional to processing times.

We differentiate two models of unavailability constraint: the breakdown model where the unavailability is due to an unexpected breakdown of the machine, and the emergent job model where the unavailability is due to occurrence of an emergent job which has to be processed immediately. The main difference is that in the latter case, the emergent job will contribute to the value of the objective function. We assume the jobs are non-resumable, i.e., when a job is interrupted by the unavailable interval, it has to be restarted when the machine is recovered later. We assume that all the jobs are known and ready to be scheduled at time 0, but the unavailable interval due to breakdown or emergent job is not known beforehand. Furthermore, when the machine becomes unavailable, the duration of the interval is also unknown. Using the 3-field notation, our problems can be denoted as $1, h_1, online \mid nr-a, w_i = p_i \mid \sum w_j C_j$ and $1, h_1, online \mid nr-a, w_i = p_i, fixed \mid \sum w_j C_j$, respectively.

1.1 Related Work

A lot of research has been done in the area of scheduling with machine unavailability constraint. This includes problems under different machine environment(single machine, parallel machine and flowshop, etc) and with different objective functions. Both resumable and non-resumable model, preemptive and non-preemptive scheduling, have been considered. The complexity results, exact algorithms and approximation algorithms in single machine, parallel machine, flow shop, open shop, job shop scheduling environment with different criteria are surveyed in a recent article [16]. The survey focuses on deterministic availability constraints. Among all the results surveyed, only one paper ([19]) studies the online problem where "online" means that the information about the jobs is incomplete, which is different from our model where the information of the unavailability of the machines is unknown beforehand. This paper studies makespan minimization on two identical machines, where each machine has a single unavailable period which does not overlap with the unavailable period of the other machine.

As for our problem in breakdown model, its offline version with arbitrary job weights, that is, $1, h_1 \mid nr - a \mid \sum w_j C_j$, has been studied a lot. The problem is obviously NP-hard. Lee ([14]) showed that the error bound of applying WSPT (Weighted Shortest Processing Time) algorithm to this problem can be arbitrarily large even if $w_i = p_i$. Recently, Kacem and Chu ([11]) show that both WSPT and MWSPT (Modified Weighted Shortest Processing Time) rules have a tight error bound of 2 under some conditions. Later, Kacem ([10]) proposes a 2-approximation offline algorithm with $O(n^2)$ time complexity and show that this bound is tight. Based on this algorithm, a Fast Polynomial Time Approximation Scheme (FPTAS) is derived in Kacem and Mahjoub [12] with a running time of $O(n^2/\epsilon^2)$. Independently, two other FPTASes for the problem are developed by Kellerer and Strusevich [13] (with a slower running time) and by Fu et al. [7] (with a faster running time). Corresponding to the emergent job model, the fixed job model studies offline problems where there is a preassigned job during some intervals. This model has been studied with the objective of makespan in [18] and [6], and with the objective of total weighted completion time in [8]. All these algorithms mentioned above except WSPT(whose error ratio can be arbitrarily large) only work offline under our online unavailability model.

For the unweighted total completion time problem, i.e., $1, h_1 \mid nr - a \mid \sum C_j$, Adiri et al. [1] and Lee and Liman [15] have shown that this problem is NP-hard, and studied the performance of SPT (shortest processing time first) for this problem. Lee and Liman [15] proved that SPT has a tight approximation ratio of $\frac{9}{7}$, instead of $\frac{5}{4}$ as shown in Adiri et al. [1]. Later Sadfi et. al [17] gave an approximation algorithm (modified SPT) with a tight approximation ratio of $\frac{20}{17}$. Breit [4] developed a parametric $O(n \log n)$-algorithm with which better worst-case error bounds can be obtained. Also for this problem, He et al. [9] presented a polynomial time approximation scheme (PTAS). Among all these algorithms, it should be noted that under our online unavailability framework, SPT rule is an "online" algorithm while others only work offline. Furthermore, we can use similar example from Lee and Liman [15] to show that the lower bound of the competitive ratio for the problem is $\frac{\sqrt{6}}{2}$. Consider three jobs J_1, J_2, and J_3 with processing time αM, M and M respectively. If an algorithm schedules J_1 first, then the adversary makes the machine unavailable during the interval $[M, M+1]$.

So in this case, the approximation ratio becomes $\frac{5+\alpha}{4+2\alpha}$ when M goes to infinity. On the other hand, if an algorithm schedules one of the longer jobs, then the adversary makes the machine unavailable during the interval $[\alpha M, M]$. In this case the approximation ratio is $\frac{6+3\alpha}{5+\alpha}$. Solving the equation, $\frac{5+\alpha}{4+2\alpha} = \frac{6+3\alpha}{5+\alpha}$, we can get the lower bound of $\frac{\sqrt{6}}{2}$.

For the online machine unavailability, a probabilistic model where at any time each machine is unavailable with a constant probability f, has been considered in Diedrich and Schwarz [5]. This paper studies scheduling problems on identical machines. Since the non-preemptive case is essentially intractable for online models, only preemptive schedules are considered. The paper shows that (1) scheduling independent jobs on identical machines with online failures to minimize the sum of completion times is $(8/7 - \epsilon)$-inapproximable, (2) SRPT (shortest remaining processing time) heuristic yields optimal results for zig-zag availability pattern.

1.2 New Contribution

From the literature review, we can see that there are no results on deterministic scheduling with online machine unavailability setting. In this paper, we first show that no $\frac{\sqrt{5}+1}{2}$-competitive online algorithm exists for breakdown model, and no $\frac{11-\sqrt{2}}{7}$-competitive online algorithm exists for emergent job model. Then we show that the simple LPT rule can give a 2-competitive ratio and 9/5-competitive ratio for breakdown model and emergent job model, respectively. We show that the ratios are tight by examples. For offline case, we show that First Fit LPT (FF-LPT) rule can give a tight approximation ratio of 2 and 4/3 for breakdown model and emergent job model respectively. Finally, our experimental results show that in practice, both LPT and FF- LPT algorithms perform very well and the performance improves when the number of jobs n increases. When $n \geq 50$, the error ratio of LPT is about 8.7%, and the error ratio of FF-LPT is about 0.7%. So in both cases, the error ratio is quite far from the theoretical bound.

1.3 Organization

Our paper is organized as follows. In Section 2, we give notations and preliminary results. In Section 3, we give the lower bounds of competitive ratios of any online algorithm for breakdown model and emergent job model, respectively. In Section 4, we analyze the competitive ratios of LPT rule for these two models. In Section 5, we analyze the approximation ratios of FF-LPT in the offline setting. In Section 6, we give our experimental results. Finally, we draw the conclusion in Section 7.

2 Notations and Preliminary Results

We consider the problems of scheduling n jobs, $J_1, J_2, \cdots, J_n$, to a single machine. Each job J_i, $1 \leq i \leq n$, has weight w_i and processing time p_i. The machine has an unavailable interval $[s, t)$ which may represent a breakdown or an emergent job, and the values of s and t are unknown before hand. We use $B = t - s$ to denote the length of the unavailable interval.

For any schedule S of the jobs, let P_1, W_1 be the total processing time and the total weight, respectively, of the jobs that finish before t; let P_2, W_2 be the total processing time and the total weight, respectively, of jobs that finish after t. Let I be the length of the idle interval before s in schedule S. The total weighted completion of schedule S is denoted by $F_w(S)$. One can easily show the following.

Lemma 1. *Let S be a schedule that schedules the jobs in the order of $J_1, J_2, \cdots, J_n$. Then $F_w(S) = \sum_{i=1}^{n} w_i p_i + \sum_{i>j} w_i p_j + W_2(B + I)$.*

When $w_i = p_i$, we have $W_2 = P_2$ and thus

$$F_w(S) = \sum_{i=1}^{n} p_i^2 + \sum_{i \neq j} p_i p_j + P_2(B + I) \ ,$$

which means that schedules are different from each other only in the third term of the above formula for $F_w(S)$, $P_2(B + I)$.

Let S^* be an optimal schedule and $F_w^*(S)$ be its total weighted completion time. Let I^* be the length of the idle interval before s, and let P_1^* and P_2^* be the total processing time of the jobs that finish before and after t, respectively. Let $\Delta = P_2 - P_2^*$. It is easy to see that we also have $\Delta = P_1^* - P_1 = I - I^*$. When $w_i = p_i$, for every i, the above formula implies that we must have $P_2 \geq P_2^*$, i.e., $\Delta \geq 0$. Thus, we have the following,

$$\begin{aligned} F_w(S) - F_w(S^*) &= P_2(B + I) - P_2^*(B + I^*) \\ &= (P_2^* + \Delta)(B + I^* + \Delta) - P_2^*(B + I^*) \\ &= \Delta^2 + \Delta(P_2^* + B + I^*) \end{aligned}$$

Corollary 1. *Let S be an arbitrary schedule for $1, h_1, online \mid nr - a, w_i = p_i \mid \sum w_j C_j$. Then $F_w(S) - F_w(S^*) = \Delta^2 + \Delta(P_2^* + B + I^*)$*

3 Lower Bounds of Competitive Ratios

Theorem 1. *It is impossible to have a $\frac{\sqrt{5}+1}{2}$-competitive online algorithm for $1, h_1, online \mid nr - a, w_i = p_i \mid \sum w_j C_j$.*

Proof. Consider two jobs J_1 and J_2 such that $p_1 = w_1 = \alpha x$ $(0 < \alpha < 1)$, and $p_2 = w_2 = x$. Any algorithm A has to decide which job should be scheduled first. If A schedules the job J_1 first, then the adversary can make machine unavailable during the interval $[x, t)$. Otherwise, if Algorithm A schedules J_2, then the adversary can make machine unavailable during the time interval $[\alpha x, t]$. So J_2 has to be restarted after t. In other words, both jobs have to be scheduled after t. When t goes to infinity, the approximation ratio becomes $\frac{1}{\alpha}$ in the former case, and $\alpha + 1$ in the latter case. Solving the equation, $\frac{1}{\alpha} = \alpha + 1$, we get $\alpha = \frac{\sqrt{5}-1}{2}$. Therefore the competitive ratio of any algorithm must be greater than $\frac{\sqrt{5}+1}{2}$.

Similarly, we can obtain the lower bound of competitive ratio for emergent job model. Due to space limit, the proof is omitted here.

Theorem 2. *It is impossible to have a* $\frac{11-\sqrt{2}}{7}$*-competitive online algorithm for* $1, h_1, online \mid nr-a, w_i = p_i, fixed \mid \sum w_j C_j$.

4 Competitive Ratio of LPT

Before we analyze the competitive ratios of LPT rule when applied to our problems, we first study some properties of LPT schedule. Let S be the schedule produced by the LPT rule. It is easy to see that S can be generated in $O(n \log n)$ time. Let J_k be the largest job that finish after t in S and J_l be the largest job finish after t in the optimal schedule S^*. We must have the following properties which will be used in our proofs.

1. $p_k > I$. This is obvious due to the nature of LPT rule.
2. $\Delta < p_k \leq p_l$. By definition and the first property, $\Delta < p_k$. Now we show that $p_k \leq p_l$. Let us assume by contradiction, $p_k > p_j$, for all J_j such that J_j finishes after t in the optimal schedule S^*. So in S^*, all the jobs with processing time greater than or equal to p_k finish before or at s. Then in the schedule S which is produced by LPT rule, all the jobs with the processing time greater than or equal to p_k should be able to be scheduled before s. This contradicts to the fact that job J_k finish after t in S. So we have $p_k \leq p_l$.
3. $\Delta < P_2^*$. This is due to $\Delta < p_l$ and J_l is a job scheduled after t in S^*.
4. $\Delta < P_1^*$. This is by definition, $\Delta = P_1^* - P_1$.

4.1 Breakdown Model

Theorem 3. *The competitive ratio of LPT is* 2 *for* $1, h_1, online \mid nr-a, w_i = p_i \mid \sum w_i C_i$.

Proof. Let S be the schedule produced by the LPT rule and S^* be an optimal schedule. By Corollary 1, we have $F_w(S) - F_w(S^*) = \Delta^2 + \Delta(P_2^* + B + I^*)$. All we need is to prove

$$\Delta^2 + \Delta(P_2^* + B + I^*) \leq F_w(S^*) = \sum_{i=1}^{n} p_i^2 + \sum_{i \neq j} p_i p_j + P_2^*(B + I^*) \ .$$

It is sufficient to prove the following inequalities: (1) $\Delta^2 < \sum_{i=1}^n p_i^2$; (2) $\Delta P_2^* \leq \sum_{i \neq j} p_i p_j$; and (3) $\Delta(B + I^*) \leq P_2^*(B + I^*)$.

Let p_k be the largest processing time among jobs that finish after t in S. By the second property of S, $p_k > \Delta$. Thus, $\Delta^2 < p_k^2 < \sum_{i=1}^n p_i^2$. Therefore Ineq. (1) is true. To prove Ineq. (2), we use property (4), $\Delta < P_1^*$. Thus, $\Delta P_2^* \leq P_1^* P_2^* \leq \sum_{i \neq j} p_i p_j$. For Ineq. (3), we use property (3).

To show the ratio is tight, consider two jobs J_1 and J_2 with processing time of x and $x+1$, respectively. The machine is unavailable during $[x, x^2)$. LPT rule will schedule both J_2 and J_1 after x^2. However, in the optimal schedule, J_1 is scheduled at 0 and J_2 is scheduled after x^2. When x is very large, the approximation ratio is close to 2.

4.2 Emergent Job Model

In this subsection, we assume that the unavailable interval is due to an emergent job which arrives at time s and its length and weight is equal to $B = t - s$. Due to its high priority, the emergent job has to be scheduled during interval $[s, t)$. In this case, the weighted completion time of the emergent job is also part of the objective value. So, in this case,

$$F_w(S) = \left(\sum_{j=1}^{n} w_j C_j \right) + B \cdot t = \sum_{i=1}^{n} p_i^2 + \sum_{i \neq j} p_i p_j + P_2(B + I) + B(P_1 + I + B) \ .$$

It is obvious that Corollary 1 still applies.

We can prove that LPT algorithm achieves a competitive ratio of 9/5. Due to space limit, the proof is omitted.

Theorem 4. *LPT rule has a* $\frac{9}{5}$*-competitive ratio for* $1, h_1, online \mid nr - a, w_i = p_i, fixed \mid \sum w_i C_i$.

To show the ratio is tight, consider the following example. The emergent job arrives at x and must be scheduled at time interval $[x, x+1)$. There are $x + 1$ jobs J_1, J_2, ..., J_{x+1} where $p_1 = w_1 = x + 1$ and $p_2 = w_2 = p_3 = w_3 = \ldots = p_{x+1} = w_{x+1} = 1$. LPT schedules all the jobs after the emergent job. However, in the optimal schedule, J_2, ..., J_{x+1} are scheduled before the emergent job and J_1 is scheduled after emergent job. When x is large, the approximation ratio is 9/5.

5 First Fit LPT as Offline Algorithms

In this section, we study the case that the unavailability information is known beforehand. We are interested in the performance of a modified LPT, First Fit LPT (FF-LPT), which schedules like LPT, but whenever a job cannot fit in before the unavailable interval, the largest job which can be scheduled before the interval will be scheduled. Apparently, FF-LPT can't be worse than LPT, but we would like to know if it has better performance than LPT asymptotically. Unfortunately, this is not the case for $1, h_1 \mid nr - a, w_i = p_i \mid \sum w_i C_i$. Consider the following example. There are 3 jobs J_1, J_2, and J_3 such that $p_1 = w_1 = x$ and $p_2 = p_3 = w_2 = w_3 = x - 1$. The machine is unavailable during the interval $[2x - 2, x^2)$. In this case, LPT and FF-LPT generate the same schedule: J_1 is scheduled before the breakdown, J_2 and J_3 are scheduled after the breakdown. The optimal algorithm, however, schedules J_2 and J_3 before the breakdown and J_1 after the breakdown. When x is large, the approximation ratio is close to 2. So we have the following theorem.

Theorem 5. *FF-LPT rule is a 2-approximation algorithm for* $1, h_1, \mid nr - a, w_i = p_i \mid \sum w_i C_i$.

For the offline version of emergent job model, we can prove that FF-LPT does have a better performance. The proof is omitted.

Theorem 6. *FF-LPT rule is a $\frac{4}{3}$-approximation algorithm for $1, h_1 \mid nr - a, w_i = p_i, fixed \mid \sum w_i C_i$.*

The following example shows that $\frac{4}{3}$ is a tight bound of FF-LPT. Suppose that there is a preassigned job that has to be scheduled during the interval $[2x, 2x+t)$, and there are three other jobs J_1, J_2 and J_3 with processing time of $x + \epsilon$, x and x respectively. FF-LPT schedules job J_1 before the preassigned job, and schedules J_2 and J_3 after the preassigned job. However, in the optimal schedule, J_2 and J_3 are scheduled before the preassigned job and J_1 is scheduled after the preassigned job. When x is large, the approximation ratio is 4/3.

6 Numerical Experiments

We have conducted computational experiments to investigate the performance of LPT and FF-LPT with randomly generated instances for both breakdown model and emergent job model. The performance is measured by the ratio of objective value of the schedule generated by LPT or FF-LPT and that of the optimal schedule. The optimal solutions can be obtained by dynamic programming, see [14].

Our data is generated as follows. We first generate jobs. We choose the number of jobs $n \in \{5, 10, 15, 50, 100, 150, 200, 300\}$. For each n, we generate two types of job instances, those consisting of "short jobs" and those consisting of "long jobs". Short jobs have processing time drawn from a uniform distribution over $[1, 10]$, while long jobs have processing time drawn from a uniform distribution over $[1, 100]$. For each type, 100 job instances are generated for each n.

Next we generate the non-available intervals $[s, t)$. For each job instance generated above, let P be the total processing time of all jobs. For each combination of α and β, where α and β are taken from $\{0, 0.2, 0.4, 0.6, 0.8\}$, we generate 100 cases of unavailable interval $[s, t)$ such that s is drawn from a uniform distribution over $[\alpha P, (\alpha + 0.2)P]$, and $t = s + B$ where B is drawn from a uniform distribution over $[\beta P, (\beta + 0.2)P]$. These unavailable intervals represent breakdowns and emergent jobs in breakdown model and emergent job models, respectively. So for both short jobs and long jobs, for each n, we generated 250,000 problem instances ($25(\alpha, \beta)$ combinations $\times$ 100 unavailable intervals $\times$ 100 job instances) in both breakdown model and emergent job model.

We investigated how the performance of LPT and FF-LPT is affected by the number of jobs n, the length of the jobs' processing time, the start time and length of the unavailable intervals which are indicated by α and β, respectively. For each combination of n, job type (short or long), (α, β), we study both the worst approximation ratio and the average ratio over 10,000 instances. Due to space limit, we only show the worst case performance below.

6.1 LPT Algorithm

For each n and job type, we list the worst case performance among 250,000 instances of LPT for both models in the following table.

Table 1. Worst performance among all instances of LPT

model	jobs	n=5	n=10	n=15	n=50	n=100	n=150	n=200	n=300
breakdown	short	1.7854	1.4874	1.3275	1.0713	1.03347	1.02254	1.01684	1.01169
	long	1.8192	1.5388	1.2852	1.0869	1.04050	1.02601	1.01942	1.01337
emergent	short	1.1294	1.0632	1.0386	1.0099	1.03342	1.02253	1.01682	1.01169
	long	1.7024	1.5341	1.2848	1.0868	1.04049	1.02594	1.01941	1.01337

From the table, we can see that for both models, LPT works very well for both short jobs and large jobs when the number of jobs is 50 or more. The worst case error ratio is less than 8.7% when $n = 50$, and it decreases as n increases. We can predict that when $n > 300$, the worst case error ratio will be less than 1.34%.

6.2 Summary of FF-LPT

For each n and job type, we list the worst case performance among 250,000 instances of FF-LPT for both models in the following table.

Table 2. Worst performance among all instances of FF-LPT

model	jobs	n=5	n=10	n=15	n=50	n=100	n=150	n=200	n=300
breakdown	short	1.2391	1.0707	1.0706	1.0071	1	1	1	1
	long	1.3227	1.0884	1.0440	1.0073	1.0016	1.0006	1.0004	1.0002
emergent	short	1.0949	1.0230	1.0167	1.0021	1	1	1	1
	long	1.2131	1.0865	1.0424	1.0070	1.0014	1.0006	1.0002	1.0001

From the table, we can see that for both models, FF-LPT works very well both short jobs and large jobs. The performance improves as n increases. The worst case error ratio is about 8.8% when $n = 10$ or 15, and becomes 0.8% when $n = 50$. When $n = 150$ or large, it is almost optimal.

7 Conclusion

In this paper we studied the problem of minimizing total weighted completion time on a single machine with single unavailable period where the start time and length of the unavailable period is not known beforehand and the scheduler has to make decision without any knowledge of unavailable period. We give the lower bound of competitive ratios for any online algorithm for both breakdown model and emergent job model. We show both theoretically and practically that LPT and FF-LPT admit very good performance for online and offline case respectively.

There is still a gap between the competitive ratio of LPT and the lower bounds in both models. For future research, one should try to develop algorithms that match the lower bounds or improve the lower bounds. Furthermore, it will be interesting to study other criteria with online unavailable periods.

References

1. Adiri, I., Bruno, J., Frostig, E., Rinnooy Kan, A.H.G.: Single machine flowtime scheduling with a single breakdown. Acta Informatica 26, 679–696 (1989)
2. Albers, S., Schmidt, G.: Scheduling with Unexpected Machine Breakdowns. In: Hochbaum, D.S., Jansen, K., Rolim, J.D.P., Sinclair, A. (eds.) RANDOM 1999 and APPROX 1999. LNCS, vol. 1671, pp. 269–280. Springer, Heidelberg (1999)
3. Arkin, R., Roundy, R.: Weighted tardiness scheduling on parallel machines with proportional weights. Operations Research 39, 64–81 (1991)
4. Breit, J.: Improved approximation for non-preemptive single machine flowtime scheduling with an availability constraint. European Journal of Operational Research 183(3), 516–524 (2007)
5. Diedrich, F., Schwarz, U.M.: A Framework for Scheduling with Online Availability. In: Kermarrec, A.-M., Bougé, L., Priol, T. (eds.) Euro-Par 2007. LNCS, vol. 4641, pp. 205–213. Springer, Heidelberg (2007)
6. Diedrich, F., Jansen, K.: Improved approximation algorithms for scheduling with fixed jobs. In: Proceeding of the Twentieth Annual ACM-SIAM Symposium on Discrete Algorithms, pp. 675–684 (2009)
7. Fu, B., Huo, Y., Zhao, H.: Exponential inapproximability and FPTAS for scheduling with availability constraints. Theoretical Computer Science 410, 2663–2674 (2009)
8. Fu, B., Huo, Y., Zhao, H.: Approximation schemes for parallel machine scheduling with availability constraints. Discrete Applied Mathematics 159(15), 1555–1565 (2011)
9. He, Y., Zhong, W., Gu, H.: Improved algorithms for two single machine scheduling problems. Theoretical Computer Science 363, 257–265 (2006)
10. Kacem, I.: Approximation algorithm for the weighted flow-time minimization on a single machine with a fixed non-availability interval. Computers & Industrial Engineering 54, 401–410 (2008)
11. Kacem, I., Chu, C.: Worst-case analysis of the WSPT and MWSPT rules for single machine scheduling with one planned setup period. European Journal of Operational Research 187(3), 1080–1089 (2008)
12. Kacem, I., Mahjoub, R.: Fully polynomial time approximation scheme for the weighted flow-time minimization on a single machine with a fixed non-availability interval. Computers & Industrial Engineering 56(4), 1708–1712 (2009)
13. Kellerer, H., Strusevich, V.A.: Fully polynomial approximation schemes for a symmetric quadratic knapsack problem and its scheduling applications. Algorithmica 57(4), 769–795 (2010)
14. Lee, C.Y.: Machine scheduling with an availability constraints. Journal of Global Optimization 9, 363–382 (1996)
15. Lee, C.Y., Liman, S.D.: Single machine flow-time scheduling with scheduled maintenance. Acta Informatica 29(4), 375–382 (1992)
16. Ma, Y., Chu, C., Zuo, C.: A survey of scheduling with deterministic machine availability constraints. Computers & Industrial Engineering 58(2), 199–211 (2010)
17. Sadfi, C., Penz, B., Rapine, C., Blazewicz, J., Formanowicz, P.: An improved approximation algorithm for the single machine total completion time scheduling problem with availability constraints. European Journal of Operational Research 161, 3–10 (2005)
18. Scharbrodt, M., Steger, A., Weisser, H.: Approximability of scheduling with fixed jobs. Journal of Scheduling 2, 267–284 (1999)
19. Tan, Z., He, Y.: Optimal online algorithm for scheduling on two identical machines with machine availability constraints. Information Processing Letters 83, 323–329 (2002)

Characterizing Mechanisms in Obnoxious Facility Game

Ken Ibara and Hiroshi Nagamochi

Graduate School of Informatics, Kyoto University, Japan
{ken.ibara,nag}@amp.i.kyoto-u.ac.jp

Abstract. In this paper, we study the (group) strategy-proofness of deterministic mechanisms in the obnoxious facility game. In this game, given a set of strategic agents in a metric, we design a mechanism that outputs the location of a facility in the metric based on the locations of the agents reported by themselves. The benefit of an agent is the distance between her location and the facility and the social benefit is the total benefits of all agents. An agent may try to manipulate outputs by the mechanism by misreporting strategically her location. We wish to design a mechanism that is *strategy-proof* (i.e., no agent can gain her benefit by misreporting) or *group strategy-proof* (i.e., there is no coalition of agents such that each member in the coalition can simultaneously gain benefit by misreporting), while the social benefit will be maximized. In this paper, we first prove that, in the line metric, there is no strategy-proof mechanism such that the number of candidates (locations output by the mechanism for some reported locations) is more than two. We next completely characterize (group) strategy-proof mechanisms with exactly two candidates in the general metric and show that there exists a 4-approximation group strategy-proof mechanism in any metric.

1 Introduction

In the *facility game*, given a set of "strategic" agents in a metric, we design a procedure, called a *mechanism*, that outputs the location of a facility in the metric based on reported locations of the agents so that the social cost (or benefit), which is defined to be the sum of individual costs (or benefits) such as the distance from the facility, is minimized (or maximized). We assume that the mechanism is known to all the agents before they report their locations and that an agent may try to manipulate outputs by the mechanism by misreporting strategically her location so that an output location of the facility will be beneficial to her (we also assume that there is no way of testing whether a reported location is a misreported one or not). A mechanism is called *strategy-proof* if no single agent can gain her benefit by misreporting her location. Moreover, a mechanism is called *group strategy-proof* if no coalition of agents can gain benefit of each member in the coalition simultaneously by misreporting the locations of the coalition. Then a (group) strategy-proof mechanism may deliver a location of the facility which is not an optimal solution in terms of the social cost (or benefit). Our game-theoretical goal is to design a (group) strategy-proof mechanism

G. Lin (Ed.): COCOA 2012, LNCS 7402, pp. 301–311, 2012.

with a good approximation ratio between locations output by the mechanism and optimal locations.

In mechanism design, we can consider mechanisms that are allowed to make payments. However, in many settings, money is not used as a medium of compensation due to ethical or legal considerations [12]. For example, in the *social choice* literature, mechanisms without payments are commonly studied. Since the facility game is rather deeply concerned in the social choice, we also design mechanisms without payments in the facility game.

Moulin [9] and Border and Jordan [3] studied a problem of the social choice in economics wherein each agent's preference is a function with a single peak at the most preferred point in a given space but no objective function to be optimized is given. Based on the median voter theorem [2], they characterized the strategy-proof mechanisms in the line and a space of a multidimensional version of the line, respectively. The traditional facility game is a problem of social choice together with a social cost that is to be minimized as an objective function. Schummer and Vohra [11] extended the mechanism [3] to the facility game on tree networks, and characterized the strategy-proof mechanisms to metrics on arbitrary networks containing at least one cycle. Recently strategy-proof *approximation* mechanisms for optimization problem have been studied extensively [1,7,8,10]. Alon et al. [1] gave a complete analysis on the approximation ratio of strategy-proof mechanisms for the facility game in metrics on arbitrary networks. Currently group strategy-proof mechanisms that attain the optimal social cost are known up to tree networks.

In this paper, we study the (group) strategy-proofness of deterministic mechanisms in the obnoxious facility game. In contrast with the traditional game, we regard the distance from each agent to the facility as the benefit of the agent in this game, and the sum of the benefits of all agents will be maximized as the social benefit. Thus each agent's preference is no longer represented as a single-peaked function. This problem setting can be interpreted as a social scenario such that the mayor of a town plans to build a garbage dump in the town according to a set of reported home addresses of the local residents, wishing to maximize the sum of the distances of all residents. Cheng et al. [4] first studied group strategy-proof mechanisms for the obnoxious facility game in the line metric. They demonstrated that a mechanism that simply outputs a socially optimal location is not strategy-proof. They designed a group strategy-proof mechanism which chooses one of two predetermined locations as an output according to the distribution of reported locations, and showed that the mechanism is a 3-approximation. They suggested that the mechanism can be extended to a 3-approximation group strategy-proof mechanism in tree networks. In this paper, we first prove that there is no strategy-proof mechanism in the line metric such that the number of *candidates* (locations output by the mechanism for some reported locations) is more than two. This suggests that we need to know the specific structure of a given metric if we wish to design a strategy-proof mechanism with more than two candidates. We next derive a complete characterization of (group) strategy-proof mechanisms with exactly two candidates in the general metric.

The paper is organized as follows. Section 2 formulates the obnoxious facility game and reviews the definition of (group) strategy-proofness. Section 3 proves that the line metric admits no strategy-proof mechanism with more than two candidates. Section 4 proposes *a valid threshold mechanism* and proves that a mechanism with exactly two candidates in the general metric is (group) strategy-proof if and only if it is a valid threshold mechanism. Section 5 then shows that there always exists a 4-approximation valid threshold mechanism in any metric. Finally Section 6 makes some concluding remarks.

2 Preliminaries

Let $\mathbb{N}$ and $\mathbb{R}_+$ be the sets of nonnegative integers and nonnegative real numbers, respectively. Let (Ω, d) be a metric such that Ω is a set of points (possibly an infinite set) and $d : \Omega \times \Omega \to \mathbb{R}_+$ is a symmetric distance function, i.e., $d(x, y) = d(y, x)$ for every two points $x, y \in \Omega$ and $d(x, y) + d(y, z) \geq d(x, z)$ for every three points $x, y, z \in \Omega$.

For a set $N = \{1, 2, \ldots, n\}$ of agents, $x_i \in \Omega$ denotes the location reported by agent $i \in N$ and the multiset $X = \{x_1, x_2, \ldots, x_n\}$ of the locations is called a *profile* of N. For a location $y \in \Omega$ of an obnoxious facility, the benefit of agent i is defined to be the distance between her location and the facility, i.e.,

$$\beta(y, x_i) = d(y, x_i).$$

The *social benefit* of a location $y \in \Omega$ of an obnoxious facility over a profile X is defined to be the total benefit of n agents

$$\mathrm{SB}(y, X) = \sum_{i=1}^{n} \beta(y, x_i).$$

For a profile X, let $\mathrm{OPT}(X)$ denote the optimal obnoxious social benefit, i.e., $\mathrm{OPT}(X) = \max_{y \in \Omega} \mathrm{SB}(y, X)$.

In the obnoxious facility game, a deterministic mechanism outputs a facility location based on a given profile X, where we do not distinguish two profiles $X = \{x_1, x_2, \ldots, x_n\}$ and $X' = \{x'_1, x'_2, \ldots, x'_n\}$ of N if there is a bijection $\sigma : N \to N$ such that $x_i = x'_{\sigma(i)}$ for all $i \in N$. We write $X = X'$ if there is such a bijection σ. A *mechanism* is defined to be a function $f : \Omega^n \to \Omega$ such that $f(X) = f(X')$ for two profiles X and X' of N with $X = X'$. We say that a mechanism f has an approximation ratio γ if

$$\mathrm{OPT}(X) \leq \gamma \mathrm{SB}(f(X), X) \quad \text{for all profiles } X \in \Omega^n \text{ of } N.$$

In the following we define the strategy-proofness and the group strategy-proofness of mechanisms. For a profile $X = \{x_1, x_2, \ldots, x_n\}$ of N and an agent set $S \subseteq N$, let X_S denote the profile of S obtained from X by eliminating locations x_i such that $i \in N - S$. We denote X_{N-S} simply by X_{-S}. In particular, for $S = \{i\}$, X_{-S} is denoted by $X_{-i} = \{x_1, \ldots, x_{i-1}, x_{i+1}, \ldots, x_n\}$. Location profile X may be written by $\langle x_i, X_{-i} \rangle$ or $\langle X_S, X_{-S} \rangle$. For simplicity, we write $f(x_i, X_{-i}) = f(\langle x_i, X_{-i} \rangle)$ and $f(X_S, X_{-S}) = f(\langle X_S, X_{-S} \rangle)$.

Definition 1. *A mechanism f is* strategy-proof *(SP for short) if no agent can benefit from misreporting her location. Formally, given an agent i, a profile $X = \langle x_i, X_{-i}\rangle \in \Omega^n$ and a misreported location $x'_i \in \Omega$, it holds that*

$$\beta(f(x_i, X_{-i}), x_i) \geq \beta(f(x'_i, X_{-i}), x_i).$$

Definition 2. *A mechanism f is* group strategy-proof *(GSP for short) if for any group of agents, at least one of them cannot benefit from misreporting their locations simultaneously. Formally, given a non-empty set $S \subseteq N$, a profile $X = \langle X_S, X_{-S}\rangle \in \Omega^n$ and a misreported profile $X'_S \in \Omega^{|S|}$ of S, there exists $i \in S$ satisfying*

$$\beta(f(X_S, X_{-S}), x_i) \geq \beta(f(X'_S, X_{-S}), x_i).$$

We remark that a stronger notation of group strategy-proofness requires that any set of misreporting agents with a strict gain contains at least one agent who strictly loses [6]. Our GSP results in this paper do not hold under the stronger definition. However, the above weaker definition of group strategy-proofness is rather common in the social choice, since in the settings without payments an agent has no incentive to misreport unless it strictly benefits.

For a mechanism $f : \Omega^n \to \Omega$, a point $y \in \Omega$ is called a *candidate* if there is a profile $X \in \Omega^n$ such that $f(X) = y$, and the set of all candidates of f is denoted by C_f. A mechanism with $|C_f| = p$ is called by a *p-candidate mechanism.* Any 1-candidate mechanism is group strategy-proof, but its approximation ratio γ can be infinitely large.

3 Mechanisms in the Line Metric

This section proves that there is a metric that admits no p-candidate SP mechanism for any $p \geq 3$. Let (I, d) be the line metric, where I denotes the 1-dimensional Euclidean space.

Theorem 1. *There is no p-candidate SP mechanism for any $p \geq 3$ in the line metric.*

We assume that f is a p-candidate SP mechanism with $C_f = \{c_1, c_2, \ldots, c_p\} \subset I$ in (I, d), where $c_1 < c_2 < \cdots < c_p$. We prove Theorem 1 via the next lemma.

Lemma 1. *Let X be a profile of N, and X'_S be a misreported profile of a coalition $S \subseteq N$. Then, for a p-candidate SP mechanism f, it holds $f(X'_S, X_{-S}) \leq c_\ell$ for an index $1 \leq \ell \leq p-1$ if $f(X) \leq c_\ell$ and $x_i < \frac{f(X)+c_{\ell+1}}{2}$ for all $i \in S$; $c_\ell \leq f(X'_S, X_{-S})$ holds for an index $2 \leq \ell \leq p$ if $c_\ell \leq f(X)$ and $\frac{c_{\ell-1}+f(X)}{2} < x_i$ for all $i \in S$.*

Proof. To derive a contradiction, assume without loss of generality that $S = \{1, 2, \ldots, k\}$ is a minimal coalition such that $f(X) \leq c_\ell < f(X'_S, X_{-S})$ and $x_k \leq \cdots \leq x_2 \leq x_1 < \frac{f(X)+c_{\ell+1}}{2}$ (the other case can be treated symmetrically). Let X^i

be the profile obtained from X by replacing $x_1, x_2, \ldots, x_i$ with $x'_1, x'_2, \ldots, x'_i$ and let $X^0 = X$. Since f is SP, it holds by Definition 1 that $\beta(f(x_i, (X^{i-1})_{-i}), x_i) \geq \beta(f(x'_i, (X^{i-1})_{-i}), x_i)$ for each agent $i \in S$ and profile X^{i-1}; i.e.,

$$|f(X^{i-1}) - x_i| \geq |f(X^i) - x_i| \text{ for } i = 1, \ldots, k. \quad (1)$$

By assumption on X and S, it holds $f(X^i) < c_{\ell+1} \leq f(X^k)$ for all $i \leq k-1$. From this and inequality (1) with $i = k$, we have $\frac{f(X^{k-1})+f(X^k)}{2} \leq x_k < \frac{f(X^0)+c_{\ell+1}}{2}$, which implies $f(X^{k-1}) < f(X^0)$ (by $c_{\ell+1} \leq f(X^k)$) and hence $k \geq 2$. Now there is an index $1 \leq j \leq k-1$ such that $f(X^j) \leq f(X^{k-1}) < f(X^{j-1})$. By inequality (1) with $i = j$, we have $\frac{f(X^{k-1})+f(X^k)}{2} \leq x_k \leq x_j \leq \frac{f(X^j)+f(X^{j-1})}{2}$, which contradicts $f(X^j) \leq f(X^{k-1}) < f(X^{j-1}) < f(X^k)$. □

Fix a candidate $c_t \in C_f - \{c_1, c_p\}$, and let $I_t = \{x \in I \mid \frac{c_1+c_t}{2} < x < \frac{c_t+c_p}{2}\}$ (where $\frac{c_1+c_p}{2} \in I_t$), $I_{t,1} = \{x \in I \mid \frac{c_1+c_t}{2} < x < \frac{c_1+c_p}{2}\}$ and $I_{t,p} = \{x \in I \mid \frac{c_1+c_p}{2} < x < \frac{c_t+c_p}{2}\}$. For a profile X of N, let $S^a(X) = \{i \in N \mid x_i \leq \frac{c_1+c_t}{2}\}$ and $S^b(X) = \{i \in N \mid \frac{c_t+c_p}{2} \leq x_i\}$. We prove Theorem 1 by deriving a contradiction.

Proof of Theorem 1. There is a profile X of N such that $f(X) = c_t \in C_f$. Let $X' \in I_t^n$ be a profile of N obtained from X by replacing each x_i, $i \in S^a(X)$ with a new location $x'_i \in I_{t,1}$ and each x_j, $j \in S^b(X)$ with a new location $x'_j \in I_{t,p}$. Assume that $f(X') \neq c_1$ (the case of $f(X') \neq c_p$ can be treated symmetrically). For $S = S^a(X)$, let $X^a = \langle X'_S, X_{-S}\rangle$ denote the profile of N obtained from X by replacing each x_i, $i \in S$ with the $x'_i \in I_{t,1}$. Since $f(X) = c_t$ and $x_i \leq \frac{c_1+c_t}{2} < \frac{f(X)+c_{t+1}}{2}$ for all $i \in S$, it holds $f(X^a) \leq c_t$ by Lemma 1.

Note that X^a is obtained from $X' = \{x'_1, \ldots, x'_n\}$ by changing the locations of the agents in $S^b(X)$. For $S^b(X)$, it holds $\frac{c_1+f(X')}{2} \leq \frac{c_1+c_p}{2} < x'_j$ for all $j \in S^b(X)$ and $f(X') \geq c_2$. By Lemma 1, we have $f(X^a) \geq c_2$.

Denote $X^a = (\tilde{x}_1, \ldots, \tilde{x}_n)$. There is a profile Z of N such that $f(Z) = c_1 \in C_f$. Then Z is obtained from X^a by misreporting all locations. We know that $\frac{c_1+f(X^a)}{2} \leq \frac{c_1+c_t}{2} < \tilde{x}_i$ for all $i \in N$ since $S^a(X^a) = \emptyset$ and $f(X^a) \leq c_t$. By $c_2 \leq f(X^a)$ and Lemma 1, we would have $c_2 \leq f(Z)$, a contradiction. □

4 2-Candidate SP/GSP Mechanisms in the General Metric

This section gives a complete characterization of 2-candidate SP/GSP mechanisms in the general metric (not necessarily on the basis of particular graphs). In the following part, we propose *valid threshold mechanisms.*

For fixed two points $a, b \in \Omega$, we partition Ω into three subspaces $\Omega_a = \{x \in \Omega \mid d(a,x) < d(b,x)\}$, $\Omega_m = \{x \in \Omega \mid d(a,x) = d(b,x)\}$ and $\Omega_b = \{x \in \Omega \mid d(a,x) > d(b,x)\}$. For a profile X of N, the set of agents and the number of agents in Ω_a are denoted by S_a and n_a, respectively, i.e., $S_a = \{i \mid x_i \in \Omega_a\}$ and $n_a = |S_a|$. Analogously, we denote $S_m = \{i \mid x_i \in \Omega_m\}$, $n_m = |S_m|$, $S_b = \{i \mid x_i \in \Omega_b\}$ and $n_b = |S_b|$.

For each integer $\ell = 0, 1, \ldots, n$, let θ_ℓ be a function that maps a profile $M \in \Omega_m^\ell$ of ℓ agents with locations in Ω_m to an integer. A mechanism f on N is called a *threshold mechanism* if there are two points $a, b \in \Omega$ and a set $\{\theta_0, \ldots, \theta_n\}$ of functions such that f returns a for all profiles X with $n_a < \theta_{n_m}(X_{S_m})$ and returns b for the other profiles X, i.e., f is given by

$$f(X) = \begin{cases} a & \text{if } n_a < \theta_{n_m}(X_{S_m}) \\ b & \text{if } \theta_{n_m}(X_{S_m}) \le n_a. \end{cases}$$

A threshold mechanism f is symmetric in terms of a and b in the sense that $f(X) = b$ if $n_b < \overline{\theta}_{n_m}(X_{S_m})$ and $f(X) = a$ otherwise for the set of complement functions $\overline{\theta}_\ell$ on Ω_m^ℓ, $\ell = 0, 1, \ldots, n$ such that

$$\overline{\theta}_\ell(M) = n + 1 - \ell - \theta_\ell(M) \text{ for } M \in \Omega^\ell \text{ and } \ell = 0, 1, \ldots, n.$$

Furthermore a threshold mechanism f is called *valid* if the set $\{\theta_0, \ldots, \theta_n\}$ of functions satisfies the two conditions: (i) $\theta_0(\emptyset) \notin \{0, n+1\}$ and $0 \le \theta_\ell(M) \le n+1-\ell$ for $0 \le \ell \le n$ and $M \in \Omega_m^\ell$; and (ii) $\theta_\ell(M) - 1 \le \theta_{\ell+1}(\langle x, M\rangle) \le \theta_\ell(M)$ for $0 \le \ell \le n-1$, $M \in \Omega_m^\ell$ and $x \in \Omega_m$. Note that the set of complement functions also satisfies the above two conditions.

We show that a 2-candidate mechanism is SP (or GSP) if and only if it is a valid threshold mechanism via the next two theorems.

Theorem 2. *Every valid threshold mechanism is a 2-candidate GSP mechanism.*

Theorem 3. *Every 2-candidate SP mechanism is a valid threshold mechanism.*

For another profile $X' = \{x'_1, \ldots, x'_n\}$ of N, we use the following notation. The set of agents and the number of agents in Ω_a are denoted by S'_a and n'_a, respectively, i.e., $S'_a = \{i \mid x'_i \in \Omega_a\}$ and $n'_a = |S'_a|$. Analogously, we denote $S'_m = \{i \mid x'_i \in \Omega_m\}$, $n'_m = |S'_m|$, $S'_b = \{i \mid x'_i \in \Omega_b\}$ and $n'_b = |S'_b|$. We first prove Theorem 2.

Proof of Theorem 2. Let f be a valid threshold mechanism. First we show that $C_f = \{a, b\}$. Clearly $C_f \subseteq \{a, b\}$. For a profile X of N with $n_a = 0$ and $n_m = 0$, it holds that $f(X) = a$ since $n_a < \theta_0(\emptyset) \in \{1, 2, \ldots, n\}$. Similarly, for a profile X with $n_a = n$ and $n_m = 0$, we have $f(X) = b$ since $n_a \ge \theta_0(\emptyset) \in \{1, 2, \ldots, n\}$. Hence $C_f = \{a, b\}$ holds and it means that f is a 2-candidate mechanism.

Let us show the group strategy-proofness of f, i.e., not all agents in any coalition S can gain simultaneously by misreporting their locations. Fix a profile $X = \{x_1, \ldots, x_n\}$ of N and a coalition $S \subseteq N$ wherein x'_i denotes the misreported location of each agent $i \in S$. A misreported profile of S is denoted by $X'_S = \{x'_i \mid i \in S\}$ and we denote $X' = \langle X'_S, X_{-S}\rangle$. We prove that there is an agent $i \in S$ such that $\beta(f(X), x_i) \ge \beta(f(X'), x_i)$. We consider the case of $f(X) = a$, i.e., $n_a < \theta_{n_m}(X_{S_m})$ (the other case can be treated symmetrically by considering the complement functions $\overline{\theta}_\ell$).

If $f(X') = a$, then $\beta(f(X), x_i) = \beta(f(X'), x_i)$ for any $i \in S$ and we are done. Assume that $f(X') = b$. If there is an agent $i \in S - S_a$, then for such an agent i it holds $d(a, x_i) \geq d(b, x_i)$, i.e., $\beta(f(X), x_i) \geq \beta(f(X'), x_i)$, and we are done. Hence it suffices to prove that $S \subseteq S_a$ implies $f(X') = a$. From $S \subseteq S_a$, we have $n'_a \leq n_a$ and $n'_m \geq n_m$. Let $k = n'_m - n_m$ ($\leq n_a - n'_a$) and we denote $S'_m - S_m = \{1, 2, \ldots, k\}$ and $M'_i = \{x'_1, \ldots, x'_i\} \in \Omega^i_m$. By repeatedly applying the second property of functions $\theta_0, \ldots, \theta_n$, we obtain $\theta_{n_m}(X_{S_m}) - k \leq \theta_{n_m+1}(\langle M'_1, X_{S_m}\rangle) - (k-1) \leq \cdots \leq \theta_{n_m+k}(\langle M'_k, X_{S_m}\rangle) = \theta_{n'_m}(X'_{S'_m})$. Then by $k \leq n_a - n'_a$, it holds that $\theta_{n_m}(X_{S_m}) - n_a \leq \theta_{n'_m}(X'_{S'_m}) - n'_a$, which implies $n'_a < \theta_{n'_m}(X'_{S'_m})$ and $f(X') = a$ since $n_a < \theta_{n_m}(X_{S_m})$ now holds by $f(X) = a$ and the assumption on f. □

We next prove Theorem 3 via the next lemma.

Lemma 2. *Let f be a 2-candidate SP mechanism with $C_f = \{a, b\}$, and X and X' be two profiles of N with $X_{S_m} = X'_{S'_m}$. Then $f(X) = f(X')$ if $f(X) = a$ and $n_a \geq n'_a$; or $f(X) = b$ and $n_b \geq n'_b$.*

Proof. For a profile X with $f(X) = a$, if $f(x'_i, X_{-i}) = b$ hold for a misreported location x'_i of some agent $i \in S_a$, then we would have $\beta(f(x_i, X_{-i}), x_i) = d(a, x_i) < d(b, x_i) = \beta(f(x'_i, X_{-i}), x_i)$, contradicting that $\beta(f(x_i, X_{-i}), x_i) \geq \beta(f(x'_i, X_{-i}), x_i)$ holds for any profile X and agent $i \in N$ in an SP f. This means that if $f(X) = a$ then $f(X'_S, X_{-S}) = a$ holds no matter how a subset $S \subseteq S_a$ misreports $X'_S \in \Omega^{|S|}$. Symmetrically if $f(X) = b$ then $f(X'_S, X_{-S}) = b$ for any $X'_S \in \Omega^{|S|}$ with a subset $S \subseteq S_b$.

To prove the lemma, we consider the case where $f(X) = a$ and $n_a \geq n'_a$ (the other case can be treated symmetrically). We construct a new profile X'' of N with $X''_{S_m} = X_{S_m}$ by changing the locations of agents i with $x_i \in \Omega_a$ as follows. We choose a set $T_a \subseteq S_a$ of n'_a agents, and let $x''_i = x'_i$ for each $i \in T_a$, x''_i be any location in Ω_b for each $i \in S_a - T_a$, and $x''_i = x_i$ for each $i \in N - S_a$. Since $f(X) = a$ and X'' is obtained from X by changing the locations of agents only in S_a, it holds $f(X'') = a$. Since X'' is obtained from X' by changing the locations of agents only in S'_b, it holds that if $f(X') = b$ then $f(X'') = b$, i.e., if $f(X'') = a$ then $f(X') = a$. Therefore we have $f(X') = a$. □

Now we give a proof of Theorem 3.

Proof of Theorem 3. Let f be a 2-candidate SP mechanism with $C_f = \{a, b\}$. For an integer $0 \leq \ell \leq n$ and a set $M \in \Omega^\ell_m$ of locations, let $\theta_\ell(M)$ be the minimum integer n_a such that there is a profile X of N such that $X_{S_m} = M$ satisfying $f(X) = b$, where if $f(X) = a$ (resp., $f(X) = b$) for all such X then we define $\theta_\ell(M) = n + 1 - \ell$ (resp., $\theta_\ell(M) = 0$). Then by Lemma 2, $f(X) = a$ holds for all profiles X such that $\theta_\ell(M) > n_a$ and $X_{S_m} = M$. Thus, f is a threshold mechanism. Now it suffices to show that f is valid, i.e., the set $\{\theta_0, \ldots, \theta_n\}$ of the above functions satisfies $\theta_0(\emptyset) \notin \{0, n+1\}$ and $0 \leq \theta_\ell(M) \leq n + 1 - \ell$ for $0 \leq \ell \leq n$ and $M \in \Omega^\ell_m$; and $\theta_\ell(M) - 1 \leq \theta_{\ell+1}(\langle x, M\rangle) \leq \theta_\ell(M)$ for $1 \leq \ell \leq n-1$, $M \in \Omega^\ell_m$ and $x \in \Omega_m$. Note that we have shown that $\theta_\ell(M) \in \{0, \ldots, n+1-\ell\}$ for $0 \leq \ell \leq n$.

We first show inequality $\theta_{\ell+1}(\langle x, M\rangle) \leq \theta_\ell(M)$ (inequality $\theta_\ell(M) - 1 \leq \theta_{\ell+1}(\langle x, M\rangle)$ follows from the inequality $\overline{\theta}_{\ell+1}(\langle x, M\rangle) \leq \overline{\theta}_\ell(M)$ on the complement functions). If $\theta_\ell(M) \geq n-\ell$, then $\theta_{\ell+1}(\langle x, M\rangle) \leq \theta_\ell(M)$ is immediate, since $\theta_{\ell+1}(\langle x, M\rangle) \leq n+1-(\ell+1) = n-\ell$ by definition. Consider the case of $\theta_\ell(M) < n-\ell$. Then there is a profile X of N such that $X_{S_m} = M$, $n_a = \theta_\ell(M)$, $f(X) = b$ and $n_b \geq 1$. We choose an agent $t \in S_b$ and change its location from x_t to an arbitrary location $x'_t \in \Omega_m$ to obtain a new profile $X' = \langle x'_t, X_{-t}\rangle$ of N. By Definition 1, it holds that $\beta(f(X), x_t) \geq \beta(f(X'), x_t)$, i.e., $d(b, x_t) \geq d(f(X'), x_t)$. Since $d(a, x_t) > d(b, x_t)$, $f(X') = b$ holds for the profile X' such that $n'_a = \theta_\ell(M)$ and $X'_{S'_m} = \langle x'_t, M\rangle \in \Omega_m^{\ell+1}$. Recall that $\theta_{\ell+1}(\langle x'_t, M\rangle)$ is the minimum integer n_a such that there is a profile X of N such that $X_{S_m} = \langle x'_t, M\rangle$ satisfying $f(X) = b$. Hence we have $\theta_{\ell+1}(\langle x'_t, M\rangle) \leq n'_a = \theta_\ell(M)$.

Finally, we prove that $\theta_0(\emptyset) \neq 0$ (property $\theta_0(\emptyset) \neq n+1$ follows from $\overline{\theta}_0(\emptyset) \neq 0$ on the complement function). If $\theta_0(\emptyset) = 0$, then inequality $\theta_{\ell+1}(\langle x, M\rangle) \leq \theta_\ell(M)$ ($0 \leq \ell \leq n-1$, $M \in \Omega_m^\ell$, $x \in \Omega_m$) inductively implies that $\theta_\ell(M) = 0$ for any $M \in \Omega_m^\ell$ and $0 \leq \ell \leq n$. This, however, means that $f(X) = b$ for all profiles X of N, contradicting that $C_f = \{a, b\}$. □

5 Approximation Ratio of 2-Candidate Mechanisms

This section analyzes the approximate ratio $\gamma = \max_{X \in \Omega^n} \frac{\mathrm{OPT}(X)}{\mathrm{SB}(f(X), X)}$ of 2-candidate SP mechanisms in the general metric.

Upper bound. We first derive an upper bound on the approximate ratio γ. Let f be a 2-candidate SP mechanism on a set N of n agents in a metric (Ω, d), where f can be given by choosing two points $a, b \in C_f$ and functions θ_i, $i = 0, 1, \ldots, n$ so that f becomes a valid threshold mechanism by Theorem 3.

Theorem 4. *Let f be a 2-candidate mechanism for a set N of n agents. If $C_f = \{a, b\}$ is a pair of most distant points in Ω and f is a valid threshold mechanism by a set $\{\theta_0, \ldots, \theta_n\}$ of functions, then the approximate ratio γ of f is less than* $\max\{\frac{2n}{\theta_0(\emptyset)}, \frac{2n}{n+1-\theta_0(\emptyset)}\}$.

Proof. Let $d(a, b) = 2r$. For a profile X of N with $f(X) = a$, we derive an upper bound on γ. We have $\mathrm{SB}(f(X), X) = \sum_{i \in N} d(a, x_i) > n_m r + n_b r = (n - n_a)r$, since $d(a, x_i) > r$ for n_b agents $i \in S_b$, and $d(a, x_i) = r$ for n_m agents $i \in S_m$. Let $c_X \in \Omega$ denote an optimal facility location, i.e., $\mathrm{OPT}(X) = \mathrm{SB}(c_X, X)$. On the other hand, we see that $\mathrm{OPT}(X) \leq 2nr$, since $d(c_X, x_i) \leq 2r$ for any location $x_i \in \Omega$ by the choice of a and b. Hence we have $\gamma < \frac{2nr}{(n-n_a)r} = \frac{2n}{n-n_a}$. Since $f(X) = a$, it holds $n_a < \theta_{n_m}(X_{S_m}) \leq \theta_0(\emptyset)$, where we use the property $\theta_{\ell+1}(\langle x, M\rangle) \leq \theta_\ell(M)$ of functions to get the second inequality. Hence $\gamma < \frac{2n}{n+1-\theta_0(\emptyset)}$. When $f(X) = b$, we apply the same argument to the complement function to obtain $\gamma < \frac{2n}{n+1-\overline{\theta}_0(\emptyset)} = \frac{2n}{\theta_0(\emptyset)}$. This proves the theorem. □

The bound $\max\{\frac{2n}{\theta_0(\emptyset)}, \frac{2n}{n+1-\theta_0(\emptyset)}\}$ in Theorem 4 is minimized and $\gamma \leq 4$ holds when $\theta_0(\emptyset) = \lceil n/2 \rceil$. In fact, such a valid threshold mechanism f for a set of n agents can be constructed as follows. For a pair $C_f = \{a, b\}$ of most distant points in Ω, let f return $f(X) = a$ if $n_a + n_m < n_b$; $f(X) = b$ otherwise.

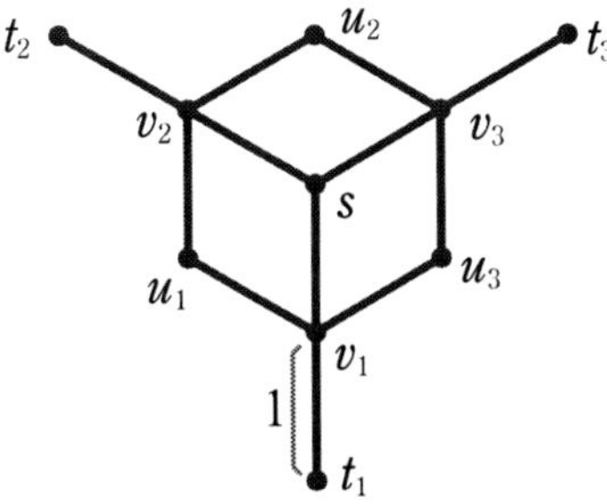

Fig. 1. An undirected graph G

Lower Bound. We have shown that every valid threshold mechanism f with $\theta_0(\emptyset) = \lceil n/2 \rceil$ has an approximation ratio $\gamma = 4$. Now we give a tight example (Ω, d) such that for every choice of $a, b \in \Omega$, the approximation ratio γ attained by a valid threshold mechanism f with $C_f = \{a, b\}$ and $\theta_0(\emptyset) = \lceil n/2 \rceil$ is at least 4. Such an example (Ω_G, d) is constructed from a graph G as follows. Let $G = (V, E)$ be the graph with a set V of ten vertices, s and u_i, v_i, t_i $(i = 1, 2, 3)$, and a set E of 12 edges, $t_i v_i, v_i s, v_i u_{i-1}, v_i u_{i+1}$ $(i = 1, 2, 3)$, where we interpret $u_4 = u_1$ and $u_0 = u_3$ (see Fig. 1). We regard each edge as a line segment of length 1, and a point x on an edge e is denoted by $x \in e$. Let Ω_G be the set of points in all edges including the end points. The distance $d(x, x')$ for two points $x, x' \in \Omega_G$ is defined to be the length of a shortest path between x and x'.

Lemma 3. *For any two points $c_1, c_2 \in \Omega_G$, there are a point $c^* \in \Omega_G$ and a shortest path P between c_1 and c_2 in (Ω_G, d) such that*

$$\rho = \frac{\max\{d(c_1, c^*), d(c_2, c^*)\} + d(x, c^*)}{d(c_1, c_2)/2} \geq 4$$

for the middle point x on P.

Proof. Given points c_1 and c_2, we choose P which does not pass through s if any. By symmetry, we only need to consider the case where x is on one of edges v_3t_3, v_3u_2 and v_3s. We show that $c^* = t_1$ suffices the lemma. Let $d(c_2, c^*) \geq d(c_1, c^*)$ without loss of generality. When $d(c_1, c_2) \leq 2$, we easily see that $\rho \leq 4$ since $\max\{d(c_1, c^*), d(c_2, c^*)\} + d(x, c^*) \geq 4$ and $d(c_1, c_2)/2 \leq 1$. In what follows, assume that $d(c_1, c_2) > 2$ and hence c_1 and c_2 are in two nonadjacent edges, respectively. Note that $x \in v_3t_3$ implies $d(c_1, c_2) \leq 2$. We distinguish two cases.

Case 1. $x \in v_3u_2$: We do not need to consider the case where one of c_1 and c_2, say c_i is on edge v_3t_3, since we can move $c_i \in v_3t_3$ to a point on edge v_3u_3 without changing both of the position of x and $d(c_1, c_2)$ or increasing $\max\{d(c_1, c^*), d(c_2, c^*)\}$. In this case, we can assume that $c_1 \in v_3u_3$ and c_2 is on one of edges v_2u_2, v_2u_1 and v_2t_2; or $c_1 \in v_1u_3$ and $c_2 \in v_2u_2$. In any case, we obtain $d(c_2, c^*) + d(c_2, x) + d(x, c^*) = 8 + 2\alpha$ and $d(c_1, c_2)/2 \leq (3 + \alpha)/2$, where $\alpha = d(c_2, v_2) \leq 1$ when $c_2 \in v_2t_2$ and $\alpha = 0$ otherwise. Hence we have $\rho \geq \frac{8+2\alpha-d(c_2,x)}{d(c_1,c_2)/2} \geq -1 + \frac{16+4\alpha}{3+\alpha} = -1 + 4 + \frac{4}{3+\alpha} \geq 4$.

Case 2. $x \in v_3 s$: We can assume that c_2 is on v_3u_3, v_3u_2 or v_3t_3 and c_1 is on sv_1 or sv_2, where c_1 is not on v_1t_1 or v_1u_1 by the choice of P. In addition, we can assume $c_1 \notin sv_2$ and $c_2 \notin v_3u_2$, since we can move $c_1 \in sv_2$ (resp., $c_2 \in v_3u_2$) to a point on edge sv_1 (resp., v_3t_3) without changing both of the position of x and $d(c_1, c_2)$ or increasing $\max\{d(c_1, c^*), d(c_2, c^*)\}$. Then we obtain $d(c_2, c^*) + d(c_2, x) + d(x, c^*) = 6 + 2\alpha$ and $d(c_1, c_2)/2 \leq (2 + \alpha)/2$, where $\alpha = d(c_2, v_3) \leq 1$ when $c_2 \in v_3t_3$ and $\alpha = 0$ otherwise. Hence we have $\rho \geq \frac{6+2\alpha-d(c_2,x)}{d(c_1,c_2)/2} \geq -1 + \frac{12+4\alpha}{2+\alpha} = -1 + 4 + \frac{4}{2+\alpha} > 4$, as required. □

By Lemma 3, we can get the following theorem.

Theorem 5. *For the obnoxious facility game on the above metric (Ω_G, d), let f be a valid threshold mechanism with $\theta_0(\emptyset) = \lceil n/2 \rceil$ of a set N of n agents. Then for any choice of $C_f = \{a, b\}$, the approximation ratio of f is not smaller than $4(1 - \frac{4}{n+2})$.*

Proof. For $C_f = \{a, b\}$, there are points $c^*, m \in \Omega_G$ such that $d(a, m) = d(b, m)$ and $\rho = \frac{\max\{d(a,c^*),d(b,c^*)\}+d(m,c^*)}{d(a,b)/2} \geq 4$ by Lemma 3. We consider the case of $d(a, c^*) > d(b, c^*)$ (the other case can be treated analogously). For a sufficiently small $\epsilon > 0$, let $m_b \in \Omega_G$ be a point that is closer to b than a in a neighbor of m within distance ϵ; $d(a, m_b) \leq d(a, m) + \epsilon$ and $d(m_b, c^*) \geq d(m, c^*) - \epsilon$ hold. Construct a profile X of N such that $\lceil n/2 \rceil - 1$ agents are situated on point a while the other $\lfloor n/2 \rfloor + 1$ agents on point m_b. Since $n_m = 0$ and $n_a = \lceil n/2 \rceil - 1 < \theta_0(\emptyset)$, we have $f(X) = a$, $\mathrm{SB}(f(X), X) = (\lfloor n/2 \rfloor + 1)d(a, m_b) \leq (\lfloor n/2 \rfloor + 1)(d(a, b)/2 + \epsilon)$ and $\mathrm{OPT}(X) \geq \mathrm{SB}(c^*, X) = (\lceil n/2 \rceil - 1)d(a, c^*) + (\lfloor n/2 \rfloor + 1)d(m_b, c^*) \geq (\lceil n/2 \rceil - 1)(d(a, c^*) + d(m, c^*) - \epsilon)$. Hence it holds $\frac{\mathrm{OPT}(X)}{\mathrm{SB}(f(X),X)} \geq \frac{(\lfloor n/2 \rfloor+1)(d(a,c^*)+d(m,c^*)-\epsilon)}{(\lceil n/2 \rceil-1)(d(a,b)/2+\epsilon)}$, which approaches to $4(1 - \frac{4}{n+2})$ when $\epsilon \to 0$. □

We remark that it is still open whether there exists an example (Ω, d) such that for every choice of $a, b \in \Omega$, the approximation ratio γ attained by a valid threshold mechanism f with $\theta_0(\emptyset) \neq \lceil n/2 \rceil$ is at least 4.

6 Concluding Remarks

In this paper, we studied SP/GSP mechanisms for the obnoxious facility game. We first showed that there is a metric that admits no p-candidate SP mechanism for any $p \geq 3$. We then proved that a valid threshold mechanism is a complete characterization of (group) strategy-proof mechanisms with exactly two candidates in the general metric. We also proved that there always exists a 4-approximation valid threshold mechanism in any metric. Note that for any integer $p \geq 3$, there is a metric (Ω, d) that admits a p-candidate GSP mechanism. For example, let (Ω, d) be a metric on a star network with a center v_c and p leaf edges v_cv_j $j = 1, 2, \ldots, p$ of length 1, and f be a p-candidate mechanism that returns $f(X) = v_k$ for a profile X such that $n_k = \min_{1 \leq j \leq p} n_j$ for $n_j = |\{i \in N \mid x_i \in v_cv_j\}|$. Then we can prove that this p-candidate mechanism is GSP and the approximation ratio is at most $2 + \frac{1}{p-1}$.

There are still several open problems on the obnoxious facility game. First, given a metric (Ω, d), it is important to know the maximum number $p(\Omega, d)$ of candidates such that there exists a $p(\Omega, d)$-candidate SP/GSP mechanism. Also for such a maximum value $p(\Omega, d)$, it is left open whether we can construct a p'-candidate SP/GSP mechanism in the metric for any $p' < p(\Omega, d)$ or not. The problem of placing an obnoxious facility when the locations of agents are fixed is called *the 1-maxian problem* [5,13]. In the 1-maxian problem, the number of optimal locations of an obnoxious facility in a network metric is known to be finite. It would be interesting to investigate the relationship between solutions of the 1-maxian problem and candidates of SP/GSP mechanisms of the obnoxious facility game. Also it is another interesting issue to derive a counterpart/extension of our arguments in randomized mechanisms in the general metric (see [4] for a randomized 2-candidate GSP mechanism in the line metric).

References

1. Alon, N., Feldman, M., Procaccia, A.D., Tennenholtz, M.: Strategyproof approximation mechanisms for location on networks. CoRR, abs/0907.2049 (2009)
2. Black, D.: On the rationale of group decision-making. Journal of Political Economy 56, 23–34 (1948)
3. Border, K.C., Jordan, J.S.: Straightforward elections, unanimity and phantom voters. The Review of Economic Studies 50(1), 153–170 (1983)
4. Cheng, Y., Yu, W., Zhang, G.: Mechanisms for Obnoxious Facility Game on a Path. In: Wang, W., Zhu, X., Du, D.-Z. (eds.) COCOA 2011. LNCS, vol. 6831, pp. 262–271. Springer, Heidelberg (2011)
5. Church, R.L., Garfinkel, R.S.: Locating an obnoxious facility on a network. Transportation Science 12(2), 107–118 (1978)
6. Jain, K., Mahdian, M.: Cost sharing. In: Nisan, N., Roughgarden, T., Tardos, E., Vazirani, V. (eds.) Algorithmic Game Theory, ch. 15. Cambridge University Press (2007)
7. Lu, P., Wang, Y., Zhou, Y.: Tighter Bounds for Facility Games. In: Leonardi, S. (ed.) WINE 2009. LNCS, vol. 5929, pp. 137–148. Springer, Heidelberg (2009)
8. Lu, P., Sun, X., Wang, Y., Zhu, Z.A.: Asymptotically optimal strategy-proof mechanisms for two-facility games. In: Proceedings of the 11th ACM Conference on Electronic Commerce, ACM-EC (June 2010)
9. Moulin, H.: On strategy proofness and single peakedness. Public Choice 35, 437–455 (1980)
10. Procaccia, A.D., Tennenholtz, M.: Approximate mechanism design without money. In: Proceedings of the 10th ACM Conference on Electronic Commerce, ACM-EC (July 2009)
11. Schummer, J., Vohra, R.V.: Strategy-proof location on a network. Journal of Economic Theory 104(2), 405–428 (2004)
12. Schummer, J., Vohra, R.V.: Mechanism design without money. In: Nisan, N., Roughgarden, T., Tardos, E., Vazirani, V. (eds.) Algorithmic Game Theory, ch. 10. Cambridge University Press (2007)
13. Zelinka, B.: Medians and peripherians of trees. Archivum Mathematicum 4, 87–95 (1968)

Efficiency of Dual Equilibria in Selfish Task Allocation to Selfish Machines*

Xujin Chen, Xiaodong Hu, Weidong Ma**, and Changjun Wang

Institute of Applied Mathematics, AMSS
Chinese Academy of Sciences, Beijing 100190, China
{xchen,xdhu,mawd,wcj}@amss.ac.cn

Abstract. In this paper we consider the task allocation problem from a game theoretic perspective. We assume that tasks and machines are *both* controlled by selfish agents with two distinct objectives, which stands in contrast to the passive role of machines in the traditional model of selfish task allocation. To characterize the outcome of this new game where two classes of players interact, we introduce the concept of dual equilibrium. We prove that the price of anarchy with respect to dual equilibria is 1.4, which is considerably smaller than the counterpart 2 in the traditional model. Our study shows that activating more freedom and selfishness in a game may bring about a better global outcome.

Keywords: Price of anarchy, Selfish task allocation, Selfish scheduling.

1 Introduction

In large-scale communication networks, it is usually impossible to globally manage network task allocation. This motivates the study on network task allocation from a game-theoretic perspective, where each network participant is viewed as a selfish player in a competitive network game. Under the standard full information setting in game theory (e.g., [1,2]), each player is aware of the situation facing all other players and tries to minimize its cost or maximize its profit, ignoring the global objective of the network system. As a typical example of *selfish task allocation* (STA) games, also known as *selfish load balancing* [10], some companies maintain a number of servers (termed as *machines* in literature) and offer content providers to store data (termed as *tasks*) [1]. The requested stream to a content provider is directed by the provider to the server that stores its data, i.e., the data requested, while this provider suffers from a *cost* of latency experienced by the request in the stream. Naturally, each content provider (*client*) would selfishly choose a *server* to minimize its own cost.

In this paper, we generalize the above classical model of STA towards more realistic settings. The classical model does not consider the welfare of individual

* Supported in part by NNSF of China under Grant No. 10771209, 11021161 and 10928102, 973 Project of China under Grant No. 2011CB80800, and CAS under Grant No. kjcx-yw-s7.

** Corresponding author.

G. Lin (Ed.): COCOA 2012, LNCS 7402, pp. 312–323, 2012.

servers which are managed by individual profit-minded companies. To make more money, each server wants to store data as much as possible, and may actively withdraw its offer to some clients it currently serves so that it has a lower latency, for a moment, to attract a more profitable client from another server. To put it differently, every client is undisciplined and seeks an latency lower than it currently experiences. The client may inquire another server, telling its data amount and current latency, about the possibility of the server offering a lower latency. The server being inquired then makes some computation to see whether or not it can make more money by discarding some clients (possibly none) it currently serves and giving better offer to the inquiring client. If the server does find a way to do so, then the corresponding data storage rejection and client migration follow to benefit both the rejecting server (inquired) and migrating client (inquiring); in the special case where the server inquired discards nothing, the migrating client is considered behaving the same as in the classical STA. We refer to this generalized model, where both clients and servers are selfish players, as *selfish task allocation to selfish machines* (STASM).

Unlike STA whose players (tasks) only have a uniform kind of action – migrating from one machine to another, our STASM model possesses two kinds of players, tasks and machines, while the machine latter players may take action of rejecting some of their tasks. Thus, it is natural to think of each action in STASM game being taken by a pair of one machine and one task, where the machine discards some (or none) of its current tasks and the task moves from another machine to the machine. In such a setting, a natural equilibrium of STASM is the state at which no pair of machine player and task player (outside the machine) has incentive to deviate, given the allocation of other tasks to machines. We refer to this kind of stable state for both task players and machine players as a *Dual Equilibrium* (DE), so as to distinguish it from the classic concept of *pure-strategy Nash Equilibrium* (abbreviated to NE throughout the paper). Any DE of STASM must be an NE of STA; in contrast, an NE of STA is not necessarily a DE for STASM. In this sense, DE is more stable than NE, and thus a more plausible outcome of the competitions among selfish tasks and selfish machines. More remarkable is the higher social welfare (lower maximum latency) at DE in comparison with that at NE, as the main result of this paper shows.

The paper is organized as follows. In Section 2, we summarize some related work and our contribution. In Section 3, we present the model for STASM on identical machines, and an example showing the lower bound 1.4 on its *price of anarchy* (PoA) – the ratio between the system objective value at the worst possible NE and the overall optimum. In Section 4, we establish the upper bound 1.4 on the PoA. In Section 5, we conclude with discussion on future work.

2 Related Works and Our Contribution

Traditionally, *task allocation*, also known as *scheduling*, was extensively studied in terms of a coordinator that, given n weighted tasks and m machines possibly with different speeds, finds an allocation of tasks to machines optimizing certain system objective function: minimizing *makespan*, the maximum completion time, or maximizing *cover*, the minimum completion time. These problems are NP-hard even for identical speeds [6], thus enormous literature are devoted to the design of approximation algorithms [7].

Recently, a great deal of research has considered task allocation from the game-theoretic perspective and modeled the problem as STA, where each weighted task is a self-interested player who tries to minimize its own latency, i.e., the latency of the machine processing it, defined as the ratio of machine load (the total weight on the machine) over speed. Koutsoupias and Papadimitriou [8] initiated the research on the PoA, and studied STA for linear latency functions with respect to mixed Nash equilibrium.

Most relevant results to ours are the existence of NE in STA [5], and the PoA of $2 - 2/(m+1)$ for STA on identical machines with respect to the system objective of minimum makespan (see [9,10]). The proof of the PoA goes back to the scheduling literature [4] where the same ratio occurs as the approximation factor of a local search heuristic. Letting m go to infinity gives the overall PoA of 2 for the problem. For the global objective of maximizing cover, the PoA of STA on identical machines was shown to falls within $[1.69, 1.7]$ [3].

The STASM model introduced in this paper presents the first study of task allocation, where tasks and machines are *both* controlled by selfish agents. We consider the STASM problem on identical machines under the system objective of minimum makespan. For the problem, we show that there is an optimal task allocation which is a DE, and *the overall value of the PoA is exactly 1.4*. This improves upon the PoAs of traditional STA – 2 for minimum makespan [4,9], and 1.7 for maximum cover [3]. The quality of stable outcomes under STASM is thus more desirable than that under traditional STA. It is an interesting phenomenon that introducing more selfishness brings about better global performance.

3 Model Specification

Given a set $\mathcal{T} = \{1, 2, \ldots, n\}$ of n tasks and a set $\mathcal{M} = \{M_1, M_2, \ldots, M_m\}$ of m identical machines, each task $j \in \mathcal{T}$ has a positive *weight* $w(j)$, and should be handled by a machine in $\mathcal{M}$. Given any nonempty subset $\mathcal{S} \subseteq \mathcal{T}$ of tasks, we write $w(\mathcal{S})$ for $\sum_{j \in \mathcal{S}} w(j)$. For convenience, $w(\emptyset)$ is set to be 0. Let $\sigma : \mathcal{T} \rightarrow \mathcal{M}$ be an allocation of tasks to machines. We abbreviate $\{\sigma(j) | j \in S\}$ to $\sigma(\mathcal{S})$, and set $\sigma(\emptyset)$ to be $\emptyset$. The *latency* (also called load) of machine $M_i \in \mathcal{M}$ under σ equals $L_i(\sigma) \equiv \sum_{j:\sigma(j)=M_i} w(j)$, and is experienced by every task allocated to M_i. The *makespan* of σ is defined as $\max_{i=1}^{m} L_i(\sigma)$, the maximum latency under σ; it measures the system performance under σ. We focus on the algorithmic game model of *selfish task allocation to selfish machines* (STASM), where

(3.1) Every task is a selfish player who wishes to select a machine of lowest latency so that it experiences latency as low as possible.

(3.2) Every machine is a selfish player who wishes to process tasks of total weight as heavy as possible so that it gets profits as many as possible.

Equilibrium Allocations. Motivated by (3.1), a task j on machine M_h would migrate to another machine M_i if M_i offers j a latency lower than j's current latency $L_h(\sigma)$. On the other hand, motivated by (3.2), a machine M_i would reject some (possibly none) of its current tasks – with total weight less than $w(j)$ – to assure attractive offer – lower than $L_h(\sigma)$ – to the more profitable task j. The task rejection by machine M_i and the migration of task j from machine M_h to machine M_i constitute an *action* (under σ) in StaSm game if and only if

(3.3) $\sigma(j) = M_h \neq M_i$, there exists $\mathcal{S} \subseteq \{k \in \mathcal{T} : \sigma(k) = M_i\}$ such that $w(\mathcal{S}) < w(j)$ and $L_i(\sigma) - w(\mathcal{S}) + w(j) < L_h(\sigma)$.

When (3.3) is satisfied, machine M_i and task j would like to take an action, which is often described as "M_i would reject tasks in $\mathcal{S}$ so that j would migrate from M_h to M_i", or simply "M_i would reject $\mathcal{S}$ to attract j" In case that no action can be taken under σ, the allocation σ is stable, and referred to as a *dual equilibrium* (DE), or simply an equilibrium allocation. The stability of DE in StaSm is stronger than that of NE (pure-strategy Nash equilibrium) in traditional Sta. By definition, a DE must be an NE (thinking of the case where machines reject nothing); on the other hand, an NE may not be a DE.

Price of Selfishness. An allocation with minimum makespan is called *optimal*. As usual, we use the ratios between the makespans of DE and optimal allocation to quantify the (in)efficiency of the equilibrium outcome. Given an instance I of the StaSm game, let opt denote optimal makespan of I, and let Π denote maximum makespan among all DEs of I. The *price of anarchy* (PoA) of I is defined as the ratio Π/OPT. The *PoA of the* StaSm *game* is defined as the supremum of PoAs over all StaSm instances.

Example 1. An infinite set $\mathscr{I}$ of StaSm instances with $\sup_{I\in\mathscr{I}}(\text{PoA of } I) = 1.4$. The set $\mathscr{I}$ consists of instances I_p for integers $p = 2, 3, 4, \ldots$. In each I_p, there are $m = 2 + 3p$ identical machines and $n = 7 + 6p$ tasks. Among these n tasks, there are seven *special* tasks that include one task of weight 5, two tasks of weight 2, and four tasks of weight 1. The remaining $6p$ tasks can be partitioned into p *groups* each consisting of 6 tasks. For $k = 1, 2, \cdots, p$, the *k-th group* consists of three tasks of weight $(3 - k/p)$ (called *leading* tasks) and three tasks of weight $(2 + k/p)$ (called *following* tasks). For I_p, Fig.1(i) and (ii) present an equilibrium allocation of makespan 7 and an allocation of makespan $5 + 1/p$, respectively. □

Lemma 1. *The price of anarchy of* StaSm *game is at least* 1.4. □

Consecutive Rejection. Given an allocation $\sigma : \mathcal{T} \to \mathcal{M}$, the limit of computational capability might need to be taken into account. Facing a query from task j who asks for a better offer, machine M_i wishes to find a subset $\mathcal{S}$ of tasks satisfying (3.3) which could be rejected for attracting the more profitable task j. This in general is a NP-complete problem [6]. On the other hand, it is easy for M_i to determine whether it can do a consecutive rejection specified as follows. A subset $\mathcal{S} \subseteq \{k \in \mathcal{T} : \sigma(k) = M_i\}$ of tasks is called *consecutive* if their weights are consecutive in an non-increasing ordering of weights of tasks allocated to M_i

by σ. In $O(n^2)$ time, M_i can check whether or not the following condition is satisfied.

(3.3)' $\sigma(j) = M_h \neq M_i$, there exists consecutive $\mathcal{S} \subseteq \{k \in \mathcal{T} : \sigma(k) = M_i\}$ such that $w(\mathcal{S}) < w(j)$ and $L_i(\sigma) - w(\mathcal{S}) + w(j) < L_h(\sigma)$.

In case that the consecutive set $\mathcal{S}$ exists, machine M_i and task j take an action – M_i "consecutively" rejecting all tasks in $\mathcal{S}$ and j migrating from machine M_h to M_i. Under the new definition of action, the *Selfish task allocation to Selfish machines with Consecutive Rejection* is abbreviated to SSCR.

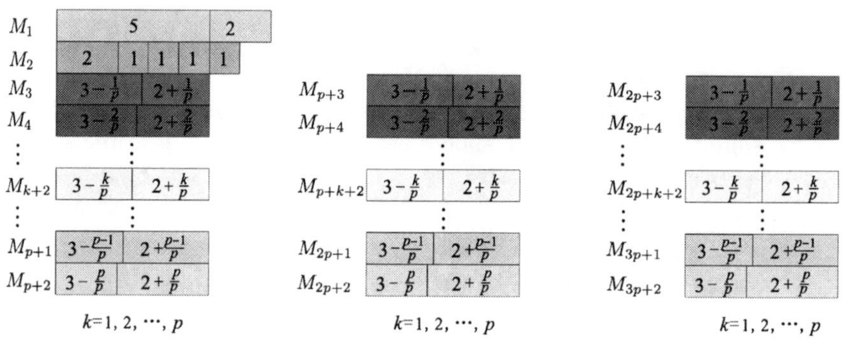

(i) An equilibrium allocation of makespan 7

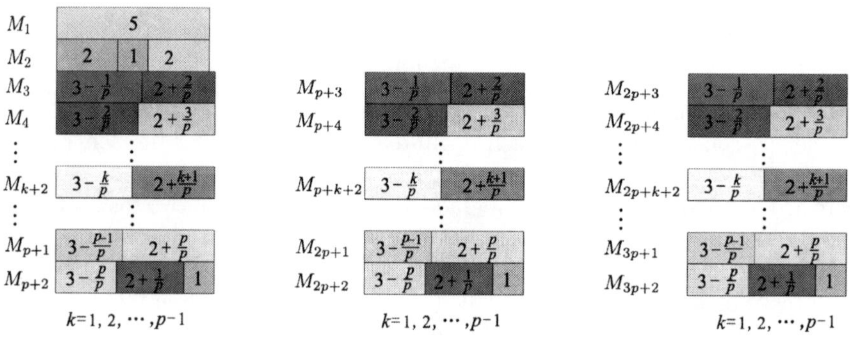

(ii) An allocation of makespan $5+\frac{1}{p}$

Fig. 1. Two allocations for StaSm instance I_p

Clearly, an action never increases the makespan, so a sequence of actions applied to the optimal solution will lead to DE that is still socially optimal.

Theorem 1. *(i) There is a DE that is optimal for* StaSm *(resp. SSCR) game. (ii) The PoA of* StaSm *game is at most the PoA of SSCR game. (iii) The PoA of* StaSm *(resp. SSCR) game is at least* 1.4.

Proof. Associate each allocation σ with a vector $\boldsymbol{v}^\sigma = (v_1, v_2, \ldots, v_m)$ such that $v_1, v_2, \ldots, v_m$ is a non-increasing ordering of $L_1(\sigma), L_2(\sigma), \ldots, L_m(\sigma)$. Let

σ and ς be two allocations. We say that σ *is smaller than* ς, denoted as $\sigma \prec \varsigma$, if $\boldsymbol{v}^\sigma$ is smaller in the lexicographic order than $\boldsymbol{v}^\varsigma$. It is easy to see that every minimum allocation under $\prec$ is a socially optimal DE, proving (i).

It is obvious that a DE of StaSm must be a DE of SSCR. This fact along with Lemma 1 implies (ii) and (iii). □

4 Efficiency Analysis

In this section, we prove the following main result of this paper. Due to limited space, we only sketch our proof. The omitted details are given in the full paper.

Theorem 2. *The price of anarchy of* StaSm *(resp. SSCR) game is* 1.4.

If the PoA of SSCR is at most 1.4, then Theorem 2 follows immediately from Theorem 1(ii) and (iii). To prove Theorem 2, we assume the contrary:

Assumption 1. There is an SSCR instance I whose PoA is greater than 1.4. □

Let *opt* denote an optimal allocation for I with makespan OPT, and let π denote a DE whose makespan Π satisfies $\Pi/\text{OPT} > 1.4$. For π, let $\mathcal{T}_i$ denote the set of tasks allocated to M_i; the load $L_i(\pi) = w(\mathcal{T}_i)$ of machine M_i is abbreviated to L_i. Suppose that M_1 and M_2 bear the maximum and minimum loads in π, respectively. Then $L_1 = \max_{i\in[m]} L_i = \Pi > w(\mathcal{T})/m > L_2 = \min_{i\in[m]} L_i$.

Property 2. (i) Under π, for any pair of distinct $h, i \in [m]$, no individual task allocated to M_h would like to migrate to M_i, i.e., $L_i \geq L_h - \min_{j\in\mathcal{T}_h} w(j)$.
(ii) The optimal makespan is at least the average load, $\text{OPT} \geq w(\mathcal{T})/m > L_2$. □

A task $j \in \mathcal{T}$ is often called a *$w(j)$-task* to indicate its weight. Given a real interval R, a task of weight in R is called a *R-task*. For nonnegative number α, an $[\alpha, +\infty)$-task is abbreviated to an *α^+-task*. Given $i \in [m]$, order the tasks in $\mathcal{T}_i$ as $1_i, 2_i, \ldots, \ell^i_i$, where $\ell^i = |\mathcal{T}_i|$, in the nonincreasing order of their weights $w(1_i) \geq w(2_i) \geq \cdots \geq w(\ell^i_i)$. The ℓ-vector $(w(1_i), w(2_i), \ldots, w(\ell^i_i))$ is called the *π-weight (vector)* of M_i. For $k = 1, 2, \ldots, \ell^i$, task k_i is referred to as the *k-th task* of M_i; and particularly, task ℓ^i_i also as the *last task* of M_i. For any real number $\alpha \in (0, L_i]$, task h_i, where $1 \leq h \leq \ell^i$, is called an *α-critical task* of M_i if $\sum_{k=1}^{h-1} w(k_i) < \alpha$ and $\sum_{k=1}^{h} w(k_i) \geq \alpha$.

Recall that M_1 experiences the maximum load Π in π. Let $a \equiv w(\mathcal{T}_1 - \{\ell^1_1\})$ denote the weight sum over all tasks but the last one allocated to M_1. Setting $b \equiv w(\ell^1_1)$, we have $a + b = \Pi$. We often call task ℓ^1_1 *the last b-task of* M_1.

Property 3. (i) $L_i \geq a$ for all $i \in [m]$.
(ii) $1 \leq \lambda \equiv a/b < 2.5$; in particular, $a \geq b$.
(iii) $\text{OPT} < 5(1+\lambda)b/7 = a + (5 - 2\lambda)b/7 \leq a + 3b/7 < 2a$. □

For any $\mathcal{L} \subseteq \mathcal{M}$, let $\mathcal{T}(\mathcal{L}) \equiv \cup_{i:M_i\in\mathcal{L}} \mathcal{T}_i$ (resp. $\mathcal{T}^o(\mathcal{L})$) denote the set of tasks allocated by π (resp. *opt*) to machines in $\mathcal{L}$. Particularly, $\mathcal{T}(\emptyset) \equiv \emptyset$ and $\mathcal{T}^o(\emptyset) \equiv \emptyset$. We categorize all machines into three types according to task allocation under π.

A machine in $\mathcal{M}$ with π-weight vector $(w_1, w_2, \ldots, w_\ell)$ is called an *A-machine* if $w_1 \geq a$, a *B-machine* if $w_1 < a$ and $\sum_{k:w_k \geq b} w_k \geq a$, and a *C-machine* if $\sum_{k:w_k \geq b} w_k < a$. In the following, let $\mathcal{A}$, $\mathcal{B}$, and $\mathcal{C}$ denote the sets of A-machines, B-machines and C-machines, respectively. Thus we have the following.

Property 4. (i) $\mathcal{M}$ is the disjoint union of $\mathcal{A}$, $\mathcal{B}$ and $\mathcal{C}$; and $M_1 \in (\mathcal{A} \cup \mathcal{B}) \setminus \mathcal{C}$.
(ii) The first *two* tasks of a B-machine are $[b, a)$-tasks.
(iii) The first task of a C-machine is a $(0, a)$-task. □

Property 3(iii) implies that no pair of a^+-task and $(3b/7)^+$-task are allocated to the same machine by *opt*. By $a \geq b$ in Property 3(ii) and $\mathcal{M} = \mathcal{A} \cup \mathcal{B} \cup \mathcal{C}$ in Property 4, we may assume without loss of generality the following.

Property 5. (i) To each A-machine $M_i \in \mathcal{A}$ (if any), the optimal allocation *opt* allocates a^+-task $1_i \in \mathcal{T}_i$, but no any other $(3b/7)^+$-task.
(ii) All $[3b/7, a)$-tasks and the last b-task ℓ_1^1 of M_1 are allocated by *opt* to machines in $\mathcal{B} \cup \mathcal{C}$. In particular, $\mathcal{B} \cup \mathcal{C} \neq \emptyset$. □

From Property 3(i), we see that every machine $M_i \in \mathcal{M}$ has load L_i at least a in π, therefore M_i must have a unique a-critical task, which we denote by t_i^i. In other words, for every $i \in [m]$, a-critical task of M_i is its t^i-th task.

Property 6. Let $(w_1, w_2, \ldots, w_\ell)$ be the π-weight vector of machine $M_i \in \mathcal{M}$. Suppose that the a-critical task of M_i is its t-th task. The following hold.

(i) $\sum_{k=1}^{t-1} w_k < a$, $\sum_{k=1}^{t} w_k \geq a$ and $\sum_{k=t}^{\ell} w_k \geq b$.
(ii) If $M_i \in \mathcal{C}$, then $w(t^i) = w_t < b$, $2 \leq t = t^i < \ell = \ell^i$ and $\sum_{k=t+1}^{\ell} w_k < b$. □

Property 7. $L_i \geq a + b/2$ for any $M_i \in \mathcal{C}$. □

Recall from Property 3(ii) that the ratio $\lambda = a/b \in [1, 2.5)$. The next two subsections deal with small $\lambda \in [1, 4/3)$ and large $\lambda \in [4/3, 5/2)$, respectively.

4.1 Small $\lambda \in [1, 4/3)$

The following upper bounds on OPT are from $a < 4b/3$ and Property 3(iii).

$$2b > a + b/2 > a + 3b/7 > \text{OPT}\,. \tag{4.1}$$

Consider *opt* allocating all tasks in $\mathcal{T}(\mathcal{C})$ (if any) to machines in $\mathcal{A} \cup \mathcal{B}$. Then by Properties 4, 5(ii) and 7, the total load of B-machines and C-machines under *opt* is at least $(2b)|\mathcal{B}| + (a + b/2)|\mathcal{C}|$. This is impossible because the average load of these machines would exceed OPT as implied by $2b > a + b/2 > \text{OPT}$ in (4.1).

Therefore, some tasks on C-machines (under π), which we study in Part I, must be allocated by *opt* to A-machines, which we study in Part II. In turn, to ensure no load exceeding OPT, the optimal allocation *opt* has to allocate an adequate number of tasks from $\mathcal{T}(\mathcal{A})$, each of which has a sufficiently "large" weight, back to B-machines or C-machines (cf. Claim 8). However, the weights brought by these tasks make the total load of B-machines and C-machines higher than $(|\mathcal{B}|+|\mathcal{C}|)$OPT, which we show in Part III. The contradiction to the optimal makespan will exclude the possiblility of $\lambda \in [1, 4/3)$.

I. Tasks on C-machines. One of our major approaches is to show that tasks from $\mathcal{T}(\mathcal{C})$ all have a considerable amount of weight such that in *opt* an A-machine accommodating any of these tasks enforces the optimal allocation to transfer enough weight from $\mathcal{T}(\mathcal{A})$ to B-machines or C-machines.

Lower Bounds on Task Weights. The next claim says that the last task (and hence all tasks) of any C-machine cannot be too light.

Claim 1. $w(\ell_i^i) > (1+\lambda)b/7 \geq 2b/7$ and $w(\ell_i^i) \geq a + b - L_i$ for any $M_i \in \mathcal{C}$. □

Another central idea is using equilibrium properties of π to show large total weight of tasks, which would imply difficulties for *opt* to average the load.

The a-critical Tasks and Their Followers. Claim 1 enables us to show that under non-increasing weight ordering of tasks allocated to any C-machine (by π), the a-critical task can be followed by at most two tasks.

Claim 2. $\ell^i \leq t^i + 2$ for any $M_i \in \mathcal{C}$. □

A crucial step in our proof is to establish the next claim, asserting the existence of a light a-critical task (of weight less than $b/2$) on some C-machine.

Claim 3. There exists $M_i \in \mathcal{C}$ such that $w(t_i^i) < b/2$. □

For each C-machine as in Claim 3, we can show that its a-critical task is followed by exactly two tasks, a kind of "best possible" in view of Claim 2.

Claim 4. $\ell^i = t^i + 2$ for every $M_i \in \mathcal{C}' \equiv \{M_i \in \mathcal{C} \,|\, w(t_i^i) < b/2\}$. □

As aforementioned, the average load over B-machines and C-machines exceeds OPT, we next expect "large" loads on A-machines. To obtain a lower bound on these loads, we think of why a C-machine does not reject its last (lightest) task for attracting a heavier task from an A-machine.

The Predecessors of the Lightest Task. Let $M_\hbar \in \mathcal{C}$ be the most loaded C-machine (in π) whose last task $\ell_\hbar^\hbar$ is lightest and has weight

$$w(\ell_\hbar^\hbar) = d \equiv \min\{w(\ell_i^i) : M_i \in \mathcal{C}\} < a\,, \tag{4.2}$$

where $a > d$ is guaranteed by Property 4. So the predecessors of the lightest task $\ell_\hbar^\hbar$ has total weight

$$c = L_\hbar - d = L_\hbar - w(\ell_\hbar^\hbar) \equiv \max\{L_i - d \,|\, M_i \in \mathcal{C} \text{ and } w(\ell_i^i) = d\}\,. \tag{4.3}$$

For any $M_i \in \mathcal{M}$, Property 2(i) says $L_i \geq c$; so M_i has a unique c-critical task, denoted by r_i^i. Using Claim 4, we can show that c is very close to the optimal makespan OPT, and thus a nice lower bound on loads of all machine.

Claim 5. (i) $\text{OPT} > c \geq \max\{a+b-2d, a+d\}$. (ii) $\text{OPT} - c < d/3$. □

II. Allocations to A-machines. In this part, we investigate the tasks allocated to A-machines. The small gap between c and OPT in Claim 5(ii) implies that all c-critical tasks on A-machines are $(d/2)^+$-tasks, as we show next.

The c-critical Tasks. Recall that every A-machine $M_i \in \mathcal{A}$ has a unique c-critical task r_i^i, the r^i-th task of M_i, satisfying $\sum_{k=1}^{r^i-1} w(k_i) < c$ and $\sum_{k=1}^{r^i} w(k_i) \geq c$.

Claim 6. (i) $w(r_i^i) \geq d/2$ for all $M_i \in \mathcal{A}$.
(ii) For any $M_i \in \mathcal{A}$, if $r^i = 2$, $w(2_i) < d$, $w(3_i) < d/2$, then $w(1_i) + d > \text{OPT}$. □

c is very close to OPT, the lower bound of $d/2$ on the weights of c-critical tasks in Claim 6(i) basically says that $(0, d/2)$-tasks count little in the total load of A-machines. We will ignore these tasks in our discussion on the optimal allocation *opt*. To get a more accurate estimation on the loads of A-machines, we divide $\mathcal{A}$ into four subsets as follows.

A Classification of A-machines. Recall from Claim 1 and (4.2) that $d \geq 2b/7$, and from (4.1) that $a + 3b/7 > \text{OPT}$. Given any $M_i \in \mathcal{A}$ with $r = r^i$, from Claim 6(i) we see that M_i is precisely of one of the following four types:

- *Type* A_1: $r \geq 2$ and $w(2_i) \geq d$;
  $\Rightarrow w(1_i) + 3 \cdot d/2 \geq a + 3d/2 > a + 3b/7 > \text{OPT}$.
- *Type* A_2: $r \geq 2$ and $d > w(2_i) \geq w(3_i) \geq d/2$;
  $\Rightarrow w(1_i) + 3 \cdot d/2 > \text{OPT}$
- *Type* A_3: $r \geq 2$ and $d > w(2_i) \geq w(3_i) < d/2$;
  $\Rightarrow r = 2$ and $w(2_i) \geq d/2$ by Claim 6(i) $\Rightarrow w(1_i) + d > \text{OPT}$ by Claim 6(ii).
- *Type* A_4: $r = 1$;
  $\Rightarrow w(1_i) \geq c \Rightarrow w(1_i) + \frac{d}{2} \geq c + \frac{d}{2} \geq a + d + \frac{d}{2}$ by Claim 5 $\Rightarrow w(1_i) + \frac{d}{2} > \text{OPT}$.

Let $\mathcal{A}_h$ denote the set of type A_h machines for $h \in [4]$. Then $\mathcal{A}$ is the disjoint union of $\mathcal{A}_1, \mathcal{A}_2, \mathcal{A}_3, \mathcal{A}_4$. Moreover, by Property 5 we have the following.

Claim 7. To any A-machine $M_i \in \mathcal{A} = \cup_{h=1}^4 \mathcal{A}_h$, except for the a^+-task 1_i of weight $w(1_i)$, the optimal allocation *opt* allocates

(i) either no d^+-task and at most two $[d/2, d)$-tasks, or exactly one d^+-task and no $[d/2, d)$-task, when $M_i \in \mathcal{A}_1 \cup \mathcal{A}_2$;
(ii) no d^+-task, and at most one $[d/2, d)$-task when $M_i \in \mathcal{A}_3$;
(iii) no $(d/2)^+$-tasks when $M_i \in \mathcal{A}_4$. □

Allocation Transformations. Consider a process in which π is transformed to *opt*. As aforementioned, some tasks in $\mathcal{T}(\mathcal{C})$ on C-machines (all of them are d^+-tasks) have to be moved to A-machines. Then by the above classification and Claim 7, an adequate number of d^+-tasks and $(d/2)^+$-tasks should be moved out of A-machines. The following claim makes the idea more precise.

Claim 8. Suppose that *opt* allocates exactly n_0 tasks from $\mathcal{T}(\mathcal{C})$ to A-machines. There exist integers n_1 and n_2 with $n_1 + n_2/2 \geq n_0$ such that *opt* allocates a number n_1 of d^+-tasks and a number n_2 of $(d/2)^+$-tasks from $\mathcal{T}(\mathcal{A})$ to machines in $\mathcal{B} \cup \mathcal{C} = \mathcal{M} \setminus \mathcal{A}$. □

III. A Partial Reallocation In this part, we allocate a subset of tasks in $\mathcal{T}^o(\mathcal{B}\cup\mathcal{C})$ to B-machines or C-machines, so that the resulting partial reallocation has average load greater than OPT, which would give a contradiction.

Let $\mathcal{K}$ ($\subseteq \mathcal{T}(\mathcal{A})$) denote the set of a number n_1 of d^+-tasks and a number n_2 of $(d/2)^+$-tasks stated in Claim 8. Then $\mathcal{K} \subseteq \mathcal{T}^o(\mathcal{B}\cup\mathcal{C}) \cap \mathcal{T}(\mathcal{A})$, $w(\mathcal{K}) \geq (n_1 + n_2/2)d \geq n_0 d$ and $w(j) \geq d/2$ for all $j \in \mathcal{K}$, which implies the following.

Claim 9. There is an allocation $\sigma : \mathcal{K} \to \mathcal{C}$ such that for any $M_i \in \mathcal{C}$, if *opt* allocates h tasks from $\mathcal{T}_i \subseteq \mathcal{T}(\mathcal{C})$ to machines in $\mathcal{A}$, then σ allocates h groups of tasks from $\mathcal{K}$ to M_i with total weight at least hd.

Let us extend σ to be an allocation from the disjoint union of $\mathcal{K}$ and $\mathcal{T}^o(\mathcal{B}\cup\mathcal{C}) \cap \mathcal{T}(\mathcal{B}\cup\mathcal{C})$ to $\mathcal{B}\cup\mathcal{C}$ by the setting in the next claim.

Claim 10. $\sigma(j) = \pi(j)$ for any task $j \in \mathcal{T}^o(\mathcal{B}\cup\mathcal{C}) \cap \mathcal{T}(\mathcal{B}\cup\mathcal{C})$. □

Recall from the definition of d in (4.2) that all tasks of a C-machine M_i are d^+-tasks. Suppose *opt* allocates h tasks from $\mathcal{T}_i$ to A-machines. Then the remaining $|\mathcal{T}_i| - h$ tasks of $\mathcal{T}_i$ belong to $\mathcal{T}^o(\mathcal{B}\cup\mathcal{C})\cap\mathcal{T}(\mathcal{C})$, and thus are allocated by σ to M_i (by Claim 10). These d^+-tasks together with the tasks from $\mathcal{K}$ of total weight at least hd (by Claim 9) give $L_i(\sigma) \geq |\mathcal{T}_i|d$.

Claim 11. $L_i(\sigma) \geq \ell^i d$ for any C-machine $M_i \in \mathcal{C}$. □

Suppose the load of machine $M_g \in \mathcal{B}\cup\mathcal{C}$ is minimum under σ. Since $\mathcal{K}\cup(\mathcal{T}^o(\mathcal{B}\cup\mathcal{C}) \cap \mathcal{T}(\mathcal{B}\cup\mathcal{C})) \subseteq \mathcal{T}^o(\mathcal{B}\cup\mathcal{C})$, it is straightforward that

Claim 12. $L_g(\sigma) = \min\{L_i(\sigma) : M_i \in \mathcal{B}\cup\mathcal{C}\} \leq \text{OPT}$. □

By Property 5(i), no b^+-task in $\mathcal{T} \setminus \mathcal{T}(\mathcal{A})$ is allocated by *opt* to any A-machine. So σ allocates every b^+-task in $\mathcal{T}(\mathcal{B}\cup\mathcal{C})$ to the same machine as π does. In particular, by Property 4, every B-machine admits at least two b^+-task under both π and σ, and therefore has load at least $2b > \text{OPT}$ under σ by (4.1). It follows from Claim 12 that $M_g \in \mathcal{C}$. We then distinguish between two cases depending on whether $w(t_g^g)$ is smaller than $b/2$ or not. In either case, we can derive a contradiction to Claim 12 and then prove the following lemma.

Lemma 2. $\lambda \notin [1, 4/3)$. □

4.2 Large $\lambda \in [4/3, 5/2)$

Similar to Claim 1, task weights on C-machine are lower bounded as follows.

Claim 13. $w(\ell_i^i) \geq (1+\lambda)b/7 = (a+b)/7$ for any $M_i \in \mathcal{C}$. □

Since $\text{OPT} < 5(1+\lambda)b/7$ by Property 2(iii), using $\lambda \geq 4/3$ and $\lambda < 5/2$, respectively, elementary mathematics gives

$$b/2 + a > (1+\lambda)b/7 + a \geq 5(1+\lambda)b/7 > \text{OPT}\,,$$
$$(1+\lambda)b/7 + 2b > 5(1+\lambda)b/7 > \text{OPT}\,.$$

Since every task on a C-machine under π has weight at least $(1+\lambda)b/7$ by Claim 13, the above two strings of inequalities imply that its weight plus that of an a^+-task or two b^+-tasks exceeds OPT, implying the following.

Claim 14. In the optimal allocation *opt*, no task from $\mathcal{T}(\mathcal{C})$ can share a machine with any a^+-task, or with two or more other b^+-tasks. □

Moreover, $\lambda \leq 5/2$ implies $b > (1+\lambda)b/7$. It follows from $(1+\lambda)b/7 + 2b > \text{OPT}$ that $3b > \text{OPT}$. Thus we have the following.

Claim 15. (i) To the same machine, *opt* does not allocate three or more b^+-tasks. (ii) For any integers $p_1 > 0$, p_2 with $p_1 \geq p_2$, $p_1(a+\frac{b}{2})+2p_2b > (p_1+p_2)\text{OPT}$. □

Suppose that in the optimal allocation *opt* there are m_1 machines each of which admits two b^+-tasks from $\mathcal{T}(\mathcal{A} \cup \mathcal{B})$. By Property 5(i), these m_1 machines all belong to $\mathcal{B} \cup \mathcal{C}$. Denote the set of these m_1 machines by $\mathcal{M}_1$. Since $\mathcal{B} \cap \mathcal{C} = \emptyset$ by Property 4, the next claim follows from Claims 14 and 15(i).

Claim 16. (i) All tasks from $\mathcal{T}(\mathcal{C})$ can only be allocated by *opt* to the *remaining* $m_2 \equiv |\mathcal{B}| + |\mathcal{C}| - m_1$ B-machines or C-machines in $\mathcal{M}_2 \equiv (\mathcal{B} \cup \mathcal{C}) \setminus \mathcal{M}_1$.
(ii) Each of these m_2 machines in $\mathcal{M}_2$ can admit at most one b^+-task from $\mathcal{T}(\mathcal{A} \cup \mathcal{B})$ under *opt*. □

Recall from Property 4(ii) that π allocates two $[b, a)$-tasks to each B-machine. Recall the last b-task ℓ_1^1 of machine $M_1 \in (\mathcal{A} \cup \mathcal{B}) \setminus \mathcal{C}$. In case of $\ell^1 \equiv |\mathcal{T}_1| = 2$, we have $w(1_1) = a$, which implies $M_1 \in \mathcal{A}$. It follows that ℓ_1^1 cannot be the first or the second task of any B-machine. Hence there are at least a number $2|\mathcal{B}| + 1$ of b^+-tasks in $\mathcal{T}(\mathcal{A} \cup \mathcal{B}) \setminus \{1_i : M_i \in \mathcal{A}\}$, none of which is allocated by *opt* to any A-machine as guaranteed by Property 5(i). By the definition of $\mathcal{M}_1$, *opt* allocates a number $2m_1$ of b^+ tasks from $\mathcal{T}(\mathcal{A} \cup \mathcal{B})$ to machines in $\mathcal{M}_1$. Therefore we have the following.

Claim 17. The optimal allocation *opt* allocates at least a number $2|\mathcal{B}|+1-2m_1$ of b^+-tasks in $\mathcal{T}(\mathcal{A} \cup \mathcal{B}) \setminus \{1_i : M_i \in \mathcal{A}\}$ to B-machines or C-machines in $\mathcal{M}_2$. □

By Claims 16(ii) and 17, we have $2|\mathcal{B}| + 1 - 2m_1 \leq m_2$, and $|\mathcal{B}| - m_1 \leq (m_2 - 1)/2 < m_2/2$. Notice from Claim 16(i) that $m_2 - |\mathcal{C}| = |\mathcal{B}| - m_1$. We thus have $m_2 - |\mathcal{C}| < m_2/2$ and $|\mathcal{C}| > m_2/2$. Applying Claim 15(ii) with $p_1 = |\mathcal{C}|$ and $p_2 = m_2 - |\mathcal{C}|$, we have

$$|\mathcal{C}| \cdot (a + b/2) + (m_2 - |\mathcal{C}|) \cdot 2b > m_2 \cdot \text{OPT}. \tag{4.4}$$

Recall from Property 7 that $L_i \geq a + b/2$ for any $M_i \in \mathcal{C}$. It is easy to see from Claims 16(i) and 17 that under *opt* the total load of the machines in $\mathcal{M}_2$ is at least $|\mathcal{C}| \cdot (a + b/2) + (2|\mathcal{B}| + 1 - 2m_1) \cdot b$. Suppose $M_g \in \mathcal{M}_2$ is the machine that has the maximum load under *opt* in $\mathcal{M}_2$. Then obviously

$$\text{OPT} \geq L_g(opt) \geq \frac{|\mathcal{C}| \cdot (a + \frac{b}{2}) + (2|\mathcal{B}|+1-2m_1) \cdot b}{m_2} > \frac{|\mathcal{C}| \cdot (a+\frac{b}{2}) + (|\mathcal{B}|-m_1) \cdot 2b}{m_2}.$$

Substituting $m_2 - |\mathcal{C}|$ for $|\mathcal{B}| - m_1$ in the above inequality, we obtain a contradiction to (4.4), which implies the following lemma.

Lemma 3. $\lambda \notin [4/3, 5/2)$. □

We are able to complete the proof of Theorem 2. Indeed, Lemmas 2 and 3 assert $\lambda \notin [1, 2.5)$, contradicting Property 3(ii). So Assumption 1 is incorrect. □

5 Concluding Remark

In this paper we extend the existing game model for the optimization problem of task allocation to machines, STA with only selfish task agents, to a more general one, STASM with not only selfish task agents but also selfish machine agents. Our study shows that activating machines to choose selfishly between accepting and rejecting tasks could considerably improve the efficiency of selfish task allocation in terms of price of anarchy. Thus it is interesting to investigate what kind of games have such a desirable property that allowing more freedom and selfishness will bring about better global outcomes.

References

1. Czumaj, A., Krysta, P., Vöcking, B.: Selfish traffic allocation for server farms. SIAM J. Comput. 39, 1957–1987 (2010)
2. Czumaj, A., Vöcking, B.: Tight bounds for worst-case equilibria. ACM Trans. Algorithms 3, Article 4 (2007)
3. Epstein, L., Kleiman, E., van Stee, R.: Maximizing the Minimum Load: The Cost of Selfishness. In: Leonardi, S. (ed.) WINE 2009. LNCS, vol. 5929, pp. 232–243. Springer, Heidelberg (2009)
4. Finn, G., Horowitz, E.: A linear time approximation algorithm for multiprocessor scheduling. BIT 19(3), 312–320 (1979)
5. Fotakis, D., Kontogiannis, S., Koutsoupias, E., Mavronicolas, M., Spirakis, P.: The Structure and Complexity of Nash Equilibria for a Selfish Routing Game. In: Widmayer, P., Triguero, F., Morales, R., Hennessy, M., Eidenbenz, S., Conejo, R. (eds.) ICALP 2002. LNCS, vol. 2380, pp. 123–134. Springer, Heidelberg (2002)
6. Garey, M.R., Johnson, D.S.: Computers and Intractabilities: A Guide to the Theory of NP-Completeness. W. H. Freeman, New York (1979)
7. Hall, L.A.: Approximation algorithms for scheduling. In: Hochbaum, D.S. (ed.) Approximation Algorithms for NP-Hard Problems, pp. 1–45. PWS Publishing, Boston (1997)
8. Koutsoupias, E., Papadimitriou, C.H.: Worst-Case Equilibria. In: Meinel, C., Tison, S. (eds.) STACS 1999. LNCS, vol. 1563, pp. 404–413. Springer, Heidelberg (1999)
9. Schuurman, P., Vredeveld, T.: Performance guarantees of local search for multiprocessor scheduling. INFORMS J. Comput. 19, 52–63 (2007)
10. Vöcking, B.: Selfish load balancing. In: Nisan, N., Roughgarden, T., Tardos, É., Vazirani, V.V. (eds.) Algorithmic Game Theory, pp. 517–542. Cambridge University Press, Cambridge (2007)

Fast-Mixed Searching on Graphs

Boting Yang

Department of Computer Science, University of Regina
boting@cs.uregina.ca

Abstract. We introduce the fast-mixed search model, which is a combination of the fast search model and the mixed search model. We establish relations between the fast-mixed search problem and other graph search problems. We also establish relations between the fast-mixed search problem and the induced-path cover problem. We present linear-time algorithms for computing the fast-mixed search number and optimal search strategies of some classes of graphs, including trees, cacti, and interval graphs. We prove that the fast-mixed search problem is NP-complete; and it remains NP-complete for graphs with maximum degree 4.

1 Introduction

Throughout this paper, we only consider finite connected graphs with no loops or multiple edges. Given a graph in which a fugitive hides on vertices or along edges, graph search problems are usually to find the minimum number of searchers required to capture the fugitive. The edge search problem and the node search problem are two major graph search problems [7,4]. Both search problems are monotonic [1,5]. Bienstock and Seymour [1] introduced the mixed search problem that combines the edge search and the node search problems. In the mixed search problem, there are three actions for searchers, that is, *placing* (place a searcher on a vertex), *removing* (remove a searcher from a vertex) and *sliding* (slide a searcher from one endpoint of an edge to the other along the edge). An edge is cleared if both endpoints are occupied by searchers or cleared by a sliding action. The mixed search problem is also monotonic. Dyer et al. [2] introduced the fast search problem, in which there are two actions for searchers, that is, *placing* and *sliding*. In the fast search problem, every edge can be traversed exactly once, and it is cleared by a sliding action. Some recent development on the fast search problem can be found in [10,12].

In this paper, we introduce the fast-mixed search model, which combines the fast search and the mixed search models. Let G be a connected graph with no loops or multiple edges. In the fast-mixed search model, initially, G contains no searchers and it contains only one fugitive who hides on vertices or along edges. The fugitive is invisible to searchers, and he can move at a great speed at any time from one vertex to another vertex along a searcher-free path between the two vertices. An edge (resp. a vertex) the fugitive may hide is said to be *contaminated*, and an edge (resp. a vertex) the fugitive cannot hide is said to be *cleared*. A vertex is said to be *occupied* if it has a searcher on it. There are two types of actions for searchers in each step of the fast-mixed search model:

G. Lin (Ed.): COCOA 2012, LNCS 7402, pp. 324–335, 2012.

- *placing* a searcher on a contaminated vertex; and
- *sliding* a searcher along a contaminated edge uv from u to v if v is contaminated and all edges incident on u except uv are cleared.

These two actions are different from those in the fast search model or the mixed search model due to the conditions on performing them. In the fast (or mixed) search model, a vertex can contain more than one searcher at any moment, but in the fast-mixed search problem, every vertex can contain at most one searcher. In the fast (or mixed) search model, if a vertex has two searchers and at least two contaminated edges incident on it, we can still slide a searcher along one contaminated edge, but this does not happen in the fast-mixed search problem.

In the fast-mixed search model, a contaminated edge becomes cleared if both endpoints are occupied by searchers, or a searcher slides along it from one endpoint to the other. An *fms-strategy* of G is a sequence of actions such that the final action leaves all edges of G cleared. The graph G is *cleared* if all edges are cleared. The minimum number of searchers required to clear G is the *fast-mixed search number* of G, denoted by $\text{fms}(G)$. An fms-strategy that uses $\text{fms}(G)$ searchers to clear G is called an *optimal fms-strategy.*

The fast-mixed search problem has a close relation with the induced-path cover problem. In [6], Le et al. proved that it is NP-complete to decide whether or not the vertex set of a connected graph can be partitioned into two subsets, each of which induces a path. In [9], Pan and Chang gave linear-time algorithms for the induced-path cover problem on block graphs whose blocks are complete graphs, cycles or complete bipartite graphs.

We use $G = (V, E)$ to denote a graph with vertex set V and edge set E, and we also use $V(G)$ and $E(G)$ to denote the vertex set and edge set of G respectively. We use uv to denote an edge with endpoints u and v. Definitions omitted here can be found in [11].

For a graph $G = (V, E)$ and a vertex $v \in V$, the *degree* of v, denoted by $\deg_G(v)$, is the number of edges incident on v. The vertex set $\{u : uv \in E\}$ is the *neighborhood* of v, denoted as $N_G(v)$. If there is no ambiguity, we use $\deg(v)$ and $N(v)$ without subscripts. Let $\delta(G) = \min\{\deg(v) : v \in V(G)\}$. For a subset $V' \subseteq V$, we use $G[V']$ to denote the subgraph induced by V', which consists of all vertices of V' and all of the edges that connect vertices of V' in G. For a vertex $v \in V$, we use $G - v$ to denote the subgraph induced by $V \setminus \{v\}$.

A *path* is a list $v_0, e_1, v_1, \ldots, e_k, v_k$ of vertices and edges such that each edge e_i, $1 \leq i \leq k$, has endpoints v_{i-1} and v_i and each vertex appears exactly once. We will denote a path by a list of vertices $v_0v_1 \ldots v_k$ or by the two ends $v_0 \sim v_k$. A path $v_0v_1 \ldots v_k$ in G is called an *induced path* if the subgraph induced by the vertex set $\{v_0, v_1, \ldots, v_k\}$ is a path.

For a graph G, a *path cover* is a set of vertex-disjoint paths that contain all the vertices of G. An *induced-path cover* is a path cover in which every path is an induced path of G. The induced-path cover problem is to find a minimum number of vertex-disjoint induced-paths that contain all the vertices of G. This minimum number is denoted by $\text{ipc}(G)$.

2 Characterizations

In this section, we give characterizations of graphs G with $\text{fms}(G) = k$. We first consider graphs that can be cleared by two searchers. For trees, we have the following characterization.

Theorem 1. *For a tree T, the following are equivalent:* (i) $\text{fms}(T) \leq 2$. (ii) *All vertices of T have degree at most 3; at most two vertices have degree 3; and if T has two vertices of degree 3, then these two vertices must be adjacent.* (iii) *T is one of the graphs in Figure 1.*

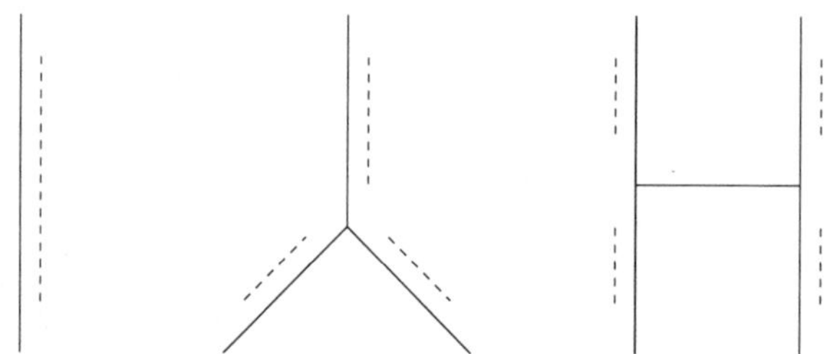

Fig. 1. Trees with fms ≤ 2, where edges marked by dashed lines can be replaced by a path of length at least one

A biconnected graph G is *outerplanar* if it has a planar embedding in which a single face includes all the vertices of G. The edges of that face are called *boundary edges*, and the remaining edges are called *chords*. An outerplanar graph is *bipolar* if there are two boundary edges ab and cd such that every cord has one endpoint on the path $a \sim c$ and the other endpoint is on the path $b \sim d$ (see Figure 2). Edges ab and cd are called *polar edges* and vertices a, b, c, d are called *polar vertices*.

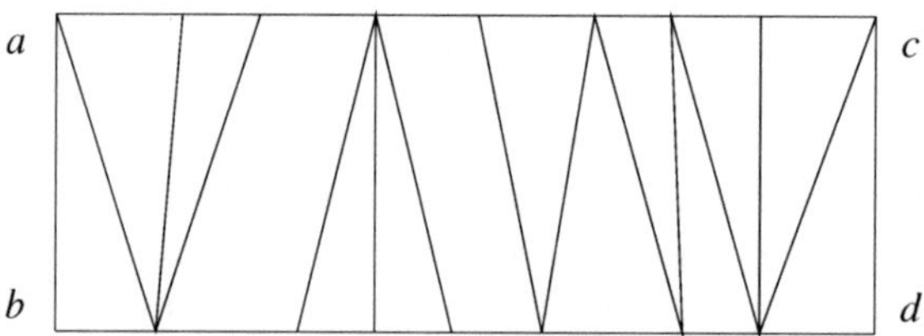

Fig. 2. A bipolar outerplanar graph with two polar edges ab and cd

Definition 1. A graph is called a *ladder* if it can be obtained from a bipolar outerplanar graph by attaching at most one path on each polar vertex (if two polar vertices coincide, then this vertex can be attached by at most two paths).

Referring to Figure 4(a), a graph is called a *standard ladder* if it can be obtained from a grid $G_{2\times n}$ by attaching four edges on the four corner vertices of $G_{2\times n}$ respectively.

Theorem 2. *For any connected graph G that is not a tree,* $\text{fms}(G) = 2$ *if and only if G is a ladder.*

Before we give a characterization of graphs G with $\text{fms}(G) = k$, we generalize ladders as follows.

Definition 2. A graph $G = (V, E)$ is called a *k-ladder* if k is the smallest integer such that G can be drawn as follows: Draw k vertex-disjoint paths $P_1, \ldots, P_k$ such that $\cup_{i=1}^{k} V(P_i) = V$ and $\cup_{i=1}^{k} E(P_i) \subseteq E$ (note that P_i may have length 0). Suppose that all paths are drawn by parallel vertical straight line segments, and let p_i, $1 \le i \le k$, be a moving point on P_i. Initially, each p_i is on the bottom vertex of P_i, and draw edges between the moving points if there is an edge between them in G. Repeat the following process until all moving point are located on the top vertices: Pick a moving point p_i, $1 \le i \le k$, which is not on the top vertex of P_i, move p_i up to the next vertex on P_i, and then draw edges between p_i and p_j ($1 \le j \le k$, and $j \ne i$) if there is an edge between them in G.

From Definitions 1 and 2, we know that a ladder is always a 2-ladder, but a 2-ladder is a ladder if it is connected and has at least one cycle.

Theorem 3. *For any connected graph G,* $\text{fms}(G) = k$ *if and only if G is a k-ladder.*

3 Relations to the Induced-Path Cover

In this section, we establish a relation between the fast-mixed search problem and the induced-path cover problem.

Lemma 1. *For a graph $G = (V, E)$ that can be cleared by k searchers in an fms-strategy S, let $V_1, \ldots, V_k$ be k subsets of V such that each vertex in V_i, $1 \le i \le k$, is visited by the same searcher in the fms-strategy S. Then $V_1, \ldots, V_k$ form a partition of V and each V_i induces a path.*

Definition 3. In Lemma 1, each induced path $G[V_i]$ is called an *fms-path* with respect to S, and the set $\{G[V_1], \ldots, G[V_k]\}$ of fms-paths is called an *fms-path cover* of G with respect to S.

Corollary 1. *For any graph $G = (V, E)$, the number of actions in any fms-strategy is $|V|$.*

We now consider the subgraph induced by the vertices on any two fms-paths. Figure 3 illustrates three families of forbidden subgraphs.

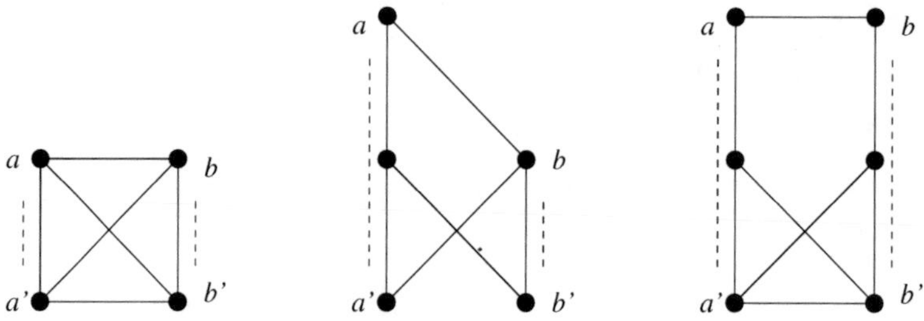

Fig. 3. Graphs with fms > 2, where edges marked by dashed lines can be replaced by a path of length at least one

Theorem 4. *For an fms-strategy S of a graph G that uses k searchers, $k \geq 2$, let P be an fms-path cover of G with respect to S. For any two paths $P_1, P_2 \in P$, let H be the subgraph of G induced by vertices $V(P_1) \cup V(P_2)$. Then the following are equivalent:* (i) *H does not contain any graph in Figure 3, where $a, a' \in V(P_1)$ and $b, b' \in V(P_2)$.* (ii) *H has one of the three patterns: (a) a forest consisting of two disjoint paths, (b) a tree consisting of two adjacent degree-3 vertices and all other vertices having degree one or two; and (c) a ladder.*

4 Lower Bounds

In this section, we give lower bounds on the fast-mixed search number.

Lemma 2. *For any graph G with ℓ leaves,* $\mathrm{fms}(G) \geq \lceil \ell/2 \rceil$.

Lemma 3. *Let G be a graph with components $G_1, \ldots, G_k$. Then* $\mathrm{fms}(G) \geq \sum_{i=1}^{k} \delta(G_i)$, *where $\delta(G_i)$ denotes the minimum vertex degree in G_i.*

Lemma 4. *For any graph G,* $\mathrm{fms}(G) \geq \mathrm{ipc}(G)$.

The minimum number of searchers required to clear a graph G in the mixed search model is called the *mixed search number* of G, denoted by $\mathrm{ms}(G)$. Since every fms-strategy is also a mixed search strategy, we have the following result.

Lemma 5. *For any graph G,* $\mathrm{fms}(G) \geq \mathrm{ms}(G)$.

It is easy to see that the family of graphs $\{G : \mathrm{fms}(G) \leq k\}$ is not subgraph-closed for any integer $k \geq 1$. Furthermore, the family of graphs $\{G$: G is connected and $\mathrm{fms}(G) \leq k\}$ is not subgraph-closed either for any integer $k \geq 2$. Although the fast-mixed search number of a graph may be less than that of its subgraph, we can show that the fast-mixed search number of a graph is always greater than or equal to that of grid subgraphs or cliques.

Corollary 2. *If a graph G contains a grid $G_{k\times\ell}$ with k rows and ℓ columns, where $2 \leq k \leq \ell$, then* $\mathrm{fms}(G) \geq k$.

Corollary 3. *If a graph G contains a clique K_k with k vertices, $k \geq 2$, then* $\mathrm{fms}(G) \geq k - 1$.

5 Relations to Fast Searching and Mixed Searching

We first consider the relationship between fast-mixed searching and fast searching. The fast search number of a graph G, denoted by $\mathrm{fs}(G)$, is the minimum number of searchers required to clear G in the fast search model.

On one hand, there are graphs whose fast-mixed search number is arbitrarily bigger than their fast search number. Consider a complete bipartite graph $K_{1,n}$, where $n \geq 2$. We can show that $\mathrm{fms}(K_{1,n}) = n - 1$ while $\mathrm{fs}(K_{1,n}) = \lceil \frac{n}{2} \rceil$. On the other hand, there are graphs whose fast-mixed search number is arbitrarily smaller than its fast search number. Consider the standard ladder L_n (see Figure 4(a)). It is easy to see that $\mathrm{fms}(L_n) = 2$, but $\mathrm{fs}(L_n) = n + 2$.

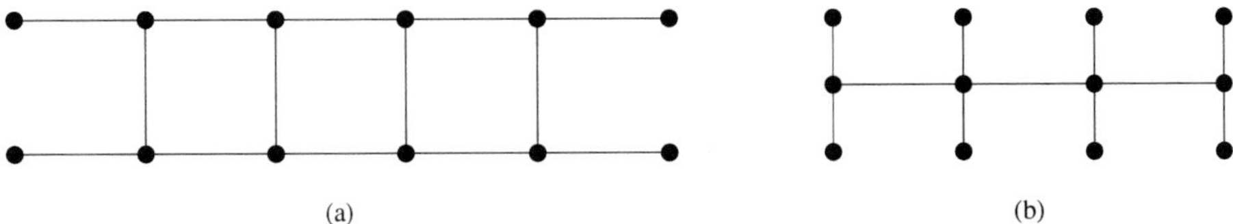

Fig. 4. (a) A standard ladder L_4 with $\mathrm{fms}(L_4) = 2$ and $\mathrm{fs}(L_4) = 6$. (b) A caterpillar H_4 with $\mathrm{fms}(H_4) = 4$ and $\mathrm{ms}(H_4) = 2$.

We now consider the relationship between fast-mixed searching and mixed searching. From Lemma 5, we know that the fast-mixed search number is always greater than or equal to the mixed search number for any graph. However, there are graphs whose optimal fms-strategies look very different from their optimal mixed search strategies. Consider a caterpillar H_n with a spine on n vertices, $n \geq 3$, such that each vertex on the spine has two pendent edges (see Figure 4(b)). The optimal fms-strategy of H_n is to place n searchers on the bottom n leaves and slide them to the n vertices on the spine, and then slide them to the top n leaves. But the optimal mixed search strategy of H_n is to place a searcher on one end vertex of the spine and move it to the other end vertex of the spine passing through all degree-4 vertices, and meanwhile, place another searcher on each leaf one by one. The following lemma gives a general case of the difference.

Theorem 5. *Given a graph G that contains at least one edge, let G' be a graph obtained from G by adding two pendent edges on each vertex. Then* $\mathrm{fms}(G') = |V(G)|$ *and* $\mathrm{ms}(G') = \mathrm{ms}(G) + 1$.

From Theorem 5, we know that the optimal mixed search strategy of G' can be obtained from that of G by using an extra searcher to clear the two pendent edges of each vertex when the vertex is occupied. However, the optimal fms-strategy of G' is always using $|V(G)|$ searchers to clear the two pendent edges on each vertex of G, which is independent of the optimal fms-strategy of G.

6 Special Classes of Graphs

First we have the following results for complete graphs, complete bipartite graphs and grids.

Lemma 6. (i) *For a complete graph* K_n $(n \geq 2)$, $\text{fms}(K_n) = n - 1$.
(ii) *For a complete bipartite graph* $K_{m,n}$ $(n \geq m \geq 2)$, $\text{fms}(K_{m,n}) = m+n-2$.
(iii) *For a grid* $G_{k\times\ell}$ *with* k *rows and* ℓ *columns* $(2 \leq k \leq \ell)$, $\text{fms}(G_{k\times\ell}) = k$.

For trees, we can show the following relation between the fast-mixed search and the induced-path cover.

Theorem 6. *For a tree* T, $\text{fms}(T) = \text{ipc}(T)$.

From [8], an optimal path cover of a tree can be computed in linear time. Note that an optimal path cover of a tree is also an optimal induced-path cover of the tree. After we obtain an optimal path cover, we can use the method on the base of the induction used in the proof of Theorem 6 to compute an optimal fms-strategy in linear time.

Corollary 4. *For any tree, the fast-mixed search number and an optimal fms-strategy can be computed in linear time.*

For some planar graphs, the optimal fms-strategy can be very different from both optimal mixed search strategy and the optimal induced-path cover. For example, the planar graph G in Figure 5 can be cleared using 6 searchers in the mixed search model, that is, using 5 searchers to clear the five rows and one more searcher to clear all degree-2 vertices inside all columns. In the fast-mixed search model, we can clear G using 9 searchers who clear the nine columns respectively. For the induced-path cover of G, the five induced paths, which correspond to the five dashed paths, contains all vertices of G.

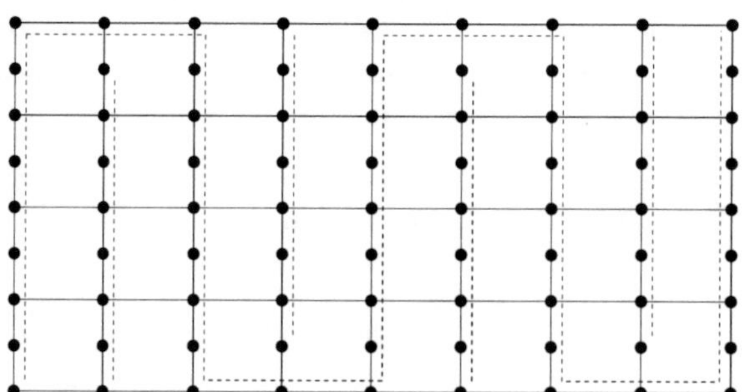

Fig. 5. A planar graph G with $\text{ms}(G) = 6$, $\text{fms}(G) = 9$ and $\text{ipc}(G) \leq 5$

We now consider cacti. A *cactus* is a connected graph in which any two simple cycles have at most one vertex in common. Thus, the subgraph induced by any pair of induced paths can contain at most one cycle. From Definition 3 and Theorem 4, we know any induced-path cover of a cactus is also an fms-path cover. From [9], we have a linear-time algorithm to compute an induced-path cover of a block graph with every block being a cycle. By modifying the algorithm in [9], we can compute the fast-mixed search number of cacti in linear time.

Theorem 7. *For any cactus, the fast-mixed search number and an optimal fms-strategy can be computed in linear time.*

A graph is *chordal* if it does not contain an induced cycle of length at least 4. A graph is an *interval graph* if it is the intersection graph of a collection of intervals on the real line. An interval graph is a special chordal graph. It is well known that a graph G is an interval graph if and only if the maximal cliques of G can be ordered $C_1, C_2, \ldots, C_m$ such that for any $v \in V(C_i) \cap V(C_k)$, $1 \leq i < k \leq m$, the vertex v is also contained in all C_j, $i \leq j \leq k$.

The fast-mixed search number can be arbitrarily larger than the induced-path number on interval graphs. For example, for an interval graph K_n, $n \geq 2$, from Lemma 6(i), we know that $\text{fms}(K_n) = n - 1$, but $\text{ipc}(K_n) = \lceil n/2 \rceil$.

Theorem 8. *Given an interval graph G, let $C_1, C_2, \ldots, C_m$ be the sequence of the maximal cliques of G such that, for any $v \in V(C_i) \cap V(C_k)$, $1 \leq i < k \leq m$, the vertex v is also contained in all C_j, $i \leq j \leq k$. If $k > 1$, then*

$$\text{fms}(G) = |V(C_1)| + \sum_{j=1}^{m-1} \max\{|V(C_{j+1})| - |V(C_j)|, 0\}.$$

Corollary 5. *For any interval graph, the fast-mixed search number and an optimal fms-strategy can be computed in linear time.*

A graph G is a *k-tree* if and only if either G is a complete graph with k vertices, or G has a vertex v such that $N(v)$ induces a k-clique, and $G - v$ is a k-tree. A vertex v of a k-tree G is called *simplicial* if $N(v)$ induces a k-clique.

Theorem 9. *For a k-tree G with more than k vertices, if G has exactly two simplicial vertices, then* $\text{fms}(G) = k$.

A graph $G = (V, E)$ is called *fms-maximal* if $\text{fms}(G') > \text{fms}(G)$ for any graph $G' = (V, E')$, where E' is a proper superset of E

Theorem 10. *Every fms-maximal graph G with* $\text{fms}(G) = k$ *is a k-tree with exactly two simplicial vertices.*

Given two graphs G and H, the *Cartesian product* of G and H, denoted by $G \square H$, is the graph whose vertex set is the Cartesian product $V(G) \times V(H)$ of the two vertex sets $V(G)$ and $V(H)$, and in which two vertices $(u, v), (u', v') \in V(G) \times V(H)$ are adjacent in $G \square H$ if and only if $u = u'$ and v is adjacent with v' in H, or $v = v'$ and u is adjacent with u' in G. We have the following result for the Cartesian product of two graphs.

Theorem 11. *For any graphs G and H,*

$$\text{fms}(G \square H) \leq \min\{|V(G)|\text{fms}(H), |V(H)|\text{fms}(G)\}.$$

7 NP-Completeness

In this section, we will show that the fast-mixed search problem is NP-complete, and it remains NP-complete even for graphs with maximum degree 4.

Let G^k be a graph with k "legs" as illustrated in Figure 6. We call graph G^k a *variable gadget*, which corresponds to a boolean variable that appears k times in a conjunctive normal form boolean formula. We first show a property of variable gadgets as follows.

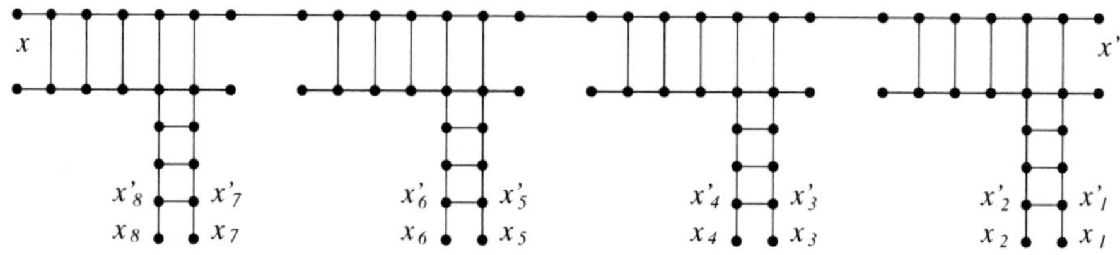

Fig. 6. A variable gadget G^k with four legs ($k = 4$)

Lemma 7. *Let G^k be a variable gadget as illustrated in Figure 6. Then* $\text{fms}(G^k) = 2k + 1$*; and furthermore, for any optimal fms-strategy of G^k, if a searcher slides from x to its neighbor, then for each leaf x_i ($1 \leq i \leq 2k$) there is a searcher sliding from x'_i to x_i; and if a searcher slides to x from its neighbor, then for each leaf x_i ($1 \leq i \leq 2k$) there is a searcher sliding from x_i to x'_i.*

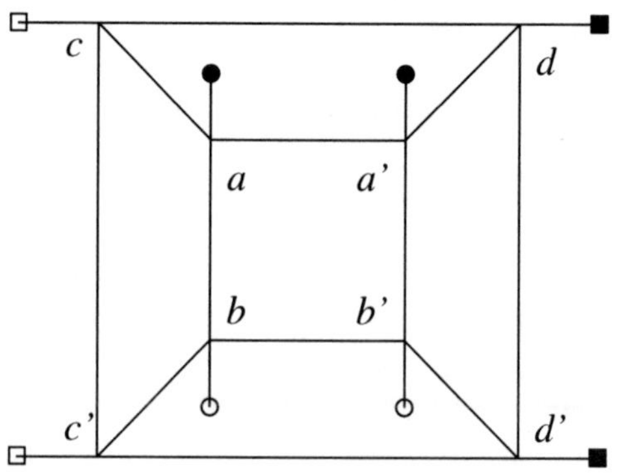

Fig. 7. A clause gadget

Let G_c be a graph illustrated in Figure 7. We call graph G_c a *clause gadget*, which corresponds to a clause c in a conjunctive normal form boolean formula. Note that G_c has four pairs of leaves that are marked using the same pattern, i.e., solid squares, hollow squares, solid circles and hollow circles. G_c has the following property.

Lemma 8. *Let G_c be a graph as illustrated in Figure 7. Then* $\text{fms}(G_c) = 4$ *and in any optimal fms-strategy of G_c, four searchers must be placed on two pairs of leaves marked with the same pattern.*

We now use variable gadgets and clause gadgets to show the NP-completeness result. The reduction is similar to the one used in [3].

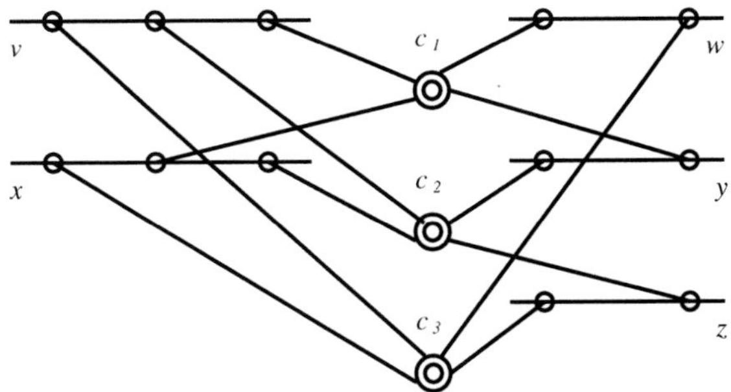

Fig. 8. The graph G_ϕ constructed for $\phi = c_1 \wedge c_2 \wedge c_3$, where $c_1 = (v \vee w \vee x \vee y)$, $c_2 = (v \vee x \vee y \vee z)$ and $c_3 = (v \vee w \vee x \vee z)$, where each single circle represents a leg of a variable gadget and each double circle represents a clause gadget

Theorem 12. *The fast-mixed search problem is NP-complete. It remains NP-complete for graphs with maximum degree 4.*

Proof. It is easy to see that the fast-mixed search problem is in NP. We will show that this problem is NP-hard by a reduction from the planar positive 2-in-4-SAT problem. From [3], we know that the planar positive 2-in-4-SAT problem is NP-complete. An instance of the planar positive 2-in-4-SAT problem is defined as follows. Let ϕ be a boolean formula in the conjunctive normal form with m clauses $\{c_1, \ldots, c_m\}$ and n variables $x_1, \ldots, x_n$ such that each clause contains exactly four variables. The incident graph of ϕ is the bipartite graph with vertex set $\{c_1, \ldots, c_m, x_1, \ldots, x_n\}$ and edge set $\{c_i x_j : \text{clause } c_i \text{ contains variable } x_j\}$. ϕ is planar positive if it contains no negations and its incident graph is planar. A truth assignment of ϕ is 2-in-4 satisfying if each clause has exactly two true variables, and ϕ is 2-in-4 satisfiable if there is a 2-in-4 satisfying truth assignment.

We now construct an instance of the fast-mixed search problem. For each clause c_j, $1 \leq j \leq m$, we construct the corresponding clause gadget G_{c_j} such that leaves marked using the same pattern (i.e., solid squares, hollow circles, etc.) are corresponding to the same literal in c_j (see Figure 7). For each variable x that appears k times in ϕ, we construct the corresponding variable gadget G_x^k

such that leaves x_{2i-1} and x_{2i} are corresponding to the i-th occurrence of the variable x (see Figure 6). If the i-th occurrence of the variable x is in clause c_j, then we use a standard ladder to connect leaves x_{2i-1} and x_{2i} to the leaves marked using the same pattern in G_{c_j}. For each clause gadget, all leaves become degree-2 vertices after they are linked by four standard ladders. In polynomial time, we can construct a graph with maximum degree 4, denoted by G_ϕ (see Figure 8). We will show that the planar positive formula ϕ is 2-in-4 satisfiable if and only if $\text{fms}(G_\phi) = n + 4m$.

Suppose that the planar positive formula ϕ is 2-in-4 satisfiable. Consider a 2-in-4 satisfying truth assignment of ϕ. For each variable x whose value is true and appearing k times in ϕ, we clear the variable gadget G_x^k by sliding a searcher from x to x' to clear all edges on the top row, in the mean time, sliding $2k$ searchers from each leaf on the second row to clear edges on the second row, edges between top and second rows, and edges on all legs, until each leaf $x_j, 1 \le j \le 2k$, is occupied by a searcher. We then slide all these $2k$ searchers along the standard ladders to all clause gadgets. After we clear all variable gadgets, which correspond to variables with true value, and all standard ladders linking to them, all clause gadgets G_c have four searchers on two pairs of leaves marked using the same pattern, which correspond to the two true literals. From Lemma 8, we can clear G_c using the four searchers such that the other two pairs of leaves marked using the same pattern are occupied when G_c is cleared. Note that these two pairs of leaves correspond to the two false literals. Then we slide all $2k$ searchers along standard ladders to all variable gadgets, which correspond to variables with false value. Finally, we clear each false variable gadget G_x^k by sliding $2k$ searchers to clear all leg edges, in the mean time, sliding a searcher from x' to x to clear all edges on the top row, edges on the second row, and edges between top and second rows. Thus, we can use $n+4m$ searchers to clear G_ϕ. On the other hand, since G_ϕ has $2n + 8m$ leaves, it follows from Lemma 2 that we need at least $n + 4m$ searchers to clear G_ϕ. Hence, $\text{fms}(G_\phi) = n + 4m$.

Conversely, suppose that $\text{fms}(G_\phi) = n+4m$. Since G_ϕ has $2n+8m$ leaves, we know that every leaf v must be associated with a searcher who either slides from v to its neighbor or slides to v from its neighbor. Thus, it follows from Lemma 7 that each variable gadget that has k legs must be cleared by $2k+1$ searchers such that either there is a searcher sliding from x to its neighbor and there is a searcher sliding from x_i' to x_i $(1 \le i \le 2k)$, or there is a searcher sliding to x from its neighbor and there is a searcher sliding from x_i to x_i' $(1 \le i \le 2k)$. Furthermore, no searcher can stay in any vertex of clause gadgets at the end of the whole searching process. Hence, from Lemma 8, each clause gadget must be cleared by four searchers and these four searchers must start from two pairs of leaves marked using the same pattern and end at the other two pairs of leaves marked using the same pattern. We can assign true to the two literals corresponding to the two pairs of leaves with four searchers starting, and assign false to the two literals corresponding to the two pairs of leaves with four searchers ending. Since $\text{fms}(G_\phi) = n+4m$, we can find a satisfying truth assignment of ϕ without conflicts. Thus, ϕ is 2-in-4 satisfiable.

From the proof of Theorem 12, we also show the following.

Corollary 6. *Given a graph G with ℓ leaves and* $\deg(G) \leq 4$, *the problem of determining whether* $\text{fms}(G) = \lceil \ell/2 \rceil$ *is NP-complete.*

8 Conclusions

In this paper, we introduced the fast-mixed search model, which is a combination of the fast search model and the mixed search model. However, the fast-mixed search number and the optimal fms-strategy can be very different from those in fast searching and mixed searching. For example, for a complete bipartite graph $K_{1,n}$ with $n \geq 2$, we know that $\text{fms}(K_{1,n}) = n - 1$, $\text{fs}(K_{1,n}) = \lceil \frac{n}{2} \rceil$ and $\text{ms}(K_{1,n}) = 2$. An obvious motivation of the new model is to capture the fugitive using the minimum number of actions. Another motivation is that the fms-paths of a graph G form an induced-path cover of G such that each pair of paths induce a ladder. We gave linear-time algorithms for computing the fast-mixed search number of some special classes of graphs such as trees, cacti and interval graphs. We proved that the fast-mixed search problem is NP-complete. We also proved that, given a graph G with ℓ leaves and $\deg(G) \leq 4$, the problem of determining whether $\text{fms}(G) = \lceil \ell/2 \rceil$ is NP-complete.

References

1. Bienstock, D., Seymour, P.: Monotonicity in graph searching. Journal of Algorithms 12, 239–245 (1991)
2. Dyer, D., Yang, B., Yaşar, Ö.: On the Fast Searching Problem. In: Fleischer, R., Xu, J. (eds.) AAIM 2008. LNCS, vol. 5034, pp. 143–154. Springer, Heidelberg (2008)
3. Kára, J., Kratochvíl, J., Wood, D.: On the complexity of the balanced vertex ordering problem. Discrete Mathematics and Theoretical Computer Science 9, 193–202 (2007)
4. Kirousis, L., Papadimitriou, C.: Searching and pebbling. Theoretical Computer Science 47, 205–218 (1986)
5. LaPaugh, A.: Recontamination does not help to search a graph. Journal of ACM 40, 224–245 (1993)
6. Le, H., Le, V., Ganjali, Y., Muller, H.: Splitting a graph into disjoint induced paths or cycles. Discrete Applied Mathematics 131, 190–212 (2003)
7. Megiddo, N., Hakimi, S., Garey, M., Johnson, D., Papadimitriou, C.: The complexity of searching a graph. Journal of ACM 35, 18–44 (1988)
8. Moran, S., Wolfstahl, Y.: Optimal covering of cacti by vertex-disjoint paths. Theoretical Computer Science 84, 179–197 (1991)
9. Pan, J., Chang, G.: Induced-path partition on graphs with special blocks. Theoretical Computer Science 370, 121–130 (2007)
10. Stanley, D., Yang, B.: Fast searching games on graphs. Journal of Combinatorial Optimization 22, 763–777 (2011)
11. West, D.B.: Introduction to Graph Theory. Prentice Hall (1996)
12. Yang, B.: Fast edge-searching and fast searching on graphs. Theoretical Computer Science 412, 1208–1219 (2011)

Inapproximability after Uniqueness Phase Transition in Two-Spin Systems

Jin-Yi Cai[1,⋆], Xi Chen[2,⋆⋆], Heng Guo[1,⋆⋆⋆], and Pinyan Lu[3]

[1] University of Wisconsin, Madison
[2] Columbia University
[3] Microsoft Research Asia

Abstract. A two-state spin system is specified by a matrix

$$\mathbf{A} = \begin{bmatrix} A_{0,0} & A_{0,1} \\ A_{1,0} & A_{1,1} \end{bmatrix} = \begin{bmatrix} \beta & 1 \\ 1 & \gamma \end{bmatrix} \quad (1)$$

where $\beta, \gamma \geq 0$. Given an input graph $G = (V, E)$, the partition function $Z_{\mathbf{A}}(G)$ of a system is defined as

$$Z_{\mathbf{A}}(G) = \sum_{\sigma:V\to\{0,1\}} \prod_{(u,v)\in E} A_{\sigma(u),\sigma(v)}. \quad (2)$$

We prove inapproximability results for the partition function $Z_{\mathbf{A}}(G)$ in the region specified by the non-uniqueness condition from phase transition for the Gibbs measure. More specifically, assuming NP $\neq$ RP, for any fixed β, γ in the unit square, there is no randomized polynomial-time algorithm that approximates $Z_{\mathbf{A}}(G)$ for d-regular graphs G with relative error $\epsilon = 10^{-4}$, if $d = \Omega(\Delta(\beta,\gamma))$, where $\Delta(\beta,\gamma) > 1/(1-\beta\gamma)$ is the uniqueness threshold. Up to a constant factor, this hardness result confirms the conjecture that the uniqueness phase transition coincides with the transition from computational tractability to intractability for $Z_{\mathbf{A}}(G)$. We also show a matching inapproximability result for a region of parameters β, γ outside the unit square, and all our results generalize to partition functions with an external field.

1 Introduction

Spin systems are well studied in statistical physics and applied probability. We focus on two-state spin systems. An instance of a spin system is a graph $G = (V, E)$. A configuration $\sigma : V \to \{0, 1\}$ assigns to each vertex one of two states. The contributions of local interactions between adjacent vertices are quantified by (1), a 2×2 matrix with $\beta, \gamma \geq 0$. The partition function $Z_{\mathbf{A}}(G)$ of a system is defined by (2), and we use $\omega(G, \sigma)$ to denote the weight of σ:

$$\omega(G,\sigma) = \prod\nolimits_{(u,v)\in E} A_{\sigma(u),\sigma(v)}$$

⋆ Supported by NSF CCF-0914969.
⋆⋆ Supported by NSF CCF-1139915 and start-up funds from Columbia University.
⋆⋆⋆ Supported by NSF CCF-0914969.

G. Lin (Ed.): COCOA 2012, LNCS 7402, pp. 336–347, 2012.

Given a fixed $\mathbf{A}$, we are interested in the complexity of computing $Z_{\mathbf{A}}(G)$, where G is given as an input. Many natural combinatorial counting problems can be formulated as two-state spin systems. For example, with $(\beta, \gamma) = (0, 1)$, $Z_{\mathbf{A}}(G)$ is exactly the number of independent sets (or vertex covers) of G. The definition of $Z_{\mathbf{A}}(G)$ in (2) can also be generalized to larger matrices $\mathbf{A}$, and the problem is known as counting (weighted) graph homomorphisms [1, 2]. On the other hand, the so-called Ising model is the special case where $\beta = \gamma$.

The *exact* complexity of computing $Z_{\mathbf{A}}(G)$ has been completely solved for any fixed symmetric $\mathbf{A}$ [3–6] and even for not necessarily symmetric $\mathbf{A}$ [7–12] as part of the dichotomy theorems for general counting constraint satisfaction problems (#CSP). When specialized to two-state spin systems, $Z_{\mathbf{A}}(G)$ is #P-hard to compute exactly, except for the two restricted settings of $\beta\gamma = 1$ or $\beta = \gamma = 0$ for which cases $Z_{\mathbf{A}}(G)$ is polynomial-time computable. Consequently, the study on two-state spin systems has focused on the approximation of $Z_{\mathbf{A}}(G)$, and this is the subject of the present paper.

Following standard definitions, a fully polynomial-time approximation scheme (FPTAS) for $Z_{\mathbf{A}}(G)$ is an algorithm that, given as input a graph G as well as a parameter $\epsilon > 0$, outputs a number Z that satisfies

$$(1 - \epsilon) \cdot Z_{\mathbf{A}}(G) \leq Z \leq (1 + \epsilon) \cdot Z_{\mathbf{A}}(G) \tag{3}$$

in time $\mathrm{poly}(|G|, 1/\epsilon)$. A fully polynomial-time randomized approximation scheme (FPRAS) is a randomized algorithm that, with probability $1 - \delta$, outputs a number Z satisfying (3) in time $\mathrm{poly}(|G|, 1/\epsilon, \log(1/\delta))$.

In a seminal paper [13] Jerrum and Sinclair gave an FPRAS for $Z_{\mathbf{A}}(\cdot)$ with $\beta = \gamma > 1$. It was then further extended to the entire region of $\beta\gamma > 1$ by Goldberg, Jerrum and Paterson [14]. We call a two-state spin system *ferromagnetic* if $\beta\gamma > 1$ and *anti-ferromagnetic* if $\beta\gamma < 1$. The approximability of $Z_{\mathbf{A}}(\cdot)$ for anti-ferromagnetic systems is less well understood. Starting with counting independent sets in sparse graphs [15], the approximability of $Z_{\mathbf{A}}(\cdot)$ in bounded degree graphs is also widely studied. Significant progress has been made recently on the algorithmic side, and approximation algorithms for anti-ferromagnetic two-state spin systems have been developed [16–19], based on the technique of correlation decay introduced by Bandyopadhyay and Gamarnik [20] and Weitz [16]. Finally a unified FPTAS was found [19] to approximate $Z_{\mathbf{A}}(\cdot)$ for all anti-ferromagnetic two-state spin systems of either bounded degree graphs or general graphs, when the system satisfies a *uniqueness condition*.

The uniqueness condition is named for, and closely related to, phase transitions that occur for the Gibbs measure. It depends on not only β, γ but also the degree of the underlying graph. Such phase transitions from statistical physics are believed to frequently coincide with the transitions of computational complexity from tractability to intractability. However, there are only very few examples where this conjectured link is rigorously proved. One notable example is for the hardcore gas model (or independent set, with $\beta = 0$ and $\gamma = 1$), for which such a conjecture was rigorously proved (for almost all degree bounds) both for the algorithmic side [16] and for the hardness side [21, 22]. As discussed above [16–19], for general anti-ferromagnetic two-state spin systems, the algorithmic part of the conjecture has recently been established. In this paper, we make substantial progress on the hardness part of the conjecture.

Our Results. For $0 \le \beta, \gamma \le 1$ except at $(\beta, \gamma) = (0,0)$ and $(1,1)$, Goldberg, Jerrum and Paterson proved that the problem does not admit an FPRAS for general graphs (when there is no degree bound), unless NP = RP [14]. In their reduction, the degrees of the hard instances are unbounded. This is consistent with the uniqueness threshold conjecture. However, for any fixed β and γ in the unit square, the uniqueness condition states that there exists a finite threshold degree $\Delta(\beta, \gamma)$ [17–19], which satisfies

$$\Delta(\beta, \gamma) > \frac{1 + \sqrt{\beta\gamma}}{1 - \sqrt{\beta\gamma}} = \frac{(1 + \sqrt{\beta\gamma})^2}{1 - \beta\gamma} \ge \frac{1}{1 - \beta\gamma} \tag{4}$$

such that the system satisfies the uniqueness condition if the degree $d < \Delta(\beta, \gamma)$ and the non-uniqueness condition if $d \ge \Delta(\beta, \gamma)$. The paper [19] gives an FPTAS for graphs with degree bounded by $\Delta(\beta, \gamma)$. The conjectured coincidence of phase transition with hardness in complexity suggests that as soon as the degree of the input graph goes beyond $\Delta(\beta, \gamma)$, $Z_{\mathbf{A}}(\cdot)$ becomes hard to approximate. Towards this direction, we show that for any fixed β, γ in the unit square, the problem does not have an FPRAS if the degree of the input graph is $\Omega(\Delta(\beta, \gamma))$, unless NP = RP. Our hardness result also holds when restricted to input graphs that are regular. Formally, we prove the following theorem:

Theorem 1. *There exists a positive constant h with the following property. For any $\beta, \gamma : 0 \le \beta, \gamma \le 1$ such that $(\beta, \gamma) \ne (0,0), (1,1)$ and for any integer $d \ge h/(1 - \beta\gamma)$, there is no randomized polynomial-time algorithm that approximates $Z_{\mathbf{A}}(G)$ in d-regular graphs G with relative error $\epsilon = 10^{-4}$, unless* NP = RP.

Note the relation between our degree bound $h/(1 - \beta\gamma)$ and $\Delta(\beta, \gamma)$ from (4).

We also make progress on β, γ outside the unit square. While the uniqueness condition is *monotone* inside the unit square, its behavior outside is significantly different. (See more discussions on this difference in the appendix of the full version [23].) Without loss of generality, we consider the region defined by $\beta\gamma < 1$ with $0 < \beta < 1 < \gamma$. There is a uniqueness curve (see Figure 1), connecting the point $(1, 1)$ and the γ-axis. Above the curve, the system satisfies the uniqueness condition for any graph [18, 19]. Hence, hardness is only possible below the uniqueness curve. Furthermore, when (β, γ) is outside the unit square but below this uniqueness curve, there is only a finite range of degrees d for which the system does not satisfy the uniqueness condition. This makes it very challenging to prove a hardness result for them. Previously, the hardness was only obtained in [14] for a very tiny square $0 \le \beta \le \eta$ and $1 \le \gamma \le 1 + \eta$ where η is roughly 10^{-7}, near $(0, 1)$ corresponding to the hardcore gas model (independent set).

We prove the following hardness result for (β, γ) outside the unit square:

Theorem 2. *Given β and γ such that $0 < \beta < 1$, $\gamma > 1$ and $\beta\gamma < 1$, let*

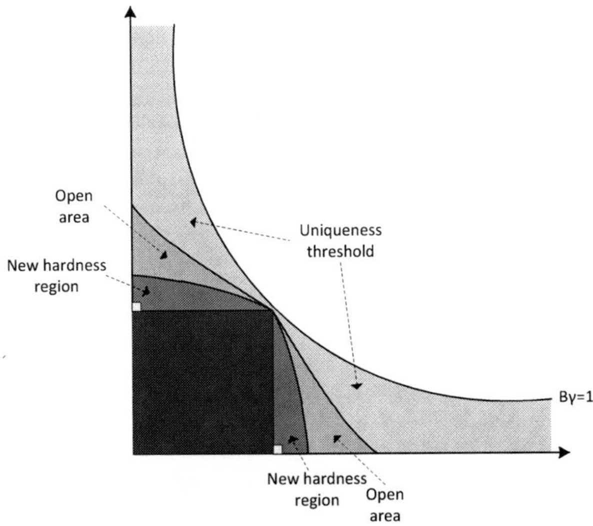

Fig. 1. The new hardness region of Theorem 2

$$\Delta' = \lceil -1/(\ln \beta + \ln \gamma) \rceil \quad and \quad \Delta^* = \lceil 1/\ln \gamma \rceil . \tag{5}$$

When $\Delta^* \geq 8000\Delta'$, there exists no randomized polynomial-time algorithm that approximates $Z_{\mathbf{A}}(G)$ in regular graphs of degree Δ^* with relative error $\epsilon = 10^{-4}$, unless NP = RP.

The new hardness region is pictured in Figure 1 [1] above. Here the two white squares are the hardness regions acquired by Goldberg, Jerrum, and Paterson [14]. Beyond the uniqueness threshold, we know that FPTAS exists. Our hardness result, Theorem 2, applies to the region between the vertical line with $\gamma = 1$ and the curve to the left of the uniqueness threshold. Let us describe this new curve in more details. We focus on the region with $0 < \beta < 1 < \gamma$ and $\beta\gamma < 1$. There is a symmetric curve for $0 < \gamma < 1 < \beta$. Near the point $(1,1)$, the condition imposed by Theorem 2 is almost linear. So the new curve is roughly a line with slope -8000 around $(1,1)$. When it approaches the line of $\beta = 0$, Δ' becomes 1 and the condition requires γ to be between 1 and roughly $1 + 1/8000$.

Moreover, using a standard translation (see the appendix of the full version [23]), we can generalize both Theorem 1 and 2 to two-state spin systems with an external field. Formally, let $\mu \geq 0$, we have the following two corollaries for

$$Z_{\mathbf{A},\mu}(G) = \sum_{\sigma:V\to\{0,1\}} \mu^{\left|\{v\in V:\sigma(v)=0\}\right|} \prod_{(u,v)\in E} A_{\sigma(u),\sigma(v)}.$$

[1] The reader should be aware that, for illustration purposes, the picture is not drawn to actual scale.

Corollary 1. *There is a positive constant h with the following property. Given any non-negative β, γ and μ with $\beta\gamma < 1$, if d is a positive integer that satisfies $\gamma \leq \mu^{\frac{1}{d}} \leq 1/\beta$ and $d \geq h/(1-\beta\gamma)$, then there exists no randomized polynomial-time algorithm that approximates $Z_{\mathbf{A},\mu}(G)$ in d-regular graphs with relative error $\epsilon = 10^{-4}$, unless* NP = RP.

Corollary 2. *Given any non-negative β, γ, μ and a positive integer d such that*

$$e^{\frac{1}{d}} \leq \gamma \cdot \mu^{-\frac{1}{d}} < e^{\frac{1}{d-1}} \quad \text{or} \quad e^{\frac{1}{d}} \leq \beta \cdot \mu^{\frac{1}{d}} < e^{\frac{1}{d-1}},$$

if d also satisfies $d \geq 8000 \lceil -1/(\ln\beta + \ln\gamma) \rceil$, then there is no randomized polynomial-time algorithm that approximates $Z_{\mathbf{A},\mu}(G)$ in d-regular graphs with relative error $\epsilon = 10^{-4}$, unless NP = RP.

Proof Outline. In the proofs of both Theorem 1 and Theorem 2, we use the phase transition that occurs in the non-uniqueness region to encode a hard-to-approximate problem. This approach has been used in previous hardness proofs for the hardcore gas model [15, 21, 24]. To this end, we reduce the approximation of E2LIN2 to the approximation of partition function in a two-state spin system. Here an instance of E2LIN2 consists of a set of variables $x_1, \ldots, x_n$ and a set of equations of the form $x_i + x_j = 0$ or 1 over $\mathbb{Z}_2$. From [25], it is NP-hard to approximate the number of satisfiable equations for E2LIN2 within any constant factor better than 11/12.

Given an E2LIN2 instance with variables $x_1, \ldots, x_n$, we use a random bipartite regular graph to encode each variable x_i. Due to the phase transition and the fact that we are in a non-uniqueness region, each of these bipartite regular graphs would be in one of two types of configurations with high probability, if sampled proportional to its weight in the partition function. This can be used to establish a correspondence between the configurations of these bipartite graphs and the assignment of the n boolean variables $x_1, \ldots, x_n$. Furthermore, we also add external connections between the random bipartite graphs according to the set of equations in the E2LIN2 instance. They contribute exponentially to the total weight in the partition function, according to the total number of equations that an assignment satisfies. Thus, a sufficiently good approximation to the partition function can be used to decode approximately the maximum number of equations that an assignment can satisfy.

Our gadget is also randomly constructed, so the probability should also be over the distribution of the gadgets. It is not hard to show that things work out beautifully if we simply substitute the expectation for the actual weight. But to make the proof rigorous, one must first obtain a sufficiently good concentration result. Such a result is unknown and could be very difficult to prove (assuming it is true), as it is already a tour-de-force in the special case for the hardcore gas model [21, 22, 24].

Instead, we use a detour: (1) We prove a lower bound for the weights of the two types of configurations we expect, guided by the phase transition; and (2) We prove that the total weight of other configurations is exponentially smaller compared to the lower bound proved in (1), with probability exponentially close to 1. The way we establish the lower bound in (1) is similar to the approach of Dyer, Frieze, and Jerrum [15]. To prove (2), they [15] used the expectation and Markov's inequality. If we use the same approach, we could not get the hardness result for bounded degree graphs in the same order of the uniqueness bound. Instead, we use a new approach for (2).

Indeed we first show a high concentration result for an expander property of the gadgets we use. Then we show that the total weight of other configurations must be exponentially small, given that the gadgets satisfy that property. This circumvented our inability to prove a complete concentration result. But we do need to prove some limited concentration results regarding the random gadget. This then led us to the hardness results for degrees in the right order conjectured according to the uniqueness threshold. It remains open whether one can use a refined version of this reduction along with the proof by Sly [21] to get the exact right bound. As discussed in the appendix of the full version [23], this random regular graph follows quite closely the property of phase transition in infinite d-ary trees, when the parameter is below or beyond the uniqueness condition.

While the high-level idea of our proofs for both Theorem 1 and Theorem 2 are quite clear and similar, it remains a challenge to work out the estimation for all ranges of parameters and at the same time, make sure that the degree is in the same order of the uniqueness bound. To this end, technically we need to use very different approaches for Theorem 1 and Theorem 2. Even within Theorem 1 itself, we need to do the estimation differently for three different subcases.

2 Proof of the Main Theorems

From now on, we will use $Z(G)$ to denote $Z_{\mathbf{A}}(G)$ whenever it is clear from the context. Given positive integers N and Δ, we use $\mathcal{H}(N,\Delta)$ to denote the following probability distribution of Δ-regular bipartite graphs $H=(U\cup V,E)$ with bipartition U,V and $|U|=|V|=N$: H is the union of Δ perfect matchings between U and V each selected independently and uniformly at random. (Because these perfect matchings are drawn independently, H may have parallel edges.)

In the proofs of both Theorem 1 and 2, we give a polynomial-time reduction from $\mathsf{E2LIN2}$ to the approximation of $Z(G)$. An instance of $\mathsf{E2LIN2}$ consists of m equations over $\mathbb{Z}_2$ in n variables $x_1,\ldots,x_n$. Each equation has exactly two variables and is of the form $x_i+x_j=b\in\{0,1\}$. Without loss of generality we may always assume $m\ge n/2$; otherwise one of the variables does not appear in any equation. Given an assignment S of the n variables $x_1,\ldots,x_n$, we use $\theta(S)$ to denote the number of equations that S satisfies and let $\theta^*=\max_S\theta(S)$. In [25] Håstad showed that it is NP-hard to estimate θ^* within any constant factor better than $11/12$.

Given an $\mathsf{E2LIN2}$ instance, we construct a random $(\Delta+\Delta')$-regular graph G as follows, with the two parameters Δ,Δ' to be specified later. This construction is used in the proofs of both Theorem 1 and 2:

Construction of G from an instance of $\mathsf{E2LIN2}$. For each variable x_i, $i\in[n]$, we let U_i and V_i denote two sets of $d_i m$ vertices each, where $d_i\ge 1$

denotes the number of equations in which x_i appears (thus, $\sum_i d_i = 2m$). Moreover, U_i and V_i can be decomposed into

$$U_i = U_{i,1} \cup \cdots \cup U_{i,d_i} \quad \text{and} \quad V_i = V_{i,1} \cup \cdots \cup V_{i,d_i}$$

where each $U_{i,k}$ and $V_{i,k}$ contains exactly m vertices. Now enumerate all the m equations in the E2LIN2 instance one by one. For each of the m equations do the following:

(1) Let $x_i + x_j = b \in \{0,1\}$ denote the current equation. Assume this is the kth time that x_i appears in an equation, and the ℓth time that x_j appears in an equation so far, where $k \in [d_i]$ and $\ell \in [d_j]$. Denote the m vertices in $U_{i,k}$ by $\{u_1, \ldots, u_m\}$, vertices in $V_{i,k}$ by $\{v_1, \ldots, v_m\}$, vertices in $U_{j,\ell}$ by $\{u'_1, \ldots, u'_m\}$ and vertices in $V_{j,\ell}$ by $\{v'_1, \ldots, v'_m\}$. All these vertices have degree 0 at this moment. If $b = 0$, we add Δ' parallel edges between (u_s, v'_s) and (v_s, u'_s), for each $s \in [m]$; or if $b = 1$, we add Δ' parallel edges between (u_s, u'_s) and (v_s, v'_s), for each $s \in [m]$.

By the end of this step, every vertex has degree Δ'. In the next step,

(2) For each $i \in [n]$, we add a bipartite graph $H_i = (U_i \cup V_i, E_i)$ drawn from $\mathcal{H}(d_i m, \Delta)$.

This finishes the construction and we get a $(\Delta + \Delta')$-regular graph G with $4m^2$ vertices.

We need the following notation. Given an assignment $\sigma : V(G) \to \{0,1\}$, we use $U_i(\sigma)$ to denote the number of vertices $u \in U_i$ with $\sigma(u) = 0$, and use $V_i(\sigma)$ to denote the number of $v \in V_i$ with $\sigma(v) = 0$.

Proof (of Theorem 1). Without loss of generality, assume $\beta, \gamma : 0 \le \beta \le \gamma \le 1$. We can also assume $\beta > 0$, as the tight hardness to the exact uniqueness bound for $\beta = 0$ has been shown in [18], by generalizing the tight hardness result for the hardcore model [21, 22].

Given an assignment S of the n variables, we let $Z(G, S)$ denote the sum of $\omega(G, \sigma)$ over assignments $\sigma : V(G) \to \{0,1\}$ that satisfy for each $i \in [n]$,

$$U_i(\sigma) \le V_i(\sigma) \quad \text{if } x_i = 0 \text{ in } S; \text{ or} \quad U_i(\sigma) \ge V_i(\sigma) \quad \text{if } x_i = 1 \text{ in } S. \tag{6}$$

From definition we have $Z(G, S) \le Z(G) \le \sum_S Z(G, S)$. We need the following key lemma. Its proof can be found in the full version [23]:

Lemma 1. *There exists a positive constant h with the following property: For any β and $\gamma : 0 < \beta \le \gamma \le 1$ with $(\beta, \gamma) \ne (1,1)$ and for any $\Delta^* \ge h/(1 - \beta\gamma)$, there are $D > 1$, $C > 0$ and positive integers Δ and Δ' with $\Delta + \Delta' = \Delta^*$, that satisfy the following property: given any input instance of* E2LIN2 *with n variables $x_1, \ldots, x_n$ and m equations, except for probability $\le \exp(-\Omega(m))$, the Δ^*-regular graph G constructed with parameters Δ and Δ' satisfies*

$$C^{m^2} \cdot D^{m\theta(S)} \leq Z(G,S) \leq C^{m^2} \cdot D^{m\left(\theta(S)+0.03m\right)}, \tag{7}$$

for any assignment S of the n variables.

Given β, γ and Δ^*, we let C, D, Δ and Δ' denote the constants that satisfy the condition in Lemma 1. Then given an input instance of E2LIN2, (7) holds with probability $1 - \exp(-\Omega(m))$.

Now assume (7) holds. We use θ^* to denote the maximum number of consistent equations and use S^* to denote an assignment that satisfies θ^* equations. We also use Y to denote an estimate of $Z = Z(G)$, where $|Y/Z - 1| \leq \epsilon = 10^{-4}$. From (7) and $Z(G,S^*) \leq Z(G) \leq \sum_S Z(G,S)$, we get

$$(1+\epsilon) \cdot 2^n \cdot C^{m^2} \cdot D^{m(\theta^*+0.03m)} \geq Y \geq (1-\epsilon) \cdot C^{m^2} \cdot D^{m\theta^*} \tag{8}$$

Using Y, we set

$$Y' = \frac{\ln Y - \ln(1+\epsilon) - n\ln 2 - m^2 \ln C - 0.03m^2 \ln D}{m \ln D}$$

and we get $Y' \leq \theta^*$ since $\ln D > 0$. We finish the proof by showing that $Y' > (11/12) \cdot \theta^*$. By (8) we get

$$Y' \geq \theta^* - \frac{\ln(1+\epsilon) - \ln(1-\epsilon) + n\ln 2 + 0.03m^2 \ln D}{m \ln D}$$

As $\theta^* \geq m/2$ and $m \geq n/2$, when m is large enough, $Y' > (11/12) \cdot \theta^*$ and the theorem is proven.

Next, we prove Theorem 2:

Proof (of Theorem 2). For β, γ with $0 < \beta < 1 < \gamma$ and $\beta\gamma < 1$, let Δ' and Δ^* be the two positive integers defined in (5) which satisfy $\Delta^* \geq 8000\Delta'$. We set $\Delta = \Delta^* - \Delta'$. Given any input instance of E2LIN2 with n variables and m equations, we let G denote the Δ^*-regular graph constructed using Δ and Δ'.

First we show that to get a good approximation of $Z(G)$, with high probability it suffices to sum $\omega(G,\sigma)$ only over assignments σ satisfying the following:

$$\min\Big(U_i(\sigma), V_i(\sigma)\Big) \leq \lambda d_i m, \quad \text{for all } i \in [n], \text{ where } \lambda = 9 \times 10^{-5}. \tag{9}$$

We let Σ denote the set of all such assignments. Formally we prove the following key lemma in Section 3:

Lemma 2. *Let G be the graph constructed from an* E2LIN2 *instance with n variables $x_1, \ldots, x_n$ and m equations, with parameters Δ, Δ'. Then with probability $1 - \exp(-\Omega(m^{1/3}))$, it satisfies*

$$\sum_{\sigma \in \Sigma} \omega(G,\sigma) \leq Z(G) \leq \big(1 + o(1)\big) \cdot \sum_{\sigma \in \Sigma} \omega(G,\sigma). \tag{10}$$

Next, given an assignment S over the n variables we use $Z_\Sigma(G,S)$ to denote the sum of $\omega(G,\sigma)$ over all assignments $\sigma \in \Sigma$ that satisfy (6) for all $i \in [n]$. We prove the following lemma in the full version [23]:

Lemma 3. *There are $C > 0$ and $D > 1$ satisfying the following property: given any instance of* E2LIN2 *with n variables and m equations, the Δ^*-regular graph G constructed with parameters Δ and Δ' satisfies*

$$C^{m^2} \cdot D^{m\theta(S)} \le Z_\Sigma(G,S) \le C^{m^2} \cdot D^{m\left(\theta(S)+0.04m\right)}, \tag{11}$$

for any assignment S of the n variables.

Let $\theta^* \ge m/2$ denote the maximum number of consistent equations, and let S^* denote an assignment that satisfies θ^* equations. From these two lemmas we have with high probability that

$$\begin{aligned} C^{m^2} \cdot D^{m\theta^*} \le Z_\Sigma(G,S^*) \le Z(G) &\le \big(1+o(1)\big) \cdot \sum\nolimits_S Z_\Sigma(G,S) \\ &\le \big(1+o(1)\big) \cdot 2^n \cdot C^{m^2} \cdot D^{m(\theta^*+0.04m)} \end{aligned}$$

Theorem 2 then follows from the same argument used in Theorem 1.

3 Proof of Lemma 2

Recall that β and γ satisfy $\beta,\gamma : 0 < \beta < 1 < \gamma$ and $\beta\gamma < 1$. Let Δ' and Δ^* be the two positive integers defined in Theorem 2, with $\Delta^* \ge 8000\Delta'$. From their definitions, we have $(\beta\gamma)^{\Delta'} \le 1/e$ and $\gamma^{\Delta^*} \ge e$. Set $\Delta = \Delta^* - \Delta' \ge 7999\Delta' \ge 7999$. By the definition of Δ^*, we have $e > \gamma^{\Delta^*-1} \ge \gamma^{\Delta}$ and thus, $\gamma < 1.001$.

Given an E2LIN2 instance with n variables $x_1, \ldots, x_n$ and m equations, we use G to denote the Δ^*-regular graph constructed with parameters Δ and Δ', where $\Delta^* = \Delta + \Delta'$. We let H_i denote the bipartite graph in G that corresponds to x_i and use $U_i \cup V_i$ to denote its vertices, with $|U_i| = |V_i| = d_i m$.

Before working on G and H_i, we start by proving a property that a bipartite graph sampled from the distribution $\mathcal{H}(N,\Delta)$ satisfies with very high probability. Let H be a bipartite graph drawn from $\mathcal{H}(N,\Delta)$ for some $N \ge 1$ and Δ defined above, with $2N$ vertices $U \cup V$. We also use $\rho : U \cup V \to \{0,1\}$ to denote an assignment and call it an (a,b)-assignment for some $a,b \in T_N$, where

$$T_N = \big\{0, 1/N, 2/N, \ldots, (N-1)/N, 1\big\}$$

if $|u \in U : \rho(u) = 0| = aN$ and $|v \in V : \rho(v) = 0| = bN$. We also use $\mathcal{I}_N(a,b)$, where $a,b \in T_N$, to denote the set of all such (a,b)-assignments, and let

$$Z_{a,b}(H) = \sum_{\rho \in \mathcal{I}_N(a,b)} \omega(H,\rho) \cdot \gamma^{\Delta'(2-a-b)N} \tag{12}$$

with Δ' as defined above. We are interested in the expectation of $Z_{a,b}(H)$ when $\min(a,b) \ge \lambda = 9 \times 10^{-5}$:

Lemma 4. *For large enough N and $a, b \in T_N$ such that* $\min(a,b) \geq \lambda$, *we have*

$$\mathbf{E}_{H \leftarrow \mathcal{H}(N,\Delta)}\Big[Z_{a,b}(H)\Big] \leq \exp\big(1.21 \cdot N\big).$$

The proof of Lemma 4 can be found in the full version [23]. By Lemma 4 we can impose the following condition on the graph G constructed from the input instance of E2LIN2: For all $i \in [n]$ and all $a, b \in T_{d_i m}$ with $\min(a,b) \geq \lambda$,

$$Z_{a,b}(H_i) \leq \exp\big(1.22 \cdot d_i m\big). \tag{13}$$

Using Lemma 4, Markov's inequality and the union bound, it is easy to show that G satisfies this condition with probability $1 - \exp(-\Omega(m))$.

In the rest of the proof, we prove that G satisfies (10) whenever it satisfies (13). Lemma 2 then follows immediately.

We assume that G satisfies (13). Then to prove (10), we randomly sample an assignment σ with probability proportional to $\omega(G, \sigma)$. (10) follows if we can show that σ satisfies (9) with probability $1 - o(1)$. For this purpose we need the following lemmas that give us properties that σ satisfies with high probability. Given any set L of vertices in G, we let

$$N_\sigma(L) = \big\{v \in L : \sigma(v) = 0\big\}.$$

Also recall the definition of $U_{i,k}$ and $V_{i,k}$ in the construction of G. Let x_i be a variable and let $k \in [d_i]$, then we have

Lemma 5. *Let σ be an assignment drawn according to its weight. Then for any $i \in [n]$ and any $k \in [d_i]$, except for probability* $\exp(-\Omega(m^{1/3}))$, *we have*

$$\big|N_\sigma(U_{i,k})\big| < \frac{1}{1+e} \cdot \Big(1 + m^{-1/3}\Big) \cdot \big|U_{i,k}\big|.$$

Proof. Pick any partial assignment σ' over vertices of G except those in $U_{i,k}$. Conditioned on σ', it is easy to see that the values of vertices in $U_{i,k}$ are independent. Each vertex in $U_{i,k}$ has $\Delta + \Delta'$ neighbors, each of which contributes a vertex weight of either β or 1 if it is assigned 0, and either 1 or γ if it is assigned 1. Since $\gamma < 1/\beta$, the total weight for assignment 1 is at least $\gamma^{\Delta+\Delta'} \geq e$ times the weight for assignment 0. The lemma follows from the Chernoff bound.

Given an assignment σ, we use σ_i to denote its restriction over vertices in H_i, and σ_{-i} to denote its partial assignment over vertices in G except H_i. We let $M_{\sigma_{-i}}(U_i)$ denote the subset of U_i whose unique neighbor outside of H_i is assigned 1. Using Lemma 5 and the union bound, we have

Corollary 3. *Let σ be an assignment drawn according to its weight. Except for probability* $\exp(-\Omega(m^{1/3}))$, *we have*

$$\big|M_{\sigma_{-i}}(U_i)\big| \geq \left(\frac{e}{1+e} - O\big(m^{-1/3}\big)\right) \cdot \big|U_i\big|, \qquad \textit{for all } i \in [n]. \tag{14}$$

It is also clear that Lemma 5 and Corollary 3 also hold for $V_{i,k}$ and V_i, respectively, by symmetry. Now we are ready to prove Lemma 2. Let $\sigma = (\sigma_i, \sigma_{-i})$ be an assignment drawn from this distribution. Recall the definition of Σ below (9). Then by Corollary 3 we have

$$\Pr\big[\sigma \notin \Sigma\big] \le \exp(-\Omega(m^{1/3})) + \Pr\big[\sigma \notin \Sigma \mid \sigma_{-i} \text{ satisfies (14) for both } U_i \text{ and } V_i \text{ and for all } i \in [n]\big] \tag{15}$$

To prove an upper bound for (15) we fix σ_{-i} to be any partial assignment over the vertices of G except those of H_i, which satisfies (14) for both U_i and V_i. Then it suffices to prove that the sum of $\omega(G, \sigma)$ over all $\sigma \in \Sigma$ that are consistent with σ_{-i}, denoted by Z_1, is exponentially larger than the sum of $\omega(G, \sigma)$ over all $\sigma \notin \Sigma$ that are consistent with σ_{-i}, denoted by Z_2.

Let $\omega(\sigma_{-i})$ denote the product of the edge weights in σ_{-i} over all edges that have no vertex in H_i. By the definition of $Z_{a,b}(H)$ in (12), we have

$$Z_2 \le \omega(\sigma_{-i}) \sum_{a,b \in T_{d_i m} :\, a,b \ge \lambda} Z_{a,b}(H_i) \le \omega(\sigma_{-i}) \cdot (d_i m)^2 \cdot \exp\big(1.22 \cdot d_i m\big) \tag{16}$$

where the second inequality follows from (13). To get a lower bound for Z_1, let

$$L = \big|M_{\sigma_{-i}}(U_i)\big| \quad \text{and} \quad R = \big|M_{\sigma_{-i}}(V_i)\big|.$$

Consider all the σ that are consistent with σ_{-i} and $U_i(\sigma) = 0$. This gives us

$$Z_1 \ge \omega(\sigma_{-i}) \cdot \gamma^{\Delta' L} \cdot (1 + \gamma^{\Delta+\Delta'})^R \cdot (\beta^{\Delta'} + \gamma^{\Delta})^{d_i m - R}.$$

Plugging in $\gamma^{\Delta+\Delta'} \ge e$, $\gamma^{\Delta} \ge e^{7999/8000}$ and the lower bound in (14), we get

$$Z_1 \ge \omega(\sigma_{-i}) \cdot \exp\big(1.22897 \cdot d_i m\big)$$

and the lemma follows from (16).

References

1. Lovász, L.: Operations with structures. Acta Mathematica Hungarica 18, 321–328 (1967)
2. Hell, P., Nešetřil, J.: Graphs and Homomorphisms. Oxford University Press (2004)
3. Dyer, M., Greenhill, C.: The complexity of counting graph homomorphisms. In: Proceedings of the 9th International Conference on Random Structures and Algorithms, pp. 260–289 (2000)
4. Bulatov, A., Grohe, M.: The complexity of partition functions. Theoretical Computer Science 348(2-3), 148–186 (2005)
5. Goldberg, L., Grohe, M., Jerrum, M., Thurley, M.: A complexity dichotomy for partition functions with mixed signs. SIAM Journal on Computing 39(7), 3336–3402 (2010)
6. Cai, J.Y., Chen, X., Lu, P.: Graph homomorphisms with complex values: A dichotomy theorem. In: Proceedings of the 37th Colloquium on Automata, Languages and Programming (2010); To appear in SIAM Journal on Computing

7. Dyer, M., Goldberg, L., Paterson, M.: On counting homomorphisms to directed acyclic graphs. Journal of the ACM 54(6) (2007)
8. Cai, J.Y., Chen, X.: A decidable dichotomy theorem on directed graph homomorphisms with non-negative weights. In: Proceedings of the 51st Annual IEEE Symposium on Foundations of Computer Science, pp. 437–446 (2010)
9. Bulatov, A.: The complexity of the counting constraint satisfaction problem. In: Proceedings of the 35th International Colloquium on Automata, Languages and Programming, pp. 646–661 (2008)
10. Dyer, M., Richerby, D.: On the complexity of #CSP. In: Proceedings of the 42nd ACM Symposium on Theory of Computing, pp. 725–734 (2010)
11. Cai, J.Y., Chen, X., Lu, P.: Non-negatively weighted #CSP: An effective complexity dichotomy. In: Proceedings of the 26th Annual IEEE Conference on Computational Complexity, pp. 45–54 (2011)
12. Cai, J.Y., Chen, X.: Complexity of counting CSP with complex weights. In: Proceedings of the 44th ACM Symposium on Theory of Computing (2012)
13. Jerrum, M., Sinclair, A.: Polynomial-time approximation algorithms for the ising model. SIAM Journal on Computing 22(5), 1087–1116 (1993)
14. Goldberg, L., Jerrum, M., Paterson, M.: The computational complexity of two-state spin systems. Random Structures and Algorithms 23(2), 133–154 (2003)
15. Dyer, M., Frieze, A., Jerrum, M.: On counting independent sets in sparse graphs. SIAM Journal on Computing 31, 1527–1541 (2002)
16. Weitz, D.: Counting independent sets up to the tree threshold. In: Proceedings of the 38th Annual ACM Symposium on Theory of Computing, pp. 140–149 (2006)
17. Sinclair, A., Srivastava, P., Thurley, M.: Approximation algorithms for two-state anti-ferromagnetic spin systems on bounded degree graphs. In: Proceedings of the 23rd Annual ACM-SIAM Symposium on Discrete Algorithms (2012)
18. Li, L., Lu, P., Yin, Y.: Approximate counting via correlation decay in spin systems. In: Proceedings of the 23rd Annual ACM-SIAM Symposium on Discrete Algorithms (2012)
19. Li, L., Lu, P., Yin, Y.: Correlation decay up to uniqueness in spin systems. arXiv:1111.7064 (2011)
20. Bandyopadhyay, A., Gamarnik, D.: Counting without sampling: Asymptotics of the log-partition function for certain statistical physics models. Random Structures and Algorithms 33, 452–479 (2008)
21. Sly, A.: Computational transition at the uniqueness threshold. In: Proceedings of the IEEE 51st Annual Symposium on Foundations of Computer Science (2010)
22. Galanis, A., Ge, Q., Štefankovič, D., Vigoda, E., Yang, L.: Improved inapproximability results for counting independent sets in the hard-core model. In: Proceedings of the 15th International Workshop on Randomization and Computation, pp. 567–578 (2011)
23. Cai, J.Y., Chen, X., Guo, H., Lu, P.: Inapproximability after uniqueness phase transition in two-spin systems. University of Wisconsin, Madison CS Technical Report (2011), http://digital.library.wisc.edu/1793/61488
24. Mossel, E., Weitz, D., Wormald, N.: On the hardness of sampling independent sets beyond the tree threshold. Probability Theory and Related Fields 143, 401–439 (2009)
25. Håstad, J.: Some optimal inapproximability results. Journal of the ACM 48, 798–859 (2001)

Dynamic Programming for a Biobjective Search Problem in a Line

Luís Paquete[1], Mathias Jaschob[2], Kathrin Klamroth[2], and Jochen Gorski[2]

[1] CISUC, Department of Informatics Engineering, University of Coimbra, Portugal
paquete@dei.uc.pt
[2] Department of Mathematics and Natural Sciences,
University of Wuppertal, Germany
klamroth@math.uni-wuppertal.de

Abstract. In this article we study the performance of multiobjective dynamic programming for a biobjective combinatorial optimization problem under several formulations. Based on our theoretical and computational results we argue that a clever definition of the recursion, allowing for strong dominance criteria, is crucial in the design of a multiobjective dynamic programming algorithm.

Keywords: Multiobjective combinatorial optimization, dynamic programming, shortest path problems, knapsack problems.

1 Introduction

Motivated by search applications like, for example, the recent search for the location of the Air France flight AF 447 underwater wreckage [12], we consider biobjective search problems aiming at the simultaneous maximization of the success probability and minimization of the search time. We assume that the search space is partitioned into square or rectangular cells and is equipped with a probability map. In this situation we consider the problem of planning a mission of an autonomous vehicle like, for example, an autonomous underwater vehicle (AUV) for detecting an underwater wreckage. We assume that the search is performed along lines, and that each line is partitioned into several segments having known success rates (or scores) according to the underlying probability map [6,11]. Note that even though this is a common approach in practical search applications, it is a simplifying assumption which *a priori* excludes more complex search patterns. Sodhi et al. [11] gives a continuous formulation of this problem that can be solved analytically; to the knowledge of the authors, our model is the first discrete formulation of this problem.

In a mission, the AUV starts from its mother ship at one end of the line, and travels along the line back and forth while scanning in a subset of the given segments. The objective is to find those missions that maximize the score and at the same time minimize the total travel time, where we assume that the travel time of an AUV is longer when scanning the seabed than without a scan. It is easy to see that we only need to consider missions having exactly one turning

G. Lin (Ed.): COCOA 2012, LNCS 7402, pp. 348–359, 2012.

point, and that the AUV performs a scan on the last (furthest) segment, i.e., it does not travel further than necessary.

In Section 2, we provide three equivalent formulations of the *biobjective search problem in a line* (BSPL) and prove intractability and NP-completeness of BSPL. The first two models are integer linear programming formulations with different constraint sets. This formulation is closely related to the 0-1 knapsack problem and is explored for the complexity proof in this article. The third model utilizes the equivalence to a biobjective shortest path problem. Dynamic programming formulations are presented for all three models in Section 3, and numerical results comparing the different approaches are described in Section 4. Our results indicate that the computational efficiency of multiobjective dynamic programming is to a large extent determined by the strength of the applied dominance operator. The paper is concluded in Section 5 with a short summary and some research ideas.

2 Problem Formulations

We define an instance of the BSPL by a quadruple (n, a, b, s), where $n \in \mathbb{N} = \{0, 1, 2, \dots\}$ is the number of line segments, $a_i \in \mathbb{N}$ is the time to travel through and search in segment i, $b_i \in \mathbb{N}$ is the time to travel through segment i, and $s_i \in \mathbb{N}$ is the score when searching in segment i, for $i = 1, \dots, n$. We assume that $n > 2$ (the cases $n = 1, 2$ are trivial), $a_i > b_i > 0$ and $s_i \geq 0$, $i = 1, \dots, n$. In the following we describe three possible formulations of this problem.

2.1 Formulation (BSPL-1)

We introduce the first integer linear programming formulation as follows:

$$\begin{aligned}
\min \quad T(x,z) &= \sum_{i=1}^{n} (a_i - b_i)x_i + 2b_i z_i \\
\max \quad S(x) &= \sum_{i=1}^{n} s_i x_i \\
\text{s.t.} \quad z_i &\geq x_i, \ i = 1, \dots, n \\
z_i &\geq z_{i+1}, \ i = 1, \dots, n-1 \\
x_i, z_i &\in \{0, 1\}, \ i = 1, \dots, n \ .
\end{aligned}$$

The agent (for example, an AUV) perfoms a search in segment i if $x_i = 1$, otherwise, $x_i = 0$, and it travels through segment i if $z_i = 1$, otherwise, $z_i = 0$. Hence, travelling through segment i yields time $2b_i$ and null score whereas searching in segment i yields time $a_i + b_i$ and score s_i. (Note that this contains the return trip on which no search is performed.) The first class of constraints $z_i \geq x_i$ ensures that the agent travels through a segment i when it is searching in it. The second class of constraints $z_i \geq z_{i+1}$ ensures that the agent has to travel through segments $1, \dots, i-1$ in order to travel through segment i.

Let X denote the set of feasible solutions according to Formulation (BSPL-1). We will mention that a pair $(x, z) \in X$ is a *mission*. The image of X in the time-score objective space is denoted by $Y \subseteq \mathbb{N}^2$. We say that a mission (x, z) *dominates* another mission (x', z') if and only if $T(x, z) \leq T(x', z')$ and $S(x) \geq S(x')$, with at least one strict inequality; if strict inequality holds for both objectives, then (x, y) *strictly dominates* (x', z'). If there is no mission that dominates (x, z), then we say that (x, z) is an *efficient mission*; in case there exists no mission that strictly-dominates (x, z), then this mission becomes *weakly efficient*. The set of all efficient missions is denoted by X^E. An efficient mission is called *supported efficient*, if it is a minimizer of a non-trivial weighted sum problem with the objectives $T(x, z)$ and $-S(x)$. We say that $y \in Y$ is *nondominated* if there is some efficient mission $(x, z) \in X^E$ such that $y = (T(x, z), S(x))$, according to Formulation (BSPL-1). We denote the set of all nondominated vectors, the *nondominated set*, by Y^{ND}. We refer to [5] for a more detailed introduction to multiobjective optimization.

2.2 Formulation (BSPL-2)

The second formulation of BSPL reduces the set of feasible missions in the formulation above, since there may exist missions where the agent travels further than necessary, as shown in the following example.

Example 1. Let (n, a, b, s) denote an instance according to Formulation (BSPL-1) with $n=3$. Then $(x, z)=((1, 1, 0), (1, 1, 0))$ dominates $(x', z')=((1, 1, 0), (1, 1, 1))$.

Therefore, we only need to consider a subset of missions whose last segment to be searched in coincides with the last segment to be travelled through. Let $\ell_z := \max\{0, i \mid z_i = 1, i = 1, \ldots, n\}$ denote the index of the last segment the agent is travelling through. The feasible set of this new formulation is defined as follows

$$\begin{aligned} \overline{X} := \{(x, z) \in X \mid \ell_z < n,\ x_{\ell_z} = 1 \text{ and } z_k = 0 \text{ for } \ell_z < k \leq n\} \\ \cup \{(x, z) \in X \mid \ell_z = n,\ x_{\ell_z} = 1\}. \end{aligned}$$

We extend Formulation (BSPL-1) with the following set of constraints:

$$\sum_{i=1}^{j} z_i \leq j - 1 + \sum_{i=j}^{n} x_i,\ j = 1, \ldots, n.$$

The nondominated set is denoted by $\overline{Y}^{ND}$. The following result establishes the relation between the two formulations.

Lemma 1. *If mission $(x, z) \in X \setminus \overline{X}$, then there exists a mission $(x', z') \in \overline{X}$, such that $T(x', z') < T(x, z)$ and $S(x', z') \geq S(x, z)$.*

Proof. It holds that $b_i > 0$, $i = 1, \ldots, n$. □

From Lemma 1, we state the following equality.

Theorem 1. $\overline{Y}^{ND} = Y^{ND}$.

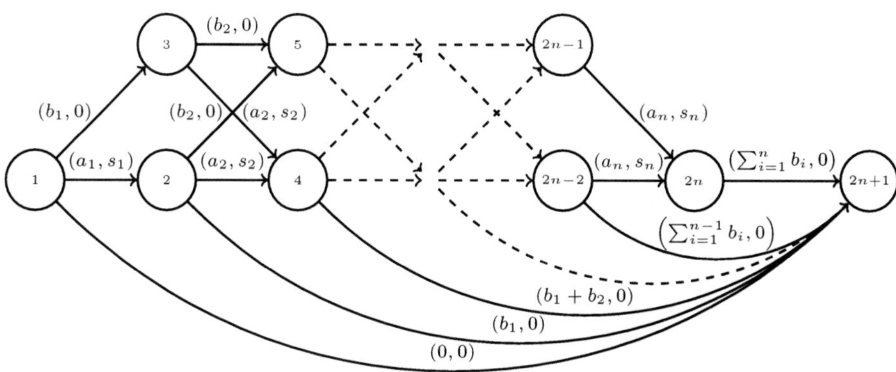

Fig. 1. The network of Formulation (BSPL-3)

2.3 Formulation (BSPL-3)

The third formulation recasts Formulation (BSPL-2) as a biobjective shortest path problem [7]. Let (n, a, b, s) be an instance of BSPL. Let G be a network with vertex set $V = \{1, 2, \ldots, 2n + 1\}$, arc set A and weight function $w = (w_t, w_s) : A \to \mathbb{N}^2$, representing time and score, respectively. The weight function for each arc in G, for $n > 1$, is defined as follows. Let $c_i := c_{i-1} + b_i$, for $i = 1, \ldots, n$ and $c_0 := 0$; for the outgoing arcs of odd nodes $3, 5, \ldots, 2n - 3$:

$$w(2i - 1, 2i) := (a_i, s_i), \quad w(2i - 1, 2i + 1) := (b_i, 0),$$

for $i = 1, \ldots, n - 1$; for the outgoing arcs of even nodes $2, 4, \ldots, 2n - 2$, $n > 2$:

$$w(2i-2, 2i) := (a_i, s_i), \quad w(2i-2, 2i+1) := (b_i, 0), \quad w(2i-2, 2n+1) := (c_{i-1}, 0),$$

for $i = 2, \ldots, n - 1$; for the remaining arcs:

$$w(2n - 2, 2n) = w(2n - 1, 2n) := (a_n, s_n),$$

$$w(2n - 2, 2n + 1) := (c_{n-1}, 0), \quad w(2n, 2n + 1) := (c_n, 0).$$

Figure 1 illustrates the resulting network, where each mission $(x, z) \in \overline{X}$ corresponds to a directed path from node 1 to node $2n+1$. We will identify a directed path p from node 1 to node $2n + 1$ in G both by a sequence of arcs and by a sequence of nodes. Note that the agent searches in a segment when an even node is used whereas it travels through it when an odd node is used. Hence, the third formulation is as follows:

$$\begin{aligned} \min \quad T_G(p) &= \sum_{\alpha \in p} w_t(\alpha) \\ \max \quad S_G(p) &= \sum_{\alpha \in p} w_s(\alpha) \\ \text{s.t.} \quad p &\in P \end{aligned}$$

where P denotes the set of all directed paths from node 1 to node $2n+1$ and $\alpha \in p$ denotes an arc in path p. We will denote the nondominated set by Y_G^{ND}. We state the equivalence between Formulations (BSPL-2) and (BSPL-3) in the following theorem.

Theorem 2. $Y_G^{ND} = \overline{Y}^{ND}$.

Proof. Let $(x, z) = (\mathbf{0}, \mathbf{0})$ and $p = (1, 2n+1)$. Then, $T(x, z) = T_G(p) = 0$ and $S(x) = S_G(p) = 0$. For the remaining cases, we prove the following statements:

$Y_G^{ND} \supseteq \overline{Y}^{ND}$: For a given $(x', z') \in \overline{X}$, let $k := \sum_{i=1}^{n} z'_i \geq 1$. Define a directed path $q = (1, v_1, \ldots, v_k, 2n+1)$ by setting $v_i := 2i$ if $x'_i = 1$ and $v_i := 2i+1$ if $x'_i = 0$, for $i = 1, \ldots, k$. By construction $T_G(q) = T(x', z')$, $S_G(q) = S(x')$, and $q \in P$ hold.

$Y_G^{ND} \subseteq \overline{Y}^{ND}$: Let $q \in P \setminus \{p\}$. Then, $q = (1, v_1, \ldots, v_k, 2n+1)$ for some $k \geq 1$. Set $\bar{x}'_i := 1$ if $v_i = 2i$ and $\bar{x}'_i := 0$ if $v_i = 2i+1$, for $i = 1, \ldots, k$, and $\bar{x}'_i := 0$ for $i = k+1, \ldots, n$. By construction $T(\bar{x}', \bar{z}') = T_G(q)$, $S(\bar{x}') = S_G(q)$ and $(\bar{x}', \bar{z}') \in \overline{X}$ hold.

□

2.4 Complexity Results

In the following, we show that the problem is intractable (see [5]):

Theorem 3. *Sets Y^{ND}, $\overline{Y}^{ND}$ and Y_G^{ND} can be exponentially large in n.*

Proof. Given Theorems 1 and 2, we only need to show it for $\overline{Y}^{ND}$. We give an instance (n, a, b, s) for which $|\overline{Y}^{ND}| = 2^n$. Let $a_i := n2^{i+1}$, $b_i := 1$, and $s_i := 2^{i-1}$, for $i = 1, \ldots, n$. Let $\overline{X} := \{(x^j, z^j)\}_{j=1}^{2^n}$, for which it holds that $\sum_{i=1}^{n} 2^{i-1} x_i^j = j - 1$. Hence, there is a one-to-one correspondence between elements in $\overline{X}$ and integers $0, 1, \ldots, 2^n - 1$. Then, the following holds:

$$S(x^j) = \sum_{i=1}^{n} s_i x_i^j = \sum_{i=1}^{n} 2^{i-1} x_i^j = j - 1$$

$$T(x^j, z^j) = \sum_{i=1}^{n} a_i x_i^j - b_i x_i^j + 2b_i z_i^j = 4n(j-1) + \sum_{i=1}^{n} 2z_i^j - x_i^j .$$

Since the following inequalities hold

$$0 \leq \sum_{i=1}^{n} 2z_i^j - x_i^j \leq 2n,$$

for $j = 1, \ldots, 2^n$, we have that $S(x^j) < S(x^{j+1})$ and $T(x^j, z^j) < T(x^{j+1}, z^{j+1})$, $j = 1, \ldots, 2^n - 1$. Hence, all elements in $\overline{X}$ are nondominated. □

In addition, we show that Formulation (BSPL-1) is also NP-complete. We introduce the decision version according this formulation.

Definition 1. *(Decision version of Formulation (BSPL-1)) For given* $n > 2, T,$ $S \in \mathbb{N}$ *and* $a, b, s \in \mathbb{N}^n$ *satisfying* $a_i > b_i > 0, s_i \geq 0, i = 1, \ldots, n,$ *does there exist* $x, z \in \{0,1\}^n$ *with* $z_i \geq x_i, i = 1, \ldots, n,$ *and* $z_{i+1} \geq z_i, i = 1, \ldots, n-1,$ *such that*

$$\sum_{i=1}^{n} (a_i - b_i)x_i + 2b_i z_i \leq T \ \ and \ \ \sum_{i=1}^{n} s_i x_i \geq S\ ?$$

We introduce the decision version of the 0-1 knapsack problem (see, e.g., [9]).

Definition 2. *(Decision version of the 0-1 knapsack problem) For given* $n,$ $D, E \in \mathbb{N}$ *and* $d, e \in \mathbb{N}^n$ *satisfying* $d_i, e_i > 0,\ i = 1, \ldots, n,$ *does there exists* $\tilde{x} \in \{0,1\}^n$ *such that*

$$\sum_{i=1}^{n} d_i \tilde{x}_i \leq D \ \ and \ \ \sum_{i=1}^{n} e_i \tilde{x}_i \geq E\ ?$$

Finally, we state the following result.

Theorem 4. *Problem Formulation (BSPL-1) is* NP*-complete.*

Proof. We give a reduction from the decision version of 0-1 knapsack problem to the decision version of Formulation (BSPL-1): Let $a_i := 1 + (2n+1)d_i$, $b_i := 1$, $s_i := e_i$, for $i = 1, \ldots, n$, $T := 2n + (2n+1)D$ and $S := E$. □

This result holds for any of the three formulations of the problem.

3 Algorithms

In this section, three algorithms are introduced that are based on multiobjective dynamic programming. This approach consists of a sequential process based on the definition of states, recurrence equations and dominance relations [3,8]. It has been mainly applied to multiobjective knapsack problems (see, e.g, [2,10,13]). For a multiobjective knapsack problem with n items, the algorithm generates, at each iteration k, a set of states, each of which representing a solution to the problem involving k first items only, for $k = 1, ..., n$. Then, dominance relations are used to discard those states that cannot lead to other states that represent efficient solutions. A similar approach will be used for solving the BSPL according to the three different formulations given in Section 2.

3.1 Notation and Definitions

The sets

$$M^i := \{(T(x,z), S(x), \ell_z) \mid (x,z) \in X, \ell_z \leq i\}$$

correspond to the missions that consist of travelling at most to the end of segment i, $i = 1, \ldots, n$. Let $\ell = (\ell_T, \ell_S, \ell_z) \in M^i$ be called a *state*. By definition,

$$\{(0,0,0)\} = M^0 \subseteq \ldots \subseteq M^n =: M.$$

For the sake of the explanation, given two states $\ell, \ell' \in M$, ℓ is *dominated* by ℓ' ($\ell \preceq \ell'$) if $\ell_T \geq \ell'_T$ and $\ell_S \leq \ell'_S$, and at least one strict inequality holds. In the following, we introduce the notion of *extension* of a state.

Definition 3. *A state* $ext(\ell) = (ext(\ell)_T, ext(\ell)_S, ext(\ell)_z) \in M$ *is an* extension *of a state* $\ell \in M^i, i < n$, *if*

$$ext(\ell) = \left(\ell_T + \sum_{k \in K}(a_k - b_k) + 2\sum_{k=\ell_z+1}^{j} b_k, \ell_S + \sum_{k\in K} s_k, j\right)$$

where $j \in \{i+1, \ldots, n\}$ *and* $K \subseteq \{i+1, \ldots, j\}$.

Index j denotes the last travelled segment and K denotes the set of indices where a search is performed. Using this definition it is possible to formalize the states that correspond to either searching or only travelling in the next segment.

Definition 4. *An extension t-succ(ℓ) of a state* $\ell \in M^i$ *is called the* travel-successor *if* $j = i + 1$ *and* $K = \emptyset$. *An extension s-succ(ℓ) of a state* $\ell \in M^i$ *is called the* search-successor *if* $j = i + 1$ *and* $K = \{i + 1\}$. *In both cases, state* ℓ *is called the* predecessor *of t-succ(ℓ) and s-succ(ℓ).*

By definition, the recursion formula

$$M^i := M^{i-1} \cup \{\text{t-succ}(\ell), \text{s-succ}(\ell) \mid \ell \in M^{i-1}\} \tag{1}$$

with $i = 1, \ldots, n$ holds and the set $M^n = M$ corresponds to the set X of feasible missions. Given Lemma 1, the smaller set $\bar{X}$ of feasible missions can be obtained by the following recursion

$$N^i := N^{i-1} \cup \{\text{s-succ}(\ell) \mid \ell \in N^{i-1}\} \tag{2}$$

with $i = 1, \ldots, n$ and $N^0 := M^0$. The following theorem justifies the use of dynamic programming since it derives the principle of optimality for this problem.

Theorem 5. *Let* $\ell, \ell' \in M^i$, $0 \leq i < n$, *for which it holds that* $\ell \preceq \ell'$. *Then, if one of the following two conditions is verified:*

1. $\ell_z \leq \ell'_z$
2. $\ell_z > \ell'_z$ *and* $\ell \preceq (\ell'_T + 2\sum_{k=\ell'_z+1}^{\ell_z} b_k, \ell'_S, \ell_z)$

then, for every extension $ext(\ell) \in M$ *there is an extension* $ext(\ell') \in M$ *for which it also holds that* $ext(\ell) \preceq ext(\ell')$.

Proof. Condition 1 follows immediately since every extension $\text{ext}(\ell) \in M$ is dominated by a corresponding extension $\text{ext}(\ell') \in M$ where the agent travels and/or searches on the same segments j for all $j \geq \ell'_z$. To show condition 2, consider an extension $\text{ext}(\ell) \in M$ and a corresponding extension $\text{ext}(\ell') \in M$ where the agent travels and/or searches on the same segments j for all $j \geq \ell'_z$. A simple calculation shows that $\text{ext}(\ell')_T \leq \text{ext}(\ell)_T$ and $\text{ext}(\ell')_S \geq \text{ext}(\ell)_S$ with at least one strict inequality, and hence $\text{ext}(\ell) \preceq \text{ext}(\ell')$. □

Algorithm 1. Algorithm for formulation (BSPL-1)

input: A feasible instance (n, a, b, s).
output: L_B^n
Initialize: $L_A^0 := L_B^0 := \{(0, 0, 0)\}$
for $i = 1$ **to** n **do**
  $L_A^i := ND\left(\{\text{s-succ}(\ell), \text{t-succ}(\ell) \mid \ell \in L_A^{i-1}\}\right)$
  $L_B^i := ND\left(L_A^i \cup L_B^{i-1}\right)$
end for

Algorithm 2. Algorithm for formulation (BSPL-2)

input: A feasible instance (n, a, b, s).
output: L^n
Initialize: $L^0 := \{(0, 0, 0)\}$
for $i = 1$ **to** $n - 1$ **do**
  $L^i := \overline{ND}\left(L^{i-1} \cup \{\text{s-succ}(\ell) \mid \ell \in L^{i-1}\}\right)$
end for
$L^n := ND\left(L^{n-1} \cup \{\text{s-succ}(\ell) \mid \ell \in L^{n-1}\}\right)$

3.2 Algorithm for Formulation (BSPL-1)

The working principle of the dynamic programming algorithm is to generate *interesting* subsets of M^i, $i = 0, \ldots, n$, by removing states whose extensions do not correspond to efficient missions. The removal of states takes into account the first condition of Theorem 5. Algorithm 1 shows the pseudo-code, where procedure ND removes dominated states. The algorithm uses two lists, L_A^i and L_B^i. At each iteration i, list L_A^i contains the nondominated states in which the last segment to be searched has index $j \geq i$; this is obtained by generating the travel and search-successors of each state in L_A^{i-1} and removing the dominated alternatives from the resulting set of states. Since all states in L_A^i have the same last travelled segment i, only the equality case in the first condition of Theorem 5 is applied. The nondominated states for the first i segments are kept into list L_B^i and are obtained by merging L_B^{i-1} with L_A^i and removing the resulting dominated states.

Some further algorithmic improvements can be obtained. Travel-successors do not need to be generated in the last iteration since they cannot lead to nondominated states. Moreover, the removal of dominated states can be performed in linear time, if the states are lexicographically ordered in both lists.

3.3 Algorithm for Formulation (BSPL-2)

Algorithm 2 gives the pseudo-code of the second dynamic programming algorithm. In this algorithm, the states are generated according to the recursion in Eq.(2) and both conditions in Theorem 5 have to be verified for the removal of states. Only one list, L^i, is needed at each iteration i, $i = 0, \ldots, n$. A search-successor state is generated at each iteration i for each state in L^{i-1}, since

Algorithm 3. Algorithm for formulation (BSPL-3).

input: A feasible instance (n, a, b, s) and induced digraph $G = (V, A)$.
output: L
Initialize: $\tau(1) := \{(0,0)\}$ and $\tau(j) := \emptyset$, for $j = 2, \ldots, 2n+1$
for $j = 2$ **to** $2n+1$ **do**
  while $(i, j) \in A$ **do**
    $\tau(j) := \tau(j) \cup \{(\tau_t + w_t(i,j), \tau_s + w_s(i,j) \mid (\tau_t, \tau_s) \in \tau(i))\}$
  end while
  $\tau(j) := ND(\tau(j))$
end for
$L := \tau(2n+1)$

travel-successors are always dominated. Furthermore, the states are removed according to both conditions of Theorem 5 at iteration i, $i = 0, \ldots, n-1$, which is performed by the procedure $\overline{ND}$. Finally, at iteration n, dominated states are removed with procedure ND, as done in Algorithm 1.

This approach has the advantage of only duplicating the current set of states at each generation. Moreover, the removal of dominated states is performed only once per iteration. However, since the second condition of Theorem 5 implies that dominated states have to be kept into list L^i, the number of states may eventually be larger and the removal cannot be performed in linear time.

3.4 Algorithm for Formulation (BSPL-3)

The shortest path algorithm extends the *pulling algorithm* [1], which is the fastest approach for finding the single-source single-sink shortest path in acyclic and topologically ordered networks, as in the graph of Formulation (BSPL-3). This extended pulling algorithm processes the nodes in the topological order. At each iteration j, $j = 1, \ldots, 2n+1$, it calculates the nondominated shortest paths from node 1 to node j by considering only the distances from each node i that is incident to j, $i < j$. Since the network is topologically ordered, the nondominated shortest paths to node i were already computed.

Algorithm 3 presents the pseudo-code of this approach. A label of a node in V is a 2-tuple $\tau = (\tau_t, \tau_s)$, where τ_t and τ_s correspond to the total time and total score stored at that label, respectively. The set of labels in a node $j \in V$ is denoted by $\tau(j)$. Procedure ND removes the dominated labels, as performed in the previous algorithms.

4 Experimental Analysis

In this section, we report the computational analysis of the three algorithms described in the previous sections. Of particular interest is to understand the difference of performance of the three approaches in a wide range of instance sizes and input data structure. For reducing memory usage, the following changes were

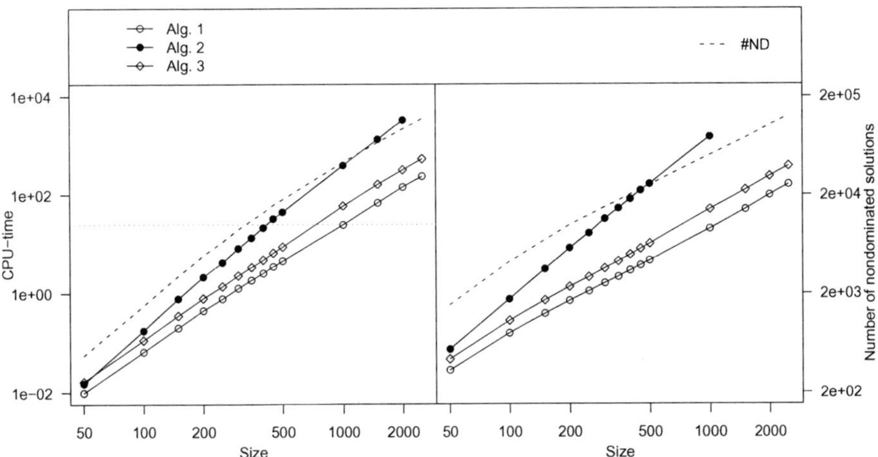

Fig. 2. CPU-time of the three algorithms (straight lines - left axis) and the number of nondominated solutions (dashed lines - right axis) for Type 1 and Type 2 instances (left and right plot, respectively)

performed: i) At the end of each iteration i of Algorithms 1 and 2, the contents of L_A^{i-1}, L_B^{i-1} and L^{i-1} are freed; ii) since keeping all labels of odd nodes in Algorithm 3 requires a large amount of memory in large instances, the contents of $\tau(j-2)$ and $\tau(j-3)$ are deleted and $\tau(2n+1)$ is updated with the contents of $\tau(j-1)$, for each odd j. Other more sophisticated speed-up techniques can also be implemented in Algorithm 3 [4] but they were not tested in this article. All implementations were coded in C and used the same data structures. They were compiled with gcc version 4.2.4. The experiments were performed in a computer cluster with 6 nodes, each with an AMD Phenom II X6 processor with 3.2GHz and operating system Ubuntu 8.04.

We generated two types of test-instances (n, a, b, s), where $a_i > b_i > 0$ and $s_i \geq 0$ for all $i = 1, \ldots, n$. All values are assumed to be integer, and are generated according to a discrete uniform distribution. For $i = 1, \ldots, n$ we defined the two following types of instances:

- Type 1: $b_i \in [1, 1000] \cap \mathbb{N}$, $a_i \in [1001, 10000] \cap \mathbb{N}$, $s_i \in [1, 1000] \cap \mathbb{N}$.
- Type 2: $b_i \in [1, 1000] \cap \mathbb{N}$, $a_i \in [1001, 10000] \cap \mathbb{N}$, $s_i \in [1, 1000] \cap \mathbb{N}$ where $a_{i+1} > a_i$ and $s_{i+1} > s_i$ for all $i = 1, \ldots, n-1$.

In instances of Type 1, the search time, the travel time, and the scores are generated randomly according to a uniform distribution. The instances of Type 2 are generated in a similar manner, but a_i and s_i are sorted in an increasing order. Then, segments with large scores also have large search time, which induces a conflict between the two objectives. The algorithms were tested for different sizes, from $n = 50$ to 2500. For each type and each size, 30 instances were generated.

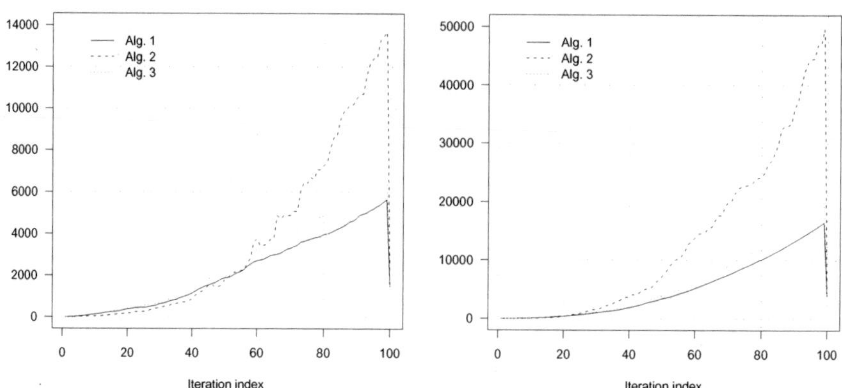

Fig. 3. Number of states generated at each iteration of the three algorithms in an instance of size 100 of Type 1 (left) and Type 2 (right)

Figure 2 shows the CPU-time in seconds taken by the algorithms to calculate the nondominated set (left axis), as well as its size (right axis). For each type and each size n, the results are averaged over 30 instances. The algorithms terminate if the processor time exceeds 3600 seconds. The experimental results indicate that the size of the nondominated set increases strongly with instance size. As expected, the size of the nondominated set is the largest for instances of Type 2. However, the difference from instances of Type 1 vanishes with growing instance sizes. The experimental results also suggest that Algorithm 1 (according to the Formulation (BSPL-1)) is the fastest among the three algorithms, independently of the instance type and size. Algorithm 2 (according to Formulation (BSPL-2)) is the slowest, not being able to solve instances of type B for $n > 1000$.

In order to investigate a possible reason for the difference of performance, the number of states that were generated at the end of each iteration of the three algorithms was recorded. At each iteration i, the number of states in both L^i_B and L^i_A for Algorithm 1, the number of states in L^i for Algorithm 2 and the number of labels related to nodes $2i$, $2i+1$ and $2n+1$ for Algorithm 3 were considered in the analysis. Figure 3 shows the results obtained in an instance of size 100 of Type 1 (left plot) and Type 2 (right plot). The results are consistent with those of Figure 2, that is, Algorithm 1 keeps the least number of states, followed by Algorithm 3, which justifies the performance ordering shown in Figure 2. Similar results were obtained in larger instances.

5 Conclusions

We present three alternative models for a biobjective search problem on a line. The problem is shown to be intractable and NP-complete. Based on the different modelling approaches three alternative dynamic programming formulations are developed. An extensive numerical analysis provides evidence that a strong dominance operator plays a central role in the design of a multiobjective dynamic programming algorithm, and that this may outweigh the computational cost of a larger number of stages. Possible generalizations to problems on general networks are an interesting topic for future research.

Acknowledgments. L. Paquete acknowledges M. Pranzo for his comments on an earlier version of this manuscript. This work was supported by the bilateral cooperation project "Connectedness and Local Search for Multi-objective Combinatorial Optimization" founded by the Deutscher Akademischer Austausch Dienst and Conselho de Reitores das Universidades Portuguesas.

References

1. Ahuja, R., Magnanti, T., Orlin, J.: Network Flows: Theory, Algorithms and Applications. Prentice-Hall (1993)
2. Bazgan, C., Hugot, H., Vanderpooten, D.: Solving efficiently the 0-1 multi-objective knapsack problem. Computers & Operations Research 36(1), 260–276 (2009)
3. Brown, T., Strauch, R.: Dynamic programming in multiplicative lattices. Journal of Mathematical Analysis and Applications 12, 364–370 (1965)
4. Delling, D., Wagner, D.: Pareto Paths with SHARC. In: Vahrenhold, J. (ed.) SEA 2009. LNCS, vol. 5526, pp. 125–136. Springer, Heidelberg (2009)
5. Ehrgott, M.: Multicriteria Optimization, 2nd edn. Springer (2005)
6. Gaver, D., Jacobs, P., Pilnick, S.: On minefield transit by detection, avoidance and demining. In: Bottoms, A., Scandrett, C. (eds.) Applications of Technology to Demining, an Anthology of Scientific Papers 1995-2005, Part 3 – Naval Mine Countermeasures, Society for Countermine Technology (2005)
7. Hansen, P.: Bicriterion path problems. In: Fandel, G., Gal, T. (eds.) Multiple Criteria Decision Making, Theory and Application. Lecture Notes in Economics and Mathematical Systems, vol. 177, pp. 109–127. Springer (1980)
8. Henig, M.: Vector-value dynamic programming. SIAM Journal on Control and Optimization 21, 490–499 (1983)
9. Kellerer, H., Pferschy, U., Pisinger, D.: Knapsack Problems. Springer (2004)
10. Klamroth, K., Wiecek, M.: Dynamic programming approaches to the multiple criteria knapsack problem. Naval Research Logistics 47(1), 57–76 (2000)
11. Sodhi, M., Swaszek, P., Bovio, E.: Stochastic line search using UUVs. In: 9th International Conference on Information Fusion (ICIF 2006), pp. 1–5 (2006)
12. Stone, L.: In search of AF flight 447. ORMS Today 38(4), 22–31 (2011)
13. Villarreal, B., Karwan, M.: Multicriteria integer programming: A (hybrid) dynamic programming recursive approach. Mathematical Programming 21, 204–223 (1981)

Characterizing Graphs of Small Carving-Width

Rémy Belmonte[1,⋆], Pim van 't Hof[1,⋆], Marcin Kamiński[2], Daniël Paulusma[3,⋆⋆], and Dimitrios M. Thilikos[4,⋆⋆⋆]

[1] Department of Informatics, University of Bergen, Norway
{remy.belmonte,pim.vanthof}@ii.uib.no

[2] Département d'Informatique, Université Libre de Bruxelles, Belgium
marcin.kaminski@ulb.ac.be

[3] School of Engineering and Computing Sciences, Durham University, UK
daniel.paulusma@durham.ac.uk

[4] Department of Mathematics, National & Kapodistrian University of Athens, Panepistimioupolis, GR-15784, Athens, Greece
sedthilk@math.uoa.gr

Abstract. We characterize all graphs that have carving-width at most k for $k = 1, 2, 3$. In particular, we show that a graph has carving-width at most 3 if and only if it has maximum degree at most 3 and treewidth at most 2. This enables us to identify the immersion obstruction set for graphs of carving-width at most 3.

1 Introduction

A *call routing tree* (or a *carving*) of a graph G is a tree T with internal vertices of degree 3 whose leaves correspond to the vertices of G. We say that the congestion of T is at most k if, for any edge e of T, the communication demands that need to be routed through e or, more explicitly, the number of edges of G that share endpoints corresponding to different connected components of $T \setminus e$, is bounded by k (we denote by $T \setminus e$ the graph obtained from T after the removal of e). The *carving-width* of a graph G is the minimum k for which there exists a call routing tree T whose congestion is bounded by k.

Carving-width was introduced by Seymour and Thomas [15] who proved that checking whether the carving-width of a graph is at most k is an NP-complete problem. In the same paper, they proved that there is a polynomial-time algorithm for computing the carving-width of planar graphs. Later, the problem of designing call routing trees of minimum congestion was studied by Khuller [10], who presented a polynomial-time algorithm for computing a call routing tree T

⋆ Supported by the Research Council of Norway (197548/F20).

⋆⋆ Supported by EPSRC (EP/G043434/1) and Royal Society (JP100692).

⋆⋆⋆ Co-financed by the European Union (European Social Fund - ESF) and Greek national funds through the Operational Program "Education and Lifelong Learning" of the National Strategic Reference Framework (NSRF) - Research Funding Program: "Thales. Investing in knowledge society through the European Social Fund."

G. Lin (Ed.): COCOA 2012, LNCS 7402, pp. 360–370, 2012.

whose congestion is within a $O(\log n)$ factor from the optimal. In [18] an algorithm was given that decides, in $f(k) \cdot n$ steps, whether an n-vertex graph has carving width at most k and, if so, also outputs a corresponding call routing tree. We stress that the values of $f(k)$ in the complexity of the algorithm in [18] are huge, which makes the algorithm highly impractical even for trivial values of k.

A graph G contains a graph H as an immersion if H can be obtained from some subgraph of G after lifting a number of edges (see Section 2 for the complete definition). Recently, the immersion relation attracted a lot of attention both from the combinatorial [1,5] and the algorithmic [8,9] point of view. It can easily be observed (cf. [18]) that carving-width is a parameter closed under taking immersions, i.e., the carving-width of a graph is not smaller than the carving-width of any of its immersions. Combining this fact with the seminal result of Robertson and Seymour in [13] stating that graphs are well-quasi-ordered with respect to the immersion relation, it follows that the set $\mathcal{G}_k$ of graphs with carving-width at most k can be completely characterized by forbidding a finite set of graphs as immersions. This set is called an *immersion obstruction set* for the class $\mathcal{G}_k$.

Identifying obstruction sets is a classic problem in structural graph theory, and its difficulty may vary, depending on the considered graph class. While obstructions have been extensively studied for parameters that are closed under minors (see [2,4,7,11,12,14,16,17] for a sample of such results), no obstruction characterization is known for any immersion-closed graph class. In this paper, we make a first step in this direction.

The outcome of our results is the identification of the immersion obstruction set for $\mathcal{G}_k$ when $k \leq 3$; the obstruction set for the non-trivial case $k = 3$ is depicted in Figure 3. Our proof for this case is based on a combinatorial result stating that $\mathcal{G}_3$ consists of exactly the graphs with maximum degree at most 3 and treewidth at most 2. A direct outcome of our results is a linear-time algorithm for the recognition of the class $\mathcal{G}_k$ when $k = 1, 2, 3$. This can be seen as a "tailor-made" alternative to the general algorithm of [18] for elementary values of k.

2 Preliminaries

We consider finite undirected graphs that have no self-loops but that may have multiple edges. For undefined graph terminology we refer to the text-book of Diestel [6].

Let $G = (V, E)$ be a graph. The set of *neighbors* of a vertex u is denoted by $N(u) = \{v \mid uv \in E\}$. We denote the number of edges incident with a vertex u by $\deg(u)$; note that $\deg(u)$ may be strictly greater than the number of neighbors of u because we allow G to have multiple edges. We let $\Delta(G) = \max\{\deg(u) \mid u \in V\}$. The n-vertex *path* is the graph with vertices $v_1, \ldots, v_n$ and edges $v_i v_{i+1}$ for $i = 1, \ldots, n-1$. If $v_n v_1$ is also an edge, then we obtain the n-vertex *cycle*. The complete graph on k vertices is denoted by K_k.

Let $G = (V, E)$ be a graph. Then G is called *connected* if, for every pair of distinct vertices v and w, there exists a path connecting v and w. A maximal connected subgraph of G is called a *connected component* of G. A vertex u is called a *cut-vertex* of G if the graph obtained after removing u has more connected components than G. A connected graph is 2-*connected* if it does not contain a cut-vertex. A maximal 2-connected subgraph of G is called a *biconnected component* of G.

The *edge duplication* is the operation that takes two adjacent vertices u and v of a graph and adds a new edge between u and v. The *edge subdivision* is the operation that removes an edge uv of a graph and adds a new vertex w adjacent (only) to u and v. A *series-parallel graph* is a 2-connected graph that can be obtained from a graph consisting of two vertices with two edges between them by a sequence of edge duplications and edge subdivisions.

The *vertex dissolution* is the reverse operation of an edge subdivision; it removes a vertex u of degree 2 that has two distinct neighbors v and w, and adds an edge between v and w. A graph G contains a graph H as a *topological minor* if H can be obtained from G by a sequence of vertex deletions, edge deletions, and vertex dissolutions. Alternatively, G contains H as a topological minor if G contains a subgraph H' that is a *subdivision* of H, i.e., H' can be obtained from H by a sequence of edge subdivisions. We mention one more equivalent definition. The graph G has H as a topological minor if G contains a subset $S \subseteq V_G$ of size $|V_H|$ that has the following property: there exists a bijection f from V_H to S such that, for each edge $e \in E_H$, say with endpoints x and y, there exists a path P_e from $f(x)$ to $f(y)$, and such that for every two edges $e, e' \in E_H$, the paths P_e and $P_{e'}$ are internally vertex-disjoint.

Let u, v, w be three distinct vertices in a graph such that uv and vw are edges. The operation that removes the edges uv and vw, and adds the edge uw (even in the case u and w are already adjacent) is called a *lift*. A graph G contains a graph H as an *immersion* if H can be obtained from G by a sequence of vertex deletions, edge deletions, and lifts. Alternatively, G contains H as an immersion if G contains a subset $S \subseteq V_G$ of size $|V_H|$ that has the following property: there exists a bijection f from V_H to S such that, for each edge $e \in E_H$, say with endpoints x and y, there exists a path P_e from $f(x)$ to $f(y)$, and such that for every two edges $e, e' \in E_H$, the paths P_e and $P_{e'}$ are edge-disjoint. Note that since any two internally vertex-disjoint paths are edge-disjoint, G contains H as an immersion if G contains H as a topological minor.

The *edge contraction* is the operation that takes two adjacent vertices u and v and replaces them by a new vertex adjacent to exactly those vertices that are a neighbor of u or v. A graph G contains a graph H as a *minor* if H can be obtained from G by a sequence of vertex deletions, edge deletions and edge contractions.

A *tree* is a connected graph with no cycles and no multiple edges. A *leaf* of a tree is a vertex of degree 1. A vertex in a tree that is not a leaf is called an *internal* vertex. A *tree decomposition* of a graph $G = (V, E)$ is a pair $(\mathcal{T}, \mathcal{X})$, where $\mathcal{X}$ is a collection of subsets of V, called *bags*, and $\mathcal{T}$ is a tree whose vertices, called *nodes*, are the sets of $\mathcal{X}$, such that the following three properties are satisfied:

(i) $\bigcup_{X \in \mathcal{X}} X = V$,
(ii) for each $uv \in E$, there is a bag $X \in \mathcal{X}$ with $u, v \in X$,
(iii) for each $u \in V$, the nodes containing u induce a connected subtree of $\mathcal{T}$.

The *width* of a tree decomposition $(\mathcal{T}, \mathcal{X})$ is the size of a largest bag in $\mathcal{X}$ minus 1. The *treewidth* of G, denoted by $\mathrm{tw}(G)$, is the minimum width over all possible tree decompositions of G.

Let $G = (V, E)$ be a graph. Let $S \subset V$ be a subset of vertices of G. Then the set of edges between S and $V \setminus S$, denoted by $(S, V \setminus S)$, is called an *edge cut* of G. Let the vertices of G be in 1-to-1 correspondence to the leaves of a tree T whose internal vertices all have degree 3. The correspondence between the leaves of T and the vertices of G uniquely defines the following edge weighting w on the edges of T. Let $e \in E_T$. Let C_1 and C_2 be the two connected components of $T \setminus e$. Let S_i be the set of leaves of T that are in C_i for $i = 1, 2$; note that $S_2 = V \setminus S_1$. Then the weight $w(e)$ of the edge e in T is the number of edges in the edge cut (S_1, S_2) of G. The tree T is called a *carving* of G, and (T, w) is a *carving decomposition* of G. The *width* of a carving decomposition (T, w) is the maximum weight $w(e)$ over all $e \in E_T$. The *carving-width* of G, denoted by $\mathrm{cw}(G)$, is the minimum width over all carving decompositions of G. We define $\mathrm{cw}(G) = 0$ if $|V| = 1$. We refer to Figure 4 for an example of a graph and a carving decomposition.

3 The Main Result

The following observation is known and easy to verify by considering the number of edges in the edge cut $(\{u\}, V \setminus \{u\})$ of a graph $G = (V, E)$.

Observation 1. *Let G be a graph. Then* $\mathrm{cw}(G) \geq \Delta(G)$.

We also need the following two straightforward lemmas. The first lemma follows immediately from the observation that any subgraph of a graph is an immersion of that graph, combined with the observation that carving-width is a parameter that is closed under taking immersions (cf. [18]). We include the proof of the second lemma for completeness.

Lemma 1. *Let H be a subgraph of G. Then* $\mathrm{cw}(H) \leq \mathrm{cw}(G)$.

Lemma 2. *Let G be a graph with connected components $C_1, \ldots, C_p$ for some integer $p \geq 1$. Then* $\mathrm{cw}(G) = \max\{\mathrm{cw}(C_i) \mid 1 \leq i \leq p\}$.

Proof. Lemma 1 implies that $\max\{\mathrm{cw}(C_i) \mid 1 \leq i \leq p\} \leq \mathrm{cw}(G)$. Now let (T_i, w_i) be a carving decomposition of C_i of width $\mathrm{cw}(C_i)$ for $i = 1, \ldots, p$. We pick an arbitrary edge $e_i = x_i y_i$ in each T_i and subdivide it by replacing it with edges $x_i z_i$ and $z_i y_i$, where each z_i is a new vertex. We add edges $z_i z_{i+1}$ for $i = 1, \ldots, p-1$. This results in a tree T. The corresponding carving decomposition (T, w) of G has width $\max\{\mathrm{cw}(C_i) \mid 1 \leq i \leq p\}$. Hence, $\mathrm{cw}(G) \leq \max\{\mathrm{cw}(C_i) \mid 1 \leq i \leq p\}$. We conclude that $\mathrm{cw}(G) = \max\{\mathrm{cw}(C_i) \mid 1 \leq i \leq p\}$. □

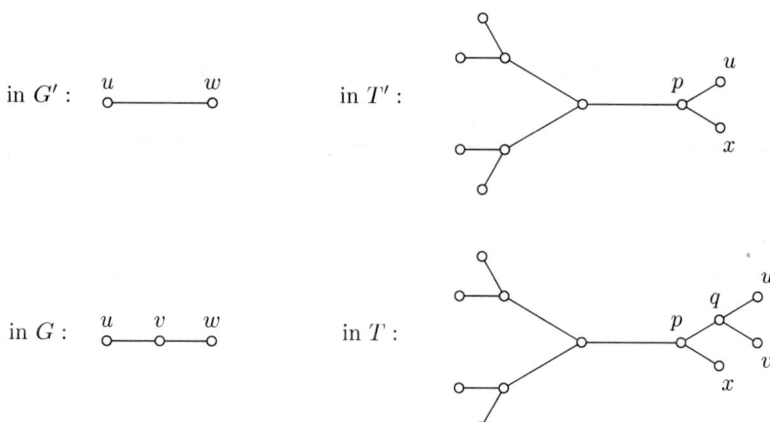

Fig. 1. A schematic illustration of how the tree T' in the carving decomposition of G' is transformed into a tree T in the proof of Lemma 3 when the edge uw in G' is subdivided. The vertex x is an arbitrary vertex of G', possibly w.

The next lemma is the final lemma we need in order to prove our main result.

Lemma 3. *Let G' be a graph with carving-width at least 2, and let uw be an edge of G'. Let G be the graph obtained from G' by subdividing the edge uw. Then* $\text{cw}(G) = \text{cw}(G')$.

Proof. Let (T', w') be a carving decomposition of G' of width $\text{cw}(G') \geq 2$, and let p be the unique neighbor of u in T'. Let v be the vertex that was used to subdivide the edge uw in G', i.e., the graph G was obtained from G' by replacing uw with edges uv and vw for some new vertex v. Let T be the tree obtained from T' by replacing the edge pu by edges pq, qu and qv for some new vertex q; see Figure 1 for an illustration. We first show that the resulting carving decomposition (T, w) of G has width at most $\text{cw}(G')$, which implies that $\text{cw}(G) \leq \text{cw}(G')$.

Let e be an edge in T. Suppose that $e = pq$. By definition, $w(e)$ is the number of edges between $\{u, v\}$ and $V \setminus \{u, v\}$ in G, which is equal to the number of edges incident with u in G. The latter number is the weight of the edge up in T'. Hence, $w(e) \leq \text{cw}(G')$. Suppose that $e = qu$. By definition, $w(e)$ is the number of edges incident with u in G, which is equal to the number of edges incident with u in G'. Hence $w(e) \leq \text{cw}(G')$. Suppose that $e = qv$. By definition, $w(e)$ is the number of edges incident with v in G, which is 2. Hence $w(e) = 2 \leq \text{cw}(G')$. Finally, suppose that $e \notin \{pq, qu, qv\}$. Let C_1 and C_2 denote the subtrees of T obtained after removing e. Let S_i be the set of leaves of T in C_i for $i = 1, 2$. Then u and v either both belong to S_1 or both belong to S_2. Without loss of generality, assume that both u and v belong to S_1. By definition, $w(e)$ is the number of edges between S_1 and S_2 in G, which is equal to the number of edges between $S_1 \setminus \{v\}$ and S_2 in G'. The latter number is the weight of the edge e in

T'. Hence, $w(e) \leq \mathrm{cw}(G')$. We conclude that (T,w) has width at most $\mathrm{cw}(G')$, and hence $\mathrm{cw}(G) \leq \mathrm{cw}(G')$.

It remains to show that $\mathrm{cw}(G) \geq \mathrm{cw}(G')$. Let (T^*,w^*) be a carving decomposition of G of width $\mathrm{cw}(G)$. We remove the leaf corresponding to v from T^*. Afterwards, the neighbor of v in T^* has degree 2, and we dissolve this vertex. This results in a tree T''. It is easy to see that the corresponding carving decomposition (T'',w'') of G' has width at most $\mathrm{cw}(G)$. Hence, $\mathrm{cw}(G) \geq \mathrm{cw}(G')$. This completes the proof of Lemma 3. □

We are now ready to show the main result of our paper.

Theorem 1. *Let G be a graph. Then the following three statements hold.*

(i) $\mathrm{cw}(G) \leq 1$ *if and only if* $\Delta(G) \leq 1$.
(ii) $\mathrm{cw}(G) \leq 2$ *if and only if* $\Delta(G) \leq 2$.
(iii) $\mathrm{cw}(G) \leq 3$ *if and only if* $\Delta(G) \leq 3$ *and* $\mathrm{tw}(G) \leq 2$.

Proof. Let $G = (V,E)$ be a graph. By Lemma 2 we may assume that G is connected. We prove the three statements separately.

(i) If $\mathrm{cw}(G) \leq 1$, then $\Delta(G) \leq 1$ due to Observation 1. If $\Delta(G) \leq 1$, then G contains either one or two vertices. Clearly, $\mathrm{cw}(G) \leq 1$ in both cases.

(ii) If $\mathrm{cw}(G) \leq 2$, then $\Delta(G) \leq 2$ due to Observation 1. If $\Delta(G) = 1$, then the statement follows from (i). If $\Delta(G) = 2$, then G is either a graph consisting of two vertices with two edges between them, or a path, or a cycle. In all three cases, it is clear that $\mathrm{cw}(G) \leq 2$.

(iii) First suppose that $\mathrm{cw}(G) \leq 3$. Then $\Delta(G) \leq 3$ due to Observation 1. We need to show that $\mathrm{tw}(G) \leq 2$. For contradiction, suppose that $\mathrm{tw}(G) \geq 3$. It is well-known that any graph of treewidth at least 3 contains K_4 as a minor (see for example [6], p. 327). It is also well-known that every minor with maximum degree at most 3 of a graph is also a topological minor of that graph (see [6], p. 20). This means that G contains K_4 as a topological minor. Then, by definition, G contains a subgraph H such that H is a subdivision of K_4. Since $\mathrm{cw}(K_4) = 4$, we have that $\mathrm{cw}(H) = \mathrm{cw}(K_4) = 4$ as a result of Lemma 3. Since H is a subgraph of G, Lemma 1 implies that $\mathrm{cw}(G) \geq \mathrm{cw}(H) = 4$, contradicting the assumption that $\mathrm{cw}(G) \leq 3$.

For the reverse direction, suppose that $\Delta(G) \leq 3$ and $\mathrm{tw}(G) \leq 2$. Bodlaender [3] showed that a graph has treewidth at most 2 if and only if all its biconnected components are series-parallel. Hence, we assume that $\Delta(G) \leq 3$ and that every biconnected component of G is series-parallel. We use induction on the number of vertices of G to prove that $\mathrm{cw}(G) \leq 3$. It is clear that this holds when $|V| \leq 2$, since we assumed $\Delta(G) \leq 3$.

Let $|V| \geq 3$. Suppose that G contains a vertex v of degree 2 that has two distinct neighbors u and w. Let $G' = (V',E')$ denote the (connected) graph obtained from G by dissolving v. Note that G' has maximum degree at most 3, and every biconnected component of G' is series-parallel. Hence, by the induction hypothesis, $\mathrm{cw}(G') \leq 3$. Because $|V| \geq 3$, we find that G' contains at least two vertices. If $\mathrm{cw}(G') = 1$, then $\Delta(G') = 1$ by Observation 1. This means that G' is a path on two vertices. Consequently, G is a path on three vertices, and hence $\mathrm{cw}(G) = 2 \leq 3$. If $2 \leq \mathrm{cw}(G') \leq 3$, then $\mathrm{cw}(G) = \mathrm{cw}(G') \leq 3$ as a result of Lemma 3.

From now on, we assume that G contains no vertex of degree 2 that has two distinct neighbors. Suppose that G contains two vertices u and v with at least two edges between them. First suppose that u and v are the only vertices of G. Then $\mathrm{cw}(G) \leq 3$, because the assumption $\Delta(G) \leq 3$ implies that u and v have at most three edges between them. Now suppose that at least one of u, v has at least one other neighbor outside $\{u, v\}$ in G, say v has a neighbor $t \neq u$. Then, because $\Delta(G) \leq 3$ and there exist at least two edges between u and v in G, we find that t and u are the only two neighbors of v in G and that the number of edges between u and v is exactly 2. Let G^* denote the graph obtained from G by deleting one edge between u and v. Let G' denote the graph obtained from G^* by dissolving v. Note that G' has maximum degree at most 3, and that every biconnected component of G' is series-parallel. Hence, by the induction hypothesis, $\mathrm{cw}(G') \leq 3$.

If $\mathrm{cw}(G') = 1$, then, for the same reasons as before, G' must a path on two vertices and G^* must be a path on three vertices, implying that $\mathrm{cw}(G^*) = 2$. Since G can be obtained from G^* by adding a single edge, $\mathrm{cw}(G) \leq 3$ in this case. Suppose $2 \leq \mathrm{cw}(G') \leq 3$. Then, by Lemma 3, $\mathrm{cw}(G^*) = \mathrm{cw}(G') \leq 3$. Moreover, from the proof of Lemma 3 it is clear that there exists a carving decomposition (T^*, w^*) of G^* of width $\mathrm{cw}(G^*)$ such that u and v have a common neighbor q in T^*. We consider the carving decomposition (T, w) of G with $T = T^*$. Let e be an edge in T. First suppose that $e = uq$ or $e = vq$. Then $w(e) \leq 3$, as both u and v have degree at most 3 in G. Now suppose that $e \notin \{uq, vq\}$. Then $w(e) = w^*(e) \leq \mathrm{cw}(G^*) \leq 3$. We conclude that the carving decomposition (T, w) of G has width at most 3, which implies that $\mathrm{cw}(G) \leq 3$.

From now on, we assume that G contains no multiple edges. Since we already assumed G not to contain any vertex of degree 2 that has two distinct neighbors, this implies that G contains no vertices of degree 2 at all. If G contains no cut-vertices, then G is 2-connected. Then G must be series-parallel, since we assumed that every biconnected component of G is series-parallel. Then, by definition, G contains either a vertex of degree 2 or two vertices with more than one edge between them. However, we assumed that this is not the case. We conclude that G contains at least one cut-vertex v.

Because v is a cut-vertex, it has degree at least 2. Since G contains no vertex of degree 2 and $\Delta(G) \leq 3$, we find that v has degree 3. Note that the graph $G - v$ has either two or three connected components. Let D_1, D_2, D_3 denote the vertex sets of the connected components of $G - v$, where D_3 is possibly empty. Let G' be the subgraph of G induced by $D_1 \cup \{v\}$. Because v is a cut-vertex of G, the set of biconnected components of G' is a subset of the set of biconnected components of G. Hence, every biconnected component of G' is series-parallel. Moreover, since $\Delta(G) \leq 3$ and G' is a subgraph of G, we find that $\Delta(G') \leq 3$. Hence, by the induction hypothesis, G' has carving-width at most 3. Similarly, the subgraph G'' of G induced by $D_2 \cup D_3 \cup \{v\}$ has carving-width at most 3. Let (T', w') be a carving decomposition of G' of width $\text{cw}(G') \leq 3$, and let (T'', w'') be a carving decomposition of G'' of width $\text{cw}(G'') \leq 3$. From T' and T'', we construct a tree T as follows (see also Figure 2). We first identify the leaves of

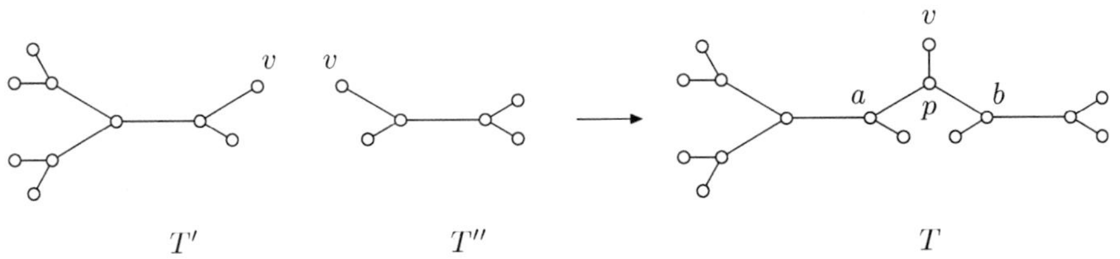

Fig. 2. A schematic illustration of how the tree T is constructed from the trees T' and T'' in the proof of Theorem 1

T' and T'' that correspond to v. Let p denote the newly obtained vertex, and let a and b be the two neighbors of p, where a belongs to T' and b' belongs to T''. We then add a new leaf adjacent to the vertex p in T, and we let this leaf correspond to the vertex v of G. This completes the construction of T. Below we show that the corresponding carving decomposition (T, w) of G has width at most 3.

Let e be an edge of T. Let C_1 and C_2 be the two subtrees of the forest $T - e$. Let S_1 and S_2 be the set of leaves of T in C_1 and C_2, respectively. We will also use S_1 and S_2 to denote the vertices of G that correspond to the leaves in S_1 and S_2, respectively. Assume that $v \in S_1$. Suppose that $e = vp$. Then $w(e) = 3$, because there are three edges incident with v in G. Suppose that $e = ap$. Due to the fact that v is a cut-vertex of G, we find that v is the only vertex in S_1 that has at least one neighbor in S_2 in G. Since v has degree 3 and D_1 is not empty, v has at most two neighbors in S_2. Hence $w(e) \leq 2$. Suppose that $e = bp$. Then $w(e) \leq 2$ by a similar argument as in the previous case. Suppose that $e \in E_{T'} \setminus \{ap, bp, vp\}$. Then $w(e) = w'(e) \leq 3$, because $\text{cw}(G') \leq 3$. Suppose that $e \in E_{T''} \setminus \{ap, bp, vp\}$. Then $w(e) = w''(e) \leq 3$, because $\text{cw}(G'') \leq 3$. We conclude that $\text{cw}(G) \leq 3$. This completes the proof of Theorem 1. □

Since graphs of treewidth at most 2 can easily be recognized in linear time, Theorem 1 implies a linear-time recognition algorithm for graphs of carving-width at most 3.

Thilikos, Serna and Bodlaender [18] proved that for any k, there exists a linear-time algorithm for constructing the immersion obstruction set for graphs of carving-width at most k. For $k \in \{1, 2\}$, finding such a set is trivial. We now present an explicit description of the immersion obstruction set for graphs of carving-width at most 3.

Corollary 1. *A graph has carving-width at most 3 if and only if it does not contain any of the six graphs in Figure 3 as an immersion.*

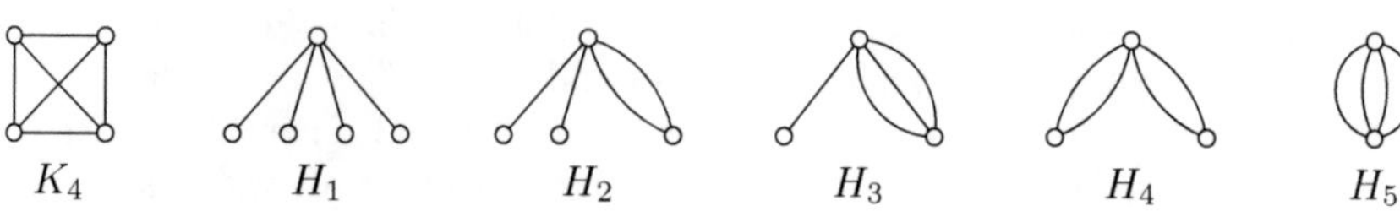

Fig. 3. The immersion obstruction set for graphs of carving-width at most 3

Proof. Let G be a graph. We first show that if G contains one of the graphs in Figure 3 as an immersion, then G has carving-width at least 4. In order to see this, it suffices to observe that the graphs $K_4, H_1, \ldots, H_4$ all have carving-width 4. Hence, G has carving-width at least 4, because carving-width is a parameter that is closed under taking immersions (cf [18]).

Now suppose that G has carving-width at least 4. Then, due to Theorem 1, $\Delta(G) \geq 4$ or $\text{tw}(G) \geq 3$. If $\Delta(G) \geq 4$, then G has a vertex v of degree at least 4. By considering v and four of its incident edges, it is clear that G contains one of the graphs $H_1, \ldots, H_5$ as a subgraph, and consequently as an immersion. Suppose that $\Delta(G) \leq 3$. Then $\text{tw}(G) \geq 3$, which means that G contains K_4 as a minor [6]. Moreover, since K_4 has maximum degree 3, it is well-known that G also contains K_4 as a topological minor [6], and hence as an immersion. □

From the proof of Corollary 1, we can observe that an alternative version of Corollary 1 states that a graph has carving-width at most 3 if and only if it does not contain any of the six graphs in Figure 3 as a topological minor.

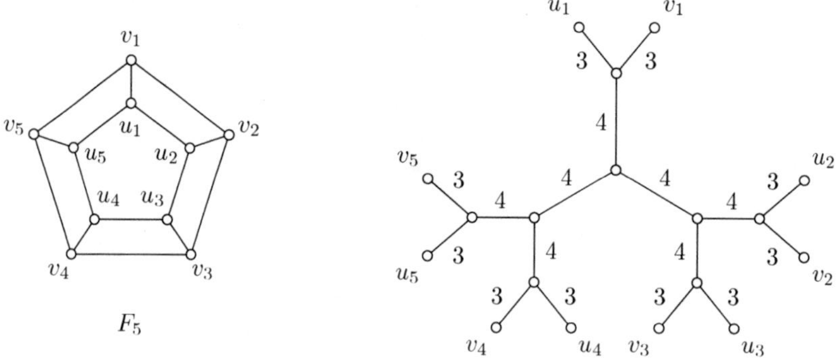

Fig. 4. The pentagonal prism F_5 and a carving decomposition (T, w) of F_5 that has width 4

4 Conclusions

Extending Theorem 1 to higher values of carving-width remains an open problem, and finding the immersion obstruction set for graphs of carving-width at most 4 already seems to be a challenging task. We proved that for any graph G, cw$(G) \leq 3$ if and only if $\Delta(G) \leq 3$ and tw$(G) \leq 2$. We finish our paper by showing that the equivalence "cw$(G) \leq 4$ if and only if $\Delta(G) \leq 4$ and tw$(G) \leq 3$" does not hold in either direction.

To show that the forward implication is false, we consider the pentagonal prism F_5, which is displayed in Figure 4 together with a carving decomposition (T, w) of width 4. Hence, cw$(F_5) \leq 4$. However, F_5 is a minimal obstruction for graphs of treewidth at most 3 [3]. Hence, tw$(F_5) = 4$.

To show that the backward implication is false, we consider the graph K_5^-, which is the graph obtained from K_5 by removing an edge. Note that $\Delta(K_5^-) = 4$ and tw$(K_5^-) = 3$. It is not hard to verify that cw$(K_5) = 6$. Since removing an edge decreases the carving-width by at most 1, we conclude that cw$(K_5^-) \geq 5$.

References

1. Abu-Khzam, F.N., Langston, M.A.: Graph Coloring and the Immersion Order. In: Warnow, T.J., Zhu, B. (eds.) COCOON 2003. LNCS, vol. 2697, pp. 394–403. Springer, Heidelberg (2003)
2. Arnborg, S., Proskurowski, A.: Characterization and recognition of partial 3-trees. SIAM J. Algebraic Discrete Methods 7(2), 305–314 (1986)
3. Bodlaender, H.L.: A partial k-arboretum of graphs with bounded treewidth. Theoretical Computer Science 209(1-2), 1–45 (1998)
4. Bodlaender, H.L., Thilikos, D.M.: Graphs with branchwidth at most three. Journal of Algorithms 32(2), 167–194 (1999)
5. DeVos, M., Dvořák, Z., Fox, J., McDonald, J., Mohar, B., Scheide, D.: Minimum degree condition forcing complete graph immersion. CoRR, arXiv:1101.2630 (January 2011)
6. Diestel, R.: Graph Theory, Electronic edn. Springer (2005)
7. Dvořák, Z., Giannopoulou, A.C., Thilikos, D.M.: Forbidden graphs for tree-depth. European Journal of Combinatorics 33(5), 969–979 (2012)
8. Grohe, M., Kawarabayashi, K., Marx, D., Wollan, P.: Finding topological subgraphs is fixed-parameter tractable. In: Proceedings of STOC 2011, pp. 479–488. ACM (2011)
9. Kawarabayashi, K., Kobayashi, Y.: List-coloring graphs without subdivisions and without immersions. In: Proceedings of SODA 2012, pp. 1425–1435. SIAM (2012)
10. Khuller, S., Raghavachari, B., Young, N.: Designing multicommodity flow trees. Information Processing Letters 50, 49–55 (1994)
11. Koutsonas, A., Thilikos, D.M., Yamazaki, K.: Outerplanar obstructions for matroid pathwidth. Electronic Notes in Discrete Mathematics 38, 541–546 (2011)
12. Robertson, N., Seymour, P.D., Thomas, R.: Linkless embeddings of graphs in 3-space. Bull. Amer. Math. Soc. 28, 84–89 (1993)
13. Robertson, N., Seymour, P.D.: Graph minors XXIII. Nash-Williams' immersion conjecture. Journal of Combinatorial Theory, Series B 100, 181–205 (2010)
14. Rué, J., Stavropoulos, K.S., Thilikos, D.M.: Outerplanar obstructions for a feedback vertex set. European Journal of Combinatorics 33, 948–968 (2012)
15. Seymour, P.D., Thomas, R.: Call routing and the ratcatcher. Combinatorica 14(2), 217–241 (1994)
16. Takahashi, A., Ueno, S., Kajitani, Y.: Minimal acyclic forbidden minors for the family of graphs with bounded path-width. Discrete Mathematics 127, 293–304 (1994)

17. Thilikos, D.M.: Algorithms and obstructions for linear-width and related search parameters. Discrete Applied Mathematics 105, 239–271 (2000)
18. Thilikos, D.M., Serna, M.J., Bodlaender, H.L.: Constructive Linear Time Algorithms for Small Cutwidth and Carving-Width. In: Lee, D.T., Teng, S.-H. (eds.) ISAAC 2000. LNCS, vol. 1969, pp. 192–203. Springer, Heidelberg (2000)

Solving the Connected Dominating Set Problem and Power Dominating Set Problem by Integer Programming

Neng Fan and Jean-Paul Watson

Discrete Math and Complex Systems Department
Sandia National Laboratories, Albuquerque, NM 87185, USA
{nnfan,jwatson}@sandia.gov

Abstract. In this paper, we propose several integer programming approaches with a polynomial number of constraints to formulate and solve the minimum connected dominating set problem. Further, we consider both the power dominating set problem – a special dominating set problem for sensor placement in power systems – and its connected version. We propose formulations and algorithms to solve these integer programs, and report results for several power system graphs.

Keywords: Connected Dominating Set Problem, Power Dominating Set Problem, Integer Programming, Spanning Tree, Connected Subgraph.

1 Introduction

The *minimum dominating set* (MDS) problem is stated as follows: For a graph $G = (V,E)$, a dominating set is a subset D of V such that every vertex not in D is linked to at least one member of D by some edge. The minimum dominating set problem is to find a dominating set with smallest cardinality. The decision version of the MDS is a classical NP-complete problem [1].

The dominating set D of a graph G is a *connected dominating set* if each vertex in D can reach any other vertex in D by a path that traverses only vertices within D. That is, D induces a connected subgraph of G. The *minimum connected dominating set* (MCDS) problem is to find a connecting dominating set with the smallest possible cardinality among all connected dominating sets of G. The concept of a connected dominating set is quite useful in the analysis of wireless networks, social networks, and sensor networks, as studied extensively by Du's group in [2–6]. The MCDS problem was recently studied in disk graphs [7] and unit ball graphs [8]. For an extensive discussion of heuristic algorithms for and applications of the MCDS problem, we refer to [9, 10].

Integer programming (IP) approaches for the MCDS problem have attracted less attention than heuristic methods. In [11, 12], although IP formulations were presented, the algorithms were still based on heuristic and simulation methods. In [13, 14], mixed integer programming (MIP) approaches were used to formulate the MCDS, while [15] introduced a MIP approach with exponential number $O(2^{|V|})$ of constraints based on spanning trees to exactly solve this problem.

In this paper, building on the IP formulation for the MDS problem, we add different kinds of constraints to ensure the connectivity of the subgraph induced by D.

G. Lin (Ed.): COCOA 2012, LNCS 7402, pp. 371–383, 2012.

Considering the fact that a graph is connected if and only if it has a spanning tree, constraints implementing sub-tour elimination, cutset, and other concepts, as reviewed in [16] for the minimum spanning tree problem, can be leveraged for IP formulations of the MCDS problem. However, because of the exponential number of constraints, the computational expense is prohibitive for large graphs. Therefore, we use a polynomial number of constraints to ensure connectivity, leveraging Miller-Tucker-Zemlin constraints, Martin constraints, and commodity flow constraints.

The *power dominating set* (PDS) problem was originally proposed for solving a sensor placement problem in power system graphs, usually referred to as the PMU placement problem [17, 18]. A *power system graph* is an undirected graph $G = (V,E)$, where the vertex set V represents a set of buses, and the edge set E represents a set of transmission lines. Additionally, there is a subset V_Z of V, which represents the set of zero-injection buses that consist of transhipment buses in the system. A power dominating set D is obtained by considering the following two physical laws: (i) if $v \in D$, then v and its neighbors (denoted by $N(v)$) are all covered (Ohm's law); (ii) if $v \in V_Z$, and all vertices within the set $\{v\} \cup N(v)$ except one are covered, then the uncovered vertex in $\{v\} \cup N(v)$ is also covered (Kirchhoff's current law). The PDS problem is to find a subset of vertices D with smallest cardinality that covers all vertices in V. This problem has been widely studied in the power systems literature, as shown in [18], and recently in the area of general combinatorial optimization [19]. The PDS problem can be extended to consider connected vertex sets, yielding the *connected power dominating set* (CPDS) problem.

The reminder of this paper is organized as follows. In Section 2, we introduce IP formulations for the MDS and MCDS problems. In Section 3, we introduce four types of connectivity constraints to ensure the connectivity of the subgraph induced by the dominating set. In Section 4, we introduce the power dominating set problem, connected power dominating set problem, and their associated IP formulations. In Section 5, we test and compare our formulations and algorithms on several power system graphs. Finally, we conclude in Section 6 with a summary of our results.

2 Dominating Set Problem

In a graph $G = (V,E)$ with $V = \{1,2,\cdots,n\}$, let $A = (a_{ij})_{n\times n}$ be the neighborhood matrix such that $a_{ij} = a_{ji} = 1$ if $(i,j) \in E$ or $i = j$, and $a_{ij} = a_{ji} = 0$ otherwise. Without loss of generality, we define the edge set E as follows: $E = \{(i,j) : a_{ij} = 1, \forall i,j \in V \text{ with } i < j\}$.

For $i \in V$, let $x_i \in \{0,1\}$ be a decision variable such that $x_i = 1$ if vertex i is included in the dominating set; $x_i = 0$ otherwise. An IP formulation of the MDS problem can then be given as follows:

$$\textbf{[MDS]} \qquad \min \sum_i x_i \tag{1a}$$

$$s.t. \sum_j a_{ij}x_j \geq 1, x_i \in \{0,1\}, \forall i \in V \tag{1b}$$

Any feasible solution to formulation (1) will form a dominating set D of G by $D = \{v_i : x_i = 1\}$. Let $G_D = (D, E_D)$ be the subgraph induced by the dominating set D, where $E_D = \{(i,j) \in E : a_{ij}x_ix_j = 1, \forall i,j \in V \text{ with } i < j\}$.

To yield an IP formulation for the related MCDS problem, we must additionally include connectivity constraints to ensure that the subgraph G_D is connected. In next section, we study four MIP approaches to model the connectivity constraints in the MCDS for subgraphs G_D.

3 Connectivity Constraints of Subgraphs

Definitionally, a graph G_D is connected if and only if it has a spanning tree. Therefore, some methods for solving the minimum spanning tree problem can be leveraged to formulate efficient MIPs for the MCDS problem.

3.1 Miller-Tucker-Zemlin Constraints

Miller-Tucker-Zemlin constraints were originally proposed for solving the traveling salesman problem in [20], and were used to eliminate sub-tours when solving the k-cardinality tree problem in [21].

Following the method proposed in [21], we let $G_d = (V \cup \{n+1, n+2\}, A)$ be a directed graph based on $G = (V,E)$, where $A = \{(n+1,n+2)\} \cup \{\bigcup_{i=1}^{n}\{(n+1,i),(n+2,i)\}\} \cup E \cup E'$ and $E' = \{(j,i) : a_{ji} = 1, \forall i,j \in V \text{ with } i > j\}$. That is, we introduce two additional vertices $n+1$ and $n+2$, add directed edges $n+1$ and $n+2$ to every $i \in V$ and $(n+1, n+2)$, and make each edge $(i,j) \in E$ bi-directional.

The idea behind Miller-Tucker-Zemlin constraints is to find a directed spanning tree $T_d = (V \cup \{n+1,n+2\}, E_d)$ of G_d such that $n+1$ is the root connecting to both $n+2$ and those vertices not in the dominating set D, $n+2$ is connected to a vertex v_r within D, and all other vertices are formed a tree with root v_r. As shown in Fig. 1, the directed spanning tree has a connected subgraph (shown within the dashed circle) whose vertices form the connected dominating set.

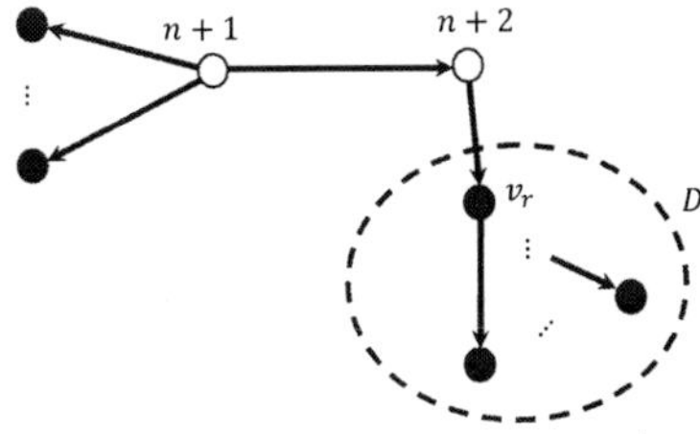

Fig. 1. The idea behind Miller-Tucker-Zemlin constraints

For $(i,j) \in A$, let $y_{ij} \in \{0,1\}$ be a decision variable such that $y_{ij} = 1$ if (i,j) is selected into the directed tree T_d and $y_{ij} = 0$ otherwise. Additionally, for $i \in V \cup \{n+1, n+2\}$, let u_i be a non-negative decision variable, as introduced in [20] to eliminate sub-tours. The Miller-Tucker-Zemlin (MTZ) constraints to ensure connectivity are formulated as follows:

[MTZ]

$$\sum_{i\in V} y_{n+2,i} = 1 \tag{2a}$$

$$\sum_{i:(i,j)\in A} y_{ij} = 1, \forall j \in V \tag{2b}$$

$$y_{n+1,i} + y_{i,j} \leq 1, \forall (i,j) \in E \cup E' \tag{2c}$$

$$(n+1)y_{ij} + u_i - u_j + (n-1)y_{ji} \leq n, \forall (i,j) \in E \cup E' \tag{2d}$$

$$(n+1)y_{ij} + u_i - u_j \leq n, \forall (i,j) \in A \setminus (E \cup E') \tag{2e}$$

$$y_{n+1,n+2} = 1 \tag{2f}$$

$$u_{n+1} = 0 \tag{2g}$$

$$1 \leq u_i \leq n+1, i \in V \cup \{n+2\} \tag{2h}$$

$$x_i = 1 - y_{n+1,i}, \forall i \in V \tag{2i}$$

In the formulation MTZ, Constraint (2a) identifies one vertex as root of the dominating set D. Constraints (2b) ensure that all vertices within V are connected to some other vertex. Constraints (2c) require that in any feasible solution either $i \in V$ is directly connected to $n+1$ or else it may be connected to other vertices in D. Without the term $(n-1)y_{ji}$, Constraints (2d) and (2e) are the original Miller-Tucker-Zemlin constraints [20] to guarantee the solutions have no sub-tours. The added term was proposed in [22] as an improvement for sub-tour elimination constraints. Constraint (2f) requires that the edge $(n+1, n+2)$ is in T_d. Constraints (2g) and (2h) present the choice of arbitrary non-negative integers for variables u_i. Finally, Constraints (2i) ensure that vertex i is either connected to $n+1$ or a vertex in the dominating set.

The constraints and variables in formulation MTZ represent a portion of a mixed-linear program. By solving MDS in conjunction with MTZ, any feasible solution x will imply a dominating set D, and form a directed spanning tree T_d of G_d. The induced subtree of T_d by D has a root, which is connected to $n+2$. Therefore, the connectivity of the subgraph G_D by D can be guaranteed. In MTZ, there are $(|V|+2) + (2|E| + 2|V|+1) = O(|E|+|V|)$ decision variables and $1+|V|+2|E|+2|E|+(2|V|+1)+1+1+|V| = O(|E|+|V|)$ constraints.

3.2 Martin Constraints

In [23], Martin presented a reformulation for solving the minimum spanning tree problem with a polynomial number of constraints instead of an exponential number of constraints. This method was also used in [24] by Yannakakis, and was recently referenced in [25, 26]. The objective is still to find a (undirected) spanning tree $T_D = (D, E_T)$ of $G_D = (D, E_D)$.

For $i, j \in V$, let $y_{ij} \in \{0,1\}$ be a decision variable such that $y_{ij} = 1$ if edge (i,j) is selected into the tree T_D and $y_{ij} = 0$ otherwise. For $i, j, k \in V$, let $z_{ij}^k \in \{0,1\}$ be a decision variable such that $z_{ij}^k = 1$ if edge (i,j) is in the tree T_D of G_D and vertex k is on side of j (i.e., vertex k is within the resulted component containing j after removal edge (i,j) from T_D), and $z_{ij}^k = 0$ if edge (i,j) is in the tree and k is not on side of j, or if edge (i,j) is not in the tree or the pair (i,j) is not an edge.

The Martin constraints to ensure connectivity of G_D are formulated as follows:
[MARTIN]

$$\sum_{(i,j)\in E} y_{ij} = \sum_{i\in V} x_i - 1 \tag{3a}$$

$$y_{ij} \le x_i, y_{ij} \le x_j, \forall (i,j) \in E \tag{3b}$$

$$z_{ij}^k \le y_{ij}, z_{ij}^k \le x_k, \forall (i,j) \in E, k \in V \tag{3c}$$

$$z_{ji}^k \le y_{ij}, z_{ji}^k \le x_k, \forall (i,j) \in E, k \in V \tag{3d}$$

$$y_{ij} - M(3 - x_i - x_j - x_k) \le z_{ij}^k + z_{ji}^k \le y_{ij} + M(3 - x_i - x_j - x_k), \forall i,j,k \in V \tag{3e}$$

$$1 - M(2 - x_i - x_j) \le \sum_{k'\in V\setminus\{i,j\}} z_{ik'}^j + y_{ij} \le 1 + M(2 - x_i - x_j), \forall i,j \in V \tag{3f}$$

$$y_{ij}, z_{ij}^k \in \{0,1\}, \forall (i,j) \in E, k \in V,\ y_{ij} = 0, z_{ij}^k = 0, \forall i,j,k \in V, (i,j) \notin E \tag{3g}$$

Constraint (3a) ensures that the number of edges within the tree T_D is one less than the number of vertices within T_D, and Constraint (3b) ensures that the selection of edges within E_D relies on the selection of its two ends.

If any one, two, or three vertices of $i,j,k \in V$ are not part of the tree of G_D (i.e., one, two, or three of x_i, x_j, x_k become 0), $z_{ij}^k = z_{ji}^k = 0$ by Constraints (3c)-(3d), and the Constraints (3e) become non-binding constraints and have no influence on the results as M is a large positive constant. Similarly, if any one or two of vertices $i,j \in V$ are not part of G_D, Constraint (3f) become non-binding.

If vertices $i,j,k \in V$ are within the proposed tree of G_D (i.e., $i,j,k \in D$ and $x_i = x_j = x_k = 1$), $z_{ij}^k, z_{ji}^k \in \{0,1\}$ by Constraints (3c)-(3d), and Constraints (3e)-(3f) become

$$z_{ij}^k + z_{ji}^k = y_{ij}, \sum_{k\in D\setminus\{i,j\}} z_{ik}^j + y_{ij} = 1, \forall i,j,k \in D.$$

This represents the original formulation of Martin's constraints, as discussed in [26].

The constraint $z_{ij}^k + z_{ji}^k = y_{ij}$ implies that (i) if $(i,j) \in E_T$ (i.e., $y_{ij} = 1$), vertex k is either on the side of j ($z_{ij}^k = 1$) or on the side of i ($z_{ji}^k = 1$); (ii) if $(i,j) \notin E_T$ (i.e., $y_{ij} = 0$), k is between i,j ($z_{ij}^k = 0, z_{ji}^k = 0$).

The constraint $\sum_{k\in D\setminus\{i,j\}} z_{ik}^j + y_{ij} = 1$ means that (i) if $(i,j) \in E_T$ (i.e., $y_{ij} = 1$), edges (i,k) who connect i are on the side of i ($z_{ik}^j = z_{ij}^k = 0$ and $z_{ij}^k = 1$); (ii) if $(i,j) \notin E_T$ (i.e., $y_{ij} = 0$), there must be an edge (i,k) such that j is on the side k ($z_{ik}^j = 1$ for some k).

The constraints and variables in formulation MARTIN represent a portion of a mixed-linear program. The number of new decision variables is $|V|^2 + |V|^3 = O(|V|^3)$, while the number of constraints to ensure connectivity is $1 + 2|E| + 4|E||V| + 2|V|^3 + 2|V|^2 = O(|V|^3)$.

3.3 Single-Commodity Flow Constraints

For $i \in V$, let $r_i \in \{0,1\}$ be a decision variable such that $r_i = 1$ if vertex i is chosen to be the root v_r of G_D for "sending" $\sum_{i\in V} x_i - 1$ unit flow to other vertices within the dominating D, and $r_i = 0$ otherwise. If each vertex in D except v_r consumes exactly one unit, and the vertices outside D consume none, the connectivity of G_D is guaranteed. This method was used to ensure subgraph connectivity in [27] for solving problems in wildlife conservation.

For each edge $(i, j) \in E \cup E'$ (see Section 3.1 for the definition of E'), let f_{ij} denote the amount of flow from vertex i to vertex j. The constraints enforcing single-commodity flow (SCF) can then be formulated as follows:

[SCF]

$$\sum_{i\in V} r_i = 1 \tag{4a}$$

$$r_i \le x_i, \forall i \in V \tag{4b}$$

$$f_{ij} \ge 0, \forall (i,j) \in E \cup E' \tag{4c}$$

$$f_{ij} \le x_i \sum_{k\in V} x_k,\ f_{ij} \le x_j \sum_{k\in V} x_k, \forall (i,j) \in E \cup E' \tag{4d}$$

$$\sum_j f_{ji} \le n(1 - r_i), \forall i \in V \tag{4e}$$

$$\sum_j f_{ji} - \sum_j f_{ij} = x_i - r_i \sum_{j\in V} x_j, \forall i \in V \tag{4f}$$

$$r_i \in \{0,1\}, \forall i \in V \tag{4g}$$

Constraints (4a) and (4b) select one vertex from the dominating set as the root to transmit the single-commodity flow. Constraints (4c) ensure the non-negativity of the flow, while Constraints (4d) require that the flow of edge (i, j) is 0 if either end of (i, j) is not selected into the dominating set. Constraints (4e) ensure that the inflow of the selected root is 0. Finally, Constraints (4f) ensure the balance of flows on each vertex. If vertex i is the selected root (i.e., $r_i = 1, x_i = 1$), the outflow of i is equal to $\sum_{j\in V} x_j - 1$, i.e., one unit is transmitted to each selected vertex. If vertex i is in the dominating set D but is not the root (i.e., $x_i = 1, r_i = 0$), the difference between the inflow and outflow will equal one, implying that vertex i consumes one unit; otherwise, vertex i is not in D (i.e., $x_i = 0, r_i = 0$), and all inflows and outflows will be 0.

Any feasible solution to the MDS problem with SCF constraints will guarantee that every vertex within the dominating set D except the selected root will consume one unit of flow transmitted from the root, and the connectivity of the subgraph induced by D will be ensured.

The quadratic terms $r_i x_j$ can be easily linearized by introducing $w_{ij} = r_i x_j$ with constraints $w_{ij} \le r_i$, $w_{ij} \le x_j$, $w_{ij} \ge r_i + x_j - 1$, and $w_{ij} \ge 0$. Similarly, the quadratic terms $x_i x_k$ can be linearized by introducing $w'_{ik} = x_i x_k$.

Additionally, the following constraints can be added such that the first appearance of $x_i = 1$ implies $r_i = 1$:

$$r_i \leq (n+1-\sum_{i'=1}^{i} x_{i'})/n, \forall i \in V. \tag{5}$$

Such constraints can reduce the degeneracy of the choice of root vertex within the dominating set. Without loss of generality, assume that i_a is first vertex with $x_{i_a} = 1$ (i.e., $x_i = 0$ for $i < i_a$) and i_b is the second one with $x_{i_b} = 1$ (i.e., $x_i = 0$ for $i_a < i < i_b$). Therefore, by (5), there are four cases: (i) for $i < i_a$, $r_i \leq (n+1-0)/n = 1+1/n$ and by (4b), $r_i = 0$; (ii) for $i = i_a$, $r_{i_a} \leq 1$; (iii) for $i_a < i < i_b$, $r_i \leq 1$ and from (4b), $r_i = 0$; (iv) for $i \geq i_b$, $r_i \leq (n+1-2)/n = (n-1)/n$ and from (4g), $r_i = 0$. Thus, by (4a), $r_{i_a} = 1$.

There are $|V|+2|E| = O(|E|+|V|)$ decision variables and $1+|V|+2|E|+4|E|+|V|+|V| = O(|E|+|V|)$ constraints in the MDS problem with SCF constraints.

3.4 Multi-commodity Flow Constraints

In Section 3.3, the connectivity of $G_D = (D, E_D)$ is enforced through a single commodity flow. In the following, the connectivity of a selected subset D is guaranteed by associating a separate commodity with each vertex. Assume that v_r is the selected root within D, such that there will be one unit of flow from v_r to each selected vertices of its own commodity type. This method was used to ensure subgraph connectivity in [27] for solving problems in wildlife conservation.

For each edge $(i,j) \in E \cup E'$ (see Section 3.1) and $k \in V \setminus \{v_r\}$, let f_{ij}^k be a decision variable such that $f_{ij}^k != 0$ if edge (i,j) carries flow of type k, and 0 otherwise. The flow outside of the dominating set should be 0, the flow of type k equals 0 if k is outside D, and the flow of type v_r should be 0, i.e.,

$$f_{ij}^k \leq x_i,\ f_{ij}^k \leq x_j,\ f_{ij}^k \leq x_k,\ f_{ij}^k \leq 1-r_k,\ f_{ij}^k \geq 0, \forall (i,j) \in E \cup E', \forall k \in V.$$

For the root v_r, there is no inflow, and the outflow of type k is x_k, i.e.,

$$\sum_{j:(j,v_r)\in E\cup E'} f_{jv_r}^k = 0, \quad \sum_{j:(v_r,j)\in E\cup E'} f_{v_r j}^k = x_k, \forall k \in V \setminus \{v_r\}.$$

For vertex $i \in V \setminus \{v_r\}$, the inflow of type i is x_i and the outflow of type i is 0, i.e.,

$$\sum_{j:(j,i)\in E\cup E'} f_{ji}^i = x_i, \quad \sum_{j:(i,j)\in E\cup E'} f_{ij}^i = 0, \forall i \in V \setminus \{v_r\}.$$

For vertex $i \in V \setminus \{v_r\}$, the flow of type $k (k \neq i)$ should be balanced at i, i.e.,

$$\sum_{j:(j,i)\in E\cup E'} f_{ji}^k = \sum_{j:(i,j)\in E\cup E'} f_{ij}^k, \forall i \in V \setminus \{v_r\}, \forall k, k \neq i.$$

For $i \in V$, let $r_i \in \{0,1\}$ be a decision variable such that $r_i = 1$ if vertex i is chosen to be the root of G_D, and $r_i = 0$ otherwise. The above constraints by multi-commodity flow (MCF) to ensure connectivity of G_D can be equivalently formulated as follows:

[MCF]

$$f_{ij}^k \le x_i,\ f_{ij}^k \le x_j,\ f_{ij}^k \le x_k,\ f_{ij}^k \le 1 - r_k,\ f_{ij}^k \ge 0, \forall (i,j) \in E \cup E', \forall k \in V \tag{6a}$$

$$\sum_{j:(j,i)\in E\cup E'} f_{ji}^k \le M(1 - r_i), \forall i,k \in V \tag{6b}$$

$$x_k - r_k - M(1 - r_i) \le \sum_{j:(i,j)\in E\cup E'} f_{ij}^k \le x_k - r_k + M(1 - r_i), \forall i,k \in V \tag{6c}$$

$$\sum_{j:(j,i)\in E\cup E'} f_{ji}^i = x_i - r_i,\quad \sum_{j:(i,j)\in E\cup E'} f_{ij}^i = 0, \forall i \in V \tag{6d}$$

$$\sum_{j:(i,j)\in E\cup E'} f_{ij}^k - Mr_i \le \sum_{j:(j,i)\in E\cup E'} f_{ji}^k \le \sum_{j:(i,j)\in E\cup E'} f_{ij}^k + Mr_i, \forall i,k \in V, k \ne i \tag{6e}$$

$$\sum_{i \in V} r_i = 1,\ r_i \le x_i, \forall i \in V \tag{6f}$$

where M is a sufficiently large positive constant.

Any feasible solution to the MDS problem with MCF constraints will guarantee that every vertex i within the dominating set D excluding the selected root will consume one unit of type i flow transmitted from the root, and the connectivity of the subgraph induced by D will be ensured. There are $|V| + 2|E||V| = O(|E||V|)$ decision variables and $8|E||V| + 2|V|^2 + |V| + |V|^2 + 1 + |V| = O(|E||V|)$ constraints in formulation (6). Similarly, Constraints (5) can be added to reduce the degeneracy of the selection for the root.

4 Power Dominating Set Problem and Connected Power Dominating Set Problem

For a power graph $G = (V,E)$, there is a given subset $V_Z \subset V$ of zero-injection vertices. As explained in [19], a power dominating set D is obtained by leveraging two physical laws: (1) if $v \in D$, then v and its neighbors (denoted by $N(v)$) are all covered (Ohm's law); (2) if $v \in V_Z$, and all vertices within the set $\{v\} \cup N(v)$ except one are covered, then the uncovered vertex in $\{v\} \cup N(v)$ is also covered (Kirchhoff's current law). The *power dominating set* (PDS) problem is to find a subset D with smallest cardinality that covers all vertices in V. The first law applies similarly as that for the dominating set problem (i.e., a selected vertex covers all neighbors of itself), while the second law can significantly reduce the dominating number for a given graph. Let the set of zero-injection vertices be denoted by $V_Z = \{v_i \in V : Z_i = 1\}$, where the parameter $Z_i = 1$ indicates that v_i is a zero-injection vertex; $Z_i = 0$ otherwise.

For $i \in V$, let $x_i \in \{0,1\}$ be a decision variable such that $x_i = 1$ if vertex i is selected into the power dominating set and $x_i = 0$ otherwise. For $i, j \in V$, let $p_{ij} \in \{0,1\}$ be a decision variable such that $p_{ij} = 1$ if Kirchhoff's current law applied to zero-injection vertex i can provide a coverage for vertex j and $p_{ij} = 0$ otherwise. Following the method in [18], the PDS problem to find a smallest dominating subset can be formulated as follows:

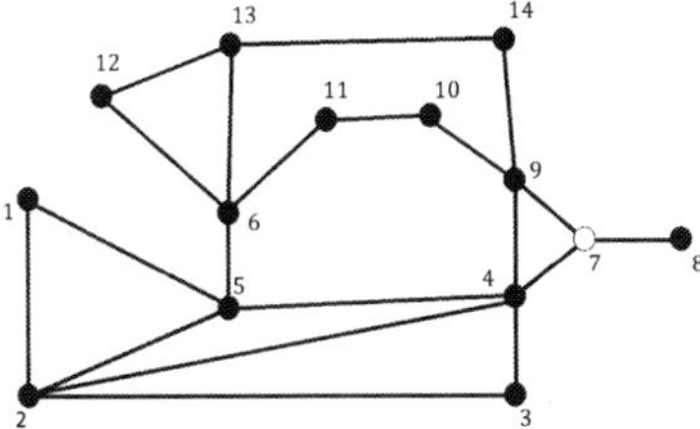

Fig. 2. An example graph with 14 vertices

$$\textbf{[PDS]} \quad \min_{x,p} \sum_i x_i \tag{7a}$$

$$s.t. \sum_j a_{ij}x_j + \sum_j a_{ij}Z_j p_{ji} \geq 1, \forall i \in V \tag{7b}$$

$$\sum_j a_{ij}p_{ij} = Z_i, \forall i \in V \tag{7c}$$

$$p_{ij} = 0, \forall i,j \text{ with } a_{ij} = 0 \text{ or } i \notin V_Z \tag{7d}$$

$$x_i, p_{ij} \in \{0,1\}, \forall i,j \in V \tag{7e}$$

The objective (7a) is to minimize the cardinality of the power dominating set. For each vertex $i \in V$, the first part of Constraint (7b) follows Ohm's law while the second part of (7b) follows Kirchhoff's current law with possible coverage from its neighbors. In the PDS problem, all vertices will be covered, and Constraints (7c) denote that every zero-injection vertex i provides coverage for itself or one of its neighbors. Constraint (7d) ensures that p_{ij} equals 0 if the pair (i,j) is not an edge or i is not a zero-injection vertex, and Constrains (7e) ensure the binary choices of the x_i and p_{ij} variables.

The PDS formulation is an integer linear program. Similarly, any feasible solution to the PDS will form a dominating set D of G by $D = \{v_i : x_i = 1\}$. Let $G_D = (D, E_D)$ be the subgraph induced by the power dominating set D, where $E_D = \{(i,j) \in E : a_{ij}x_ix_j = 1, \forall i,j \in V \text{ with } i < j\}$. For the *connected power dominating set* (CPDS) problem, we have to add connectivity constraints to ensure that the subgraph G_D is connected, following the methods introduced in Section 3.

5 Numerical Experiments

All MIP formulations are implemented in C++ and solved using CPLEX 12.1 via IBM's Concert Technology library, version 2.9. All experiments were performed on a Linux workstation with 4 Intel(R) Xeon(TM) CPU 3.60GHz processors and 8 GB RAM. The optimality gap was set to be 1%.

First, we consider an illustrative example using a graph with 14 vertices and 20 edges (as shown in Fig. 2). By solving the MDS problem formulated in (1), a minimum

Table 1. Minimum objective function values for power graphs

Graph				Dominating Set Problems		Connected Dominating Set Problems	
Name	$\|V\|$	$\|E\|$	$\|V_Z\|$	MDS	PDS	MCDS	CPDS
IEEE-14-Bus	14	20	1	4	3	5	4
IEEE-30-Bus	30	41	6	10	7	11	9
IEEE-57-Bus	57	78	15	17	11	31	24
RTS-96	73	108	22	20	14	32	28
IEEE-118-Bus	118	179	10	32	28	43	39
IEEE-300-Bus	300	409	65	87	68	129	112

Note: The column $|V_Z|$ denotes the number of zero-injection vertices.

Table 2. Comparison of formulation sizes

Number of	MDS(1)	PDS(7)	MTZ(2)	MARTIN(3)	SCF(4)	MCF(6)
decision var.	$\|V\|$	$2\|E\|+\|V\|$	$O(\|E\|+\|V\|)$	$O(\|V\|^3)$	$O(\|E\|+\|V\|)$	$O(\|E\|\|V\|)$
constraints	$\|V\|$	$2\|V\|$	$O(\|E\|+\|V\|)$	$O(\|V\|^3)$	$O(\|E\|+\|V\|)$	$O(\|E\|\|V\|)$

dominating set is $\{2,6,7,9\}$, with cardinality 4. By solving the MCDS problem as formulated in (1) coupled with any one type of connectivity constraints (2), (3), (4), or (6), a minimum connected dominating set is $\{4,5,6,7,9\}$ with cardinality 5.

For the power dominating set of the graph in Fig. 2, assume that the set of zero-injection vertices is $V_Z = \{7\}$. A minimum power dominating set is $\{2,6,9\}$ with cardinality 3, obtained by solving the formulation (7). In this dominating set, vertices $2,1,3,4,5$ are covered by vertex 2; vertices $6,5,11,12,13$ are covered by vertex 6; and vertices $9,4,7,10,14$ are covered by vertex 9. By Kirchhoff's current law, vertex 8 is covered because vertices in $\{7\} \cup N(7) = \{7,4,9,8\}$ are all covered with the exception of vertex 8. Similarly, the minimum connected power dominating set $\{4,5,6,9\}$ with cardinality 4 can be computed using formulation (7) with any one of the constraints (2), (3), (4), or (6).

Next, we test our models on the six power graphs considered in [28]. First, we removed all parallel edges in these graphs. The objective values for the minimum dominating set, minimum connected dominating set, minimum power dominating set, and minimum connected power dominating set problems are shown in Table 1, while the wall clock run-times (in seconds) are reported in Table 3. In Table 1, we also present statistical information for the test instances, including the number of vertices, edges, and the number of zero-injection vertices in the case of the power dominating set problem.

From Table 1, we observe that the cardinality of the minimum power dominating set is less than the cardinality of minimum dominating set for a given graph. Application of Kirchhoff's current law to zero-injection vertices can reduce the dominating number of a graph. Additionally, minimum connected dominating sets have larger cardinality than their non-connected counterparts.

In Table 2, we present the number of decision variables and constraints for each formulation. The four types of constraints we used to ensure set connectivity have at most $O(|V|^3)$ decision variables, and at most $O(|V|^3)$ constraints. In contrast to the formulation for sub-tour elimination in [15], which has exponential number $O(2^{|V|})$ number of constraints, our proposed methods should yield more tractable computation for even larger graphs.

Table 3. Solution times for formulations with different connectivity constraints

Graph Name	Dominating Set Problems		Minimum Connected Dominating Set Problems				Minimum Connected Power Dominating Set Problems			
	MDS	PDS	MTZ	MARTIN	SCF	MCF	MTZ	MARTIN	SCF	MCF
IEEE-14-Bus	0	0	0.02	0.14	0.15	0.14	0.04	0.12	0.49	0.15
IEEE-30-Bus	0	0	0.22	2.39	299.39	0.89	0.32	1.72	265.10	1.17
IEEE-57-Bus	0.01	0.01	200.59	14671.70	64641.60	6738.05	60.81	5309.74	12448.10	2579.66
RTS-96	0.01	0.03	445.69	>24h	>24h	47236.40	55266.10	>24h	>24h	53752.20
IEEE-118-Bus	0.01	0.04	699.83	85455.70	>24h	36263.10	50.67	>24h	>24h	78715.40
IEEE-300-Bus	0.01	0.27	5033.97	>24h	>24h	>24h	72437.40	>24h	>24h	>24h

From Table 3, we observe that it is quite fast to compute optimal solutions to the dominating set problem without connectivity constraints. In contrast, the imposition of connectivity constraints significantly impacts computational tractability. The MTZ constraints (2) yield the best performance. Comparing the two methods with the same number $O(|E|+|V|)$ of decision variables, MTZ and SCF, MTZ (with fewer constraints) yields significantly better performance. The other connectivity formulations are quite slow requiring more than 24 hours for solving problems arising in large graphs. For example, there are $|V| = 73$ vertices and $|E| = 108$ edges in the RTS-96 graph, yielding more than 73^3 binary variables in formulation (3).

6 Conclusions

We presented four optimization models to ensure the connectivity of the subgraph induced by the dominating set of a graph. All models are formulated as mixed integer programs with a polynomial number of constraints, and were tested on many representative graphs. Among these models, the one with Miller-Tucker-Zemlin constraints to ensure connectivity has the best performance. We further note that the MIP formulations we examine here can be easily extended to solve the minimum spanning tree problem, the maximum leaf spanning tree problem, the k-cardinality tree problem, and the Steiner tree problem.

Future research directions include using efficient branch-and-cut methods to further reduce the computational complexity, and comparing the results obtained by formulations considering an exponential number of constraints. To improve the efficiency of the methods described in this paper, more valid inequalities should be further studied and high-performance computing methods should be leveraged. For some graphs, for example $1 \times n$ grid graphs, the dominating set problem can in theory be solved in polynomial time and tests should be performed on these cases to verify computational complexity results.

Acknowledgements. Sandia National Laboratories is a multi-program laboratory managed and operated by Sandia Corporation, a wholly owned subsidiary of Lockheed

Martin Corporation, for the U.S. Department of Energy's National Nuclear Security Administration under contract DE-AC04-94AL85000.

References

1. Garey, M.R., Johnson, D.S.: Computers and Intractability: A Guide to the Theory of NP-Completeness. W. H. Freeman & Co., New York (1979)
2. Cheng, X., Huang, X., Li, D., Wu, W., Du, D.-Z.: A polynomial-time approximation scheme for the minimum-connected dominating set in ad hoc wireless networks. Networks 42(4), 202–208 (2003)
3. Li, Y., Thai, M.T., Wang, F., Yi, C.W., Wan, P.J., Du, D.-Z.: On greedy construction of connected dominating sets in wireless networks. Wirel. Commun. Mob. Comp. 5, 927–932 (2005)
4. Zhu, X., Yu, J., Lee, W., Kim, D., Shan, S., Du, D.-Z.: New dominating sets in social networks. J. Global Optim. 48(4), 633–642 (2010)
5. Wu, W., Gao, X., Pardalos, P.M., Du, D.-Z.: Wireless networking, dominating and packing. Optim. Lett. 4(3), 347–358 (2010)
6. Ding, L., Gao, X., Wu, W., Lee, W., Zhu, X., Du, D.-Z.: An exact algorithm for minimum CDS with shortest path constraint in wireless networks. Optim. Lett. 5(2), 297–306 (2011)
7. Thai, M.T., Du, D.-Z.: Connected dominating sets in disk graphs with bidirectional links. IEEE Commun. Lett. 10(3), 138–140 (2006)
8. Kim, D., Zhang, Z., Li, X., Wang, W., Wu, W., Du, D.-Z.: A better approximation algorithm for computing connected dominating sets in unit ball graphs. IEEE Trans. Mob. Comp. 9(8), 1108–1118 (2010)
9. Blum, J., Ding, M., Thaeler, A., Cheng, X.: Connected dominating set in sensor networks and MANET. In: Du, D.-Z., Pardalos, P. (eds.) Handbook of Combinatorial Optimization, pp. 329–369 (2004)
10. Liu, Z., Wang, B., Guo, L.: A Survey on connected dominating set construction algorithm for wireless sensor networks. Informa. Technol. J. 9, 1081–1092 (2010)
11. Mnif, K., Rong, B., Kadoch, M.: Virtual backbone based on mcds for topology control in wireless ad hoc networks. In: Proceedings of the 2nd ACM International Workshop on Performance Evaluation of Wireless Ad Hoc, Sensor, and Ubiquitous Networks, Quebec, Canada (2005)
12. Yuan, D.: Energy-efficient broadcasting in wireless ad hoc networks: performance benchmarking and distributed algorithms based on network connectivity characterization. In: Proceedings of MSWiM, Quebec, Canada (2005)
13. Morgan, M.J., Grout, V.: Finding optimal solutions to backbone minimisation problems using mixed integer programming. In: Proceedings of the Seventh International Network Conference (INC 2008), Boston, MA, pp. 53–63 (2008)
14. Wightman, P.M., Fabregasy, A., Labradorz, M.A.: An optimal solution to the MCDS problem for topology construction in wireless sensor networks. In: 2010 IEEE Latin-American Conference on Communications (LATINCOM), Belem, Brazil (2010)
15. Simonetti, L., da Cunha, A.S., Lucena, A.: The Minimum Connected Dominating Set Problem: Formulation, Valid Inequalities and a Branch-and-Cut Algorithm. In: Pahl, J., Reiners, T., Voß, S. (eds.) INOC 2011. LNCS, vol. 6701, pp. 162–169. Springer, Heidelberg (2011)
16. Pop, P.C.: A survey of different integer programming formulations of the generalized minimum spanning tree problem. Carpathian J. Mathematics 25(1), 104–118 (2009)
17. Haynes, T.W., Hedetniemi, S.M., Hedetniemi, S.T., Henning, M.A.: Domination in graphs applied to electric power networks. SIAM J. Disc. Math. 15, 519–529 (2002)

18. Aminifar, F., Khodaei, A., Fotuhi-Firuzabad, M., Shahidehpour, M.: Contingency-constrained PMU placement in power networks. IEEE Trans. Power Syst. 25, 516–523 (2010)
19. Aazami, A.: Domination in graphs with bounded progagation: algorithms, formulations and hardness results. J. Comb. Optim. 19, 429–456 (2010)
20. Miller, C.E., Tucker, A.W., Zemlin, R.A.: Integer programming formulation of traveling salesman problems. J. Assoc. Comp. Mach. 7, 326–329 (1960)
21. Quintao, F.R., da Cunha, A.S., Mateus, G.R., Lucena, A.: The k-cardinality tree problem: reformulations and lagrangian relaxation. Disc. Appl. Math. 158, 1305–1314 (2010)
22. Desrochers, M., Gilbert, L.: Improvements and extensions to the Miller-Tucker-Zemlin subtour elimination constraints. Oper. Res. Lett. 10, 27–36 (1991)
23. Martin, R.K.: Using separation algorithms to generate mixed integer model reformulations. Oper. Res. Lett. 10, 119–128 (1991)
24. Yannakakis, M.: Expressing combinatorial optimization problems by linear programs. J. Comp. Syst. Sci. 43(3), 441–466 (1991)
25. Conforti, M., Cornuéjols, G., Zambelli, G.: Extended formulations in combinatorial optimization. 4OR (8), 1–48 (2010)
26. Kaibel, V., Pashkovich, K., Theis, D.O.: Symmetry Matters for the Sizes of Extended Formulations. In: Eisenbrand, F., Shepherd, F.B. (eds.) IPCO 2010. LNCS, vol. 6080, pp. 135–148. Springer, Heidelberg (2010)
27. Dilkina, B., Gomes, C.P.: Solving Connected Subgraph Problems in Wildlife Conservation. In: Lodi, A., Milano, M., Toth, P. (eds.) CPAIOR 2010. LNCS, vol. 6140, pp. 102–116. Springer, Heidelberg (2010)
28. IEEE reliability test data (2012), `http://www.ee.washington.edu/research/pstca/`

Measuring Structural Similarities of Graphs in Linear Time

Zheng Fang, You Li, and Jie Wang

Department of Computer Science, University of Massachusetts, Lowell, MA 01854

Abstract. Measuring graph similarities is an important topic with numerous applications. Early algorithms often incur quadratic time or higher, making it unpractical to use for graphs of very large scales. We present in this paper the first-known linear-time algorithm for solving this problem. Our algorithm, called Random Walker Termination (RWT), is based on random walkers and time series. Three major graph models, that is, the Erdős-Rényi random graphs, the Watts-Strogatz small world graphs, and the Barabási-Albert preferential attachment graphs are used to generate graphs of different sizes. We show that the RWT algorithm performs well for all three graph models. Our experiment results agree with the actual similarities of generated graphs. Built on stochastic process, RWT is sufficiently stable to generate consistent results. We use the graph edge rerouting test and the cross model test to demonstrate that RWT can effectively identify structural similarities between graphs.

1 Introduction

Detecting similarities between graphs of very large scales, e.g., between online social networks and knowledge networks, asks for linear-time algorithms. Classical graph similarity algorithms that measure isomorphism or maximum (or minimum) subgraphs (or supergraphs) are not suitable for large graphs because they incur high computational complexity. Approximate measurement on certain graph signatures (e.g., degree distribution), although efficient, may not be able to truly represent structural similarities.

To overcome these problems we devise an effective, linear-time algorithm to detect structural similarities between graphs. Our algorithm, called Random Walker Termination (RWT), is based on random walkers and time series. Three major graph models, namely the Erdős-Rényi random graph, the Watts-Strogatz small world graph and the Barabási-Albert preferential attachment graph are used to generate synthetic graphs for performance testing. Experiment results show good agreements with the actual similarities of generated graphs. To study the stochastic nature of the algorithm, we also carry out a set of self-similarity experiments to show that RWT is stable.

The rest of the paper is organized as follows. In Section 2 we briefly review previous work on measuring graph similarities. In Section 3 we present a linear-time algorithm called Random Walker Termination (RWT) based on random walkers and time series for finding structural similarities for graphs of very large scales. In Section 4 we show a number of numerical experiment results of the RWT algorithm. In Section 5 we conclude the paper.

G. Lin (Ed.): COCOA 2012, LNCS 7402, pp. 384–395, 2012.

2 Related Work

Measuring graph similarities has numerous applications, and a number of algorithms have been proposed. Isomorphism is the strictest measure of graph similarity. Two graphs are considered similar if they share a sufficiently large isomorphic component, or they have a sufficiently small super graph in common. The difference between two graphs can be represented as edit distance, that is, the number of steps transforming one graph to another, with minimum editing costs. However, since computing the exact solution of isomorphism requires exponential time unless P = NP, such algorithms are not applicable to large-scale graphs of interests.

Another class of similarity algorithms takes a local perspective on similarity, based on the philosophy that "two nodes in different graphs are considered similar if their neighboring nodes are also similar". This recursive definition on similarity naturally leads to an iterative updating process, in which similarity scores between nodes propagate to neighboring nodes at each time step. Melnik et al. [7] applied a similarity flooding algorithm in matching of database structures, and attempted to find the correspondence between the nodes of two given graphs. The SimRank algorithm, described by Jeh and Widom [4], computes the self-similarity of a graph, that is, the similarities between all pairs of nodes in one graph. A recursive method proposed by Zager and Verghese [9] introduces the idea of coupling the similarity scores of both nodes and edges.

Extensive research has been conducted to identify the essential features of graphs and use them to measure similarities. The key idea is that the features of given graphs, also referred to as the "signatures", are expected to be similar if graphs are pairwise similar. Common graph features include degree distributions, diameters, and graph spectrum eigenvalues [2]. These methods are usually quite powerful and scale well for large inputs, for they map the graphs to several statistical measures that are much smaller in size than the original graphs. However, such features may not produce desirable results. For example, the degree distribution, if solely used as the similarity measure, is considered too loose to truly represent a structural similarity. It is also possible that two graphs with high similarity may have different numbers of nodes and edges.

3 Random Walker Termination Similarity Measurement

In a thermodynamic system, the expansion of heat content depends on various factors coming from both internal and external, but mostly from the structure (or shape) of the object. Consider a cooling process for two objects of the same material with similar shapes in the same environment. They are expected to have a similar heat transfer pattern. On the other hand, if the two objects have major different structures, the cooling processes are expected to be significantly different.

Based on this observation, if we assign each node in a graph with an initial energy associated with its degree, then the whole graph can be thought of as a thermodynamic object. Energy units can exchange freely between nodes along the edges. When a certain node reaches the lower bound of energy threshold, it stops emitting energy while starting to absorb energy. In this manner, the graph will stochastically "cool down" in time. If two graphs are structural similar, they are expected to follow a similar cooling process and the time series for such cooling process can be used for detecting similarities. Random walkers and stochastic walking would be a natural fit for simulating such process.

3.1 Random Walks on Graphs

Random walks are used in many models in mathematics and physics. The classical theory of random walks deals with random walks on simple, but infinite graphs, such as grids, and studies their qualitative behaviors: Does a random walk return to its starting point with probability one? Does it return infinitely often? More recently, random walks on general and finite graphs have received much attention. Extensive studies have focused more on quantitative properties. A step in a random walk can be defined as follows:

Definition 1. Random Walk
A one step random walk on an undirected graph $G = (V,E)$ is the movement from node i to j with probability $p(j|i) = 1/d_i$, where $(i,j) \in E$ and d_i is the degree of node i.

A random walk is a finite Markov chain that is time-reversible, and the theory of random walks on graphs and the theory of finite Markov chains are quite similar [6]. The studies on the probability of random-walk paths indicate that such a stochastic process may have deep relations to the spectrum of graphs, especially to a class of natural kernels based on heat equations, called diffusion kernels [5]. The random-walk approach has shown great power and flexibility to quantitative properties of graphs.

3.2 RWT Algorithm

We present a new approach to measuring graph structural similarity using a number of random walkers and time series. In particular, instead of studying the probability of random-walk paths, we are concerned about the termination rate of random walkers on a graph with dynamically generated sink nodes (A random walker dies when it reaches a sink node).

Given an undirected graph $G = (V,E)$, we initially assign d_i random walkers on each node i, where d_i is the degree of node i. Define a sink threshold ϕ, so that if the number of walkers on node i is not greater than ϕ, then node i becomes a sink node. Note that once a node becomes a sink, it remains as a sink. During each iteration, each walker randomly chooses one of its neighboring nodes. If the chosen node is not a sink, then the walker will move to it; otherwise the walker dies. The overall termination rate is calculated at the end of each iteration. Let k be the number of iterations. Then a k-Random Walker Termination (k-RWT) algorithm is obtained, as shown in the following pseudocode.

```
CALC_RWT (G,k,φ)
    ▷ G : input graph, k : number of iterations, φ : sink threshold
for each node i in G
    do walkers[i] ← i.degree
       walkers_next[i] ← 0
total_walkers ← walkers.sum
while (k)
    do for each node i in G
           do for each walker on node i
                  do j ← randomly pick a neighbor of node i
                     if (walkers[j] > φ)
                        then walkers_next[j] increases by 1
                        else total_walkers decrease by 1
              ▷ end of for
       ▷ end of for
       zero(walkers[])
       swap(walkers, walkers_next)
       rate ← calculate the termination rate
       RWT.append(rate)
       k ← k − 1
▷ end of while
return RWT
```

We use the k-element vector RWT of termination rates obtained for each graph as a signature that represents the structure of the graph, and calculate RWT_{Score} with two RWTs to determine the similarity score between two graphs.

Definition 2. RWT Score
Let $U = \langle u_1, \cdots, u_l \rangle$ and $V = \langle v_1, \cdots, v_m \rangle$ be two RWTs for graphs G and H. Define a normalization factor $N(u,v)$ as follows: $N(u,v) = u+v$ if $u+v \neq 0$ and 1 otherwise. Let

$$RWT_{Score}(U,V) = \sum_{i=1}^{k} \frac{|u_i - v_i|}{N(u_i, v_i)}, \text{ where } k = \min\{l,m\}.$$

Note that in the case when U and V have different dimensions, we should calculate RWT_{Score} by aligning the two with the minimum dimension of U and V. The range of RWT_{Score} is $[0,k]$, and a smaller score means a better similarity of the two graphs.

Definition 3. α-Similarity
Two graphs G and H are α-similar if $RWT_{score}(G,H) \leq \alpha$.

While the selection of the α value would depend on the underlying graphs, in general, setting α to 0.1 is deemed appropriate.

Note that the similarity measured by RWT scores contains the case of "scale similarity". That is, two graphs are considered scaly similar if they are both generated using the same model with similar growth parameters, regardless their current sizes.

Although this kind of similarity is quite useful in some scenarios, e.g., in finding objects that follow the same growing dynamics but in different time phases, in most cases, the similarity measurement requires that two similar graphs be of about the same size. Therefore, we should also apply a secondary filter of edge ratio and vertex ratio on top of RWT scores, to eliminate any similar pairs with different sizes.

The complexity of the RWT algorithms depends on the total number of random walkers and the number of iterations (which is often a small constant). Thus, the time complexity is linear in terms of the size of the input graph. The linear-time complexity guarantees that the RWT algorithm is efficient even on large-scale inputs. The RWT algorithm may also be applied in a distributed environment, since the random walking of each walker is independent from each other. Through appropriate communications within a distributed system, the time needed for calculating RWTs can be further reduced.

4 Numerical Experiments

In this section, we use randomly generated synthetic graphs to test the accuracy of the RWT algorithm. Three major graph models, namely, Erdős-Rényi random graphs, Watts-Strogatz small world graphs, and Barabási-Albert preferential attachment graphs are used to generate graphs of different sizes. The number of iterations of RWT algorithm is set to 10. To determine the threshold, we first sort node degrees in increasing order, and pick the first 20% position's degree as the threshold. Data experiments are focused on algorithm's stability, ability to identify similar graphs. Experiment results show good agreements with the actual similarities of generated graphs. Regarding the stochastic nature of the algorithm, we also conduct self-similarity experiments to show that RWT is stable.

4.1 Graph Models

Erdős-Rényi Random Graphs. The Erdős-Rényi model [3] (ER) generates random graphs with equal probability of creating new edges between nodes, independently of the other edges. There are two parameters n and p, where the graph initially has n nodes and each edge is introduced with probability p for each pair of nodes. The expected number of edges is $\binom{n}{2}p$.

Let D be a random variable that represents the degree of a node in an Erdős-Rényi graph. Then we have $E(D) = (n-1)p$ and $P(D=d) = \binom{n-1}{d}p^d(1-p)^{n-1-d}$. Keeping the expected degree constraint as $n \to \infty$, D can be approximated with a Poisson random variable with $\lambda = (n-1)p$, and $P(D=d) = e^{-\lambda}\lambda^d/d!$. This degree distribution falls off faster than an exponential in d, and so is not a power-law distribution. Figure 1 shows an Erdős-Rényi graph of 60 nodes.

Watts-Strogatz Small-World Graphs. The Watts-Strogatz model (WS) [8] generates graphs with small-world properties, including short average path lengths and high clustering.

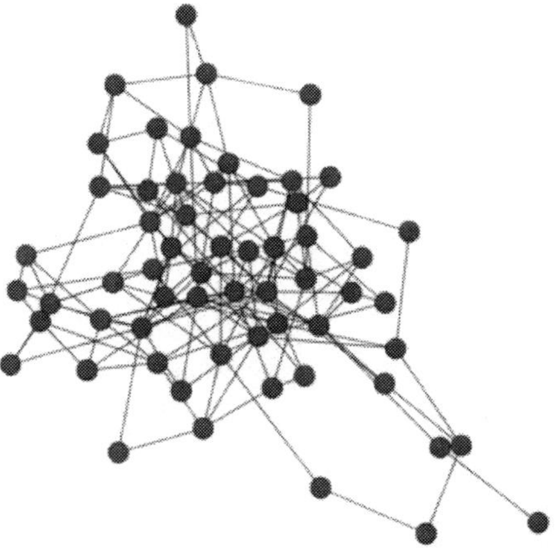

Fig. 1. Erdős-Rényi random graph

Given the number of nodes N, the mean degree K and rewiring probability β, where $N \gg K \gg \ln(N) \gg 1$, the model constructs an undirected graph with N nodes and $NK/2$ edges in the following way: first construct a regular ring lattice, a graph with N nodes each connected to K neighbors $K/2$ on each side. Then, for every node $n_i = n_0, \ldots n_{N-1}$ take every edge (n_i, n_j) with $i < j$, and rewire it with probability β. Figure 2 shows a Watts-Strogatz graph of 60 nodes.

Fig. 2. Watts-Strogatz small world graph

Barabási-Albert Preferential Attachment Graphs. The Barabási-Albert model [1] (BA) generates at random, scale-free networks using a preferential attachment mechanism. By preferential attachment it means that the more connections a node has, the more likely it is to receive new links. That is, nodes with higher degrees have stronger ability to attract new connections.

To construct a Barabási-Albert graph, the graph should initially contain m nodes, where $m \geq 2$ and the degree of each node in the initial graph should be at least 1. New nodes are added to the network one at a time. A new node is connected to the existing nodes with a probability proportional to the number of links that already exist in the graph. The probability that a new node is connected to node i can be calculated as follows:

$$p_i = \frac{k_i}{\sum_j k_j},$$

where k_i is the degree of node i and the divisor is the summation of degrees of all existing nodes. Heavily linked nodes, a.k.a. hubs, tend to accumulate more links quickly, while nodes with small degrees are unlikely to be chosen to form a new edge. The new nodes have a preference to attach themselves to the already heavily linked nodes. Figure 3 shows a Barabási-Albert graph of 60 nodes.

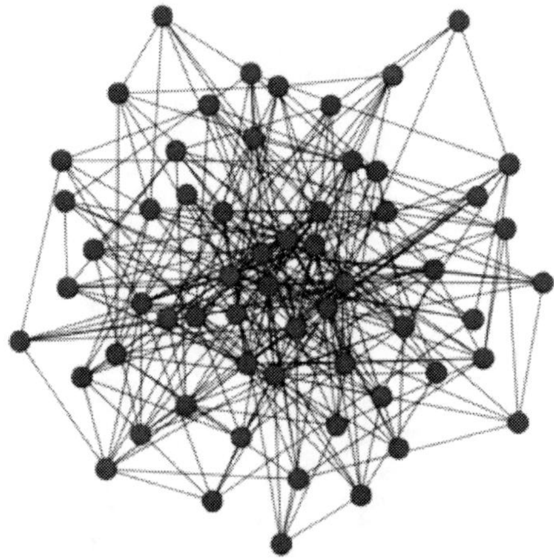

Fig. 3. Barabási-Albert preferential attachment graph

4.2 Self-similarity Stability

Since the RWT algorithm is based on random walks with multiple walkers, the stochastic behavior of the algorithm requires validation of the stability of outputting consistent results. The most effective way to test the stability is through self similarity.

In a self-similarity test, we generate a graph and apply RWT on it twice. Since the graphs for comparison are exactly the same, the RWT score is expected to be close to zero, but not exactly zero because of stochastic behaviors. Using the three graph models, we randomly generate graphs of nodes ranging from 100 to 3,000, and test the similarity of each graph against itself. Figure 4.2 shows the average self-similarity score and error ranges versus node sizes for all three graph models.

The self-similarity RWT scores in all three models are sufficiently small to conclude that each graph is similar to itself. The RWT score gradually approaches to zero as the size of graph increases, which makes sense for the stochastic behavior becomes smoother on bigger systems. At a 1,000-nodes graph, the average score can be as small as 0.0036. The consistency of the results shows that the RWT algorithm is stable in terms of graph types and sizes, and the large input size increases the stability of the algorithm.

4.3 Graph Edge Rerouting

In this experiment, we test the RWT algorithm for the ability to identify small changes in a graph. We randomly generate graphs of edges ranging from 50 to 1,000 for each of the three graph models, and reroute selected edges at random, but maintain graph properties (such as preferential attachment). The numbers of selected edges ranged from 1 to 50. The rerouted graph is then compared with the original graph, and the RWT score is calculated. Figure 4.3 shows the RWT score plotting of edge rerouting on each of the three models.

Edge rerouting similarity essentially tests the algorithm's ability to detect gradually changes in graph structures. In all three models, the two graphs for comparison become less similar as the number of rerouted edges increases. We also note that edge rerouting has much bigger impact on Barabási-Albert graphs, with significantly less influence on Erdős-Rényi graphs.

In an Erdős-Rényi graph, edges are randomly added uniformly between nodes. That is, every node is in an equivalent state and no edge is more special than the others. Rerouting edges on such graph is nothing more than removing the edges first and re-generating edges using the normal generating process. Therefore, the rerouting does not seem to affect the graph structure too much. Although large portion of edge rerouting could still result in structural changes, the impact can be expected to be small compared to the other models.

On the other hand, in a Barabási-Albert graph, edges are created following the preferential attachment, and so the nodes each have different priorities to form new edges. Since nodes are not in equivalent state, rerouting on a Barabási-Albert graph may put more impact for low degree nodes and maybe even cause detachment creating orphan nodes. Certain techniques can be applied during rerouting to avoid creating orphan nodes, but after certain number of edge reroutings, a significant structural change is inevitable. Thus, it is not surprised that a large difference exists between the original graph and its edge rerouted graph.

4.4 Cross Model Test

Finally, we randomly generate graphs of up to 1,000 nodes, and calculate the RWT score on each graph generated under different models (ER, WS, BA). Table 1 records the similarity among all generated graphs. We can see that all graph pairs have high RWT scores and none of them seem similar. This result agrees with the observation on

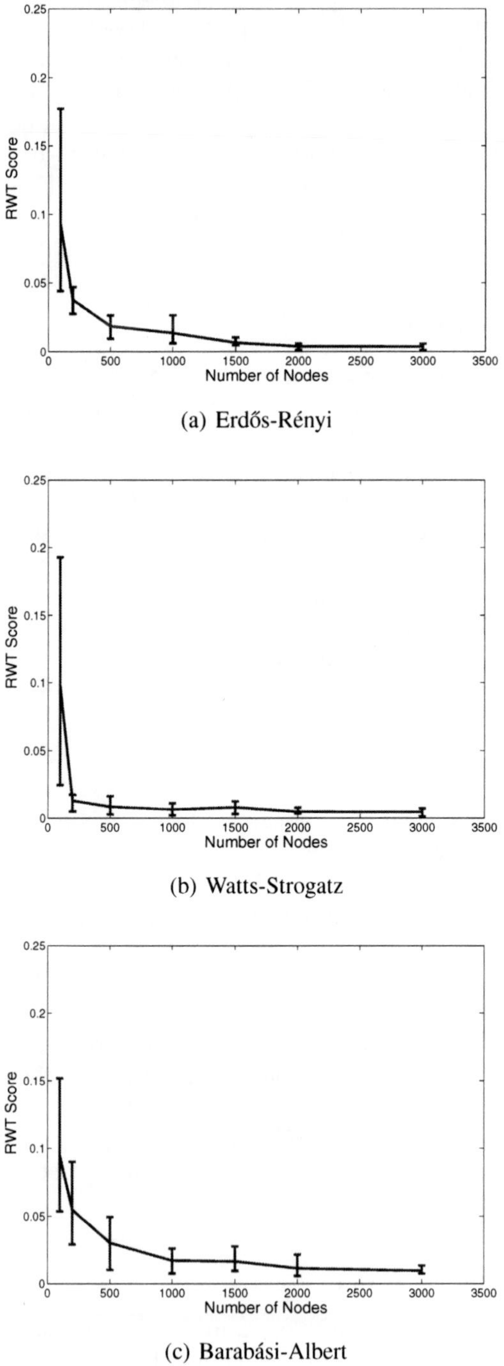

(a) Erdős-Rényi

(b) Watts-Strogatz

(c) Barabási-Albert

Fig. 4. Self-similarity and error ranges for three graph models

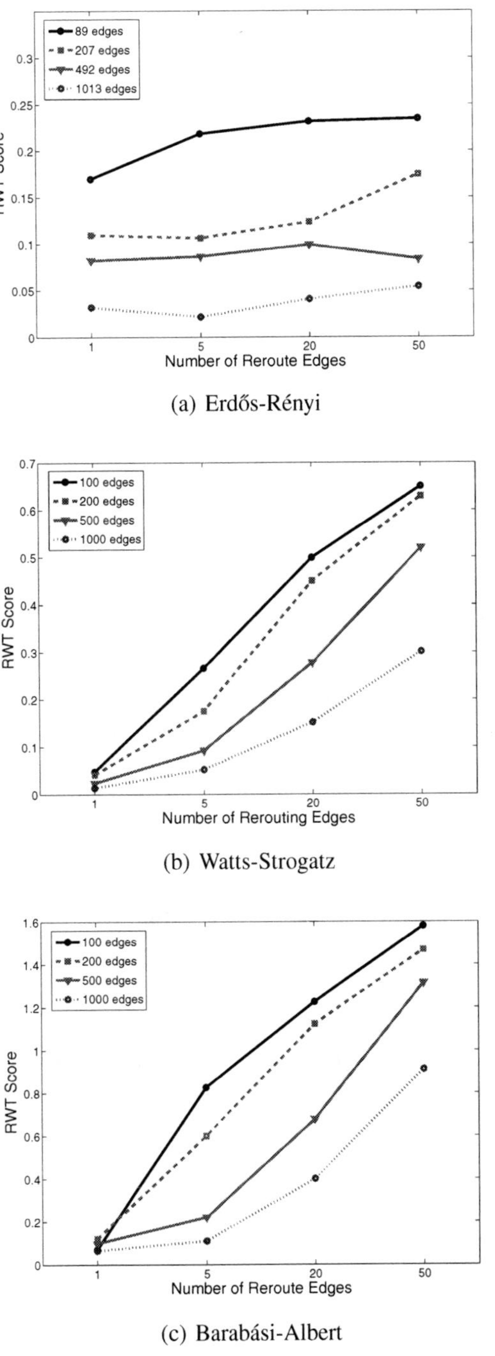

(a) Erdős-Rényi

(b) Watts-Strogatz

(c) Barabási-Albert

Fig. 5. Edge rerouting similarity for three graph models

Table 1. RWT Scores of Cross Model Test

Model 1 vs 2	Vertices	Edges 1	Edges 2	RWT_{Score}
ER vs BA	100	263	475	1.1147
	200	1046	1900	1.1690
	500	6259	11875	1.6337
	1000	24872	47500	1.9921
ER vs WS	100	263	200	1.3383
	200	1046	400	1.5537
	500	6259	1000	1.3569
	1000	24872	2000	1.2260
BA vs WS	100	475	200	2.3014
	200	1900	400	2.3271
	500	11875	1000	2.7603
	1000	47500	2000	2.9754

the actual graphs, that none of the graphs from different models should be considered similar to each other, regardless of their sizes.

5 Conclusion

We devised an effective, linear-time algorithm called Random Walker Termination (RWT) for finding structural similarities in graphs of very large scales based on random walkers and time series. We showed that the algorithm is scalable for a wide range of input sizes on various randomly generated synthetic graphs. Three major graph models, namely, the Erdős-Rényi random graphs, the Watts-Strogatz small-world graphs, and the Barabási-Albert preferential attachment graphs are used to generate graphs of different sizes. The RWT algorithm performs well on all three graph models, and the experiment results agree with the actual similarities of generated graphs. Built on stochastic process, the algorithm is sufficiently stable to generate consistency results. The graph edge rerouting test and the cross model test also demonstrate good performance of identifying similar graphs.

Acknowledgements. We thank Prof. Weibo Gong at University of Massachusetts Amherst for providing motivation of the RWT algorithm. We thank Jian Lu for helping us to organize the data and figures.

Z. Fang and Y. Li were supported in part by the NSF under grant CCF-0830314. J. Wang was supported in part by the NSF under grants CCF-0830314, CNS-0958477, and CNS-1018422. Points of view in this document are those of the authors and do not necessarily represent the official position of the NSF.

References

1. Albert, R., Barabási, A.L.: Statistical mechanics of complex networks. Reviews of Modern Physics 74(1), 47–97 (2002)

2. Cha, S.H.: Comprehensive survey on distance / similarity measures between probability density functions. International Journal of Mathematical Models and Methods in Applied Sciences 1(4), 300–307 (2007)
3. Erdős, P., Rényi, A.: On the evolution of random graphs. Publication of the Methematical Institute of the Hungarian Academy of Sciences, 17–61 (1960)
4. Jeh, G., Widom, J.: Simrank: a measure of structural-context similarity. In: Proceedings of the Eighth ACM SIGKDD International Conference on Knowledge Discovery and Data Mining, KDD 2002, pp. 538–543. ACM (2002)
5. Kondor, R.I., Lafferty, J.D.: Diffusion kernels on graphs and other discrete input spaces. In: Proceedings of the Nineteenth International Conference on Machine Learning, ICML 2002, pp. 315–322 (2002)
6. Lovasz, L.: Random walks on graphs: A survey. Combinatorics, Paul Erdos is Eighty 2, 1–46 (1993)
7. Melnik, S., Garcia-Molina, H., Rahm, E.: Similarity flooding: a versatile graph matching algorithm and its application to schema matching. In: Proceedings on 18th International Conference on Data Engineering, pp. 117–128 (2002)
8. Watts, D., Strogatz, S.: Collective dynamics of small-world networks. Nature 393(6684), 440–442 (1998)
9. Zager, L.A., Verghese, G.C.: Graph similarity scoring and matching. Applied Mathematics Letters 21(1), 86–94 (2008)

Author Index